Abbreviations for Fibers Used in this Book[a]

Abbreviation	Explanation
CA	Secondary cellulose acetate
CTA	Cellulose triacetate
CO	Cotton
PAN	Polyacrylonitrile
PA 66	Polyamide 66, Nylon 66
PA 6	Polyamide 6, Nylon 6
PA 11	Polyamide 11, Nylon 11
PET	Polyester (Polyethylene terephthalate)
PP	Polypropylene
PU	Polyurethane
PVA	Polyvinyl alcohol
PVC	Polyvinyl chloride
CV	Viscose
WO	Wool

[a] The word *fiber* is used for any textile material such as staples, filaments, yarns, fabrics, and so on.

CHEMICAL PROCESSING
OF SYNTHETIC FIBERS
AND BLENDS

CHEMICAL PROCESSING OF SYNTHETIC FIBERS AND BLENDS

KESHAV V. DATYE, Director

and

A. A. VAIDYA, Assistant Director

Sir Padampat Research Centre
J. K. Synthetics Ltd.
Kota, Rajasthan, India

A Wiley-Interscience Publication

JOHN WILEY & SONS

New York, Chichester, Brisbane, Toronto, Singapore

Library of Congress Cataloging in Publication Data:

Datye, Keshav V.
   Chemical processing of synthetic fibers and blends.

   "A Wiley-Interscience publication."
   Bibliography: p.
   Includes index.
   1. Dyes and dyeing—Textile fibers, Synthetic.
2. Textile finishing. I. Vaidya, A. A. II. Title.
TP904.D37 1984   667'.34   83-19809
ISBN 0-471-87654-2

Printed in the United States of America

10 9 8 7 6 5 4 3 2 1

PREFACE

Synthetic fibers occupy a prominent position in the multifiber industry of today. The large-scale introduction of synthetic fibers in the textile industry has created a number of problems for the textile dyer and finisher because these fibers require special techniques for chemical processing. A large number of operations, such as heat setting, high-temperature dyeing, thermosoling, transfer printing, antistatic and antipilling finishing, and so on were developed following the advent of synthetic fibers. Blends of synthetic fibers with natural fibers such as cotton and wool have become a common textile material. The wet processing of these blends is complicated, since the dyeing and finishing of the two fibers in a blend has to be carried out by different techniques.

Because of the persistent efforts of synthetic fiber producers, textile-machine designers, dye and auxiliary manufacturers, and wet processors, most of the initial problems of chemical processing of new fibers and their blends are solved and satisfactory methods for dyeing, printing, and finishing are now available. In the present book, an attempt is made to review various aspects of chemical processing of synthetic fibers and their blends. The book is written for the use of technologists engaged in the dyeing and finishing of synthetic fibers, and contains a high proportion of comprehensive process instructions. The theoretical aspects of the chemical processing are introduced in brief. The chemistry of dyes and textile chemicals is not discussed fully since both these aspects have been thoroughly described elsewhere. References are given for those who are interested in pursuing the subject in more detail.

The properties of synthetic fibers are decided by the manufacturing processes and can influence the chemical processing. The factors that are of

v

importance are considered in detail. This book may be of interest to those who are producing these fibers and who are engaged in technical service departments.

The research scientists in fiber science and textile chemistry will find many new ideas in our book that can be developed further. We have gone into some detail in describing mass coloration, modified fibers, new forms of nonionic dye formulations, faults in fibers generated in the manufacturing operations, their identification and remedial measures, and so on, so that research workers will find this book interesting and useful.

The chemical processing scene has changed recently because of the energy crisis, stringent pollution control measures, and automation. All these aspects are described in the present volume.

We gratefully acknowledge the authors who allowed us to use their work. We do not mean to claim any credit for the original work and request that the readers see the original references listed in these papers and articles. We thank the various journals for allowing us to reproduce figures and tables and the machine manufacturers for supplying us with photographs. Sources are given at the end of each chapter.

We would like to thank all those who have helped us directly or indirectly in completing this book, in particular, S. C. Nanavati for critically reading the chapters on printing, S. N. Mishra for reading all of the manuscript, H. M. Raje and C. K. Datye for preparing the drawings, Urmila Sharma, P. S. Eswaran, and especially M. D. Bhat for typing the manuscript.

The technology of chemical processing is changing very rapidly and we may need to revise the book in future. We, therefore, welcome any criticism, suggestions, and comments on the present volume.

KESHAV V. DATYE
A. A. VAIDYA

Kota, Rajasthan, India
February 1984

CONTENTS

14 PRINTING OF POLYESTER AND ITS BLENDS   **375**

15 TRANSFER PRINTING   **396**

CHEMICAL PROCESSING
OF SYNTHETIC FIBERS
AND BLENDS

1 | INTRODUCTION

While working with bifunctional compounds that can react at both ends to form long molecules, Carothers synthesized a polymeric ester, that is, a polyester from hexamethylene glycol and adipic acid. The molten polyester gave a long filamentlike material that could be stretched several times its original length. The stretched material did not return to its original length on release of stress. This was probably the first synthetic fiber.[1]

The initial fibers made from aliphatic polyesters were not satisfactory. Subsequently, Carothers synthesized a polyamide, nylon 66, that was found to have good textile properties.[2] The commercial production of nylon 66 started in 1938 in a small pilot plant at DuPont's Wilmington factory. The success of nylon 66 exceeded all expectations. Other fibers followed one after another during the next four decades. Nylon 6 was synthesized by Schlack of IG Farbenindustrie in 1937 and was commercially produced in West Germany in 1948.[3] Aromatic polyester fiber was first synthesized by Whinfield and Dickson of Calico Printers Association (U.K.).[4] It was commercially produced by Imperial Chemical Industries (ICI) in 1950.[5] The first acrylic fiber, Orlon, was produced by DuPont in 1945 on a pilot plant scale. In 1954, Natta and Ziegler patented a process of producing polypropylene.[6] Commercial production of the fiber was started by Montecatini Societa Generale in Italy in 1957. The first polyvinyl chloride fiber, PECE, was developed in Germany during the Second World War.[7] Polyvinyl alcohol fiber was produced from 1957.[8] Polyurathane elastomeric fibers appeared in 1958.

Synthetic fibers of interest to the textile industry are polyamides, polyesters, acrylics, polyhydrocarbons, and substituted polyhydrocarbons. Most new fibers fall into one or the other of the above groups and no new class of

1

commercial textile fibers has been introduced on a large scale in recent years. Polymer alloys and copolymers with two or more types of bonds in polymer chains have been explored recently. Replacement of the aliphatic chain with an aromatic ring in polyamides, use of aromatic dihydroxy compounds in place of aliphatic diols, and use of substituted monomers are some other routes to finding fibers with desirable new properties. It appears very unlikely that any basically new chemical structure will be introduced in synthetic textile fibers. On the other hand, nontextile fibers may be developed with entirely new chemical structures. Modified cellulosic fibers, such as cellulose acetate and triacetate, are still produced commercially. They are also discussed along with the synthetic fibers, wherever necessary.

The first synthetic fibers were rodlike, long filaments with metallic luster. The fibers were dulled (made opaque) with the addition of titanium dioxide. Their cross-sections were altered to improve the glitter or to impart dullness and soft feel. They were crimped and permanently textured. The ionic character of the polymer was modified to impart or eliminate affinity for a given class of dyes. Hollow fibers, bicomponent fibers, mass-colored fibers, fibers with microcrators on the surface, elastomeric fibers, and fibers with specific elongation at break were produced to achieve the desired textile effects.

1.1 GROWTH OF SYNTHETIC FIBERS

There has been a phenomenal growth in the production of synthetic fibers during the last three decades while the growth in regenerated cellulose fiber has remained stagnant (Table 1.1, Figure 1.1).[9] Of the total fiber production, synthetic fibers accounted for a meagre 0.73% in 1950, but rose to 37% in 1980. Polyester, nylon, and acrylic fibers are produced in major quantities

TABLE 1.1 Production of Textile Fibers (Thousand Metric Tons)[9]

Year	Cotton	Wool	Synthetic Filament	Staple	Regenerated Cellulosic Filament	Staple
1900	3162	730	—	—	1	
1910	4200	803	—	—	5	
1920	4629	816	—	—	15	
1930	5870	1002	—	—	205	3
1940	6907	1134	1	4	542	585
1950	6647	1057	54	15	871	737
1960	10113	1463	417	285	1131	1525
1970	11784	1602	2391	2831	1397	2187
1980	14137	1581	4731	5756	1159	2085

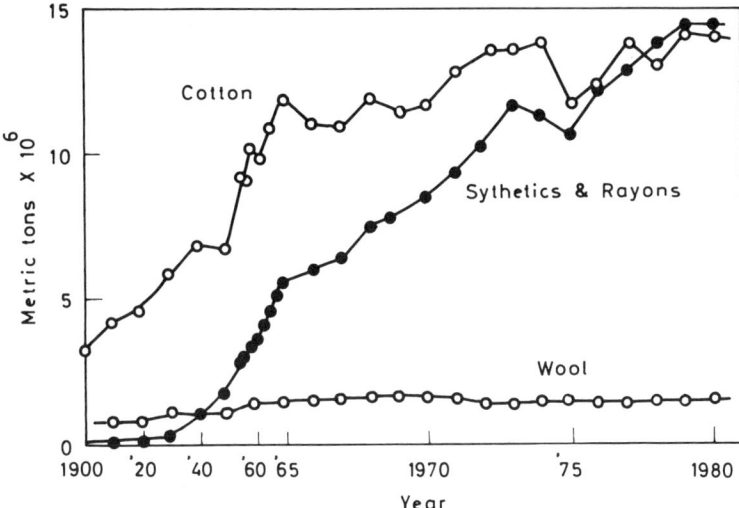

FIGURE 1.1.   World production of fibers.[9]

(Table 1.2); polyester fiber shows the highest rate of growth (Figure 1.2)[10] and its total production has surpassed that of nylon since 1972. Polypropylene fiber is slowly gaining ground (Table 1.3), although it is now consumed mainly in nontextile end uses. The production of other fibers such as polyvinyl alcohol, polyvinyl chloride, and polyurathane for textile use is still very limited.

The production of synthetic fibers is increasing despite the big price hike in 1973 of crude oil, the raw material for synthetic fibers. The volume of oil that goes into the synthetic fiber industry is very small and is efficiently used compared to generating energy by combustion. Synthetic fibers are making a substantial contribution to the overall improvement in the quality of life through their diverse characteristics. Their production can be easily increased, unlike cotton, regenerated cellulose, or wool.

The synthetic fibers can be tailormade in their properties, have a longer life, and are cheaper to use than other fibers.[11] Attempts have been made to overcome their limitations so that they will satisfy the demands of comfort, appearance, and ease of wear, washing, and processing. The future of synthetic fibers appears to be quite bright. By the turn of this century, they are likely to account for 60% or more of total fiber, as opposed to cotton and wool (24%) and the rest (16%).

A far-reaching development in the synthetic fiber industry is the production of staples of short length that can be blended with natural, animal, or other synthetic fibers. Filament materials have metallic luster, unpleasant handle, lack of hairiness, low bulk, and high cost because of a low rate of production. Blended materials cannot easily be made with filament yarns. It is, therefore, a common practice to have the fiber in staple form and to spin a

TABLE 1.2 **Production of Synthetic Fibers[9] (Except Polyolefins)**

Fiber	1971	1972	1973	1974	1975	1976	1977	1978	1979	1980
				Thousand Metric Tons (by Year)						
				Acrylic + Modacrlyic						
Yarn	4	5	5	4	3	3	3	3	4	3
Staple	1164	1266	1575	1445	1388	1739	1784	2014	2061	2080
				Nylon + Aramid						
Yarn	1859	2042	2272	2193	2083	2342	2379	2506	2636	2591
Staple	299	395	455	431	405	509	559	622	648	534
				Polyester						
Yarn	956	1127	1510	1545	1641	1744	1892	2045	2227	2093
Staple	1169	1396	1665	1722	1728	2143	2401	2641	2907	3039
				Other Fibers						
Yarn	41	39	46	43	36	39	44	41	40	44
Staple	111	107	112	104	71	75	79	74	91	103
				World Total						
Yarn	2860	3213	3833	3785	3763	4128	4318	4595	4907	4731
Staple	2743	3163	3807	3702	3590	4466	4323	5351	5707	5756
Total	5603	6377	7460	7484	7353	8494	9141	9946	10614	10187

yarn from these staples, either alone or in combination with other fibers. Polyester and acrylic staples are commonly used with cotton, wool, and viscose to get blended yarns. Nylon staple material is not commercially popular because of its high rigidity and high denier (usually not less than 4).

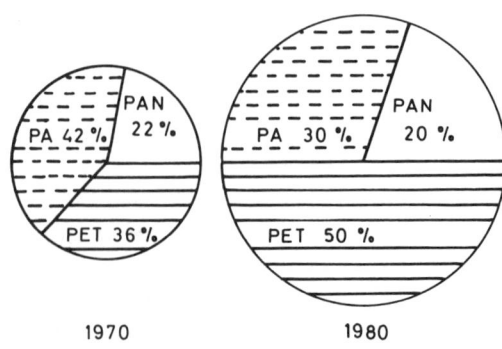

FIGURE 1.2. Production (%) of polyester, polyamide, and acrylic fibers in 1970 and 1980.[10]

TABLE 1.3 Production of
Polyolefin Fiber
(Thousand Metric Tons)[9]

Year	Production
1971	449
1972	562
1973	683
1974	712
1975	711
1976	738
1977	841
1978	912
1979	1352
1980	1319

The weak cotton fiber is cut by the nylon staple during wear, which thus makes the blend unsuitable for textile use.

The properties of component fibers in a blend compliment each other. Many drawbacks of synthetic fibers such as low-moisture absorption, accumulation of static charge, easy soiling, metallic luster and feel, and lack of ideal warmth, feel, comfort, drape, fall and so on can be overcome by blending them with other fibers. Cellulosic fibers have good moisture absorption, low accumulation of static charge, and comfort, even though they have limited strength, poor dimensional stability, and low crease recovery. Wool has good moisture absorption and aesthetic properties, but it is a delicate fiber with a tendency towards felting. Fabrics from blended yarn exhibit properties superior to those from a single fiber. Blending is also done to replace part of the expensive fiber by a cheaper one in order to bring down the price without affecting the quality of the resultant fabric. The blending of fibers gives a material that can be dyed in solid, contrast, or tone-in-tone shades to produce multicolored fancy textiles. Because of such reasons, the production of staples and blends of different fibers is growing rapidly (Table 1.4). Besides blends of polyester with cotton, viscose, and wool, acrylic/ wool, nylon/wool, acrylic/cellulosic, synthetic fiber/silk, jute, cellulose acetate and cellulose triacetate, and polyester/nylon are also produced to a small extent.[12] The cationic dyeable polyester is blended with normal polyester and cotton or wool to get fancy color effects or frost-free dyed fabric. Blends or mixtures of a differentially dyeable fiber with the normal fibers are produced to get multicolored effects. The processing of blends involves treatments for component fibers having entirely different chemical properties, and has been described in the present book along with the synthetic fibers.

TABLE 1.4   Production of Filament, Staple, Blended Yarn, and Fabric

| Year | World Production[9] (kg × 10⁶) | | Production in India[12] | |
	Filament	Staple	Blended Yarn (kg × 10⁶)	Fabric (m × 10⁶)
1973	5206	6116	—	—
1974	5098	5984	25.48	113.94
1975	4098	4970	40.49	233.97
1976	5327	6569	74.96	330.35
1977	5496	6960	188.14	890.77
1978	5780	7600	224.88	983.15
1979	6095	7998	187.88	942.31
1980	5932	7954	—	—

1.2   COMMERCIAL SYNTHETIC FIBERS

Synthetic fibers are produced all over the world by many manufacturers. They are marketed under various trade names as staples and filaments of different deniers, staple length, filaments per yarn and so on. Typical trade names and manufacturers for polyester, polyamide 66, polyamide 6, and acrylic (including modacrylic) fibers are given in Table 1.5. The raw materials and manufacturing process of a given fiber vary from producer to producer. The mechanical and thermal history of a fiber vary from source to source and fibers produced by different manufacturing processes invariably have differences in their properties. Any change in the parameters of a given manufacturing process may yield fibers with slightly or even significantly different properties. It is essential for the success of chemical processing to have a fiber with the specified properties. Fiber manufacturers usually give a merge number for a fiber lot with identical specified properties. The mixing of fibers from different manufacturers or from fiber lots with different merge numbers from a given manufacturer invariably endangers the success of the chemical processing of mixed materials. Problems created by the mixing of lots are tough to solve and, at times, unsurmountable.

Block copolymers containing polyurethane linkages are called elastomeric fibers, polyurethane elastomers, or Spandex. The first elastomeric fiber was introduced in 1958 and the field expanded rapidly during the 1960s. Elastomeric fiber consumption is not very large compared to other textile fibers. The polyurethane technology has been reviewed by Gregg.[13]

1.3   PRODUCTION OF SYNTHETIC FIBERS

Synthetic fibers are produced from highly pure monomers under very strictly controlled standard conditions. The following steps are generally involved in the manufacturing process:

TABLE 1.5 Some Commercial Synthetic Fibers

Manufacturer		Trade Name of Fiber		
	Acrylic	Nylon	Polyester	PP, PVA, PVC, CA, and CTA
Akzo N. V. (Netherlands, Enka Co.)	—	Enkalon	Diolen and Terlenka	—
American Cyanamid Co. (U.S.A.)	Creslan	—		—
Asahi Chemical Ind. Co., Ltd. (Japan)	Cashmilon	Asahi kasei nylon	Asahi kasei ester	Roica (PU)
Badische (BASF Amer.) Corp. (U.S.A.)	Zefran	Zeftran	—	—
Bayer AG (W. Germany)	Dralon	Perlon	—	Dorlastan (PU)
Celanese Fibers Corp. (U.S.A.)	—	—	—	CA and CTA
Courtaulds Ltd. (U.K.)	Courtelle	—	—	CA, CTA
E. I. DuPont De Nemours and Co. (U.S.A.)	Orlon	Antron	Dacron	Typar (PP) Lycra (PU)
Eastman Kodak Co. (U.S.A.)	—	—	Kodel	—
Hercules Incorporated (U.S.A.)	—	—	—	Herculon (PP)
Hoechst A. G. (W. Germany)	Dolan	—	Trevira	—
Imperial Chemical Ind., Ltd. (UK)	—	Bri Nylon	Terylene	Ulstron (PP)
J. K. Synthetics Ltd. (India)	Jaykrylic	Jaykaylon	Jaykaylene	—
Kuraray Co., Ltd. (Japan)	—	—	Kuraray ester	Kuralon (PVA)
Mitsubishi Rayon (Japan)	Vonnel	—	Soluna	Mitsubishi Pylen (PP), CA and CTA
Monsanto Co. (U.S.A.)	Acrilan	Blue C, Cadon	Blue C	—
Montefibre SPA (Italy)	Leacril	Helion Nailon	Terital	Meraklon (PP), CA
Rhone-Poulenc S. A. (France)	Crylor	Nylfrance	Tergal	Rhovyl (PVC)
Snia Viscosa (Italy)	Velicren	Lilion	Wistel	—
Teijin Ltd. (Japan)	—	Teijin Nylon	Teijin Tetoron	Teviron (PVC) CA
Toray Industries (Japan)	Toraylon	Amilan Promilan	Toray Tetoron	Toray Pylen (PP)

1. Conversion of monomers into a polymer.
2. Conversion of the polymer into a suitable form for spinning.
3. Pressing polymer melt or dope through spinning jets to form filaments.
4. Development of the morphological fine-structure of the fiber.
5. Modifying the filaments to suit the end use.

The methods of making special fibers such as bicomponent fibers, fibers with modified cross-sections, hollow fibers and so on are described in the literature.[11] The production may be carried out in a batchwise manner or in a continuous manner by combining a few steps. Similarly, each step may be carried out in a batch process or a continuous process. Each step in fiber manufacture has a considerable influence on the chemical and physical structure of the synthetic fiber, which, in turn, decides the properties of the yarn and the fabric. It is, therefore, essential to consider briefly these production steps from a wet processer's viewpoint, so that the coloring and finishing of these fibers can be subsequently well-understood. Monographs on each fiber and reviews may be referred to for details of their production. Only typical examples will be described here.

1.3.1 Polyester

The monomers for the production of polyesters are diols and dibasic acids or their esters. Polyethylene terephthalate, the most common polyester, is produced from ethylene glycol (EG) and dimethyl terephthalate (DMT) or terephthalic acid (TPA) using transesterification and polycondesation catalysts (Scheme 1.1).

The melting point of the monomer (DGT) is 101°C, the dimer ($n = 1$) 167°C, the trimer ($n = 2$) 200–202°C, and the tetramer 220°C. If DMT is used instead of TPA, the first stage is the transesterification with the evolution of

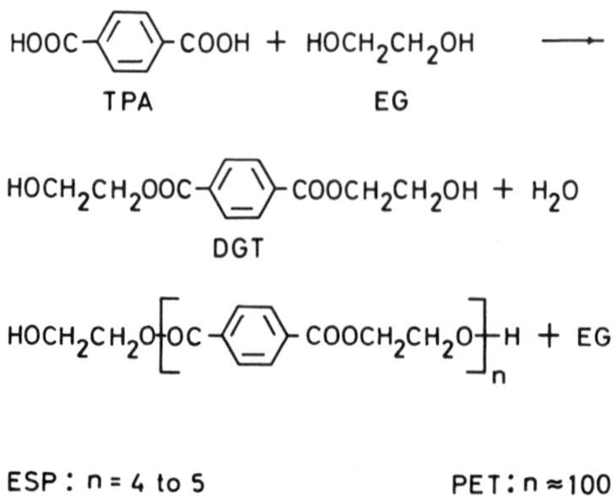

HOOC—⟨☐⟩—COOH + $HOCH_2CH_2OH$ ⟶

  TPA   EG

$HOCH_2CH_2OOC$—⟨☐⟩—$COOCH_2CH_2OH$ + H_2O

          DGT

$HOCH_2CH_2O$[OC—⟨☐⟩—$COOCH_2CH_2O$]$_n$H + EG

ESP : $n = 4$ to 5   PET : $n \approx 100$

SCHEME 1.1

methanol instead of water. This step may be carried out separately to get ESP followed by the polycondensation step to get polyester. DMT has some advantages over TPA. It can be easily purified by distillation or crystallization, can be added to the reactor in molten form, and yields PET with a low carboxyl group content. The catalyst for transesterification is an acetate of sodium, manganese, zinc, or cobalt, or a mixture thereof. Antimony trioxide or acetate is used as a polycondensation catalyst. A heat-stabilizer is also added. The delustering agent, titanium dioxide, is finely dispersed before it is added to the molten DMT. The ratio of DMT:EG is about 1:2. Agitation, increased surface area, generation of fresh surface, and a vacuum of 0.1–0.5 torr increase the rate of reaction in the final stages. Total time of transesterification and polycondensation is 8–15 h. Apart from the polymerization reaction, a number of side reactions take place. Most of these side reactions are undesireable since they lower the quality of the polymer, impart color to the fiber, may develop cross-bonded polymer gels and so on. The reaction between two hydroxyl groups to form ether linkages (diethylene glycol residue) has to be minimized since these linkages are sensitive to UV light and the fiber loses its strength on exposure to light. The melting point of the polymer is also lowered by diethylene glycol residue. The formation of vinyl groups, aldehydic groups, and other unsaturated structures impart color to the polymer. Naps and gels in the fiber, broken filaments, and poor whiteness are produced by these undesireable side reactions. The proper selection of reaction conditions, purity of monomers, and design of the polymerization unit are therefore of critical importance. The formation of low-molecular-weight oligomers, particularly, cyclic oligomer, is an unavoidable part of the polymerization reaction (see Section 1.7).

The polymer melt is spun into ribbons or bands that are granulated. The total time of spinning ribbons has to be as low as possible to get chips of uniform viscosity. The chips are thoroughly mixed to minimize batch-to-batch variations as well as in a given batch from the beginning to end of the spinning of ribbons. The raw chips must meet specifications, such as:

Intrinsic viscosity	0.65–0.66
Melting point	255–260°C
Carboxyl group content	30 ± 5 meq/kg
Diethylene glycol residue content	Not more than 1.5%

The color of the chips is white. These chips have a tendency to soften and to stick together at 130°C. Therefore, the drying of the chips is carried out under tumbling or agitation. During drying, the crystallinity of the chips increases up to about 40% after which the chips do not soften below 200°C. The moisture content of the chips after drying is very low and is never higher than 0.005%, since polyester is easily hydrolyzed by traces of water during melt-spinning. In a continuous polymerization–melt-spinning process, the polymer melt is used for spinning the filaments; thus, ribbon spinning, chips cutting, drying, and remelting steps are avoided.

The conversion of polyester chips or melt into filaments is carried out by the melt-spinning process. The polymer chips are melted in an extruder, the melt is filtered and fed to the spinnerets by a metering pump. The temperature of the melt is adjusted to get the desired melt viscosity and is about 30–40° higher than the melting point of the polymer, that is, around 290–300°C. The polymer emerges from the spinnerets as filaments that are solidified by blowing cold air. An appropriate spin finish is applied and the filaments are wound on take-up bobbins. The number of holes in a spinneret may be as low as 1–4 or as high as several hundred. For staple fibers, the tow of filaments is collected in a drum.

The spin-finish has three main functions: (*a*) surface lubrication, (*b*) proper filament cohesion, and (*c*) imparting antistatic properties to provide protection from static charge.[14,15] It is an emulsion of oil, antistatic agent, and surface active agent in water. The filaments are wound on a bobbin under slight tension at a speed of 800–1000 m/min. Speeds of 2500–4000 m/min are used when moderately or partially oriented yarns (MOY or POY) are obtained. At this stage, the filaments have low strength and high extensibility and cannot be used for weaving or knitting textiles. The filaments are subjected to a stretching or drawing process in order to impart high tenacity and desired extensibility. In this process, the filaments are stretched to 3–4 times their original length or 1.2–1.5 times the original length in the case of MOY–POY. The drawing process involves orienting polymer molecules in the direction of the fiber axis (see Chapter 2). Temperatures higher than the glass-transition temperatures of PET are employed so that the polymer chain molecules are free enough to orient. Thus, the undrawn yarn is heated, usually in two stages—90°C and 150–170°C. It is then stretched between the rollers (godets). The desired draw ratio decides the ratio of the surface speed of the two godets. The upper (first) godet is at about 90°C. There is a heating rail in between the two godets that is maintained at about 160°C. The lower godet is at the ambient temperature. After drawing, the yarn is wound by the conventional ring traveller spindle mechanism when a certain amount of twist is inserted into the yarn. Simultaneously, the yarn is allowed to relax so that the stability of the yarn is improved.

The technology of staple manufacture differs slightly from that for filaments mainly after the melt-spinning operation. A large number of filaments are brought together to form a thick tow. The tow is passed through water to remove the spin-finish, squeezed, heated, and stretched on a series of stretching godet rollers that revolve at increasing speed depending on the desired draw ratio. The godets are heated from 140°C to finally 180–190°C to set the fiber. The stretched tow is wetted, crimped, dried, heat set, and cut into staple fibers of desired length. The stretched tow can be directly converted into tops that are used for yarn spinning. The staple fibers are pressed into bales and packed. Generally, the bales are marketed for blending and spinning into yarns. The final structure of the fiber builds up during the draw-twisting (D/T) process for filaments and the drawing–heat-setting–crimping, process for the staples. From the viewpoint of processors, the drawing operation is the most important one since many of the yarn characteristics, defects, and faults are developed during this operation. It is discussed in more detail at a later stage (see Section 1.4).

1.3.2 Polyamide 66

A variety of polyamides have been synthesized and spun into fibers. However, nylon 66 and nylon 6 are the main polyamide textile fibers. Their manufacturing processes are described here.

The monomers for nylon 66 are hexamethylene diamine (HMD) and adipic acid (AA) which are converted into nylon 66 salt and then polycondensed into the polymer with the elimination of water molecules (Scheme 1.2).

Hexamethylene diamine and adipic acid are dissolved separately in methanol. When the two solutions are mixed, the neutralization reaction takes place and the nylon 66 salt is formed. The salt is relatively insoluble in methanol and crystallizes out on cooling. The crystals are separated by centrifuging and washing with methanol. The product contains only a small amount of methanol and can be used directly for the polycondensation reaction. Nylon 66 salt is dissolved in water to get a 50–60% solution. Acetic acid (0.5–1%) is then added as a viscosity stabilizer. The dispersion of titanium dioxide is added if dulled nylon is to be produced. The solution is then heated in an autoclave when a pressure of 15–20 kg/cm^2 builds up. The pressure is slowly released and heating is continued to distill off all the water. The autoclave is evacuated to facilitate the removal of water from the highly viscous molten polyamide mass. The polymer melt is then spun into ribbons and granulated into chips. The chips are dried, mixed together, and stored in silos under nitrogen. These chips are melt-spun in a manner similar to that described earlier for PET. The filaments on take-up spools are cold-drawn to 3–4 times their original length in order to build up the fine structure of the fiber. Staple fibers of nylon 66 are also produced from the tow that is cold-drawn, washed, crimped, heat set, and cut as usual.

$$H_2N(CH_2)_6NH_2 + HOOC(CH_2)_4COOH \longrightarrow$$
$$\quad\quad HMD \quad\quad\quad\quad\quad AA$$

$$H_2N(CH_2)_6NHCO(CH_2)_4COOH + H_2O$$
$$\quad\quad Nylon\ 66\ salt$$

$$H\!-\!\!\left[HN(CH_2)_6NHCO(CH_2)_4CO\right]_n\!\!-\!OH + H_2O$$
$$\quad\quad Polyamide\ 66$$

SCHEME 1.2   Nylon 66 polymer.

1.3.3   Polyamide 6

6-Amino caproic acid or caprolactam is the raw material for the production of nylon 6. The polymerization of caprolactam takes place in two stages (Scheme 1.3).

The polymerization of caprolactam is carried out either in a batch process or a continuous process. Water and acetic acid are added to the molten caprolactam. In a batch process, the molten caprolactam, water, acetic acid, and a slurry of titanium dioxide are heated in an autoclave to 220–240°C under pressure up to 12–15 kg/cm^2. The steam is slowly released, nitrogen is flushed, and a vacuum is applied to remove the last traces of water. Unlike nylon 66, the reaction never reaches a state of completion. The melt is kept under constant temperature when the reaction reaches an equilibrium. The molten polymer is spun into ribbons and cut into chips.

In the continuous polymerization process,[11] the molten caprolactam, water, acetic acid, and delustrant are passed through a reaction (VK) tube that is about 8–10-m high. The inner part of the tube contains several fittings such as perforated plates at varying inclinations to keep the moving flux as flat as possible. The temperature of the tube is kept at 250–270°C. The retention time of the polymerizing mixture is 20–24 h. The polymer is continuously extruded from the bottom of the tube into a water bath in order to get ribbons that are then cut into chips.

The chips of nylon 6 contain about 7–10% of unconverted monomer, which is extracted by hot water (90–100°C) so that the concentration of extractables in the chips drops below 1%. The washwater is collected so that the caprolactam can be recovered. Along with the caprolactam, oligomers are also removed from the chips. The chips are dried in rotating drum dryers at 100–120°C and a final reduced pressure of 1–2 torr. A time of 20–50 h is required to get chips with less than 0.1% moisture. The dried chips from different batches are mixed together to get a uniform product and are stored in silos under nitrogen. The melt-spinning, cold-drawing, and other operations are similar to those described earlier. Since staple fibers of nylon 6 are not used commercially, they are rarely produced.

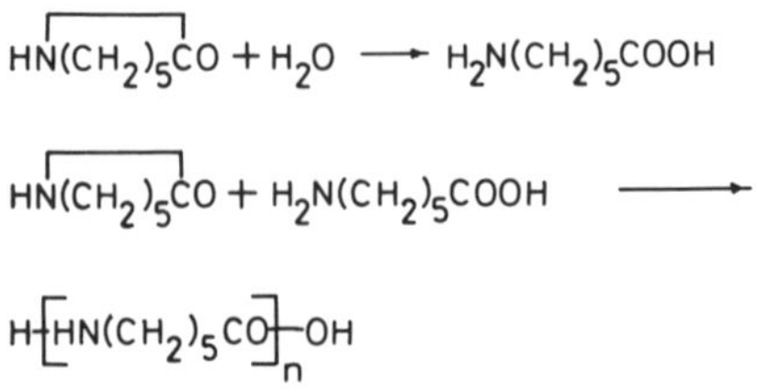

$$HN(CH_2)_5CO + H_2O \longrightarrow H_2N(CH_2)_5COOH$$

$$HN(CH_2)_5CO + H_2N(CH_2)_5COOH \longrightarrow$$

$$H[HN(CH_2)_5CO]_n OH$$

Polyamide 6 $n \approx 200$

SCHEME 1.3

1.3.4 Acrylic Fiber[16]

The worldwide demand for woollike synthetic fibers at a reasonable price can be satisfied by acrylic fibers. In 1940, DuPont produced the first acrylic fiber by polymerizing acrylonitrile ($CH_2 \cdot CH \cdot CN$). The fiber had no affinity for any dye and was difficult to color by any known method, which retarded the growth of the fiber for quite some time. After considerable research, a commercially acceptable fiber was developed.[16] The development of improved dyeing technology and dyes greatly influenced the market potential for acrylic fibers. The major markets for acrylic fiber did not develop until after 1954 when cationic dyeable fibers were introduced. The world production of acrylic fiber then grew rapidly from 109 kton in 1960 to 2080 kton in 1980 and accounted for 18% of the total production of synthetic fibers and 35% of the staple fibers.

Almost all the acrylic fibers that are produced today are copolymers of acrylonitrile and other vinyl monomers. The acrylonitrile content of the polymers is at least 85%. Those fibers that contain less than 85% but more than 35% acrylonitrile are called *modacrylic fibers*. The comonomers most commonly employed in the production of acrylic fiber are of three types. The addition of about 5–10% of neutral comonomers such as vinyl acetate, methyl acrylate, and methyl methacrylate, reduces the glass-transition temperature from 104°C of homopolymer of acrylonitrile to about 80–90°C, thereby facilitating the dyeing to be carried out at boiling. It also reduces the fiber compactness and improves the diffusion of dyes in the fiber. The acidic comonomer such as acrylic acid, allyl sulfonic acid, and itaconic acid imparts cationic dyeability to acrylic fibers. Cationic dyeability is also imparted to the acrylic fiber if potassium persulfate–sodium bisulfite redox catalyst system is used for polymerization. This system introduces sulfate and sulfonate groups at the end of polymer chains that impart substantivity for cationic dyes to the fiber. The fibers containing basic comonomers such as vinyl pyridene or ethylene imine are dyeable with acid dyes. However, acid dyeable acrylic fibers have not become very popular and are mainly used as differentially dyeable fibers for getting multicolored effects. The commercial acrylic fiber is noted for its woollike feel, bulking, inertness to chemicals, molds, and bacteria, and resistance to weather. A variety of acrylic and modacrylic fibers are produced that differ in their properties and dyeability, and are used in a specific field of textile application.

The polymerization of acrylonitrile along with other comonomers can be carried out by aqueous slurry polymerization techniques, either by a batch process or a continuous process. In this process, the monomers are polymerized in aqueous dispersions with free radicals generated by a redox–catalyst system such as persulfate–bisulfite–iron sulfate; the polymer formed is isolated and dried before being redissolved in a spinning solvent. In a typical batch process, ammonium persulfate catalyst and sodium bisulfite as an activator are dissolved in demineralized water at 40°C, and a mixture of about 90 percent acrylonitrile and 10 percent other ethylenic monomer is slowly stirred in over a period of 2 h. The copolymer of acrylonitrile with other monomers precipitates out. It has a molecular weight of around 60,000. The precipitated copolymer is filtered, washed, and dried.

A continuous polymerization is carried out from an aqueous dispersion (suspension) or alcoholic solution of calcium or sodium thiocyanate, zinc chloride, and sodium perchlorate. In a solution polymerization, dimethyl formamide is used as a solvent medium. After polymerization, the solvent is partially removed by distillation and the concentrated dope is used for spinning instead of using dry polymer powder.

Acrylic polymers do not melt before decomposition and are not melt-spun. They are spun from a dope of high viscosity by either a dry or wet-spinning process. The polymer powder is dissolved in a solvent such as dimethyl formamide, dimethyl acetamide, dimethylsulphoxide, or nitric acid. The polymer dope is carefully prepared to avoid the formation of gels, filtered, deaerated, and is immediately used for spinning. If the dope is stored for a long time, the polymer becomes oxidized and gives a dull yellow fiber. The dope is maintained at a very low temperature if nitric acid is used and at about 80–150°C in the case of organic solvents. The dope is fed to the metering pumps, which, in turn, feed it to the spinnerets. In dry spinning, filaments coming from the spinneret pass through a column in which hot air is circulated at 230–260°C. Since the boiling point of DMF is 153°C, it evaporates rapidly to cause solidification of the polymer dope into filaments that are collected on a bobbin at 100–300 m/min. For the wet-spinning process, the polymer dope is fed to the spinnerets having 1000–60,000 holes. Streams of the polymer solution emerging from the spinneret holes coagulate into filaments when they come in contact with the spinning bath that consists of the dilute aqueous solution of the solvent. The major fraction of the solvent diffuses out of the filaments during the coagulation process while the residual solvent is washed with hot water with simultaneous stretching of the filaments (to build-up the fiber structure). The latter is carried out by passing the filaments over rollers rotating at increasingly higher speeds when filaments are stretched 4–10 times their original length. The stretching is carried out at 70–100°C in water or steam or at 80–110°C in hot air. The water-swollen structure of filaments collapses during stretching and drying. After applying softener, antistatic agent, and other finishing chemicals, the filament tow is crimped; heat set (annealed), preferably by steam; and finally cut into staple fibers. To get high bulk, prerelaxed staples and stretched staple fibers are mixed together when a yarn is spun. On steaming, the unrelaxed staple fibers shrink while the others do not. They therefore form loops around the shrunk fibers. The bicomponent fibers are also used to get high-bulk acrylic yarn.

1.3.5 Polypropylene

Polypropylene polymers in which all side-chain methyl groups are located on the same side of the paraffinic backbone, are used for fiber spinning. Sterio regular, high-molecular weight, linear, fiber-forming polypropylene is synthesized using a Ziegler–Natta catalyst by an anionic mechanism. These catalysts are based on some form of TiCl₃ and an aluminium alkyl or alkyl halide. Mixed metal-alkyls such as lithium aluminium alkyls and various metal hydrides are also commercially used.

There are five key steps involved in the manufacturing process:

1. Catalyst preparation.
2. Polymerization.
3. Purification.
4. Recovery of solvents.
5. Compounding and finishing.

Both batch and continuous polymerization plants are under operation. In batch operation, the reactor is filled with the monomer and a diluent, catalyst and related components are added, conditions are maintained for a given period of time with periodic additions of more propylene as the reaction proceeds and the material is discharged into the flash tank. In the continuous process, fixed pressure is employed and the production rate is controlled by the rate of addition of catalyst and indirectly by the rate of heat removal.

The polymer slurry is passed into the flash tank, which is maintained at a relatively low pressure. Most of the monomer, along with some of the diluent, is flashed away. The slurry leaving the flash chamber contains solid polymer, soluble amorphous polymer, inert diluent, some residual monomer, and an active catalyst. The catalyst is deactivated with an alcohol and is removed by separation processes. The soluble polymer is filtered or centrifuged away and residual soluble polymer is washed with the diluent. After centrifuging, the polymer cake contains considerable quantities of other volatiles. Both rotary dryers and spray dryers are used to remove the volatile compounds.

The use of an inert medium in the reactor helps the proper integration of propylene and the catalysts, and facilitates the removal of heat and the transfer of the solid polymer. Paraffins, cycloparaffins, or mixtures thereof may be used as the inert medium.

The filaments are produced from the polymer by melt-spinning. Unlike polyester or nylon, polypropylene is spun at 100–150°C higher than its melting point because of the high melt viscosity of the polymer melt. The spinneret jet holes are also longer (higher L/D ratio) than those used for PET or nylon spinning. To minimize degradation, the filaments are quenched fast.

The spun fiber has a crystalline content of 33 percent. The fiber is subsequently stretched under various conditions of temperature and stretch ratio. The fiber is then annealed to improve the dimensional stability.

Polypropylene (PP) is a melt-spun fiber with excellent physical properties. Tenacities range from 3–5.5 gpd for staple fiber and 5–8 gpd and possibly higher for continuous filament yarns. Since the fiber has extremely low-moisture absorption, wet and dry strengths are the same. The loop strength is 80–85% of the regular strength and the modulus of polypropylene filament yarns falls between those of polyester and nylon. The same is true for staples at low elongations, but polypropylene has a higher modulus at higher elongations.

Polypropylene has the lowest fiber density of any commercial fiber. A specific gravity of 0.9 gives this fiber an advantage in bulk over all other fibers. It has high flat-and-flex abrasion resistance. The pilling tendency of PP is very low because of its high abrasion resistance. This property, high strength, and high bulk have made PP fiber very attractive for use in ropes, cordage, netting, industrial fabrics, and carpets. In blends with cotton, rayon, or wool, PP appears to be an excellent fortifying fiber.

The main shortcoming of PP is its low softening point (160°C). Fabrics containing more than 30% polypropylene are not exposed to temperatures higher than 125–130°C as they become harsh due to shrinkage. Polypropylene is completely free from polar groups, such as the amide function in nylon and the ester group in PET. This makes the fiber stain-resistant and easy to clean. It also makes it difficult to dye.

The melting point of PP is 176°C and the sticking temperature of the fiber is 145 ± 3°C. The molecular weight varies from about 100,000 to 1,000,000. The crystallinity is about 60% and the specific gravity 0.9–0.91. The moisture regain of the fiber, due to its paraffinic nature, is almost zero. Resistance to most chemicals acids, alkalis, and oxidizing agents is excellent. PP is soluble in hot chlorinated and aromatic hydrocarbons.[17]

The low melting point of PP–polymer is a drawback that affects the ironability and hot processing. If the pendant methyl group is replaced by isopropyl or isobutyl, the melting point of the new polymer is greatly elevated. Both these polyolefins can be spun into fiber. However, the monomers are not available at a low enough price for commercial production.[18]

The resilience of PP fiber compares unfavorably with that of other fibers. However, in newer PP fibers, this limitation has been overcome and it is claimed that polypropylene carpet yarns equal or exceed nylon in crimp development.[18]

The dyeing of PP fiber is a serious problem since there are no polar groups in the fiber to act as dye-receptor sites. As a result, there is no driving force to bring the dye into the fiber and no binding force to keep them in. The general techniques that have been used to impart dyeability to polypropylene are (a) additives, (b) graft copolymers, (c) chemical modification,[19] and (d) combination of additives with posttreatments.[20,21]

Permeability of PP Fibers.   Lack of dye receptors and lack of permeability to dyes are both responsible for the undyeable nature of polypropylene. A fiber with basic dye receptors in sizable quantity but without permeability remains essentially undyeable. The fiber behaves as though the dye receptors were encapsulated in a waxy hydrocarbon matrix through which the dyes cannot penetrate. Thus, the graft loses its effect on dyeability, if the fiber is remelted and spun into a fiber again. A basic polymer dye-receptor in the PP polymer can be made effective by a simple chemical treatment. Without the after-treatment, the fiber is either undyeable or dyeable only to a pale shade.

Static Electricity.   Polypropylene has a low tendency to build-up static electric charge, even though it is hydrophobic fiber with no moisture. Only in very dry weather when the relative humidity falls below 20% is static build-up of low order felt on PP carpets.

1.3.6   Elastomeric Fibers[11,13]

Elastomeric fibers are synthetic segmented copolymers containing polyurethane linkages. Lycra was the first commercial polyurethane elastomer fiber marketed by DuPont. It is based on polyethylene glycol and toluene-2-4-diisocyanate. The polymerization is carried out in a water containing a small amount of acid chloride.

   Elastomeric fibers achieve their stretch and recovery through a combination of high-melting (hard) and low-melting (soft) segments joined in the same molecular chain. The lateral bonding forces, physical size, and molecular shape of the hard segments contribute to the modulus and thermal stability. They are important to recovery because they help limit the plastic flow. The most common hard segments in commercial use are aromatic co-polyureas linked to the soft segments by urethane linkages. The hard segment is formed in the polymer chain by the reaction of a diisocyanate prepolymer with a diamine chain extender. Various techniques and modifications are developed to get elastomers with increased elastic power.[11]

   Elastomeric fibers are commonly formed by dry-spinning into a heated gas or by wet-spinning into a dilute solvent bath. In a technique known as *reaction spinning,* the isocyanate-terminated prepolymer is extruded into a solution containing the chain extender which builds the final polymer as it penetrates the surface of the extrudate. Melt-spinning of elastomeric polymers with good fiber properties has proved difficult because of their poor stability at extrusion temperatures. Hard segments in the elastomeric fibers are believed to associate in sufficient number for the resulting discrete domains to be large enough to act as reinforcing filler particles, much as carbon black does in rubber. The size of the domains is in the range of 2–10 nm.[11]

   Such fibers are used in various applications as is or in combination with other fibers. The blends may be core-spun or core-plied. They are used in foundations, swimwear, stretch fabrics, hosiery, belts, gridles and so on.

The stretch before break of an elastomeric yarn can be 500–600 percent. At the breaking point, the strength is 0.7 g/d (which is equivalent to about 4.5 g/d before stretching). At a given stretch, the recovery of Spandex (Lycra) fiber is usually 93–96% while rubber exhibits 100% recovery from 100% elongation. Lycra is white in color and can be dyed. It exhibits good resistance to chemicals but turns yellow when it is exposed to hypochlorite.

1.3.7 Other Fibers

Polyvinyl Chloride. There are a number of commercial fibers based on the polyvinyl chloride polymer. These fibers are produced from the polyvinyl chloride polymer by wet-spinning.

These fibers have a limited textile use. They are used as filter cloths, diaphragms, and industrial packing. Other uses are fishing gear, ropes, sailcloth, and swimsuits.

Fibers Based on Vinylidene Chloride. Fibers are melt-spun from polymers based on vinylidene chloride and other comonomers such as vinyl chloride and acrylonitrile (vinyl cyanide). A typical example is the Saran fiber, which is spun from a copolymer of 85 parts by weight of vinylidene chloride, 13 parts vinyl chloride, and 2 parts acrylonitrile. The copolymer is melt-spun at about 180°C and quenched by a current of cold air. The filaments are cold-drawn and wound on suitable packages.

Vinylidene chloride polymer is pale gold or straw colored and thus is at a disadvantage compared to other textile fibers. It has very good photostability. These fibers do not support combustion and are self-extinguishing. They are not affected by acids and alkalis except ammonium hydroxide.

The textile use of this fiber is prohibited by its low ironing temperature, higher specific gravity, low moisture content, and unacceptable color limiting the subsequent coloration. The main uses of the fiber are insect screens, filter cloths, and fishing ladders. It is also used for upholstery, belts, suspenders, and braids.

Polyvinyl Alcohol. Polyvinyl alcohol is obtained by saponification of polyvinyl acetate. There are two methods of manufacture of vinyl acetate: (*a*) addition reaction of acetic acid and acetylene and (*b*) reaction of acetaldehyde and acetic anhydride. The most widely used polymerization process is the free-radical initiation of vinyl acetate. Since high-molecular weight polyvinyl alcohol is required for fiber spinning, care is taken to minimize termination of reacting chains during polymerization. Methanol is commonly used as a solvent during polymerization, because the chain transfer content with methanol is small and the alcoholic system is preferable for the process of hydrolysis.

Polyvinyl alcohol is predominantly of 1,3-glycol type. End groups are mainly primary alcohol groups. Because of chain transformation and chain termination, ketonic, carboxyl, and carbonyl groups are also present in the polymer.

Fibers are produced by the wet-spinning process. The viscosity of the aqueous solution of the concentrated polyvinyl alcohol (PVA) depends on the method of preparation of the polymer, the degree of polymerization (DP), the concentration, and the temperature. Aqueous salt solutions are usually used for the coagulation bath.

Wet-spun PVA fiber dissolves in hot water. At an early stage of development, formaldehyde treatment was carried out to make the fiber insoluble in hot water. This is not practiced now since the shrinkage of the fiber is large. To counteract the disadvantage, the fibers are heat-treated.

Hot-drawing is usually carried out simultaneously with heat treatment to improve the tensile properties of the fiber. The relaxation heat-treatment after hot-drawing decreases the shrinkage of the fiber at high temperatures or in hot water.

PVA fibers are used in many textile applications such as school uniforms, raincoats, protective clothing, umbrella cloth, surgicals, filter cloths, and fishing nets. Other uses include suitings, linings, stockings, socks, gloves, hats, and sewing threads. It has also been blended with cotton or rayon.

Polyblend Fibers.   Attempts have been made to balance or improve the properties of a fiber by producing them from polymer blends. Fibers produced from polymer blends can be broadly classified as (a) bicomponent, in which the fiber is made by physically joining together the different polymers and (b) biconstituent, in which two polymers are intimately mixed.

Two polymers can be arranged (a) side-by-side, (b) in a sheath-core, and (c) in a matrix-fibril. The side-by-side structure is generated to give self-crimping. In a sheath-core, one polymer is exposed and one polymer is encased. This structure is employed when it is desirable for the surface to have the property of one of the polymers such as luster, dyeability, or stability, while the core may contribute to strength, reduced cost, and wearlife. Variations in spatial arrangements are many, for example, an eccentric core leads to self-crimping in addition to the usual advantages of the sheath-core structure.[11]

When two polymers are mixed intimately and spun, a matrix-fibril-type structure is obtained. The major component forms the matrix and the fibrils of the minor component are dispersed in the matrix. The size and number of fibrils depend on the proportions and the rheological conditions of the fiber formation.

Bicomponent fibers are spun either by melt or wet-spinning by specially designed spinning heads. A common feature of the process is two different streams of melt or solution of polymers that meet each other just behind the spinneret hole.

Biconstituent fibers are produced by intimately mixing the polymers in the melt or the solution by standard mixing processes and spinning the fiber from the resultant mixture.

Attempts have been made to improve fiber properties by polymer blending. Examples are permanent antistatic fiber, improving flat-spotting behavior of nylon tire cord, improved hot-wet mechanical properties, improving flame retardancy or dyeability, making ultra-low denier fiber, synthetic fibrils, and other fibrillated products and so on.

This blending technique has been used to incorporate dye-sites into a fiber and to achieve some special effects. On the other hand, the dyeing of a polyblend is of interest since the addition of even a small amount of another polymer may affect the dyeability of the product. For example, in the case of sheath-core fibers, a sheath of polyester and a core of nylon will be difficult to dye with acid dyes. Whereas, if the situation is reversed and nylon is in the sheath, the fiber is easily dyeable. In the case of side-by-side polyblend fibers, special dyeing effects may be obtained since the two components are differentially dyeable or have an affinity for different classes of dyes. A polypropylene blend with polymers based on vinylpyridine gives a dyeable polypropylene fiber. Similarly, polyethylene terephthalate can be blended with nylon to enhance the dyeability.

1.4   DRAWING OF FILAMENTS

1.4.1   Draw-Twisting Process

The filaments formed from the polymer melt or dope by cooling or coagulation have little mechanical strength. Their storage stability is also very poor unless they are further stretched. In the process of drawing, the filaments are stretched up to their natural draw ratio. The drawing is carried out either after collecting the filaments, that is, after spinning or simultaneously during spinning. Simultaneous melt-spinning and partial drawing processes give POY and MOY. These yarns are fully drawn during the final texturing operation. In the draw-twisting (D/T) process, the conditioned-as-spun filament yarn is fed to the godet rollers, which rotate at different speeds and thus, have different linear (surface) speeds. The yarn passes over each godet several times in order to avoid slippage before the stretched yarn is collected. The first godet may be heated as in the case of polyester and the yarn may be further heated in-between the godets. Heavy denier nylon such as sewing-thread monofilaments may be drawn in the presence of steam. The typical load–elongation curves of undrawn PET fibers are shown in Figure 1.3.

1.4.2   Simultaneous Spin-Drawing Process

The melt-spinning and drawing operations are combined in spin-draw machines.[22,23] The extruded filaments from the spinning head are not wound on

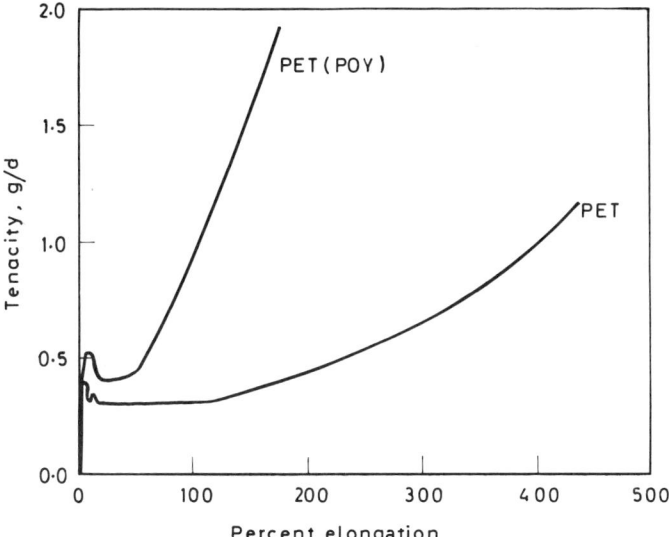

FIGURE 1.3. Tenacity vs. elongation (%) of undrawn and POY polyester at 20°C. Tenacity (g/d) on the basis of raw titre. The peaks show the necking point.

the usual take-up unit running at 1000–1300 m/min, but are passed through a drawing zone that is defined by a pair of godet rollers of which the output roller runs at a speed of 3500–5000 m/min. Thus, the yarn gets drawn between the two godet rollers and the intermediate take-up operation is eliminated. The drawn yarn is then wound on a high-speed winding head running at almost the same speed of 3500–5000 m/min. The properties of these yarns are similar to those produced by the two-stage spinning D/T process described above. A typical machine is shown in Figure 1.4.[23]

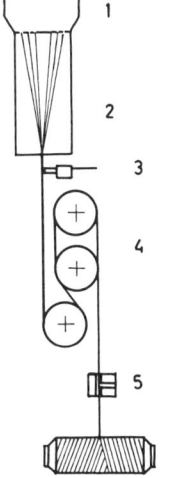

FIGURE 1.4. Simultaneous spinning and drawing process.[23] 1: Melt-spinning; 2: Cooling chimney; 3: Spin-finish applicator; 4: Stretching godets; 5: Steam setting; 6: Take-up unit.

These machines have high initial and maintenance costs, high noise-levels, and specialized technology without a substantial gain in productivity. Development of high-speed take-up units to produce POY and MOY materials are more attractive and are preferred, particularly, for yarn that is to be further texturized.

The position of the neck below the first godet must remain stable in order to produce a yarn of uniform properties. For polyester filaments, if the temperature of the godet and heated rail is not proper, fluctuates for a given machine, or is not constant from machine to machine, it will be adversely reflected in the yarn properties. For example, the rate and extent of dyeing will vary for yarns from machine to machine. If the stretch or draw ratio varies because of a faulty feed mechanism, faults in godet movements, improper spin-finish oil, or variations in spin-finish on filaments, uneven godet surface, poor quality feed yarn and so on, the stretched yarn exhibits unevenness (high Uster value), periodic variations in filament diameter, unstretched portions in the yarn, high or uneven elongation, low strength and so on. Poor quality textiles will be produced from such yarns. It is impossible to get a uniform solid shade on such a material; it will exhibit the *weft bar* or *ladder* effect on dyeing. The variation in the yarn diameter influences the dye uptake not only because of the variations in the surface area through which the dye is absorbed by the fiber but also because of differences in the molecular orientation (which is easily quantified as the birefringence value). The fabric will exhibit thin and thick portions that are dyed light and heavy in depth. Some of these defects in the yarn are noticed only after weaving and dyeing. One pern of such a defective yarn will spoil an otherwise good fabric. One package of faulty yarn will give several perns and the weft bar defect will be introduced in the fabric whenever such a pern is used. Uneven dyeing in woven fabric may thus be due to faulty D/T working.

The spinning of filaments at speeds higher than 1000–1400 m/min has been thought of for many years. However, the commercial development of high-speed spinning, particularly of PET, is only about a decade old.[24] As the spinning (or take-up winding) speed increases, the throughput increases (Figure 1.5).[25] The filament yarn produced exhibits high storage stability, orientation (birefringence), crystallinity (Figure 1.5), density, Young's modulus, and melting point (see Section 2.2.1). In practice, speeds of up to 3500 m/min are commonly used to produce two types of yarns.

MOY: Moderately oriented yarn (speed 2500 m/min).
POY: Partially oriented yarn (speed 3500 m/min).

Since these yarns are spun at high speeds, they get partially stretched and have to be drawn to a lesser extent (about 120–150%) in order to reach their final length. Speeds beyond 4000 m/min are uneconomical since the price of the winding unit rises exponentially with the spinning speed. In fact, the speed of 3500 m/min gives sufficiently preoriented yarns with the desired

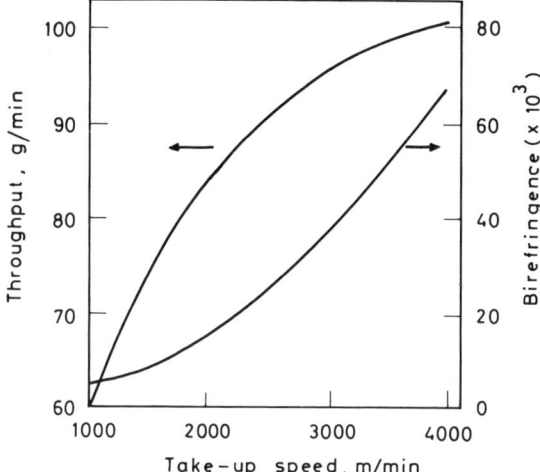

FIGURE 1.5.   The effect of winding speed on the birefringence and the throughput per spinneret in the production of polyester fibers.[25]

storage stability. Thus, such yarns remain stable independent of climatic conditions for several months (Table 1.6)[22] and can be textured away from the spinning units. Throsters and crimpers can therefore use these yarns to produce textured yarns. Since their residual draw ratio is very low, they are

TABLE 1.6   Storage Stability of Undrawn PET Yarns[22] (33°C/30% RH)

	Tenacity (g/dtex)		
Storage Days	Normal	MOY	POY
1	2.9	3.2	3.3
4	2.8	3.2	3.3
7	1.3	3.2	3.3
13	0.9	3.3	3.3
27	—	3.3	3.3
35	—	3.2	3.2
55	—	3.0	3.2
64	—	3.0	3.3
77	—	3.0	3.2
90	—	2.9	3.2
Spinning speed (m/min)	1300	2500	3500
Birefringence $\times 10^3$	5	25	50

drawn and textured in a single step. The uniformity of the preoriented yarns is also better. The omission of the draw-twist operation involves additional heating–cooling operations and improves the dyeability of MOY/POY. The ultimate tensile properties of stretched yarns spun at different speeds are, however, similar.[26]

The production of POY and MOY materials is rapidly increasing because of the economical attractions and the storage properties of the yarns. High-speed spinning has eliminated the separate process of D/T by dividing it into two parts, one achieved in the filament spinning and the other in the texturing. The yarn produced by the high-speed spinning process is thus always textured before use.

1.4.3 Drawing and Fiber Properties

Molecular orientation in the fiber develops essentially during drawing and can be traced by measuring the birefringence and X-ray diagrams. The mechanical stress applied to the fiber during drawing acts at first to align polymer segments along the direction of force until a certain degree of orientation is established and then it gradually increases the lateral order. Thus, there is greater increment in birefringence in the low draw ratio range while the increase in density occurs mainly at the higher draw ratio range.[27] The moisture regain decreases with an increase in the lateral order rather than with orientation and thus with higher draw ratios (Figure 1.6).[27]

Breaking stress and strain of fibers as a function of the draw ratio are shown in Figure 1.7.[27] The stress at break increases while the strain at break decreases with the drawing process. The dependence of the breaking stress on the draw ratio is essentially linear over the entire draw ratio range but the decrease in the breaking strain is particularly pronounced in the initial stages of drawing.[27] The dynamic elastic modulus of nylon 66 fiber is larger than the

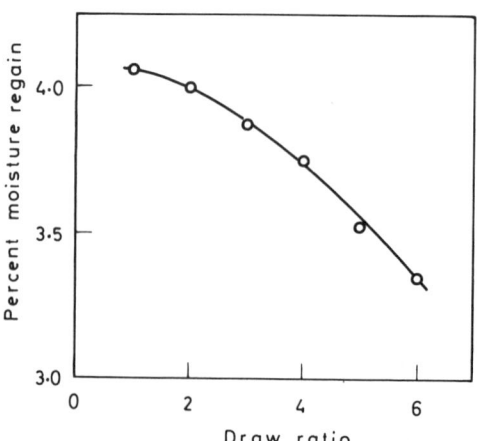

FIGURE 1.6. Equilibrium moisture regain (21°C, 65% R.H.) of PA 66 as a function of the draw ratio.[27]

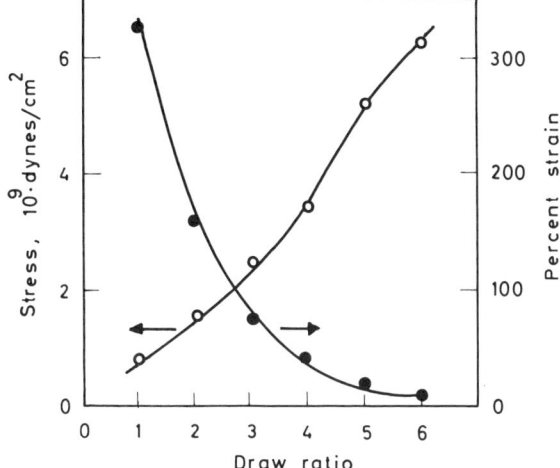

FIGURE 1.7.   Stress and strain at break of PA 66 as a function of the draw ratio.[27]

static one, with both moduli increasing linearly with the draw ratio. During free shrinkage of an oriented fiber, there is a considerable disorientation of the primary axes of the polymer crystallites and the material eventually reverts to the amorphous state on melting. Dismore and Statton,[28] however, found no change in the orientation of highly drawn nylon 66 yarns after they were allowed to shrink freely at temperatures up to 250°C indicating that the thermal shrinkage of oriented polymers involves a structural transformation rather than disorientation of a crystalline phase. Thus, intermolecular bonds are loosened thermally, thereby freeing chains to move into new conformations with lower bond energies. Thermal shrinkage involves a phase transformation between two crystal forms—the extended-chain and folded-chain conformations (see Chapter 2).

A fourfold decrease in dye diffusion in nylon 6 is found as the draw-ratio is increased from 1 to 3. The orientation of molecular chains and the structural disruption of the undrawn fiber play opposite roles in the microstructure of a yarn during drawing. The orientation increases the stability which overrides the stabilizing effect of orientation at certain points in the drawing. This structure break probably takes place in the necking region.[29] Many workers have studied the effect of drawing on the properties of PET fibers; the higher the draw ratio, the lower the accessibility of fibers to the dye molecules.[30] The diffusion coefficient of acid dyes into nylon 66 decreases markedly as the draw ratio increases.[27]

1.5   TEXTURING

Continuous filament yarns do not have the same hairiness, bulk, and warmth of handle as those made from natural fibers or synthetic staples. Some of these properties are imparted by a process known as *texturing* or *texturizing*. In this process, the filament yarn is permanently deformed to impart bulk. The process alters the drape and handle of the fabric by developing air spaces, and, in turn, increases warmth, comfort, and ease of care. Reversible stretch is introduced in the yarn by certain texturing processes. New and improved properties are developed in the yarn by the permanent introduction of crimps, coils, loops, curls, and/or crinkles into the filaments by various texturing techniques that take advantage of the thermoplasticity of synthetic filaments and their ability to permanently deform and set in a deformed state by a thermal treatment. Thus, all the texturing processes are mechanical or thermomechanical in nature and do not involve any chemical treatment.[31]

The following techniques are available for producing textured yarns:

1. False-twist process.
   (*a*) Spindle technique.
   (*b*) Friction-texturing technique.
2. Air-texturing process.
3. Knit–deknit process.
4. Gear crimping process.
5. Knife-edge crimping process.
6. Stuffer-box crimping process.

Texturing is possible with all synthetic fiber, multifilament yarns. Each method imparts a specific quality to the yarn. Each and every method is therefore not applicable to every type of yarn and every fiber material.

1.5.1 False-Twist Texturing Process

The false-twist texturing process is the process of introducing deformations in the synthetic filament yarns and releasing the stresses imposed by the deformations by heat treatment so the deformations are set permanently in the yarn. Thus, the stress development and the stress release are the two major parameters of the texturing. The basic principle of the false-twist texturing process is that the filament yarn is twisted, heat set, and finally untwisted. The filaments are set in twisted form and are unable to regain their original flat form on unwinding; hence, they buckle and loop. If a stationary multifilament yarn is held at both ends and twisted in the center by a hollow false-twist spindle, an equal amount of twist in opposite directions will be imparted on each side of the spindle. In a moving yarn, the twist produced on lower portion will be removed on the upper portion of the false-twist spindle. In the false-twist texturing process, the spindle is kept between the heater and the output roll. The temperature of the heater is maintained at 200–220°C for polyester. Some of the machines are equipped with a second heater, which is used for heat setting the textured yarn (for stabilization).

In these texturing techniques, the stress development is decided by the torque produced in the yarn by twisting and tension, while the stress release is controlled by the heat treatment given to the yarn. The rotating spindle acts as a twist-introducing mechanism as well as a twist-migration inhibitor. The rotational speed of the yarn given by the rotational speed of the spindle cannot go above 0.8 million revolutions/min and thus, limits the yarn speed to 350 m/min. This led to the invention of friction texturing, in which the yarn is rotationally driven by discs (instead of spindles) with rotational speeds of up to 2.9 million turns/min and theoretical speeds of above 1000 m/min.

Friction-texturing offers the advantage of high speed.[32] It is suited to MOY/POY and undrawn yarns. Spin-finish, friction disc surface, basic yarn properties, the geometry of yarn within friction aggregates, and the pressure of the yarn against the discs are some of the factors that decide the quality of the texturing. Since the grip provided by the friction discs on the yarn is not as positive as that provided by the pin in spindle texturing, the control of the parameters in friction texturing has to be very close in order to get reproducible results. A migration of twist and, in turn, absence of texturing is a frequent fault in friction-textured yarns.

1.5.2 Air-Texturing Process

The air-texturing process is a mechanical process in which a multifilament yarn is overfed into a nozzle with an air jet. The current of compressed air impinges on the yarn, whips the filaments, causing them to buckle and to form loops, curls, and crimps so that the yarn contracts in length and increases in bulk.[33–36] In the nozzle, the air velocity is increased by using the principles of an air jet and the air pressure is turned into air velocity to whip the filaments.[36] On emerging from the jet, the tangled filaments are held in place by the friction between the curled filaments. A small amount of twist inserted in the yarn helps to increase the stability of the yarn. Heat setting of the air-textured yarn is also used to improve the stability. Yarns that are bulked in this way do not need to be thermoplastic since the technique does not rely on any heat-setting treatment. The yarn loses in strength since the crimped, curled, and looped filaments do not bear the load. However, on weaving or knitting, the fabric exhibits the normal strength. In the air-texturing process, the filaments mingle with each other. If two yarns of different properties such as denier of filament or yarn, number of filaments, modulus, color and so on are used, a blended, mingled textured yarn is produced.[36] For example, polyester and nylon filament yarns may be used to get a blended yarn that can be dyed with acid/disperse dyes to produce fancy effects. Mingling of filaments by the air-texturing process is the only technique we have to produce blended, textured filament yarns. However, extreme care is required to get reproducible, uniform blended yarn. It is also possible to produce skin-core, blended yarn in which one yarn will be in the core and the other on the surface. Finer denier, less rigid fibers fed with low tension form the loops and outer part of the blended yarn.[36] The feel of the air-bulked yarn is nearer to that of a spun yarn. The yarn does not exhibit reversible stretching. The advantage of polyester yarn of this type over the spun yarn is that it exhibits low pilling tendency. This technique has a potential that still needs to be fully explored for commercial textile end uses.

1.5.3 Crimping Processes

Knit–Deknit Process.   This is a three-stage batch process: knitting, setting, and deknitting or unravelling. A filament yarn is knitted into a fabric with a plain weave. The fabric is heat set and deknitted. The yarn unravelled after the fabric is heat set retains the memory of the knitted stitch shape and bulks. This is a rather expensive process and may be combined with dyeing to produce colored textured yarns.

Gear Crimping Process.   A thermoplastic yarn can be crimped by passing the flat filament yarn between two heated intermeshing cog-wheels or some similar device. Yarns produced in this way and by the knit–deknit process have a wavy configuration and do not have a tendency to twist. They are referred to as crinkle yarns. An advantage of gear crimping is that it is a continuous process.

Knife-Edge Crimping.   Polyester yarn is heated and drawn over a knife edge while cooling. This distortion causes one side of the individual filaments to shorten relative to the other so that they coil into a helical shape.[37] The yarn is then relaxed in hot water in order to achieve the full bulking potential. These yarns are suitable for such end uses as stockings.

Stuffer-Box Crimping.   The Banlon process or the stuffer-box crimping process has been developed by the Bancroft Corporation.[38] The yarn is packed into a small heated box known as a stuffer box. The filaments are folded into zigzags as they are compressed into a limited volume and are heat set so that the folded filaments are set in that form. Thermoplastic yarns can be modified in this way. Dyeing of such a crimped yarn can be a problem since heated and unheated portions of the filaments exhibit different rates of dyeing. On the other hand, dope-dyed or mass-colored filaments may be processed by this technique.

1.5.4 Draw-Texturing Process[39–41]

In draw-texturing processes, the operation of drawing is shifted to the texturing machine. This is possible by two different methods. In a sequential two-stage process, drawing takes place in a separate drawing zone allotted to each spindle. In a simultaneous one-stage process, the drawing is carried out directly between the feed unit and the texturing spindle (Figure 1.8). The latter method has a number of advantages such as high productivity, greater efficiency, and uniform dyeability of the resultant yarn. Most of the high-speed spun PET yarns—POY and MOY—are processed by this method. The draw-twisters are dispensed with in the draw-texturing processes but the cost of the sequential draw-texturing units are considerably higher than that of the texturing units using drawn yarn. In the one-stage, simultaneous-texturing process, this is not the case. In addition, POY/MOY polyester shows favorable crimp-elastic properties as a result of the residual drawing operation moving in parallel with texturing. As a result, the filaments take-up the twist more easily and uniformly in the simultaneous texturing process. A draw-twisted PET filament yarn loses about 10% of its initial tenacity during texturing operations because of the increase in crystallinity of the fiber and a simultaneous decrease in the orientation in relation to the fiber axis. Thus, deformation caused by twisting or rotation and setting brings about a partial destruction of the structure. Since POY/MOY achieves the final orientation only during draw-texturing, the influence of the original structure of POY/MOY is not significant.[42]

1.5.5   Further Processing

Heat setting, twisting-doubling, and cone winding are the other processes that the yarns may have to pass through before marketing.

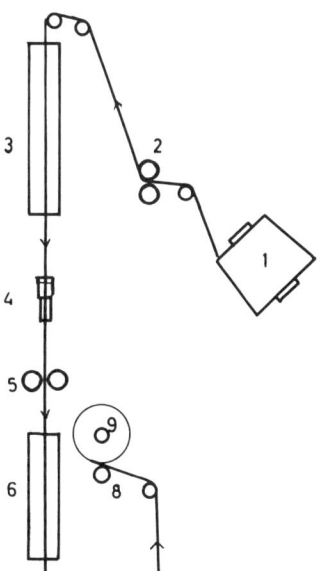

FIGURE 1.8. Schematic diagram of the simultaneous draw-texturing process for POY yarn (draw ratio 1.7). 1: Supply package; 2: Input rollers (relative speed 50); 3: First stage heater; 4: False-twist spindle (560,000 rpm); 5: Intermediate rollers (relative speed 85); 6: Second-stage heater; 7: Output rollers (relative speed 70); 8: Package drive roller (relative speed 77); 9: Take-up package, [From *Draw Textured Yarn Technology* Monsanto Textiles, (1974).]

Heat Setting.   This operation imparts the dimensional stability to the PET yarn. Besides controlling the residual shrinkage of the yarn, it minimizes the tendency of the flat yarns to curl during processing. The heat-setting operation is carried out by winding the yarn on rigid or flexible tubes and subjecting them to steaming in an autoclave. If the yarn is textured on double-heater texturing machines, a heat-setting treatment is not required because the second heater performs the setting. Similarly, if the goods are to be dyed in the yarn form, a separate heat-setting operation is not required since the hot dye-liquor sets the yarn.

Twisting-Doubling.   Since the continuous filaments on draw-twist bobbins have a low twist (about 10–12 tpm) that is inadequate for further operations, it is necessary to increase the twist. This uptwisting operation is carried out either on single-end twisting machines or on two-for-one twisting machines.[43,44] Textured yarns with *S* and *Z* twists are doubled so that there is improved twist stability. Twisted yarns are set to minimize the twist migration.

Coning.   Two functions are carried out in the cone winding: (*a*) application of coning oil and (*b*) winding marketable cheeses. The coning oil is a mixture of a wetting agent, softener, and antistatic agent. It acts as a lubricant during the handling of the yarn in the knitting or weaving. However, it is removed before dyeing and therefore must be readily washable. It should be stable to oxidation and heat and should not turn yellow during storage.

1.6   PROPERTIES OF SYNTHETIC FIBERS

A few properties of synthetic and other fibers are summarized in Table 1.7. In general, synthetic fibers are characterized by high strength, superior abrasion resistance, very good crease recovery property, excellent biological resistance, and low density. Many properties such as tenacity, elongation (%), cross-section and so on can be adjusted to the end-use requirements. Usually, the durability of synthetic fibers is very high compared to viscose, cotton, or wool, but there is a problem of pilling.

The moisture regain of many synthetic fibers is very low and in extreme cases, such as polypropylene fibers, almost nil. These fibers dry easily after washing. Low moisture regain creates a number of problems such as the development of static charge and its accumulation to dangerous levels, easy soiling and difficulties in soil removal, poor comfort, and loss of aesthetic properties during their use.[45]

Unlike cellulosic and animal fibers that char on heating and during burning, many synthetic fibers melt before burning. The melting point varies from as low as 100°C for polyvinyl chloride fiber to as high as 255°C for PET. Important properties of synthetic fibers are thermoplasticity and setting abil-

TABLE 1.7   Properties of Typical Textile Fibers

Fiber	Density (g/ml)	Tenacity (g/denier)	Elongation (%)	Moisture Regain (%)	Melting Point (°C)
Polyester	1.38	3.5–7	12–55	0.4	255
Nylon 66,	1.14	3–7	18–75	4	255
Nylon 6, and					215
Nylon 11					205
Acrylic	1.17	2–4.5	15–50	1	d–200[a]
Polypropylene	0.9–0.92	4.5–7	15–100	0	165–176
Polyvinyl chloride	1.39	1.8–3.5	25–40	0	100
Polyvinyl alcohol	1.26	3.5–6.5	15–30	4–5.5	200–240
Polyurethane	1.2–1.25	0.6–1.0	400–700	1–1.2	230–290
Sec. Acetate	1.32	dry 1.4	25–45	6.5	230
and Triacetate		wet 0.9		4.5	290
Viscose	1.52	dry 2.6	15–30	12–13	chars above 400
		wet 1.5	20–40		
Cotton	1.52	2.5–5	6–10	6–8	chars above 400
Wool	1.32	1.5–2	25–40	16	d–350

[a] d = decomposes.

ity. These fibers recover well from creases formed during wear. The garments keep their elegant crease-free appearance during wear. Practically, no ironing is required if heat set garments are carefully washed and dried. On the other hand, pleats and creases set in the fibers at high temperatures remain almost through the life of the garments.

The resistance of synthetic fibers to the action of acids in general is quite good. Polyester and acrylic fibers are attacked by hot alkaline solutions. Nylon exhibits better resistance to the action of alkali. Synthetic fibers degrade on prolonged exposure to light. However, the degradation is much slower than cotton. Nylon 66 has a tendency to yellow during heat setting or on exposure to UV light. Nylon 6 is in this respect slightly better than nylon 66.

Synthetic fibers have excellent biological resistance. This is because, unlike natural and animal fibers, synthetic fibers do not serve as a food for bacteria. Even after being buried in soil for three months, the synthetic fibers were not attacked by bacteria.

Synthetic fibers have a tendency to shrink in hot water. This tendency can be completely eliminated by a heat-setting process. After being heat set at an appropriate temperature and time, nylon and polyester fibers will not shrink on subsequent hot treatments at lower temperatures. Acrylic fiber is not responsive to dry heat-setting treatments. Its' shrinkage is controlled by the annealing process using steam.

Synthetic fibers are produced as staples and continuous filament yarns. The number of filaments and their denier vary over a wide range depending on the end use. Staples of synthetic fibers vary in their denier and staple length depending on the spinning process used to produce the spun yarn. Variable cut-lengths of staples are common in acrylic staples that are blended to produce the bulked yarn. The cross-section of the fibers can be circular or multilobal. Essentially triangular cross-section gives glittering yarn. Multilobal cross-sections give dulled yarns with soft feel. Normally dulled yarn is produced by incorporating a dulling agent, titanium dioxide. Its concentration varies from fiber to fiber and can be up to 2.5% for full-dulled material. The presence of a dulling agent makes the fiber opaque and white due to the scattering of light. This influences the color of the dyed material[46] (see Figure 3.16).

The fine structure of synthetic fibers will be described in Chapter 2. Other properties of synthetic fibers will also be discussed in Chapter 2.

1.7 OLIGOMERS

When fiber-forming polymers are prepared, low-molecular weight constituents of the same chemical composition are also formed during the polymerization. These are differentiated essentially by their chain lengths, molecular weight, and physical characteristics, and are called *oligomers*.[47] PET contains 2–3% oligomers that cannot be removed before spinning the fiber, as they are difficult to extract and are likely to reform during melt-spinning.[48,49] Among the various oligomers in PET fibers, cyclic trimers (Figure 1.9) occur in the largest amounts. This oligomer gets extracted on heating and in the wet processing. It is characterized by a high crystallization rate and extremely difficult and poor water solubility, even above 100°C (1–3 mg/liter in a HT bath). The cyclic trimer gets deposited in large amounts as dirt in the polyester dyeing plant and on textiles, causing considerable production

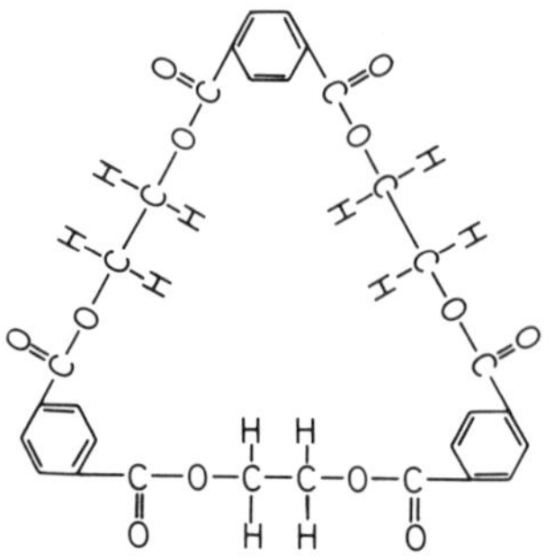

FIGURE 1.9. Cyclic oligomer.

problems[50] (Figure 1.10). The textured PET filaments create an acute oligomer problem since the oligomer migrates to the surface because of the heat treatment during crimp-setting in the texturizing process. (see Section 8.7). Nylon 6 oligomers are extracted from the polymer before melt-spinning and pose no serious problems in the processing of PA 6 fibers.

1.8 IDENTIFICATION AND ANALYSIS[51,52]

Synthetic fibers are marked in different deniers, filament cross-sections, and staple lengths, with or without a dulling agent. The manufacturing process and raw materials may vary for a given type of a fiber from different sources. Molecular-weight distribution, fiber crystallinity, tensile behavior, chemical properties, properties of sorption of moisture, dyes, and chemicals, solubility, and other properties may be deliberately varied. Addition of a small fraction of a comonomer may alter the fiber response to various agencies. For example, nylon 6 can be made normal dyeable, deep dyeable, or undyeable with acid dyes or cationic dyes. It may have a circular or trilobal cross-

FIGURE 1.10. Cyclic oligomer on dyed polyester fibers.

section. Thus, it is essential to identify an unknown fiber and a blend by using various techniques and to confirm them by comparing them with authentic samples.

Various types of fibers from natural and synthetic origin are blended together. The component fibers of a blend have to be identified individually after separation. The wet processing of such a textile material is undertaken after the quantitative analysis of the material. The proportion of each component fiber is determined by the selective removal of the other fiber by chemical, physical, and/or mechanical means. Quantitative methods of analysis involve the preferential selective dissolution of a component and the gravimetric estimation of the residue. The selected solvent dissolves the component completely without affecting the other fibers. Typical blends with cotton as one component are given in Table 1.8.[53] For unknown blends, different solvents are used in sequence, since many of the solvents will remove more than one type of a fiber.[54] The qualitative and quantitative analysis of blends using cadoxen of suitable composition has been described by Achwal and co-workers.[55] According to Langley,[54] a microscopic analysis of the fibers in a blended fabric should be made prior to the chemical removal of fibers. If the blend contains two or more fibers with similar chemical structures, the solvent method does not succeed.

The physical and other methods involve microscopic examination, measurement of density, melting point, moisture regain, staining or tinting and so on. For this purpose, the fibers are separated by cutting and powdering followed by some flotation techniques.[56] A polarizing microscope is preferred for microscopic examination. Under crossed polarizers, each fiber generates a specific color depending on the extent of orientation or birefringence (Δn) (Table 1.9). Acetate and acrylic fibers have a low birefringence and appear almost grey-white. Nylon exhibits bright multicolored stripes along the fiber axis ($\Delta n = 0.046$–0.06). PET has the highest birefringence ($\Delta n = 0.175$) which is easily measured. It gives pale-colored stripes. Knowing the stripes and their color, the fibers can be identified in combination with other test methods. The shape of the cross-section, the longitudinal

TABLE 1.8 Quantitative Analysis of Blends Containing Cotton[53]

Blend	Solvent	Solute	Conditions
PET/CO	H_2SO_4	Cotton	50°C/1h
PA6/CO	Formic Acid (85%)	PA6	rt/10 min
Acrylic/CO	DMF	Acrylic	60°/30 min
PVA/CO	Formic Acid (85%) + m cresol	PVA	80°/15 min
Acetate/CO	Acetone	Acetate	25°/40 min
Cellulosic/synthetic fiber	Cadoxen I[a]	Cellulosic	25°/2 h

[a] Cadoxen: I = Cadmium: 5%, ethylene diamine: 30% + 0.5N NaOH

appearance, the presence of voids and so on may help to confirm the nature of the fiber. The density of a fiber can be easily measured by the flotation technique using the density gradient tube. The density gradient column is prepared and calibrated by a standard method. Xylene and tetrachloroethylene are mixed in different proportions (0 : 100 to 100 : 0) and are added to the column either automatically or manually. For quick identification, dyed reference fibers are tied in a knot and the loose ends are snipped off. The reference fibers are boiled for about two minutes in xylene to remove moisture and air, and then are introduced into the density gradient column. After about half an hour, the fibers come to a rest at a level representing their densities. Unknown fibers prepared in a similar manner are dropped into the column. The level at which they float gives their density and identification.

The melting point (m.p.) is a characteristic property of a pure compound. The polymeric fiber materials have a molecular weight distribution and their fine structure varies in crystallinity and crystal size. All these factors are reflected in the melting behavior of fibers as is observed on a differential scanning calorimeter. However, in practice, the melting point of a fiber can be a guide to its identification. The fiber shrinks and softens prior to melting which should not be confused with the final melting. Some fibers do not melt before decomposition. The flame test of the fiber yields valuable information about the nature of the fiber. A small tuft of fibers (held in tweezers) is placed close to the side of a small flame. The fibers are then moved into the flame to note whether the fiber burns when held in flame. They are then removed from the flame very slowly and carefully to note whether they continue to burn outside the flame. The flame of the burning fiber, if any, is then blown out and the smoke is smelled. The odor is noted and the color and nature of any ash/residue is observed. The behavior of the test specimen is then compared with known reference fibers. The fibers with flame-retardant finishes show different behavior and can give misleading results. Staining of fibers is a simple technique to differentiate the fibers. Fibers can be exclusively dyed

TABLE 1.9 Typical Optical Properties of Fibers

Fiber	Refractive Indices		Birefringence $n_e - n_w = \Delta n$
	n_e	n_w	
Cellulose acetate	1.478	1.473	0.005
Acrylic fiber	1.515–1.520	1.517–1.525	$-0.002--0.005$
Nylon	1.580	1.520	0.060
Polyester	1.710	1.535	0.175
Cotton	1.578	1.532	0.046
Viscose rayon	1.547	1.521	0.026
Silk	1.591	1.538	0.053
Wool	1.556	1.547	0.009

with a specific dye. Such dyes are used for identifying the fibers. Special staining agents are based on this principle.

1.9   FAULTS IN SYNTHETIC FIBERS

The structure of a synthetic fiber builds up through the various stages of fiber manufacture. Any variations in the conditions and parameters at any stage or simultaneously at a number of stages are reflected in the properties of the fiber. The origin of many problems faced by the processor of synthetic fibers may be in the thermal and mechanical history of the fiber. It is, therefore, essential to know the faults that are introduced during their production.

1.9.1   Polymerization

The chemical constitution of the polymer is decided essentially during polymerization. The type of monomers used to get a given polymer are usually not altered. However, the source of the monomers may vary from time to time. Regenerated or recovered monomers are used, either as is or in combination with the virgin monomers. Such monomers may not match in all respects the virgin monomers and can introduce added properties to the polymer; the color may be affected, the end-group concentrations may alter, polymer gels are formed that may create problems in fiber-spinning and drawing and may give faults in dyeing. Variations in the concentrations of monomers, temperatures, and the time of polymerization can also alter the end-group concentration. Side reactions such as aldehyde group and diethylene glycol (DEG) formation in the polymerization for PET fiber influence the final product. A higher concentration of DEG in the final PET fiber increases the dyeability of the fiber. However, diethylene glycol ether lowers the m.p. of the polymer (Table 1.10) and gives an inferior polymer, with poor strength and high sensitivity to UV light. If the PET polymer waste is recycled or glycolized, the diethylene glycol ether content increases, thus, giving a weaker fiber.[57] The composition of the chain molecules and oligomers should not change from lot to lot. Some of the oligomers, for example, cyclic trimer in PET, create problems in dyeing. Many times, particularly, in the wet-spinning process, polymers are recycled to produce fibers. Such fibers are usually of inferior quality with low strength, color, and uneven dyeing properties. If such fibers get mixed in with good quality material, they yield a product of uneven properties.

The carboxyl group content of PET fiber varies from manufacturer to manufacturer (Table 1.11).[58] The number of end groups in nylon 6 and nylon 66 decides their behavior in dyeing with acid and metal-complex dyes. It is carefully controlled by the fiber manufacturer, but may vary from one merge number to another or from one source to another (Table 1.12).[58] Nylon is prone to oxidation and the cross-linked polymer thus formed gives gels in

TABLE 1.10 Diethyleneglycol Content in
PET and the Melting Point[57]

	Diethylene Glycol (%)	Melting point (°C)
1	0.60	270
2	0.76	266
3	0.83	265
4	1.40	262
5	1.60	261
6	2.06	259
7	2.90	255
8	3.01	254
9	3.25	253
10	3.64	250

the fiber that may exhibit different dyeing behavior and the fiber exhibits lower tenacity and extension at break. Similarly, the dyeability of acrylic fibers depends on the number of available sulfonic and sulfate groups in the fiber. Various manufacturing conditions can influence these values. A mixing of fibers differing in end group concentration creates problems in dyeing because of the difference in their rate of dye uptake and saturation values. How to identify the cause of uneven dyeing of textured nylon is described in Section 9.10.

1.9.2 Spinning

A fiber-forming polymer melt/dope is continuously extruded with a constant intensity (throughput) through a circular spinneret orifice of small diameter.

TABLE 1.11 Carboxyl Group Concentration in
PET Fibers[58]

Manufacturer and Trademark	Carboxyl Group meq/kg
ICI: Terylene W11	32
W16	37
Enka: Diolen	40
Diolen FL	59
Hoechst: Trevira 220	20
Rhodiaceta: Tergal	25
J. K. Synthetics: Jaykaylene[a]	40

[a] Unpublished SPRC work.

TABLE 1.12   End-Group Data, Number Average Molecular Weight, and Breaking Stress of Commercial Nylon Fibers[58]

Manufacturer and Trademark	Amino Groups (meq/kg)	Acylamino Groups (meq/kg)	Carboxyl Groups (meq/kg)	M_n	Breaking Stress (kg/mm²)
Bayer: Perlon	25	26	50	19,800	43
Enka: Perlon	49	6	53	18,500	57
Phrix: Perlon	40	18	53	18,000	41
Emser: Werke Grilon	36	17	57	18,200	45
Rhodiaceta: Nylon	45	26	54	16,000	53
Monsanto Nylon (deep dyeing)	65	11	68	13,500	27
JK Synthetics: Jaykaylon[a]	35	25	60	18,000	50

[a] Unpublished SPRC work.

On the spinning way, the liquid stream becomes fiber when the velocity increases and the diameter decreases from the spinneret to the take-up device in order to reach the take-up speed. During this flow, the liquid stream gets solidified into a filament in a cooling chimney or in a coagulating–washing bath and receives spin-finish. The variations in the fiber diameter introduced in the spinning process cannot be corrected later and remain or even enlarge during the drawing process. This variation influences the dyeing behavior and the final appearance of the finished goods. Various factors contribute to the diameter differences, such as:

1. Spinneret hole diameter variations because of partial clogging, poor cleaning, chocking, wear, and tear.
2. Poor quality melt/dope of the polymer with big particles of additives, dulling agents, oxidized or cross-linked polymers and foreign bodies.
3. Turbulence in the solidification zone, draft in the cooling chimney, variations in the cooling air-flow rates, inefficient cooling, variations in the coagulation bath concentrations and so on.
4. Poor spinneret design, crowding of holes.
5. Fluctuations in throughput.
6. Presence of gels, naps, cross-linked polymers, dust of polymer in filtered melt-dope that pass through the spinnerets.
7. Mixing of incompatible polymers, polymers with different melt rheology, redried polymer and so on.

The partial chocking of jet holes gives a filament with low denier (small diameter) and high orientation. The turbulence during cooling or coagulation increases the short-distance unevenness in the fiber diameter. Such turbulence is caused by a variety of factors. The take-up unit vibrations also impart unevenness to the fiber diameter. Drawing accentuates many of these faults. The denier variation in a filament and from filament to filament (or fiber to fiber) in a multifilament yarn is reflected in the properties of the yarn. The variation in diameter on take-up should be as low as possible and never more than 2–3% to get uniform textile yarn. The wet-spun fibers have voids in their structure that are formed during the coagulation–drawing–drying process. The extent of the void volume, their shape, and size decide the appearance and dullness of such fibers. Variations in voids may give differences in appearance or color for acrylic fibers.

1.9.3   Stretching

The as-spun yarns are stretched to build-up the final structure of the fiber. Variations in the temperature and the stretch, the movement of the necking position on the draw-twisting or any other stretching machine, changes in the extent of relaxation during and after stretching and so on give fibers that vary in their physical structures. Ladder effect, weft bars, streaky dyeing and so on have their origin in the defective stretching process. Broken filaments, undrawn filaments or fibers, and unevenly heated filaments are some other faults introduced in the yarn on the drawing machines. The combination of faulty spinning and faulty drawing gives yarns with unacceptable unevenness in properties and many faults. A slight variation in the draw ratio may understretch or overstretch the yarn. Such a yarn may exhibit different shrinkage and dyeing properties. It may respond to the annealing, texturing, and other processes in a different way than the normal yarn. Changes in the orientation and crystallinity of the fiber due to stretching variations makes the yarn faulty. Denier variation is also a fault introduced in the drawing process. The slippage of filaments on the stretching godet, faulty spin-finish composition that fails to keep the filament together and to give sufficient lubrication, and improper temperature of the godets give faulty stretching of a tow and, in turn, faulty fibers. If a few fibers in a yarn are dyed deeper, they are probably not drawn fully and their diameters will be higher than the normal fibers. The orientation (birefringence) of such PET staple fibers will be lower than that of the normal fibers. Achwal and Rao have suggested solvent staining for a visual assessment test, iodine absorption to quantify the difference, and trichloroethylene swelling to confirm the differences in PET fiber due to variations in the parameters of the drawing process.[59]

1.9.4   Texturing

The main aim of texturing is to deviate from the uniform structure of flat filament yarns and to approach nonuniform geometry in a controlled manner. While the yarn is to be made nonuniform along its contour, the extent of nonuniformity must be controlled to be uniform over the length of yarn from package to package and from day to day. This uniformity of quality depends upon the uniformity of texturing parameters as well as the feed yarns. A variety of defects are observed in textured yarns which spoil the appearance and handle of the fabric apart from defective dyeing. The common defects in textured yarns and their main causes are as follows.

Tight Spots. The melting of filaments and their sticking together give tight spots—the portions of the yarn from which the twist has not been removed. High-texturing temperature, low ratio of tension before and after spindle, and a high rate of twist are the causes of this defect.[60]

Untextured Yarn. Bad, overstored feed yarn gives this defect. The uneven structure is due to improper texturing parameters. The migration of twist over the pin of the spindle or friction discs also gives untextured yarn. Improper spin-finish, improper frictional properties of yarns, and poor cleaning of friction discs are some other causes.

Weak Yarn. Excessive tension or temperature could cause a drop in strength. The number of load-bearing filaments in the yarn decreases on air-texturing, giving a low-breaking load.

Poor Package. Hard package, bulging, or hard edge of packages and so on are caused because of the improper setting of machine and rough handling. The poor package gives uneven rewinding tensions during knitting which causes uneven yarn density in the fabric which causes a Barre' effect and thick and thin portions. The tension in the layers of tightly wound package sets the yarn in that condition if the package is stored for long. Differential hardness in the package sets the yarn to different residual shrinkage. Use of such a yarn gives puckering, uneven dyeing, loss of texturing and so on.

Nonuniform Temperature of Texturing. The temperature of the heaters in a texturing machine should be the same at all the positions. A slight variation in temperature will produce a yarn having different dyeability.[61,62] It can be seen from Figure 1.11 that as the texturing temperature increases, the dyeability decreases up to 215°C.[62]

Nonuniform Tension. The dye uptake of PET yarns textured at low tensions is higher than that of the untextured yarn (Figure 1.12).[63] At high tensions, the textured sample has a relatively lower dye uptake.

The defects in the textured yarn are seen in the knitted or woven fabric, particularly, after dyeing. Measurements of filament diameters and birefringence, the impression of the fabric on a plastic sheet, and a microscopic

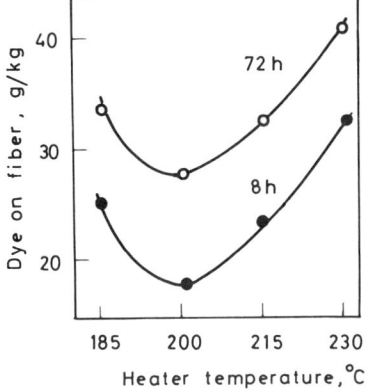

FIGURE 1.11. Effect of texturing heater temperature on the dye uptake of polyester.[62] Dye: Cibacet Turquoise Blue G. Dyeing temperature: 100°C; L:G Ratio 500:1.

examination are some of the methods used to find out whether texturing is the cause of these defects. Visual inspection of the fabric (against light) brings out tight spots, thick/thin yarns and so on, but dyeing confirms the nature of faults.

1.9.5 Coning

Many times the textured nylon yarn turns yellow during storage. This yellowness cannot be removed by repeated scouring. There are two possible reasons for the yellowing of nylon yarn. One is the thermal decomposition of nylon and the second is the degradation of coning oil into yellow products.

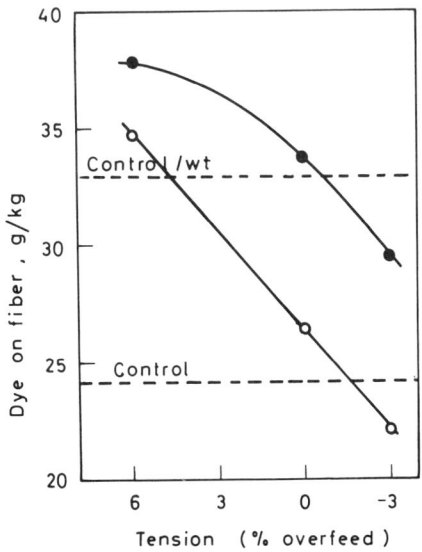

FIGURE 1.12. The dye uptake of textured polyester as a function of tension (or overfeed) during texturing.[63] Dye: C.I. Disperse Red 17. Dyeing temperature: 100°C, *MLR:* 1:500. ○: Original textured. ●: Water-treated (80°C, 2 h). The dotted line represents the dye uptake of the flat yarn.

The degradation of nylon can be detected by recording infrared or UV spectra of the yarn. A peak at 1738 cm^{-1} is obtained in IR spectra if thermal degradation has taken place. Similarly, in the case of UV spectra, a peak of 2400 Å is observed. The viscosity as well as tensile strength of the yarn may show quite small changes. A manufacturer of the fiber tests the coning oil by an accelerated ageing test. In this test, the coning oil is heated at 70°C for three days. A good coning oil does not turn yellow during the accelerated ageing test. The amount of coning oil applied to the yarn is not more than 2% and, under ideal conditions, the oil is uniformly distributed throughout the length of the yarn. Furthermore, the oil is such that it is easily removed during the scouring treatment before dyeing. If the coning oil remains on the fiber, particularly, PET, after dyeing, the disperse dye migrates and dissolves in the oil during heat setting of the dyed fiber. This impairs the fastness properties, soils the setting machines, and stains the undyed fiber. Improper selection of a coning oil is thus a cause of faulty dyeing.

1.9.6 Doubling

If yarns from two different merge numbers are doubled, they exhibit differential dyeing and appear as a heather mixture. This fault cannot be corrected in dyeing. If the twist of yarns to be doubled is the same (both S or both Z), the doubled yarn will appear different from the normal S–Z twisted doubled yarn. The migration of twist also gives differential appearance after dyeing.

REFERENCES

1. Carothers, W. H., and Hill, J. W., *JACS,* **54** (1932), 1559.
2. DuPont, USP 2,130,523 (1938); 2,130,947 (1938); 2,130,948 (1938), and 2,163,636 (1939).
3. Schalack, P., USP 2,241,321 (1941).
4. Whinfield, J. R., and Dickson, J. T., BP 578,079 (1941).
5. Whinfield, J. R., *TRJ,* **23** (1953), 289.
6. Thompson, A. B., *J. Roy, Inst. Chem.,* **85** (1961), 293.
7. Rein, H., *MTB,* **22** (1941), 5.
8. Wells, R. D., and Morgon, H. M., *TRJ,* **30** (1960), 668.
9. *Text. Organon* (June 1978, June 1979, June 1980).
10. *Guide to Man Made Fibers,* British Man Made Fibers Federation (1980).
11. The Production of Synthetic Fibers, *Text. Prog.,* **3**(1), 1971, 1.
12. Gulrajani, M. L., Ed., *Blended Textiles,* The Textile Association, India (1981).
13. Gregg, R. A., in *Polyurethane Technology* Bruins, P. F., Ed., Interscience, New York (1969), p. 257.
14. Redston, J. P., Burnholz, W. F., and Schlatter, C., *TRJ,* **43** (1973), 325.
15. Postman, W., *TRJ,* **50** (1980), 444.
16. Falkai, B. U., International Conference on Man-Made Fibers for Developing Countries, SASMIRA, Bombay (1982).

17. Farber, M., *ADR*, **55** (1966), 537.
18. Campbell, T. W., *JAPS*, **5** (1961), 184.
19. Sawaya, A. R., USP, 3,039,840 (1962).
20. Shore, J., *Rev. Prog. Color.*, **6** (1975), 7.
21. Montecatini, BP, 942, 131 (1963).
22. Riggert, K., *Chemiefasern/Textilindustrie*, **23/75** (1973), 543; **29/81** (1979), 683.
23. Bush, M., and Kemp, U., *Fiber Prod.*, **9**(2) (1981), 26.
24. DuPont, BP, 106,810 (1972).
25. Hughes, A. J., McInntyre, J. E., et al., *Text. Prog.*, **8**(1) (1976), 36.
26. Geller, V. E., Baranova, S. A., Zhestkova, E. I., Aizenstein, E. M., and Shablygin, M. V., *Fiber Chem.*, **10** (1978), 124.
27. Sakuma, Y., and Rebenfeld, L., *JAPS*, **10** (1966), 637.
28. Dismore, P. F., and Statton, W. O., *JPS*, **Part C** (13) (1966), 133.
29. Sweet, G. E., and Bell, J. P., *TRJ*, **46** (1976), 477.
30. Dumbleton, J. H., Bell, J. P., and Murayama, T., *JAPS*, **12** (1968), 2491.
31. Gulrajani, M. L., Ed., *Texturing*, IIT, Delhi, India (1977).
32. Dupeuble, J. C., *Chemiefasern/Textilindustrie*, **29/81** (1979), 104.
33. Aken, F. J., *Chemiefasern/Textilindustrie*, **29/81** (1979), 108.
34. Hoffsommer, K. P., *Fiber Prod.*, **8**(3) (June 1980), 22.
35. Rilling, R., *Chemiefasern/Textileindustrie*, **31/83** (1981), 298.
36. Datye, K. V., and Bose C., in *Blended Textiles*, Gulrajani, M. L., Ed., The Textile Association, India (1981), p. 42.
37. Weller, A. F., in *Modern Yarn Production*, Columbine Press, Manchester, England (1960), Chapter 8.
38. McCormick, W. H., in *Modern Yarn Production*, Columbine Press, Manchester, England (1960) Chapter 7, p. 111.
39. Tucker, P., Johnson, D., Dobb, M., and Sikorski, J., *TRJ*, **47** (1976), 29.
40. Luneschloss, J., Sulkowaski, A., and Bock, G., *Chemiefasern/Textilindustrie*, **30/82** (1980), 119.
41. Ahrendt, D., and Dietrichs, H., *Chemiefasern/Textilindustrie*, **31/83** (1981), 207.
42. Schatzel, H., *Chemiefasern/Textilindustrie*, **23/75** (1973), 295.
43. Friedrich, A., *Text. Prax.*, **7** (1952), 433.
44. Gerhardt, W., *Deutsche Textiltech.*, **9**(3) (1959), 127.
45. Thomas, L. B., Seffers, M., and Broome, E., in *Blended Textiles*, Gulrajani, M. L., Ed., The Textile Association, India (1981), p. 385.
46. Mishra, S., PhD Thesis I.I.T. (Delhi) (1983).
47. Goodman, I., and Nesbitt, B. F., *JPS*, **48** (1960), 423.
48. Peebles, L. H., Huffman, M. W., and Ablette, C. T., *JPS*, **Part A-1, 7** (1969), 479.
49. Valk, G., Loers, E., and Kiipers, P., *MTB*, **51** (1970), 504.
50. Valk, G., *MTB*, **59** (1978), 843.
51. *AATCC Technical Manual*, U.S.A., AATCC Test Method 20-1973, *Fibers in Textiles—Identification* (1975), p. 52.
52. Jones, E. B., *Text. Prog.*, **10**(4) (1981), 2.
53. Achwal, W. B. and Vaidya, A. A., *Text. Praxis.*, **25** (1970), 48.
54. Langley, K. D., in *Blended Textiles*, Gulrajani, M. L., Ed., The Textile Association, India (1981), p. 210.

55. Achwal, W. B., in *Blended Textiles,* Gulrajani, M. L., Ed., The Textile Association, India (1981), p. 216.

56. Datye, K. V. and Pitkar, S. C., Seminar, UDCT Bombay (1967), 48.

57. Janssen, R., Ruysschaert, H., and Vroom, R., *Die Micromolecular Chemie,* **77** (1964), 153.

58. Zahn, H., Kusch, P., Muller-Schulte, D., Nissen, D., and Rossbach, V., *TRJ,* **43** (1973), 601.

59. Achwal, W. B., and Rao, C. S., Resumé of Papers 23rd Tech. Conf., The Northern India Textile Research Association, Gulrajani, M. L., Ed., (Feb 12–13, 1982), p. 101.

60. Stein, W., and Heiser, H., *Chemiefasern/Textilindustrie,* **29/81** (1979), 470.

61. Warwicker, J. O., Shirley Institute Publication, **S25** (1976), 49.

62. Gupta, V. B., Kumar, M., and Gulrajani, M. L., *TRJ,* **45** (1975) 463.

63. Gupta, V. B., and Amrithraj, J., *TRJ,* **46** (1976), 785.

2 STRUCTURE OF SYNTHETIC FIBERS

A synthetic filament yarn is produced from a polymer made up of long chain molecules by a succession of treatments using a combination of temperature, tension, and residence time. For a given polymer, the arrangement of the chain molecules in the fiber is decided by a combination of process conditions that, in turn, decide the thermal and mechanical properties of the yarn produced.[1,2] Thus, fibers with different properties can be prepared from the same polymer by altering the spinning and drawing parameters or by a heat treatment. The relationship between the process conditions and the fine structure of the fibers is of particular importance to the processor as well as to the producers of the fibers.

2.1 FINE STRUCTURE OF FIBER

2.1.1 Miceller Theories

Various theories for the supermolecular physical structure of synthetic fibers have their origin in the earlier theories proposed for the structure of cotton fiber. One may broadly group these theories as miceller theories.[3] The micelle was assumed to be the smallest structural aggregate of a macromolecule. These discrete bricklike submicroscopic particles are oriented with respect to the axis of the fiber and account for the crystallinity of the fiber. The remaining matter is just amorphous matter separating the micelles. Since the polymer molecules are much longer than a micelle, they can run through successive crystalline and noncrystalline regions giving fringed micelles (Figure 2.1). Thus, crystalline micelles are completely embedded in

FIGURE 2.1. Fringed micelle structure.[8]

the amorphous matrix of the noncrystalline polymer.[4-7] Hearle[8] modified these theories by assuming the crystalline regions were continuous fringed fibrils composed of molecules diverging from the microfibrils at different positions along their length. Owing to the partially intertangled arrangement of polymer macromolecules in the viscous melt/dope during fiber spinning, it is not possible for a given chain molecule to be incorporated indefinitely in the same microfibril. It will diverge and pass through noncrystalline regions before being included in another microfibril (Figure 2.2).

2.1.2 Chain Folding

Studies of single crystals of polyethylene with thickness of about 100 Å show that these crystals were formed by the folding of long chain molecules.[9-12] This phenomenon of chain folding at the crystal surface in order to reenter the crystal is observed with all the crystallizable polymers including polyacrylonitrile, the crystallinity of which was doubted for a very long time.[13-15] This led to a modified theory of the physical structure of fibers. In the chain-folding model,[14] the longitudinal period corresponds to the folding length.

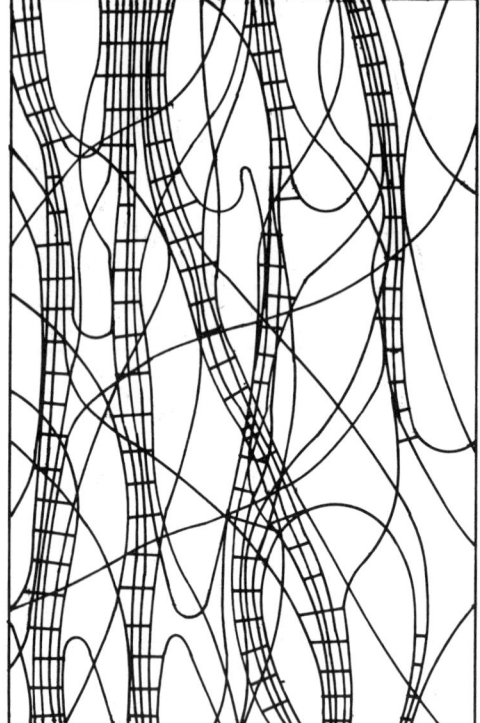

FIGURE 2.2.   Fringed fibril structure.[8]

2.1.3   Crystal Defects

Polymer molecules are not of the same lengths but have a size distribution. The random placement of the chain ends creates defects in the crystalline structure.[16,17] The presence of these defects results in an apparent limitation of crystal size. For a hypothetical polymer of molecular weight 10^4, there are 6×10^{19} molecules in a gram of polymer, that is, 12×10^{19} molecular ends. If the density is 1 g/cc, each molecular end occupies a volume of $(1/1.2) \times 10^{-20}$ cc, which is equivalent to a cube with a 20 Å side. Thus, the crystal size is of the order of 20 Å before the distortion produced by the chain end appears. This is probably the cause of the amorphous halo observed in the X-ray diffraction patterns of fibers. It can also explain the mechanical properties of polyethylene crystals.[12,18] The movement of molecules during the annealing of the synthetic fibers enables the molecular ends to aggregate in order to reduce the distortion in the structure and hence, the approach is energetically favored. The process of annealing produces larger crystallites by reducing the chain end distortions. (Figure 2.3).[16]

Statton[18] has observed the presence of microvoids with dimensions of the

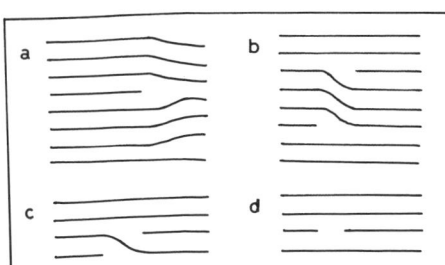

FIGURE 2.3. The progressive decrease in distortion produced by a chain- end in a crystalline defect model.[16]

order of 10 Å within polymer structures. He suggests that these microvoids are formed by imperfections in the packing together of folded and extended crystalline structures.

The crystal defect model for the synthetic fiber structure has been successful in explaining a number of physical and mechanical observations that previous models based on the two-phase crystalline-amorphous concept could not handle.

2.1.4 Chain Folding and Fiber Annealing

The folding of molecular chains during annealing gives a stable configuration.[19–21] Annealing of drawn fibers at temperatures near the melting point results in a mechanical weakening of the fibers. The process of drawing the filaments is supposed to orient the crystallites in the direction of the fiber axis and to increase the mechanical strength of these fibers. However, Statton[18,21] regards the increase in the mechanical strength of nylon fibers after drawing as a process of alignment of polymer molecules in an extended rather than a folded configuration. He also regards the process of annealing as the movement of extended chains to a state of folded configuration. Thus, drawing results in strengthening and annealing results in the mechanical weakening of the fiber. The crystallization of nylon chain molecules in a folded form has been demonstrated by X-ray analysis[22] and electron micrography.[23] Dismore and Statton[24] observed that heating without tension increases the intensity of the small-angle X-ray diffraction crystallinity without significantly affecting the crystalline orientation and defect mobility (or mobile segment fraction), and decreases the sonic modulus to a value lower than the theoretical one. Simple relationships are found for the long-period intensity vs. fiber shrinkage or fiber tenacity. In the oriented, as-drawn yarn, the nylon 66 molecules are highly extended and are essentially parallel to each other for the bulk of their length. There are relatively few chain folds enclosing highly crystalline regions which are probably remnants of unfolded lamellae. However, the overall statistical order is quite low (34%) so that there is a broad distribution of hydrogen and van der Waals bond energies

ranging from the nearly ideal in highly crystalline regions between the folds to a very low value in the surrounding regions. The intermolecular bonds with the lowest energy will give way or melt on heating. This allows short lengths of polymer molecule to move and form more stable bonds. In the terminology of the fringed micelle model, the polymer crystallizes. If enough adjacent intermolecular bonds are melted during heating, entropy will cause the segments to approach the random state of the melt. New crystallization will then take place in the folded form, using the folds that are already present as oriented nuclei.

The properties of PET yarns of low crystallinity (less than 2%) and low orientation (birefringence of 0.002) which were drawn to different extents on a hot pin at 80°C and annealed in relaxed conditions at various temperatures of up to 240°C by immersing in silicone oil for 1 min are given in Table 2.1.[20] Dumbleton explained these results in the manner similar to that described above for nylon 66. The chain folding during annealing gives a large increase in intensity in the small-angle X-ray diffraction maximum, a change in crystal size, high shrinkage, and a minimum change in crystalline orientation.

The tenacity decrease with the increase in annealing temperature may be explained as a consequence of the formation of regular fold surfaces that allow easier slip between crystals. At high annealing temperatures, the tenacity will also decrease because of the decrease in crystallinity.

The main difference between the behavior of nylon 66 and PET on annealing is in the much larger increase of the small-angle X-ray diffraction intensity for PET. Also, the intensity decreases at the highest temperature for PET. The decrease is probably caused by the decrease in crystallinity which itself is a sign of melting in the crystalline regions. The change of long period with temperature of annealing is somewhat different for PET, since there is no break in the curve at high temperatures as is found in nylon 66. This implies that there is a wide distribution of fold lengths to act as nuclei for folding in PET. The chain-folding model is also suggested for acrylic fibers.[19]

2.1.5 Three-Phase Model

The three-phase model of the fiber structure has been suggested by Prevorsek[25] for melt-spun and drawn fibers such as nylon, polyester, polyethylene, and polypropylene (Figure 2.4). The three phases are amorphous domains of microfibril, crystalline domains of microfibril, and intermicrofibril regions. In the microfibrils, the sequence of crystalline and amorphous regions is regular and the chains are folded to form the crystalline regions. The microfibrils form an essentially endless interwoven structure with branching and fusion without an abrupt end.[26] Microfibrils have a well-defined element structure with a width of 60–200 Å. The microfibrils have long-period characteristics that represent the spacing between the adjacent crystallites (Figure 2.4). Typical dimensions of microfibrils are given in Table 2.2.[27]

TABLE 2.1 Properties of Annealed PET Yarns[20] (Stretch Ratio : 5)

Annealing temperature (°C)	20	100	150	175	200	225	240
X-ray (long-period)	—	124	122	124	147	167	195
Intensity in small-angle maximum	—	0.8	47	76	147	214	132
Percent crystallinity	35	38	34	39	39	39	33
Crystallite size (010) (Å)	40	44	54	57	57	80	84
Birefringence	0.206	0.200	0.174	0.164	0.159	0.112	0.040
Percent shrinkage	—	8	23	31	43	60	75
Orientation function							
Crystallite	0.943	0.943	0.910	0.910	0.858	0.858	—
Amorphous	0.790	0.811	0.709	0.668	0.680	0.480	
Tenacity (g/d)	5.25	6.50	5.75	4.75	3.75	2.25	2.20

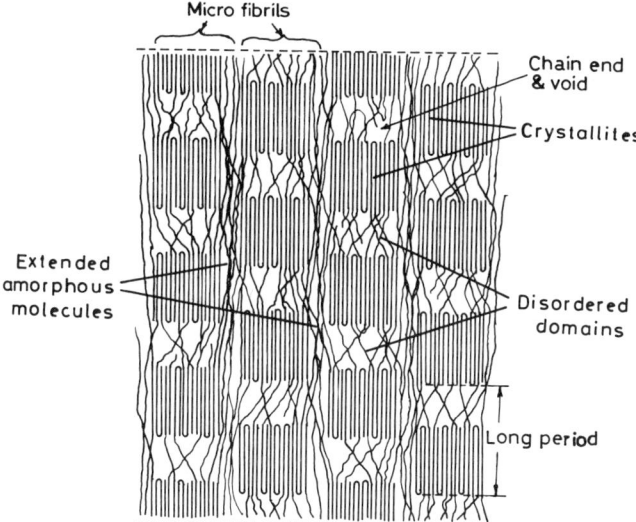

FIGURE 2.4.   Three-phase structural model of PET fiber.[25]

The crystalline length along the microfibril axis is always about two-thirds of the long period. The longitudinal dimensions of the microfibril are not well-defined. In the highly extended intermicrofibrillar regions, the chain molecules are oriented but the crystallinity is absent. These regions are called *oriented amorphous regions*. The presence of these regions is also proposed by Peterlin[28] for PE and PP fibers. He reported that at least some of the interfibrillar space is filled with highly extended interfibrillar tie molecules that are the main factor in fiber shrinkage at temperatures below their melting point. These interfibrillar domains do not contribute to the fiber strength which is mainly due to the strength of microfibrils. In contrast, Prevorsek et al.[29,30] consider that extended-chain interfibrillar domains are the strongest elements of fiber structure and have an important effect on the fiber strength of PET and nylon. They consider the amorphous domains of the microfibril as the weakest element of the fiber structure.[27] The increase in fiber strength on drawing is attributed to an increase in the volume fraction of the extended-chain molecules that are formed as a result of the relative

TABLE 2.2   Typical Dimensions (Å) of Microfibrils[27]

	PE	PP	PA 66	PET
Long period	—	90	90	150–180
Crystalline length	—	60	60	—
Amorphous length	—	30	30	—
Diameter	160	120–90	—	330–160

displacements of microfibrils. In this process, the molecules from the surface of the microfibrils are sheared off and stretched. The dimensional stability at elevated temperatures is provided by microfibrils. The melting point of the microfibrils is considerably above the softening point of the extended interfibrillar regions, whose order and density are intermediate between those of the crystalline and amorphous domains of the microfibrils. This model of the physical structure is supported by the following observations.

The room-temperature drawing of amorphous PET film leads to the formation of oriented amorphous regions. The absence of crystallization in highly extended chains is due to the high surface to volume ratio, which results in significant supercooling.[31] Boyer[32] has observed the phenomena of double glass transition in PET that has survived several drawings and annealing processes. Chao[33] suggested that one of the two transitions may be due to the oriented amorphous regions. The lower of the two transition temperatures increases with the draw ratio. Valk et al.[34] reported a double relaxation mechanism close to the glass-transition temperature for carrier-treated PET samples. He assigned relaxation to isotropic amorphous parts around 35°C, whereas the relaxation above 80°C was assigned to anisotropic amorphous regions.

2.1.6 Macrofibrils

Nylon and PET fibers have considerable interfibrillar strength, that is, the strength perpendicular to the fiber axis. Transmission electron microscopic studies on these fibers indicate the presence of another element of fiber structure whose shape is ribbonlike and is about 300–400 Å thick.[26] These elements are called *macrofibrils* and are presumed to consist of several microfibrils and intermicrofibril regions.[31,35] The size of these units is highly dependent on the thermomechanical treatment. Further information is necessary to understand these super structures.

2.1.7 Present Status of Theories

The overall picture that emerges from the voluminous experimental data and theories is:[36]

1.   Crystalline (ordered) regions alternate with amorphous (less ordered) domains. A molecule can run through several crystalline and amorphous regions. On annealing, new crystallization occurs in the folded form.[37]

2.   A molecule may fold back on the surface of a crystal to reenter it. Thus, highly crystalline regions are enclosed in the chain folds.[22,38]

3.   The polymer molecules are more or less oriented along the fiber axis to form the microfibrils, in which crystalline and amorphous regions are arranged in a regular sequence. These microfibrils also have a lateral extension.

4.   An oriented amorphous region is enclosed between the microfibrils.

2.2 FINE STRUCTURE AND FIBER PROPERTIES

2.2.1 Crystallinity

The characterization of the fine physical structure of synthetic fibers based on an amorphous–crystalline, two or three-phase model is now universally accepted. PET does not develop crystallinity beyond 1% under normal slow-speed spinning conditions.[39–41] The winding speed influences the fiber structure immensely. The relationships of fiber properties and winding speed are shown in Figure 2.5.[42] The elongation at break gradually decreases with the winding speed and the orientation[33] and, in turn, the modulus increases with the speed. There is a sudden decrease in the boiling water shrinkage that drops to its lowest value at the highest winding speed. The boiling water shrinkage also depends on the average molecular weight of PET;[42] the higher the molecular weight, the smaller the shrinkage (Figure 2.6). The density of the fiber starts increasing sharply at 3500 m/min. A change from amorphous to partly crystalline material is suggested by this sharp increase. This is clearly brought out in the differential scanning calorimetry (DSC) diagrams.

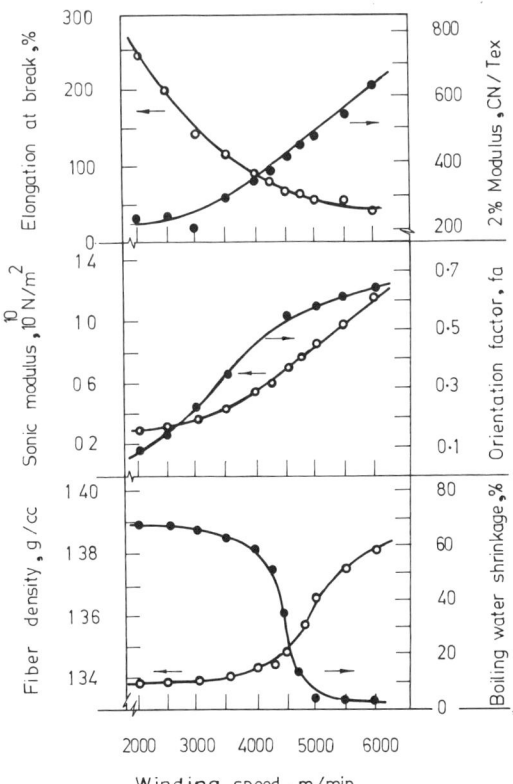

FIGURE 2.5. Properties of polyester fiber spun at different winding speeds.[42]

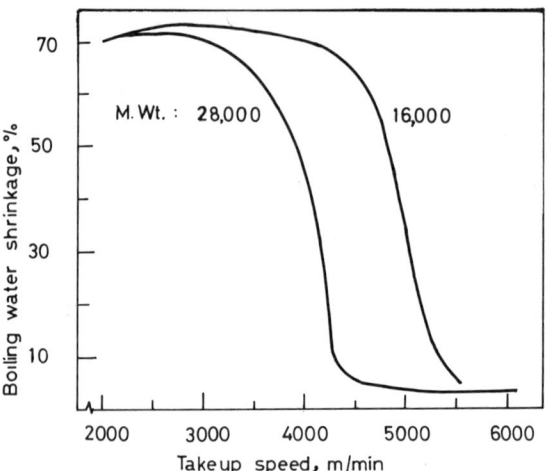

FIGURE 2.6.   Boiling water shrinkage of polyesters spun at different winding speeds.[42]

For the lower winding speed, the glass transition takes place at 76°C followed by the crystallization exotherm with a peak temperature near 132°C and a melting endotherm with its minimum at about 253°C. From this DSC diagram, it is concluded that the original fiber wound at 2000 m/min, is largely amorphous, and the crystalline material that melts finally formed during the DSC run. In sharp contrast to this DSC trace is the one brought about by the yarn wound at 6000 m/min. This curve only shows a sharp melting endotherm that indicates that this material originally was already partially crystalline. The situations in-between these two extremes show a gradual change (Figure 2.7). The physical quantities related to orientation are shown in Table 2.3. The steep decrease in boiling water shrinkage is caused by the crystallization process taking place during winding at higher speeds.

The effect of stretching the as-spun yarn on the physical structure of the fiber is to increase the crystallinity in the initial stages of drawing (Figure 2.8).[33] X-ray diagrams of undrawn, drawn, and set PET are shown in Figures 2.9(a) and 2.9(b).

Nylon 6 spun in moderately dry air or in a saturated steam environment does not exhibit detectable crystallinity.[43] The as-spun yarn, however, gradually crystallizes if it is left in the air on the take-up bobbins. The major development of crystalline structure occurs during drawing.

Acrylic fiber exhibits a type of crystalline order that extends laterally from one molecular chain to another, but little or none along the chain axis.[1,2,44] The crystallinity in acrylic fiber is low. Polypropylene is an isotactic polymer with all of its methyl groups having the same orientation in space. About 96% of the structure is isotactic, which enables the polymer chains to fall into a helical configuration and crystallize, thus, giving it

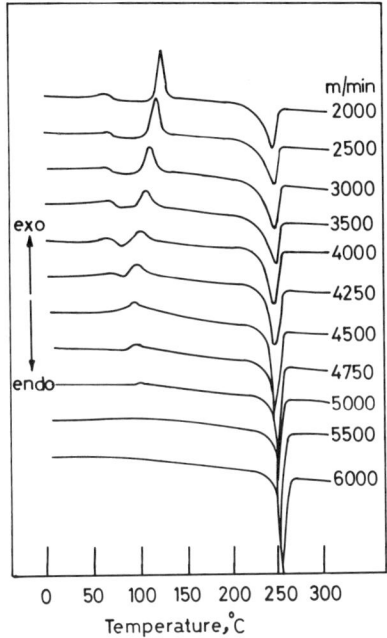

FIGURE 2.7. DSC thermograms of polyester yarn spun at different winding speeds.[42]

strength and a high melting point for a paraffinic hydrocarbon. In the atactic form, where the methyl groups are randomly oriented up and down, not only it is not crystalline, but it is a rubber.

Crystallinity of PET increases with the heat-setting temperature under relaxed or taut conditions[37] (Figure 2.10). Similarly, in the case of nylon, the crystallinity increases with the heat-setting temperature (Figure 2.11).[24]

TABLE 2.3   Physical Quantities Related to Orientation[41,42]
Fiber: PET

Winding Speed	Sonic Modulus (GPa)	Crystalline Orientation Factor (fc)	Amorphous Orientation Factor (fa)	Volume Fraction Crystallinity
2000	3.01	—	0.093	0
3000	3.62	—	0.246	0
3500	4.31	—	0.367	0
4500	6.91	0.76	0.589	0.056
5000	8.57	0.972	0.626	0.155
5500	9.78	0.976	0.650	0.209
6000	11.60	0.979	0.693	0.242

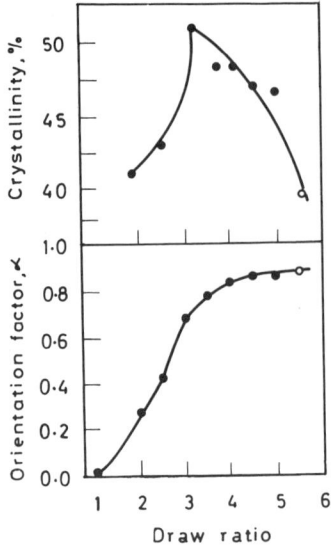

FIGURE 2.8. Crystallinity and orientation of PET fiber stretched to different draw ratios.[33] Draw temperature: ●: 100°C; ○: 125°C.

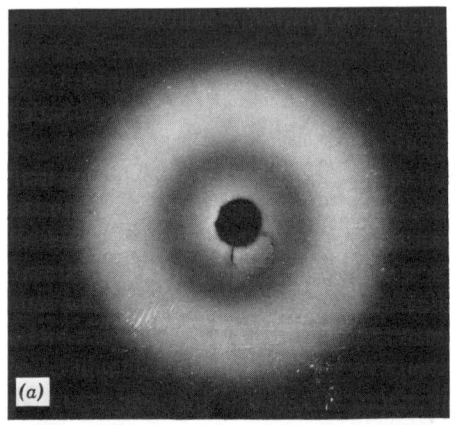

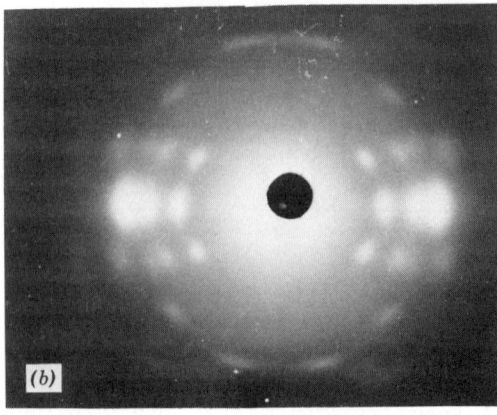

FIGURE 2.9. X-ray diagram of PET fiber. (*a*) Undrawn; (*b*) Drawn and set.

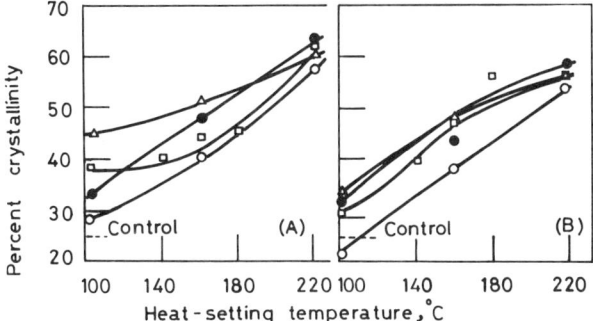

FIGURE 2.10. Crystallinity of polyester after heat setting without and under tension.[37] Heat-setting time: ○: 1 min; △: 15 min; □: 30 min; ●: 60 min. Heat-setting condition: (A): without tension; (B): under tension.

The melting point of the fiber is attributed to the crystalline part and is size-dependent; a smaller crystal has a lower melting point because of its high surface area per unit size which contributes a relatively large amount of energy to the system. The crystal size can be determined from the radial half-width of the X-ray diffraction. The polymer fiber does not have a fixed melting point. During melting, the heating rate is low with respect to the quick recrystallization process. The small crystal melts and recrystallizes into a bigger one with a higher melting point. In this way, during DSC trace, the endothermic effect of melting is compensated for by the exothermic effect of recrystallization. If the recrystallization during melting is prevented by cross-linking the molecules in the amorphous regions, thus blocking the necessary segmental movements for recrystallization, the real melting point

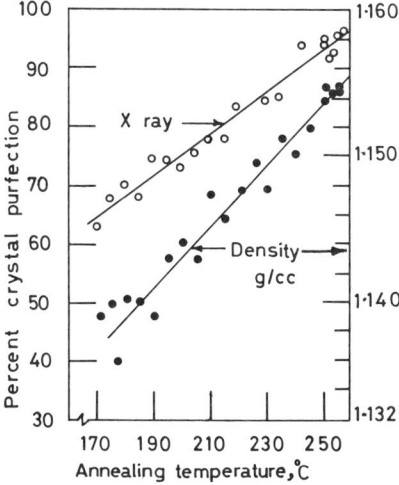

FIGURE 2.11. X-ray crystallinity and density of PA 66 vs. annealing temperature.[24]

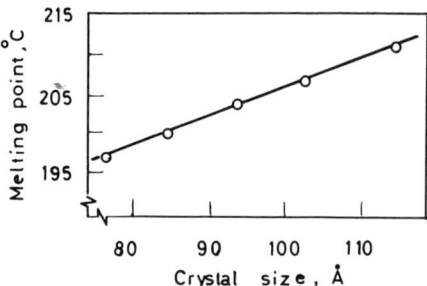

FIGURE 2.12. Melting of PA 6 as a function of crystal size.[42]

of a crystal can be determined. The size of the crystal can be estimated by X-ray diffraction. The relation between the size of the crystal and the melting point for nylon is shown in Figure 2.12.[42]

False-twist texturing processes lead to an increase in crystallinity that increases with temperature and time of heating (Figure 2.13)[45] but decreases with tension and twist in the yarn. However, the degree of crystallinity of textured yarn is always higher than that of untextured yarn. If the texturing temperature is higher than the melting point of the crystals, the crystals will melt which results in the sticking together of the filaments. The optimal texturing is therefore performed at a temperature where only partial melting takes place. All the crystals that have a melting point below the texturing temperature will melt and recrystallize into new larger crystallites in the newly crimped shape. This is schematically illustrated by Huisman and Heuvel,[42] as shown in Figure 2.14. To develop the latent crimp into an actual one, a tensionless heat treatment above the glass-transition temperature is sufficient. If the crimp development is hindered by tension, the crystals weaken or melt, orientation differences disappear, and the latent crimp is

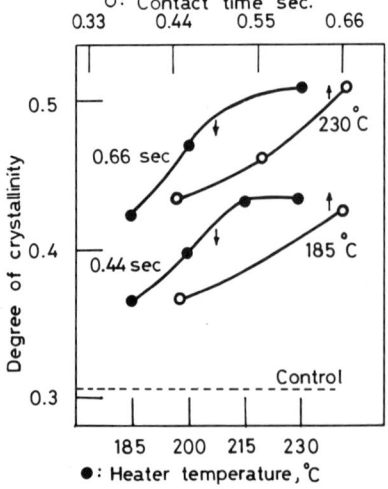

FIGURE 2.13. Degree of crystallinity as a function of temperature and time of false-twist texturing.[45]

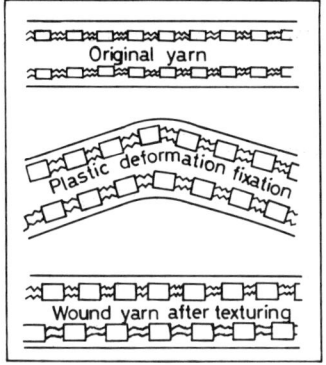

FIGURE 2.14. Physical structure of feeder yarn, yarn during texturing, and wound yarn after texturing.[42]

lost. Thus, the crystal size (as obtained during texturing) is large enough to be stable during treatments under tension such as dyeing. The larger the crystal size, the smaller will be the change of crystal size during dyeing (Figure 2.15).[42] The thermal crimp stability and the crystal size of the textured yarn also show a similar relationship (Figure 2.15).[42] The effect of temperature on crystallinity of fiber may not be the same in all planes; the growth of the crystals is more in one direction than in the other. (Figure 2.16).[42]

Various methods for measuring the crystallinity of a fiber are based on density, X-ray diffraction, infrared absorption spectra and so on.[18,46,47] Since these methods are based on different principles, the values of crystallinity obtained by these methods need not be the same. (Table 2.4).[48]

2.2.2   Orientation

In the case of synthetic fibers, it is often possible to control the degree of orientation by applying a stretch of desired intensity to the spun filaments.

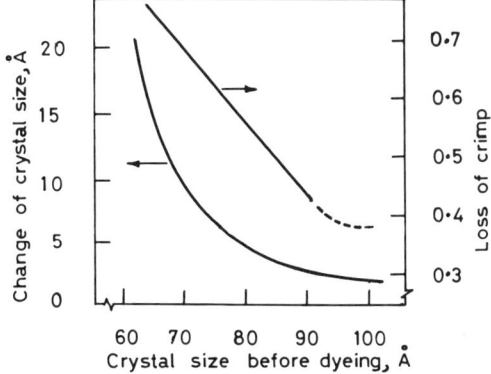

FIGURE 2.15. Change of crystal size and loss of crimp during dyeing as a function of the original crystal size for PA 6.[42]

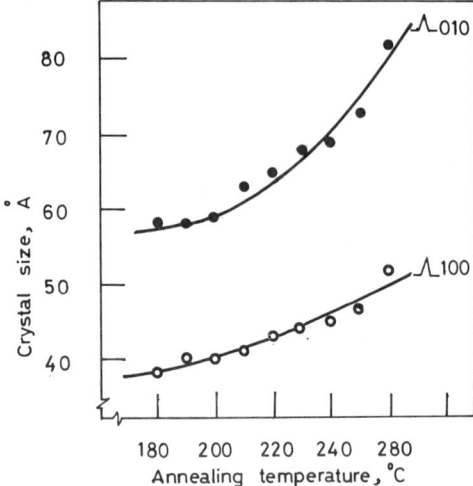

FIGURE 2.16. Crystal size in different planes as a function of the annealing temperature.[42] ○: 100 plane; ●: 010 plane.

Consider a filament that has just been extruded and not stretched. The molecules will probably orient themselves (to a slight degree) parallel to the fiber axis. This is caused by the direction of flow of the spinning solution or melt through the spinneret jet. The degree of orientation due to this effect will not be high and the molecules will approximate a random arrangement. The birefringence of the undrawn fibers is very low, which confirms that the molecules are not highly oriented. If, a fiber consisting of randomly arranged molecules is stretched, the molecules tend to orient themselves in the direction of stretching. Thus, the birefringence of the fiber increases on stretching.

There are two types of orientations present in synthetic fibers: (a) crystalline orientation and (b) amorphous orientation. The orientation of the crystals can be determined directly by means of azimuthal scanning of the X-ray

TABLE 2.4 Polyester Fiber Crystallinity (%) by Different Methods[48]

Sample	Density	Infrared	X-rays
1	20	61	29
2	18	41	31
3	18	50	nil
4	42	58	27
5	56	81	39
6	61	75	40

reflection. Birefringence measures the orientation in both crystalline and amorphous domains. Other methods are light scattering, IR dichroism, NMR, fluorescence polarization,[49] and spectroscopic[50] and dielectric anisotropy techniques.[51]

By measuring the sonic modulus by means of a pulse propagation, the overall orientation of crystalline and amorphous regions can be evaluated.[20] If the sonic modulus, the crystallinity, and crystalline orientation factor are known, the value of the amorphous orientation factor can be calculated (Table 2.3).

The molecular orientation develops during various stages of fiber manufacture. Thus, orientation increases with the draw-ratio[33,43] (Figure 2.8). In the case of acrylic fiber, the orientation in the fiber depends on the coagulation conditions during wet-spinning, draw ratio, and drawing speed (Figure 2.17).[52–54] Gupta and Kumar[37] observed that an increase in the annealing temperature increases the crystalline and amorphous orientation factors, except when the fiber sample is annealed without tension and when the crystalline orientation decreases.

The false-twist texturizing causes a decrease in both the crystalline and amorphous orientation factors, indicating that the texturing results in a disorientation of crystallites.[45] Although crystalline orientation decreases on texturizing, the decrease is larger at lower temperatures. The reverse is the effect of texturizing temperature on amorphous orientation.[45]

2.2.3 Dyeability

The dyeability of synthetic fibers depends, apart from the chemical nature, on the fine structure of the fiber. The dye uptake of a given fiber is remarkably affected by the thermal and mechanical history of the fiber. Marvin[55]

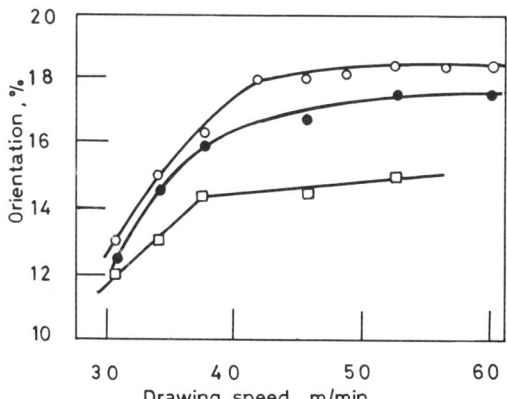

FIGURE 2.17. Relationship between orientation and drawing speed of acrylic fibers spun under different conditions.[54] Coagulation conditions: ●: DMF : H_2O; 30 : 70; 10°C. ○: DMF : H_2O; 53 : 47; 23°C. □: DMF : H_2O; 70 : 30; 45°C.

reported that PET annealed at different temperatures shows a dye uptake minimum at 200°C when dyed by aqueous dyeing methods. A typical relationship between dye uptake vs. annealing temperature is shown in Figure 2.18.[56] The fibers annealed at a low temperature show a low crystallinity build-up of many small crystals, while those annealed at high temperature have a high crystallinity and are composed of fewer larger crystals together with large adjacent amorphous regions. The crystals can be considered completely inaccessible to the dye molecules and dyeing takes place only in the amorphous regions. If the total volume of PET amorphous regions is considered for the calculation of dye uptake, different fibers show similar dye uptake (see Figure 7.1). The mobility of the molecular segments in the amorphous region decides the rate of dyeing. If the crystallites are considered physical cross-links decreasing the segmental mobility, many small crystals have a greater cross-linking effect than fewer big ones, and the accessibility of the amorphous regions can be related to their volume. At a high temperature, a large increase in the amorphous volume because of the decrease in the number of small crystals (which are converted into a few big crystals) is found.[42] Accordingly, the accessibility to the dye molecule is high. Wide-line NMR signals are observed[57] in the yarns annealed at elevated temperatures, indicating increased segmental mobility. The effect of increasing the annealing temperature is made up of two opposite factors; (*a*) a decrease in amorphous content and (*b*) an increase in the accessibility, facilitating the diffusion of dye. The balance of these two effects accounts for the observed minimum in Figure 2.18.

The size of the dye molecule influences the dyeing behavior of a given

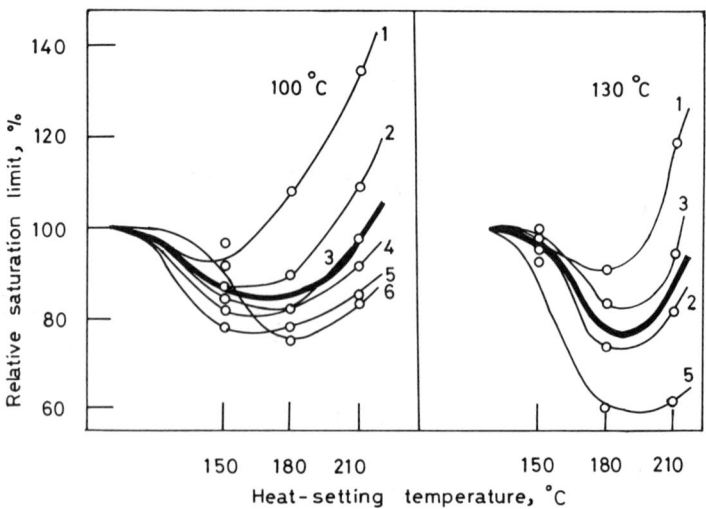

FIGURE 2.18. Dependence of the dye-saturation limit on the heat-setting temperature.[56] C.I. Disperse: 1: Red 7; 2: Orange 13; 3: Blue 14; 4: Blue 23; 5: Red 13; 6: Blue 3. Thick line (—), average.

fiber. The relationship of the molecular weight of dyes and their diffusion coefficients for PET from perchloroethylene (121°C) has a coefficient of correlation of 0.9380[58] (Figure 2.19). The dependence of the activation energy for diffusion in PET on the molecular weight of dyes in thermofixation or sublimation transfer is clearly established (see Chapter 7).[59] With a small dye molecule, the accessibility of the amorphous region will be a less important factor in the control of the dyeing process. As the size of the dye molecule increases, the structure of the fiber will play a more and more decisive role. With an increase in the dyeing temperature, the segmental mobility improves and the dyeing rates will be less influenced by the fine structure of the fiber.

The amorphous orientation contributes adversely to the accessibility. It may be expected on physical grounds that the diffusion rate will be higher in a less-oriented medium. Figure 2.20 gives dye uptake curves as a function of the volume of amorphous regions for different amorphous orientations.[42] (In this picture, highly stabilized yarns are used; therefore, only the high-temperature side of the dyeing picture is involved.) The figure clearly shows the effect of orientation.

Thus, the dyeing behavior of a fiber can be described in terms of the total volume of amorphous material, the volume of individual amorphous regions, and the orientation of the molecules in the amorphous domains. The influence of the molecular size of a dye and temperature on dyeing is explained on the basis of ease of segmental mobility of polymer in the amorphous regions. More detailed discussion will be found in later chapters.

2.3 TRANSITIONS IN FIBER STRUCTURE

One of the most important properties of synthetic fibers is their thermal behavior. Knowledge of this behavior enables one to select the proper pro-

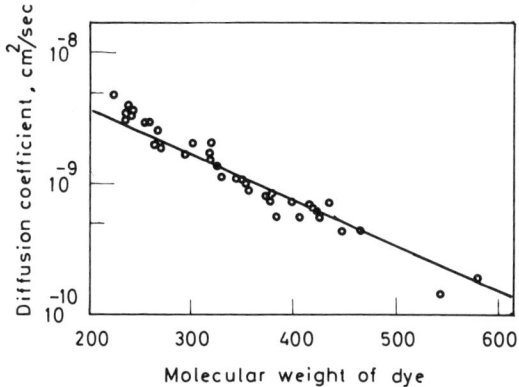

FIGURE 2.19. Relationship of the molecular weight of dyes and their diffusion coefficient in PET from perchloroethylene at 121°C.[58]

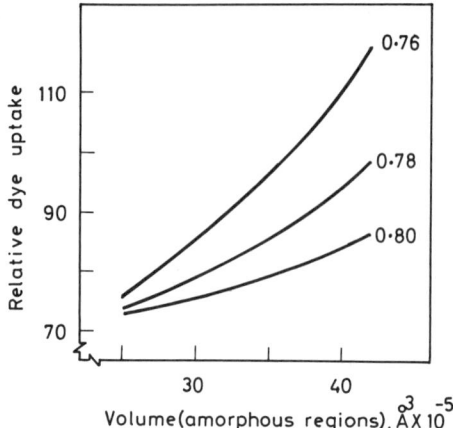

FIGURE 2.20. The dye uptake of polyester yarn as a function of the volume of the individual amorphous regions for different levels of amorphous orientation.[42] (The values in the figure are for amorphous orientation factors.)

cessing conditions and to characterize the physical and mechanical properties for the appropriate end use. The temperature-dependent properties of the fibers undergo major changes at one of two transition points: first-order transition at the crystalline melting point and second-order transition at the glass-transition temperature. The latter is the more important transition in connection with the processing of fibers.

The glass-transition temperature T_g may be defined as the temperature below which the amorphous polymer is glassy (or frozen) and above which it is rubbery. T_g is the temperature of the onset of large-scale motion of molecular chain segments. At very low temperatures, chain atoms undergo low-amplitude vibratory motion around fixed positions. As the temperature is raised, both the amplitude and the cooperative nature of these vibrations among neighboring atoms increase. When the thermal energy is sufficient to overcome the barriers for movement, segmental movement becomes possible and the material becomes rubbery. Thus, above T_g, the chain segments undergo concerted rotational, translational, and diffusional motions. As the temperature is raised further, larger and larger chain segments undergo motions until the amorphous material becomes like a liquid of high viscosity. After melt-extrusion or after any treatment involving high temperature, the fiber is cooled rapidly. The fiber is therefore not in thermodynamic equilibrium at T_g. The thermal history of the fiber, heating rate, and the presence of low-molecular-weight compounds, dyes,[60] liquids[61] and so on influence the T_g. According to relaxation or hole theory, a chain segment passes from one energy state to another. In order for this to happen, an empty space or free volume or a hole for the segment to move into is required. The creation of this hole requires hole energy to overcome the cohesive forces of the surrounding molecules and an activation energy to overcome the potential barrier associated with the arrangement, by way of an activated state. As the hot fiber is cooled, the T_g depends only on the rate of cooling. In contrast, as the fiber is heated, T_g depends not only on the heating rate but on the thermal history of the sample, since the segmental movement is frozen below T_g and the fibers that have been cooled differentially represent different starting materials that have different enthalpies and different time-dependent heat capacities in the glass-transition range. The conditions of setting or annealing of fibers remarkably influence the T_g.

2.3.1 Molecular Structure and T_g

The basis for considering the effects of molecular structure variables on T_g is the onset, as T_g is approached, of concerted motion of chain segments larger than the monomer residue. As the temperature increases, the number of these groups becomes larger and larger until about T_g; the entire polymer coil in the amorphous region is the elastic unit. Thus, any molecular parameter affecting the chain mobility can be expected to influence T_g. The chain microstructure, such as chemical type of monomer units, copolymerization effects, molecular architecture variables such as MWD, average molecular weight, branching, and cross-linking, and the presence of low-molecular-weight compounds such as oligomers, plasticizers, diluents, solvents, dyes, water and so on are such factors. The T_g can be measured in a variety of ways, not all of which yield the same value.[61] The T_g of some synthetic fibers are given in Table 2.5. Since the fiber is not all amorphous polymer and contains a substantial proportion of crystalline fraction, the T_g of the fiber is influenced by its crystallinity, crystal size, and orientation. For example, the T_g of PET fiber increases with crystallinity from 80°C to 120°C. The amor-

TABLE 2.5 Typical Glass
Transition Temperatures of
Synthetic Fibers

Fiber	$T_g(°C)^a$
Nylon 6	50–60
Nylon 66	47
Polyester	80
Acrylic	85
Polypropylene	−35
Polyethylene	−100
Polyvinylchloride	−75

[a] These values are only indicative and vary with the thermal history and the method of determination. For an example, see Ref. 60.

phous PET chips soften and stick together if they are heated together at 100–120°C. On crystallization, the same chips do not soften, even up to 170°C.

REFERENCES

1. Hearle, J. W. S., and Greer, R., *Text. Prog.*, **2**(4) (1970), 1.
2. Meredith, R., *Text. Prog.*, **7**(4) (1975), 1.
3. Mark, H., and Meyer, K. H., *Z. Physik. Chem.*, **B2** (1929), 115.
4. Kratky, O., *Kolloid, Z.*, **68** (1934), 347 and **70** (1935), 14.
5. Gerngross, O., Hermann, W., and Abitz, W., *Z. Physik. Chem.* (*Frankfurt*), **10** (1939), 371.
6. Hess, K., Mahl, H., and Gutter, E., *Kolloid, Z.*, **155** (1957), 1.
7. Centola, G., *JTI*, **59** (1968), 445.
8. Hearle, J. W. S., *JPS*, **28** (1958), 432; **Part C.**, **20** (1967), 215; *JAPS*, **7** (1963), 1175.
9. Keller, A., *Phil. Mag.*, **2** (1957), 117.
10. Hill, P. H., *JPS*, **24** (1957), 301.
11. Fischer, E. W., *Z. Naturforsch.*, **12** (1957), 753.
12. Lindenmeyer, P. H., *JPS*, **Part C**, **15** (1966), 109; and **20** (1967), 145.
13. Holland, V. F., Mitchell, S. B., Hunter, W. H., and Lindenmeyer, P. H., *JPS*, **62** (1962), 145.
14. Szaboles, O., and Szaboles, I., in *Man Made Fibers*, Mark, H. F., Atlas, S. M., and Cernia, E., Eds., **1** (1967), 341.
15. Geil, P. H., *Polymer Single Crystals*, Interscience, New York (1963).
16. Feughelman, M., in *Applied Fiber Science*, Vol. 1, Happey, F., Ed., Academic, New York (1978), p. 43.
17. Statton, W. O., *JPS*, **Part C, 18** (1967), 33.
18. Statton, W. O., *JPS*, **Part C, 20** (1967), 117; *J. Applied Phys.*, **38** (1967), 4149; **41** (1970), 4290.
19. Kobayashi, Y., Okajima, S., and Kosuda, H., *JAPS*, **11** (1967), 2525.
20. Dumbleton, J. M., *JPS*, **Part A2, 6** (1968), 795; **7** (1969), 667.
21. Kelly, K. M., and Statton, W. O., *TRJ*, **39** (1969), 662.
22. Dryfuss, P., and Keller, A., *JPS* (*Polymer Phys.*), **11** (1973), 193.
23. Pennings, A. J., Proc. Int. Conf. on Crystal Growth, Boston. Pergamon Press, Oxford (1966), p. 389.
24. Dismore, P. F., and Statton, W. O., *JPS*, **Part C, 13** (1966) 133.
25. Prevorsek, D. C., *JPS*, **Part C, 32** (1971), 343.
26. Reimsohnessel, A. C., and Prevorsek, D. C., *JPS*, (*Polym. Phys.*), (**14**) (1976), 485.
27. Prevorsek, D. C., Butler, R. H., Kwon, Y. D., Lamb, G. E. R., and Sharma, R. K., *TRJ*, **47** (1977), 107.
28. Peterlin, A., *JPS*, **Part C, 32** (1971), 297; *J. Macromol. Sci., Phys.*, **B6**, (4) (1971), 583.
29. Prevorsek, D. C., Harget, P. J., Sharma, R. K., and Reimschnessel, A. C., *J. Macromol. Sci.*, **B8**(1–2) (1973), 127.
30. Prevorsek, D. C., Tirpak, G., Harget, P. J., and Reimschnessel, A. C., *J. Macromol. Sci.*, **B9**(4) (1974), 733.
31. Deopura, B. L., Sinha, T. B., and Verma, D. S., *TRJ*, **47** (1977), 267.
32. Boyer, R. F., *J. Macromol. Sci., Phys.*, **B8** (1973), 503.

33. Chao, N. P. C., Cuculo, J. A., and George, T. W., *Appl. Polym. Symp.*, **27** (1975), 175.
34. Valk, G., Jellinek, G., and Schroider, U., *TRJ*, **50** (1980), 46.
35. Gupta, V. B., *JAPS*, **20** (1976), 2005.
36. Deopura, B. L., in *Polyester Textiles*, Gulrajani, M. L., Ed., The Textile Association of India (1980), 317.
37. Gupta, V. B., and Kumar, S., *JAPS*, **26** (1981), 1865.
38. Parker, J. P., and Lindenmeyer, P. H., *JAPS*, **21** (1977), 821.
39. Amano, T., *JAPS*, **16** (1972), 1072.
40. Wasiak, A., and Ziabicki, A., *Appl. Polym. Symp.*, **27** (1975), 111.
41. Sotton, M., Arniand, M., and Rabourdin, C., *JAPS*, **22** (1978), 2585.
42. Heuvel, H. M., and Huisman, R., *JAPS*, **22** (1978) 2229; SASMIRA Conf, Bombay, India, March 29–April 12 (1982).
43. Spruiell, J. E., and White, J. L., *Appl. Polym. Symp.*, **27** (1975), 121.
44. Holme, I., *Chimia*, **34**(3) (1980), 110.
45. Gupta, V. B., Book of Papers, First Annual Symposium on Texturing, Gulrajani, M. L., Ed., IIT, Delhi, India, September (1977), p. 35.
46. Warwicker, J. O., Shirley Inst. Publication, **S25** (1976), 49.
47. Dumbleton, J. H., and Bowler, B. B., *JPS, Part A-2*, **4** (1966), 951.
48. Ferrow, G., and Ward, I. M., *Polymer*, **1** (1960), 330.
49. Ruland, W., and Weigand, W., *JPS, Polym. Symp.*, **58** (1976), 43.
50. Ward, I. M., *JPS, Polym. Symp.*, **58** (1976), 1.
51. Gupta, A. K., Chand, N., and Man Singh, A., *TRJ* **50** (1980), 328.
52. Craig, J. P., Knudsen, J. P., and Holland, V. F., *TRJ*, **32** (1962), 435.
53. Takahashi, M., Nukushina, V., and Kosugi, S., *TRJ*, **34** (1964), 87.
54. Stoyanov, A. I., and Krustev, V. P., *JAPS*, **26** (1981), 1813.
55. Marvin, D. N., *JSDC*, **70** (1954), 16.
56. Merian, E., Carbonell, J., Lerch., U., and Sanahuja, V., *JSDC*, **79** (1963), 505.
57. Statton, W. O., Keenig, J. L., and Hannon, M., *J. Appl. Phys.*, **41** (1970), 4290.
58. Datye, K. V., Pitkar, S. C., and Purau, U. M., *Textilveredlung*, **6** (1971), 600.
59. Datye, K. V., *Textilveredlung*, **4** (1969), 562.
60. Saxena, A. K., and Gulrajani, M. L., *JSDC*, **95** (1979), 330.
61. Collins, E. A., Bares, J., and Billmeyer, F. W., *Experiments in Polymer Science*, John Wiley and Sons, New York (1973), p. 217.

3 | THEORETICAL ASPECTS OF COLORATION

The mysterious way in which dye from the dyebath becomes concentrated in the substrate has been the subject of many theoretical speculations and there have been many attempts to explain the affinity that a dye appears to have for a fiber. The rate of dyeing depends not only on the affinity of dyeing but also upon the degree of resistance that the dyeing system offers to the dyeing process, that is, the rate of dyeing depends on two distinct factors: the rate parameter and the affinity of dyeing.

Dyeing will tend to occur spontaneously only if the Gibbs free energy G of the whole dyeing system decreases during the dyeing process as a consequence of the transfer of dye molecules from the solution phase (σ) to the fiber phase (ϕ).

3.1 AFFINITY OF DYEING[1]

The rate at which the Gibbs free energy G of the whole system changes during dyeing is given by the difference ($\Delta\mu$) between the chemical potentials of the dye in the fiber (μ^ϕ) and the bath. (μ^σ) The chemical potential is $\mu = \Delta\, \delta G/\delta n$, where n is a mole of dye in the fiber or solution.

The standard affinity of dyeing ($-\Delta\mu^0$) for a simple discontinuous model of dyeing is given by

$$-\Delta\mu^0 = RT \ln \frac{(C^\phi)_\infty}{(C^\sigma)_\infty} = RT \ln K$$

where $(C^\phi)_\infty$ and $(C^\sigma)_\infty$ are the equilibrium concentrations of dye and K is the partition coefficient. For instantaneous affinity of dyeing $(-\Delta\mu)$ from an infinite bath (when C^σ remains constant)

$$-\Delta\mu = RT \ln C^\phi$$

where C^ϕ is the relative dye concentration in the fiber, and

$$-\Delta\mu = -\Delta\mu^0 + RT \ln C^\phi/C^\sigma$$

The instantaneous rate of dyeing may be assumed to be directly proportional to the instantaneous affinity of dyeing $(-D\mu)$, that is,

the rate of dyeing = (rate parameter) $(-\Delta\mu)$

Thus,

$$\frac{dC^\phi}{dt} = BRT \ln C^\phi$$

where B is the rate parameter.

Thus, the instantaneous affinity of dyeing $(-\Delta\mu)$, the Gibbs free energy of the whole discontinuous system, the instantaneous rate of dyeing, and the relative dye concentration in the fiber phase are intimately interconnected. A single curve in a three-dimensional space determines the path that is taken by a dyeing system that conforms to this simple model.[1] It is assumed that the concentration of dye within the fiber phase is uniform and independent of position, which is rarely the case. In most dyeing systems, the diffusion process within the fiber phase plays an important role. The concentration of dye in the fiber phase at the surface (interface) is always the equilibrium concentration $(C^\phi)_\infty$ that corresponds to the constant solution concentration $(C^\sigma)_\infty$. The dye concentration is minimum at the center of the fiber and zero at the beginning of the dyeing process.

The dyeing process is controlled by the diffusion of dye from a constant surface concentration under the influence of the concentration gradient into the polymer. When surface equilibrium exists, the chemical potentials μ^σ and μ^ϕ must be the same at the interface and no change in chemical energy or the Gibbs free energy of the system occurs when a dye molecule passes from the solution into the fiber phase through the interface.

The instantaneous affinity of dyeing $(-\Delta\mu)$ is therefore zero for the interfacial transport. The driving force within the fiber phase is the gradient of the chemical potential in the system, or, roughly, the concentration gradient. It is the steepness of the variation of the chemical potential with distance that provides a measure of the driving force within the fiber phase.

3.2   EQUILIBRIUM ADSORPTION ISOTHERMS

The adsorption isotherms are constructed from equilibrium sorption of dye on fiber in relation to equilibrium concentration in solution (or vapor). Depending on the nature of a dye–fiber–solvent system, these isotherms are either linear or nonlinear. When the dye is not adsorbed on specific sites in the fiber, but is present in free volume in the fiber, the nonlinear isotherm is the Freundlich type, that is, the dye may form multimolecular layers on the fiber without any apparent saturation value. On the other hand, when the dye is adsorbed on specific sites in the fiber and when the site is occupied by the dye, it is no longer available for further adsorption; the adsorption has a limiting value—a saturation point when all the sites are blocked. The dye in this system is present on the fiber as a monolayer. The nonlinear adsorption isotherm under these conditions is the Langmuir type. When the dye is taken up by the fiber by the solid-solution mechanism and the partition between two immiscible solvents, the isotherms are linear. Many times, combinations of these mechanisms may be operating simultaneously and give complex isotherms. The standard affinity $-\Delta\mu^0$ under solid-solution conditions is given by

$$-\Delta\mu^0 = RT \ln \frac{(C^\phi)_\infty}{(C^\sigma)_\infty} = RT \ln K$$

The equilibrium sorption isotherm in the simplest possible dyeing system is linear since the equilibrium partition coefficient K is constant and is given by the slope of a straight line (Figure 3.1).[2] The isotherm generally terminates at the point where the fiber phase and the solution phase are saturated with the monomeric dye.

For a Freundlich-type isotherm of a nonionic dye

$$-\Delta\mu^0 = RT \ln \frac{(C^\phi)_\infty}{V} - RT \ln (C^\sigma)_\infty$$

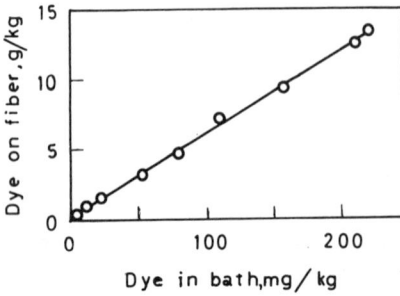

FIGURE 3.1.   Simple linear sorption isotherms at constant temperature and pressure.[2]

where V is the volume (liter/kg) in which the dye is adsorbed on the fiber. Similarly, for a Langmuir-type isotherm

$$-\Delta\mu^0 = RT \ln \frac{\Theta}{1 - \Theta} - RT \ln (C^\sigma)_\infty$$

where Θ is the fraction of sites occupied by the dye in the fiber.

The ionic dyes—acid, cationic, direct, reactive, and so on—are present in the ionized state, both in the dyebath and the fiber phase. For a Freundlich type isotherm, if there are no added electrolytes in the solution, then for anionic dyes

$$-\Delta\mu^0 = (Z + 1) RT \ln \frac{(C^\phi)_\infty}{V^{(Z+1)}} - (Z + 1) RT \ln (C^\sigma)_\infty$$

In presence of an electrolyte (NaCl),

$$-\Delta\mu^0 = \frac{RT \ln (C^\phi)_\infty (Na^\phi)_\infty^Z}{V^{(Z+1)}} - RT \ln (C^\sigma)_\infty (Na^\sigma)_\infty^Z$$

where $(Na^\phi)_\infty$ and $(Na^\sigma)_\infty$ are sodium ion concentrations and Z is the valency of the dye ion. For a Langmuir type isotherm of ionic dyes, in the absence of an electrolyte

$$-\Delta\mu^0 = (Z + 1) RT \ln \frac{\Theta}{1 - \Theta} - (Z + 1) RT \ln (C^\sigma)_\infty$$

If the sodium ions are not adsorbed on specific sites, but are diffusely adsorbed

$$-\Delta\mu^0 = RT \ln \left(\frac{\Theta}{1 - \Theta}\right) \left(\frac{(Na^\phi)_\infty^z}{V^z}\right) - RT \ln (C^\sigma)_\infty (Na^\sigma)_\infty^z$$

If the system contains added electrolyte, the appropriate values of $(Na^\phi)_\infty$ and $(Na^\sigma)_\infty$ are substituted. Similarly, when a cationic dye is used, the concentration of anions must be considered.

It is seen from the above that the standard affinity of dyeing $-\Delta\mu^0$ has a direct relationship to the way the dye is distributed between the dyebath and the fiber at equilibrium under given pressure and temperature. For more complicated ionic dyeing systems, $(-\Delta\mu^0 = RT \ln F)$ may be used where F represents the function of concentrations in the fiber and the dyebath (obtained from equilibrium sorption measurements). Thus, $-\Delta\mu^0$ values are no more reliable than the theoretical model on which the calculation is based.

3.3   THERMODYNAMIC PARAMETERS

The standard affinity $-\Delta\mu^0$ is related to the standard heat of dyeing ΔH^0 and the standard entropy of dyeing ΔS^0 by

$$-\Delta\mu^0 = -\Delta H^0 + T\Delta S^0$$

where T is the absolute temperature.

The standard heat of dyeing ΔH^0 at a given temperature and pressure is obtained from the slope of a plot $\Delta\mu^0/T$ vs. $1/T$ (Figure 3.2).[3] ΔH^0 can be negative, zero, or positive when the equilibrium dye sorption decreases, remains constant, or increases, respectively, as T increases. ΔH^0 gives a different value for each temperature except where it is independent of temperature. In practice, however, one single value of ΔH^0 can usually give the effect of temperature on the equilibrium adsorption isotherm.

From the standard affinity of dyeing $-\Delta\mu^0$ and the standard heat of dyeing ΔH^0 at a given temperature T and pressure P, the standard entropy of dyeing ΔS^0 can be calculated. For any spontaneous process, the entropy increases and the degree of randomness or disorder in the environment increases.

The standard heat of dyeing ΔH^0 reveals the interplay of forces between all the components in the system, for example, between the dye and the fiber. It represents the total change in enthalpy of the system during dyeing. Both the solution phase and the fiber phase contribute to ΔH^0 which tells the manner in which the forces of attraction between the dye and the fiber are influenced, for example, by the chemical structure.

The standard entropy of dyeing ΔS^0 reflects the manner in which the various components become intermixed within the system during dyeing. It reflects the loss or gain in freedom of the molecules.

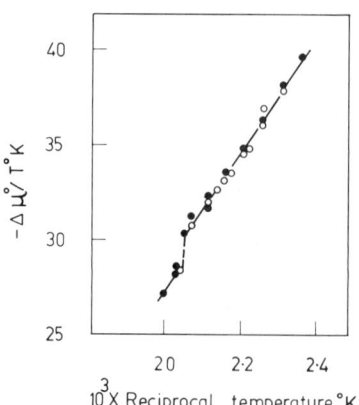

FIGURE 3.2. $1/T$ vs. $\Delta\mu^0/T$. Data on two polyester fibers.[3]

3.3.1 Vapor-Phase Dyeing

Synthetic fibers and cellulose acetates are dyed with disperse dyes from the vapor phase in the thermofixation process or the sublimation transfer process.[2-5] The experimental sorption isotherms for the dyeing of synthetic fibers with nonionic dyes are linear, so the chemical potentials may be related to the concentrations.[1-7]

The typical results of equilibrium dyeing at 180°C are given in Table 3.1 and at different temperatures in Table 3.2. The relationship between $-\Delta G^0/T$ vs. $1/T$ for a number of dyes on PET and PA 66 is shown in Figure 3.3.[4]

In vapor-phase dyeing, dye molecules must leave the vapor phase before they can enter the fiber, which contributes to the values of ΔH^0 and ΔS^0. If the enthalpy changes during dyeing are considered, Figure 3.4 can be constructed.[5]

TABLE 3.1   1-4 Diaminoanthraquinone on Fibers by Vapor Phase Dyeing at 180°C[4,8]

| Fiber | Saturation Value $(C^\phi)_\infty$ | | $(K^b \times 10^5)$ | $-\Delta G^0$ (kcal/mole) |
	(g/kg)	(g/liter)^a		
PA 66	30.5	35.6	9.78	12.41
PA 6	38.8	49.9	12.88	12.67
PET	37.9	52.8	14.51	12.77
CTA	38.8	51.2	14.07	12.75

^{a} Fibers are assumed to be completely accessible to the dye vapor.
^{b} Partition coefficient $K = (C^\phi)_\infty/(C^\sigma)_\infty$, where the concentration of dye vapor is $(C^\sigma)_\infty$ at 180°C = 3.64 × 10^{-5} g/liter.

TABLE 3.2   1-4 Diaminoanthraquinone on Nylon 66 by Vapor Phase Dyeing at Different Temperatures[4,8]

| Temperature (°C) | Saturation Value $(C^\phi)_\infty$ | | Dye in Vapor (g/liter × 10^5) | $K \times 10^{-5}$ | $-\Delta G^0$ (kcal/mole) |
	(g/kg)	(g/liter)			
170	24.6	28.0	1.63	17.19	12.64
180	30.5	34.7	3.64	9.54	12.39
190	37.1	42.3	7.88	5.36	12.14
200	43.3	49.4	16.45	3.00	11.85
210	56.0	63.8	33.34	1.92	11.67

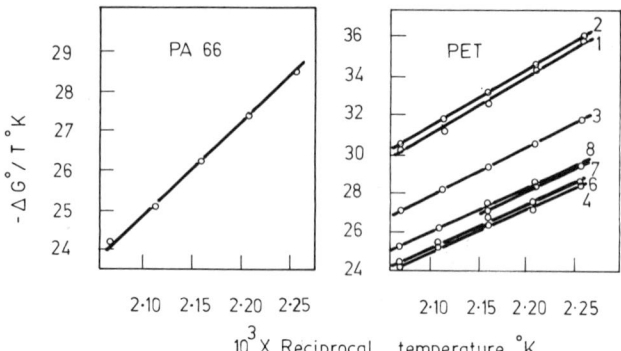

FIGURE 3.3. The apparent Gibbs free energy $-\Delta G^0$ as a function of temperature.[4] The numbers in the figure are for the dyes in Reference 4.

Since removing dye from the vapor phase is simply the opposite of introducing dye into the vapor phase, the enthalpy change during the vaporization of the solid dye ΔH_v^0 and the enthalpy change during dyeing from the vapor phase ΔH^0 can give the enthalpy change (ΔH_D^0) that would occur if the dye molecules were taken straight out of the solid dye and placed directly in the fiber. Because the enthalpy is a thermodynamic state variable, its value depends only on the state of a system and not upon the way that state is reached. Thus,

$$\Delta H_D^0 = -\Delta H^0 + \Delta H_v^0$$

ΔH_D^0 values can be experimentally determined from the fiber saturation values of a dye since the concentration of a solid (dye) in the external phase is always constant. Typical calculated values for the differences in heat of vaporization and heat of dyeing from the vapor phase are given in Table 3.3.

The enthalpy change ΔH_D^0 is usually small, which implies that the forces of attraction that bind the dye molecules within the fiber, which may be dye–dye interactions or dye–fiber interactions or both, are very similar in strength to the forces of attraction between the dye molecules in the solid

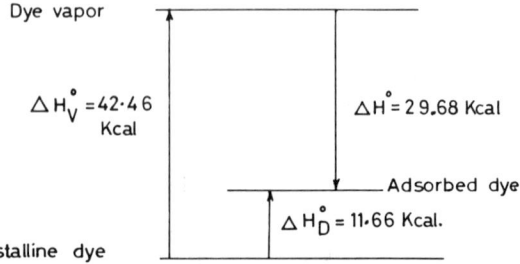

FIGURE 3.4.   Enthalpy changes during the vapor-phase dyeing of polyester.[5]

TABLE 3.3   Enthalpy Changes in Vapor Phase Dyeing (kcal/mole)[3] [a]

| | PET | | Nylon 66 |
	Dye 3	Dye 6	Dye 1
ΔH_v^0	42.46	40.04	33.0
$-\Delta H^0$	29.68	30.91	24.2
ΔH_D^0	11.66	8.60	8.17
$\Delta H_D^0 = (\Delta H_v^0 - \Delta H^0)$	12.78	9.13	8.8

[a] For dyes formulas, see Ref. 3.

dye. The heat of dyeing from the vapor phase (ΔH^0) is a direct measure of the dye–fiber bonds. When ΔH^0 is large and negative, dyeing is favored.[7] On the other hand, the entropy change ΔS^0 during vapor phase dyeing is also negative, that is, entropy is decreasing during dyeing. This reflects the need to crowd the dye molecules together within the fiber in an ordered arrangement after removing them from their state of comparative disorder and freedom of movement in the vapor phase.

3.3.2 Dyeing from Solution

Linear isotherms are obtained for a system—nonionic dye–solvent (or water)–synthetic fiber—particularly, when the dye solutions are dilute. Datye et al. calculated the thermodynamic parameters for such a system for a large number of dyes.[8] The comparative data for vapor-phase dyeing and the solvent dyeing for PET for a few dyes are shown in Table 3.4 and Table 3.5.

For solvent dyeing systems, the change in Gibb's free energy ($-\Delta G^0$) or standard affinity $-\Delta\mu°$ is about 10–15% of that for vapor-phase dyeing (Table 3.4). The partition coefficient K for the latter is much bigger than that for a solvent system. Similar results have been quoted by McGregor for azobenzene (nonionic dye)–water–cellulose acetate.[1] A correction for the temperature difference between vapor-phase dyeing and solvent dyeing accentuates the differences in K and $-\Delta G^0$ values. The values of the standard heat of dyeing ΔH^0 for the vapor-phase system and the solvent system are also very different (Table 3.5).

It is interesting to note that the heats of solution of a solid dye in PER and PET are of similar orders. (Table 3.5). This may be due to the similar solubility parameter of PET and PER (see Section 3.5).

The solvent or water plays a positive role in modifying the behavior of the dyeing system. It suggests the formation of dye–solvent (or water), dye–dye, and solvent–solvent complexes that are associated with the enthalpy change in both the fiber and the solvent system. The molecular interactions

TABLE 3.4 Dyeing PET with Disperse Dyes from the Vapor Phase and a Solution in Perchloroethylene (PER)[8]

Dye[a]	Vapor Phase Dyeing			Solution Dyeing		
	°C	$K \times 10^5$	$-\Delta G^0$ (kcal/mole)	°C	K	$-\Delta G^0$ (kcal/mole)
3	170	88.13	14.08	91	9.45	1.625
	180	46.02	13.81	95	8.26	1.544
	190	25.68	13.59	101	7.55	1.502
	200	14.42	13.33	110	6.11	1.377
	210	8.14	13.07	115	5.75	1.349
				121	5.43	1.325
4	170	15.94	12.58	91	9.81	1.670
	180	8.79	12.33	95	9.31	1.614
	190	5.26	12.48	101	7.18	1.465
	200	3.34	11.96	110	5.50	1.298
	210	1.97	11.71	115	5.41	1.301
				121	4.52	1.181
8	170	28.64	13.10	91	10.41	1.695
	180	17.03	12.92	95	8.53	1.568
	190	9.45	12.66	101	7.95	1.541
	200	5.57	12.43	110	6.22	1.391
	210	3.40	12.22	115	5.77	1.324
				121	5.75	1.369

[a] For dye formulas, see Ref. 8.

in the solution tend to retain the dye in the solution phase and, as a result, the heat of dyeing is not as large as that for vapor-phase dyeing. The molecular interactions within the dyebath must be overcome before the dye can be removed from the dyebath; this tends to oppose dyeing. This is the reason

TABLE 3.5 Changes in Enthalpy in Dyeing PET with Disperse Dyes from Perchloroethylene (PER) and from Vapor Phase (kcal/mole)[8]

Heat of	Dye 3	Dye 4	Dye 8
Vaporization of solid dye	36.3	34.2	33.0
Solution in PER	11.0	11.6	10.8
Solution in PET			
of solid dye	10.3	11.1	11.0
from dye solution in PER	5.9	3.9	6.4
Dyeing from Vapor Phase	25.5	21.7	23.5
Dyeing from PER	5.1	7.7	4.4

why $-\Delta G^0$ values for dyeing from solvent are so much smaller than those for vapor phase dyeing.

The vapor phase and solvent dyeings are used mainly for synthetic fibers. The other dye–fiber systems are not described here since they have been particularly developed with reference to cotton–water–ionic dye systems and have been exhaustively described in various books, monographs, and reviews.

3.4 MOVEMENT OF DYE THROUGH FIBER PHASE

In aqueous dye solutions, the hydrophilic fibers swell markedly, yielding an open, water-filled structure. The water-filled channels or pores provide a route for the dye to reach its adsorption site. This pore model presents the fiber as a network of interconnecting channels or pores through which the dye diffuses. Many of the channels in as-spun fibers are lost for dyeing when the fiber is dried in its virgin state. The dye diffuses much more rapidly through never-dried cotton or wet-spun fibers than the dried (commercial) fibers. For example, as-spun acrylic tow before drying can be dyed within a few seconds at 50°C, but it takes several minutes at 100°C once the fiber is dried before dyeing.

Hydrophobic fibers such as PET do not take up large quantities of water. The dye must, therefore, find a path through the polymer matrix itself. This is the case in dry dyeing systems such as thermofixation[3] or sublimation dye-transfer processes.[4] Such a diffusion process is explained by a free-volume model of diffusion.

3.4.1   Free-Volume Model

In the free-volume model, it is assumed that the thermal motion of atoms generate a volume associated with molecules that is not occupied by the constituent atoms. This volume increases with increasing temperature.

In the dyeing system, the dye molecules are visualized as moving through the fiber in a series of jumps from one location to another that can accommodate the dye molecule. This depends on the development of a free volume due to the segmental movement of intervening polymer chains adjacent to the position of the dye molecule.[9] The dyeing properties and the viscoelastic properties are therefore related because they are both governed by the segmental mobility.[10]

Below a certain temperature, the polymer chains are frozen into a position and the only motion they undergo are thermal vibrations. But an increase in temperature eventually provides sufficient energy for bond rotation in the backbone of the polymer chain. At this point, a whole segment of the polymer chain between two simultaneously rotating bonds changes its position by rotation (a segmental jump) until it is hindered from moving further by other polymer molecules. Such motions may occur only when sufficient free volume has been created to provide a space large enough to accommodate the polymer segment. Once the segment has moved, the space is free to be occupied by another segment or a diffusing dye and the process is repeated throughout the whole of the polymer structure. Consequently, there is a sudden marked increase in free volume over a narrow temperature range and an associated increase in segmental mobility. This temperature is called the glass-transition temperature (T_g) (Cf. Section 2.3).

The influence of temperature on diffusion or viscoelastic properties is represented by the Williams, Landel, and Ferry (WLF) equation.[11]

$$\frac{\log D_T}{\log D_{T_g}} = \log \frac{T_g}{T} = \log \frac{1}{a_T}$$

where D_T is the diffusion coefficient at temperature T or T_g and a_T is the shift factor of the WLF equation, where

$$\log a_T = \frac{-A(T - T_g)}{B + (T - T_g)}$$

where A and B are constants.

The shift factor represents the fraction by which the property is changed in going from T_g to the ambient temperature T. This could represent a change in the modulus of a fiber or the diffusion coefficient of a dye through the fiber. The fractional change depends only on the difference between T_g and T, which reflects the effect of temperature on the frequency of the segmental jump. Thus, using correct values of A and B and multiplying by an appropriate value of the shift factor, the results for one temperature can be superimposed on those for another (Figure 3.5).[9]

In the equation, the constants A and B appear in terms of the critical free volume needed to permit a segmental jump and the free volume at a given temperature. A large number of polymers have been found to conform with the WLF relationship over the temperature range from T_g to $T_g + 120°$. The equations fit well with amorphous, rubbery polymers. Dyeing is governed by the response of the small number of amorphous domains present in the fiber and the latter are in the rubbery state above T_g.

Rosenbaum[10] has successfully used the WLF equation to express the relationship between the diffusion coefficients of cationic dyes in acrylic fibers and the variations in properties with temperature. Similar arguments are used to explain the effect of temperature on the diffusion of acid dyes into nylon 66[12] and the diffusion of disperse dyes into polyester.[2–5,13]

The action of carriers in dyeing is explained by the free-volume theory. The carrier plasticizes the fiber and thus brings down the T_g. The effect of carriers on acrylic fiber dyeing is shown in Figure 3.6.[9] The rate of diffusion is a direct function of the difference ($T - T_g$), where T is the dyeing tempera-

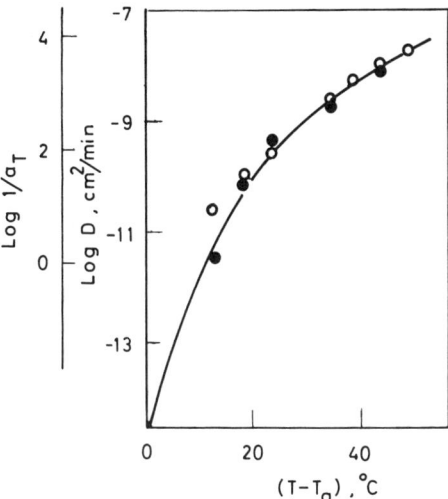

FIGURE 3.5. Matching diffusion data for C.I. Disperse Red 1 on Acrilan filament and the physical properties to the WLF equation.[9] O: Diffusion; ●: Data with shift factor.

ture. The greater the value of $T - T_g$, the more rapid is the diffusion of dye.[14] Thus, a reduction in T_g could have the same effect as an increase in the temperature of dyeing. Use of carriers in the bath leads to a decrease in T_g and, in turn, to an increased rate of dyeing.[15] Dumbleton et al.[13] have shown that diffusion is controlled by the variation of the glass-transition tempera-

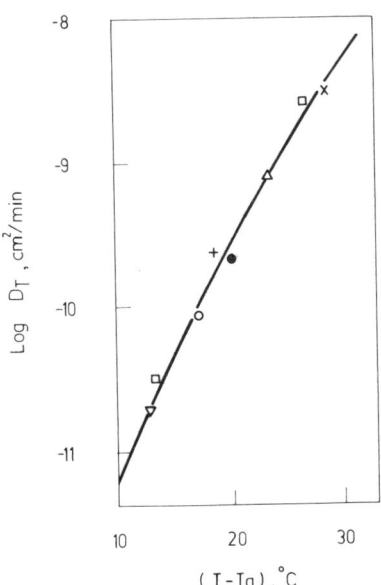

FIGURE 3.6. Effect of carriers on the diffusion coefficient (D_T) of paraminoazobenzene in Acrilan fiber at 70°C as a function of ($T - T_g$).[9] ▽: Water; +: Phenol 5 g/liter; △: Phenol 10 g/liter; ×: Phenol 15 g/liter; ●: Toluidene 2–3 g/liter; O: Nitrobenzene 1 g/liter; □: Dimethylformamide 10 g/liter.

ture T_g, which depends on the mechanical and thermal history of the fiber and on dyeing conditions. The dynamic loss modulus measured under dyeing conditions is related to the dye diffusivity, which confirms that the diffusion is controlled by the mobility of the polymer chain segments.

3.4.2   Other Factors

The orientation of amorphous chains, the tortuosity of the internal voids, and the porosity in water of PET are all related to the rate of dyeing. Sufficient shrinkage must be allowed during heat setting of polyester fabrics to smoothen out the structural difference caused by the tension applied to the yarn during mechanical textile processes.[16] A disorientation of the amorphous regions leads to an increased dye uptake.[17] It is assumed that the dye penetrates into the amorphous regions of PET that behaves kinetically as a viscous liquid. However, synthetic polymer fibers are too far from homogeneous to be able to represent them as solid solvents. Thus, Okijima et al.[18] studied the dependence of the dichroism of a PET–disperse dye system on chain orientation and concluded that (a) there exists a distribution of lateral order in the amorphous regions between adjacent polymer chains, (b) the dye molecules exist only within the regions of the lowest lateral order, and (c) the dye molecules interact with the polymer chains by a definite geometric relation. The angle of the dye axis with the chain axis remains unchanged during uniaxial deformation and heat treatments.

The visible dichroism D developed by disperse dyes changes reversibly with temperature, so the dye molecules seem to return to the original state of combination, which is defined by the structure of the polymer and the dye molecules, even if the dye molecules are detached slightly by the thermal oscillation from polymer chains at high temperatures.[19]

The dye molecules are absorbed only in the amorphous phase and dichroism D developed by the dyed polymer is relevant to the amorphous orientation Δn_a that exhibits a linear relationship with D.[18,20]

The dyeing properties of the polymer are modified by a heat treatment. A dry-heat treatment causes a reduction in the rate of diffusion of dyes into nylon, whereas a steam treatment increases the rate.[21] An increase in the molecular size of a dye decreases the diffusion coefficient of dry-set material but not wet-set material.[22] The increased segmental mobility in the presence of moisture leads to increased crystalline perfection and the rejection of chain ends from the ordered regions. Such chain ends are surrounded by the void space through which diffusion of dyes may be enhanced, since dye passes not only through the network of molecules but also through the void spaces.

It is obvious that dyeing properties are controlled by the fiber structure that is a very intricate system with single and clustered fibrils, single crystals, fault regions, voids, an amorphous phase, tie molecules, *para* crystalline layer-lattice, and chain entanglements. One important structural feature is the arrangement of microfibrils that produce lamellae oriented more or less perpendicularly to the fiber axis. During dyeing, diffusion into fibers takes place in the radial direction, at right angles to the major axis of the fiber and hence to the polymer molecules. However, the major axis of the dye is oriented parallel to the fiber axis.[9]

3.5 SOLUBILITY PARAMETER CONCEPT

Inherent in the molecular mechanism for polymer–solvent interactions is the concept of mutual solubility of a polymer–solvent or polymer–dye pair, which is frequently expressed by the solubility parameter principle of Hildebrand and Scott.[23] The solubility parameter δ is defined as the square root of the cohesive energy density or molar energy of vaporization per unit volume. This is the amount of energy that is required to separate all molecules into a gas phase. Thus, δ is a measure of the intermolecular forces that hold molecules together in the liquid and solid states and it describes the forces that must be overcome in the polymer before mixing can occur. When the difference in δ-values is zero, solute and solvent are said to be perfectly compatible. However, the mutual solubility is not assured even with the solubility parameters being equal unless the contributions of polar and specific group interactions are considered.

These interactions collectively give rise to the cohesive energy that is overcome during the evaporation of a liquid. The first of these interactions is common to all molecules and the only interaction available to simple saturated hydrocarbons, known as nonpolar dispersion (London) forces (d). The second mode of interaction arises from the permanent dipoles or permanent and induced dipoles in interacting molecules. These are generally called dipole–dipole or polar interactions (p) and are characterized by polar molecules. The third mode of interaction contributing to the cohesive energy density is hydrogen bonding (h). Thus, the total solubility parameter may be expressed as

$$\delta^2 = \delta_d^2 + \delta_p^2 + \delta_h^2$$

The mutual compatibility is given by the three-dimensional difference terms.

$$(\delta_2 - \delta_1)_d^2 + (\delta_2 - \delta_1)_p^2 + (\delta_2 - \delta_1)_h^2$$

The solubility parameter due to polar forces is called δ_A (A = association), which is

$$\delta_A^2 = \delta_p^2 + \delta_h^2$$

The solubility parameter concept can be used to interpret liquid miscibility, polymer solubility, polymer compatibility, adsorption on solid surfaces, dispersion phenomena, solubility of inorganic and organic substances in organic liquids, and salting phenomena. The division of the solubility parameter into dispersion, and polar and hydrogen-bonding components helps in understanding the above phenomena clearly.

The solubility parameter concept has been applied to the PET–solvent interaction.[24] For solubility purposes, PET may be treated as an $(AB)_x$ copolymer, where A is a semirigid aromatic residue, CO—C_6H_4, with a value of 9.8 and B is the flexible aliphatic ester residue, O—CH_2CH_2—O—CO, with a value of 12.1. Solvents or chemicals can be effective carriers if their δ-value is between 10 and 11 [which is close to that of PET (10.7)] (Table 3.6).[25–27] However, not all compounds whose δ-values are of this magnitude are effective as carriers. The individual components δ_d, δ_p, and δ_h should be close to the corresponding values of PET.[28] Thus, benzyl alcohol acts as a carrier for PET because its δ_d is close to that of PET. Similarly, it acts as a carrier for PAN (acrylic) fibers because its δ_A is similar to that of PAN.[29] Thus, the polar forces of benzyl alcohol govern the plasticization of PAN, which is polar, while the dispersion forces of benzyl alcohol govern the plasticization of PET, which is much less polar (Table 3.6).

The solubility parameter concept offers a logical link between the theory of solutions and nonionic (disperse) dyeing.[25] The concept correlates the solubility of disperse dyes in secondary cellulose acetate, cellulose triacetate, polypropylene, and polyester with the calculated solubility parameters of both the polymer and the dyes. The concept is particularly applicable for vapor-phase dyeing. However, with the hydrophilic polymers, the correlation is not close since the dyeing of hydrophilic polymers consists of the dye displacing water molecules in the water-swollen fibers.

TABLE 3.6 Comparison of the Solubility Parameters (δ) of Solvents with Their Solubility in PET at 95°C[25] [a]

Carrier	δ	δ_A	Solubility (g/100g)
Ethyleneglycol monomethyl ether	11.9	8.55	3.7
Dimethyl formamide	12.14	8.5	6.4
Anisole	9.5	3.9	7.4
Chlorobenzene	9.57	2.4	8.4
Benzyl alcohol	11.97	7.9	8.5
Benzaldehyde	10.40	4.9	9.7
Acetophenone	9.68	4.7	10.0

[a] PET δ = 10.70 $(cal/cm^3)^{1/2}$
δ_A = 4.80 $(cal/cm^3)^{1/2}$

The solubility parameter δ and its polar fraction δ_A for a number of dyes have been correlated with their fiber saturation values (solubility in polymer) at 80°C in Table 3.7.[25] Dyes for cellulose acetate and cellulose triacetate have δ-values within ± 1.4 units of the polymer. The δ of polypropylene is very close to that of dry-cleaning solvents and therefore the dye is easily extracted from the fiber during dry cleaning. Gerber[28] studied the dyeing of PET with 60 azo dyes and confirmed that the dye must have a solubility parameter very close to PET. When the solvent, polymer, and dye have about the same solubility parameter, the dye has little preference for either phase.[30] Thus, the partition coefficient K of disperse dyes for the PET–PER (perchloroethylene) system is very low,[8] since both PET and PER have similar solubility parameters. On the other hand, the disperse dye–PET–water system exhibits very high K values, since the solubility parameter of water is 20.3 $(cal/cm^3)^{1/2}$ while that of PET is 10.7 $(cal/cm^3)^{1/2}$; dyes that have δ nearer to PET prefer the fiber over the water.

TABLE 3.7   Solubility Parameter (δ), Solubility Parameter Component due to Polar Force (δ_A), and Solubility of Nonionic Dyes in Polymers at 80°C[25]

Dye	Constitution	δ $(cal/cm^3)^{1/2}$	δ_A $(cal/cm^3)^{1/2}$	Solubility in Polymer (g/kg)
	Cellulose acetate: $\delta = 11.0$, δ_A - 5.3			
1	4-NO$_2$-aniline $\rightarrow$ N-ethyl-N hydroxyethylaniline	10.51	4.3	18
2	4-NO$_2$-4'-NH$_2$-azobenzene	11.40	6.8	16
3	N,N-Dimethyl-p-NH$_2$ azobenzene	9.6	3.8	19
4	1-Methylamino-4-β-hydroxy ethyl anthraquinone	11.92	4.9	22
5	2-Cl-4–NO$_2$–aniline $\rightarrow$ N-ethyl-N-β-hydroxyethyl aniline	9.77	5.9	22
6	4-NH$_2$-acetanilide $\rightarrow$ o-Cresol	11.10	4.5	28
7	Azobenzene	9.85	4.0	43
8	p-aminoazobenzene	11.10	4.4	141
9	p-nitroaniline	13.40	5.4	150
	Cellulose Triacetate: $\delta = 9.1$, $\delta_A = 4.4$			
3	Same as above	9.6	3.8	10
7	Same as above	9.85	4.0	49
9	Same as above	13.40	5.4	150
	Polypropylene: $\delta = 8.0$			
3	Same as above	9.6	—	6.7
10	p-Aminoazobenzene $\rightarrow$ o-cresol	10.0	—	1.4

3.6   DIFFUSION

Diffusion in a liquid solution is the spontaneously occuring and thermo-dynamically irreversible process by which a solution approaches uniformity of concentration without any convective flow. The equalization of concentration throughout the system is not instantaneous because of the resistance offered by the solvent environment to the migration of solute particles. Thus, the concentration (C) varies with distance (x) and the concentration gradient dC/dx acts as a driving force that causes the solute to diffuse through the medium. The transfer of the ions of an electrolyte by diffusion is affected not only by the driving force and the frictional resistance of the solvent, but by Coulomb forces among the ions. Under the concentration gradient, both cations and anions experience the driving force. However, if their sizes are different, the larger ions are retarded by a greater frictional resistance than the smaller ions, causing a separation of charges that is opposed by the Coulombic electric forces between opposite types of charge on the two types of ions. These Coulombic forces retard the motion of small ions and accelerate the motion of the big ones. Thus, an electric potential gradient in the opposite direction to the concentration gradient retards the diffusion rate. No ultimate separation of charge occurs after the establishment of a constant rate of diffusion. Equivalent amounts of both types of ions pass any cross-section of the diffusion column in any interval of time, maintaining electrical neutrality. The diffusivity coefficient D, which is a measure of the mobile molecules in the solvent, depends on the mean free-path length (λ) between collisions, that is,

$$D = \frac{\lambda^2}{2t}$$

where t is the mean time between two collisions. For diffusion in solution, D is constant and more or less independent of the solute concentration. For aqueous systems, D is 10^{-6}–10^{-7} cm^2/sec.

The resistance to the transport of a dye molecule through the fiber arises from two principal causes. One cause is because of the strong forces of adsorption that exist between the dye molecule and the polymer chains. The other cause is due to the restrictions imposed on the movement of dye by the structure of polymers.

The forces of adsorption are interactions that occur between polar and polarizable molecules, such as dipole interactions, hydrogen bonds, van der Waal forces, dispersion forces and so on. Because of their influence, a fraction of the total amount of dye present in the fiber at any instant is always immobilized and is not free to diffuse. Consequently, the diffusion coefficient of dyes in polymers is much smaller than those in aqueous solution. The adsorption process does not modify the characteristics of the diffusion process and the diffusion coefficient of nonionic dyes in fibers is independent of dye concentration. The diffusion coefficient of ionic systems is dependent on the concentration of ions due to the electrical potential gradients, which arise from the different mobilities of simultaneously diffusing ions. However, the concentration dependence of diffusion coefficients of ionic dyes is not influenced by the form of the substrate; for example, acid dyes on nylon 6 and nylon 66 have similar concentration dependence of diffusion coefficients.[31]

During the dyeing and finishing of synthetic fibers, chemical reactions may also take place involving the fiber. For chemical processes inside a textile substrate, a flow of reagents or reactive dyes from or to the interface and the interior of the substrate is necessary. In such a process, the combined effect of the diffusion, adsorption, and chemical reaction may decide the kinetics of the process. Adsorption and reaction immobilize the reacting species and lowers the diffusion rate. On the other hand, the diffusion rate decides the position where the reaction of the dye or finishing agent will take place within the fiber.

The determination of a diffusion coefficient in dyeing is a complex process.[32,33] The dye is first adsorbed at the fiber surface if the fiber is hydrophobic and free of water-filled channels. In the early stages of dyeing, the concentration builds up at the surface and falls away rapidly to the center of the fiber. During the course of dyeing, the dye adsorbed on the surface of a fiber diffuses inwards (or outwards during migration–desorption) until the concentration is uniform in the fiber. Most of the mathematical models of diffusion are derived from Fick's equation of diffusion:[32]

$$\frac{\text{Rate}}{\text{Gradient}} = \text{constant} \quad \text{or} \quad F = -D\frac{dC}{dx}$$

where F is the flux or rate of diffusion at any point and time and dC/dx is the concentration gradient at the same point and time. For steady-state diffusion, the gradient remains constant and is given by $\Delta C/L$, where ΔC is the concentration difference at the boundaries that are at a distance L away from each other. During dyeing, the concentration gradient is not constant but decreases steadily throughout the dyeing. Under the nonsteady state, the overall diffusion coefficient is given by

$$\frac{C_t}{C_\infty} = \left(\frac{Dt}{4\pi r^2}\right)^{1/2}$$

where C_t is the concentration at time t and C_∞ at equilibrium (after long time) and r is the radius of the fiber. Thus, C_t/C_∞ vs. $\sqrt{t}$ are linear plots. By determining the dye-concentration profile from the surface to the center of the fiber by the densitometer technique,[34] it is possible to determine diffusion coefficients at various points and concentrations in the fiber. A similar distribution of concentration in the layers of a cylindrical roll of PET fabric[35] during desorption at 210°C is shown in Figure 3.7.[3]

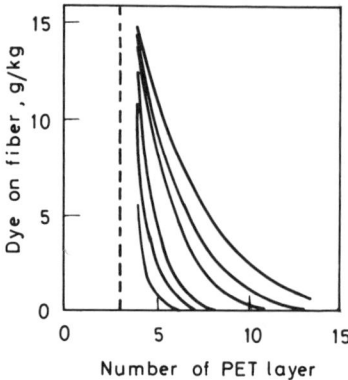

FIGURE 3.7. Dependence of dye concentration on the PET layer in the roll.[3] From left-to-right heating time: 30 min, 60 min, 120 min, 240 min, 480 min, and 960 min.

3.6.1   Diffusional Boundary Layer

The dyeing of hydrophobic fiber with a nonionic dye from water has been described by a model in which the dye in an aqueous solution diffuses across a boundary layer followed by adsorption at the fiber surface from which diffusion into the polymer proceeds.[36-38] The shape of the rate of dyeing curve in such a model depends on the parameter L as defined by

$$L = D_\sigma a / K D_\phi \, \delta D$$

where D_σ and D_ϕ are diffusion coefficients of the dye in a diffusional boundary layer of thickness δD and in a fiber of radius a. K is the partition coefficient of the dye in the two phases. The thickness δD of the diffusional boundary layer near the fiber surface decreases with vigorous stirring or a high rate of liquor flow, creating a turbulence that increases the rate of dyeing. In solvent dyeing, the δD is either very small due to the lower viscosity of solvents or the rate of diffusion in the boundary layer does not influence the rate of dyeing because of small K values. Thus, the rate of dyeing from PER is not appreciably influenced while that from aqueous media under similar conditions is significantly increased with the increase in the rate of flow or stirring.[8] The design of a dyeing machine, the flow rates, and the movement of the material and the liquor during dyeing influence the dyeing rate.

The rate of sorption of an acid dye by nylon is strongly influenced by the mass-transfer process in the liquid phase, even at rates of flow far exceeding those to be expected in the existing commercial dyeing equipment. The concentration gradients exist in the solution within the yarn elements of the fabric and the individual fibers in the yarn elements differ in their accessibilities to the dye; the yarn cross-sections are ring-dyed. This can be explained by equivalent diffusional boundary resistance to mass transfer represented by δD. Within the yarn elements, dye transport is by convective diffusion in the liquid phase, which limits the overall dye-transfer rate.[38]

3.6.2 Influence of Fiber Denier

The surface area of the fiber through which dye diffuses into the fiber phase increases as the fiber denier decreases. Thus, if the cross-section, crystallinity, and amorphous orientation of the fiber are similar, the rate of dyeing increases with decreasing denier (Figure 3.8),[8] exhibiting a linear relationship between the dyeing rate and reciprocal-square-root denier (Figure 3.9).[38] The synthetic fibers attain their final denier through the drawing process. If the denier is high due to undrawn or partially drawn conditions, the role of the surface area and the ease of diffusion in the undrawn fiber phase are counteractive in deciding the rate of dyeing. Usually the dye diffuses in the undrawn or less crystalline fiber at a very fast rate, even if the surface area is smaller.

3.6.3 Activation Energy of Diffusion

The mechanism of dye diffusion in a fiber is conceived as a rate process that depends on the spontaneous appearance at a suitable position in the amorphous regions of a network of molecular chains of a hole large enough to accommodate the dye molecule. Above the glass-transition temperature, the restriction on the molecular motion of the polymer chain segment disappears and the necessary energy to produce such holes will be possessed by an increasingly large number of chain segments. Consequently, diffusion will be more rapid. According to the theory of rate processes, there is a close analogy between the viscosity and the diffusion process. The activation energy for viscous flow will be similar to that for diffusion. For PET, the former is 40 kcal/mole[39] while that for diffusion is between 20–40 kcal/mole, depending on the molecular weight of the dye[38] (Figure 3.10). In the close-packed structure of polymer, the production of a hole sufficiently large enough to accommodate a dye molecule involves the movement of several

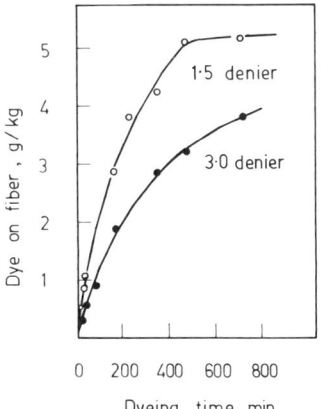

FIGURE 3.8. Adsorption of a nonionic dye on PET from perchloroethylene at 77°C.[8]

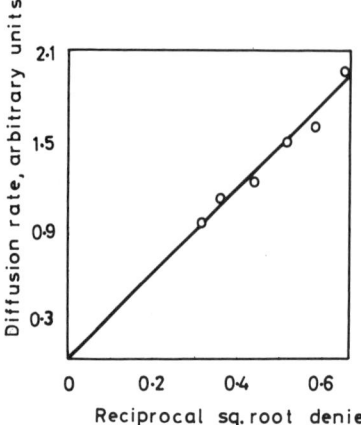

FIGURE 3.9. Rate of diffusion of a disperse reactive dye on nylon 6 fibers with varying deniers.[38]

surrounding polymer chains and the consequent disarrangement of a considerable zone in the structure, thus giving high entropy of diffusion.[40] Thus, the diffusion coefficient at a given temperature decreases with increasing molecular weight of the dye.[8]

The diffusion behavior of a dye in fiber is influenced by the transitions in the polymer substrate. In turn, the activation energy of diffusion in the fiber shows abrupt changes when there is a transition in the fiber structure.[41]

From the sublimation rate of disperse dyes from dyed polymers, electrical conductivity, and dilatometry of PET film, the transitions in the amorphous regions and crystalline transition points are noted. The apparent activation energy of diffusion decreases from 100 kcal/mole for the glass state to 22–24

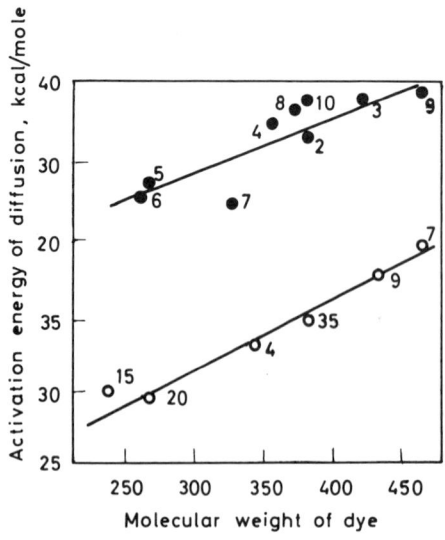

FIGURE 3.10. Activation energy of diffusion ΔE vs molecular weight of dyes.[38] ○: From perchloroethylene (91–121°C). ●: In thermofixation process (170–210°C).

kcal/mole for the region above 180°C, which changes slightly with the size of the dye molecule and the crystallinity and orientation of PET film.[42]

In the case of polypropylene, the diffusion transition points are indicated by the breaks in log D vs. $1/T$ at about 70°, 90°, and 115°C, which are related to the structural changes in polypropylene polymer.[43]

3.7 COLOR AND COLOR MEASUREMENT

We can say that all textile goods are colored materials if white and black are considered colors. These colors are produced to match a specimen. It is a routine practice to compare a sample of dyed goods with a given specimen to match the shade and to evaluate the differences. This is done visually by an experienced colorist who can detect differences in depth of approximately 5% if the hue is similar. However, with small differences in hue, the assessments become less accurate. Even though visual comparison of colors is always made and relied on, there is a variability of color vision among observers and the existence of a large number of defective observers.

Any color can be clearly identified by three parameters: hue, saturation level, and subjective brightness. Hues are yellow, red, green, blue and so on. Colors that have hues are called *chromatic colors*. There are also *achromatic colors*—physically neutral white, grey, and black. They have no hue and consequently, no saturation, since saturation simply means degree of chromaticity. The pure colors of the spectrum have the highest saturation. The only distinguishing factor for achromatic colors is their differing degrees of subjective brightness. The ideal black is of zero brightness, that is, all the visible light striking the sample is absorbed. The ideal white reflects 100% of all light. Such a white has a subjective whiteness of 100. It is almost attained by barium sulfate and magnesium oxide. Between the ideal white and the ideal black, there are ideal greys that reflect all incident light in an even manner. However, they reduce the intensity of light without changing its composition. The brightness-of-grey range lies between 0 and 100. The ideal white and black are the limits of the grey range. In practice, white is a color of great subjective brightness and it is chromatic, but with such a low saturation as to be almost achromatic. The hue may have a positive or negative influence on the perceived whiteness. If saturation and subjective brightness are simultaneously increased, (for example, as a result of applying an optical brightener with a suitable tint), the whiteness level continues to increase.

3.7.1 Color Parameters[44]

The human eye is sensitive to visible light, that is, electromagnetic radiations of wavelengths 380–780 nm. Color is the interpretation by the observer's mind of the response of the retina to stimulation by light. The color vision of the human eye is decided by three types of receptors, each possessing sensitivity maxima in different regions of the visible spectrum. The *Commission Internationale de L'Eclairage* (CIE) established a standard for the spectral characteristics of the human eye. Figure 3.11 represents the sensitivity of three different color receptors in the retina of the eye to blue, green, and red light. The CIE spectral tristimulus values define the color parameters for a human observer. The red sensitivity curve $\bar{x}(\lambda)$ has a secondary peak in the blue $\bar{z}(\lambda)$ region. The curve $\bar{y}(\lambda)$ for the receptor sensitive to green also represents the eye's impression of subjective brightness, that is, wavelength 554 nm light is experienced as the brightest.

The sensory impression, color, is created by the combined effect of the spectral characteristics of the lighting, the object illuminated, and the human eye. Any change or difference in these three components alters the color impression. The color of a textile depends on the nature of the radiations leaving it and entering the eye of the observer. The source of light is an important component in deciding the color. The source of radiation provides a characteristic energy distribution throughout the spectrum. Tungsten-filament lamps emit very little uv radiation, that is, wavelengths shorter than 380 nm. They provide a high and regular emission of energy in the visible region (380–780 nm) and in the near-IR region (>780 nm). Daylight has a relatively greater ultraviolet and blue content than a tungsten source. Some sources of radiation, such as mercury or sodium vapor discharge lamps, emit strongly only at certain wavelengths.

Thus, the radiation that stimulates the eye is given by the curve representing color stimulus $R(\lambda) \times S(\lambda)$, where $R(\lambda)$ is the spectral reflectance of an object and $S(\lambda)$ is the spectral energy distribution of the illuminating light source. This color stimulus is multiplied by each of the three CIE distribution coefficients. These products are worked out for individual wavelengths and receptors. The summations of these products give the three CIE tristimulus values X, Y, and Z:[45]

$$X = \sum_{400}^{700} R(\lambda) \cdot S(\lambda) \cdot \bar{x}(\lambda)\Delta\lambda$$

$$Y = \sum_{400}^{700} R(\lambda) \cdot S(\lambda) \cdot \bar{y}(\lambda)\Delta\lambda$$

$$Z = \sum_{400}^{700} R(\lambda) \cdot S(\lambda) \cdot \bar{z}(\lambda)\Delta\lambda$$

If the tristimulus values of two objects under a given illuminating condition are identical, the color will look the same. If the reflectance curves of the two objects are identical, a change in the light source will not affect the match or develop a color difference. If the reflectance curves of two samples are quite different, they may still give identical XYZ values under certain illuminating conditions (since tristimulus values are summations) and may

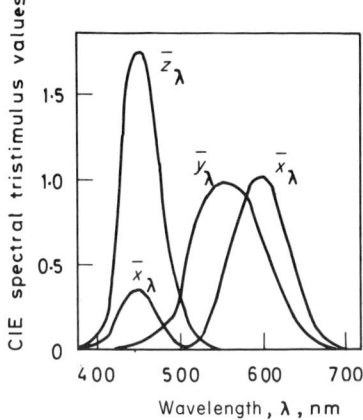

FIGURE 3.11. CIE spectral tristumulus values.

exhibit identical color. However, if light with another energy distribution is used for illumination, the tristimulus values will show differences. Colors that look alike under a given illumination, although their reflectance curves differ are known as *metameric colors* (or dyeings). Colorists naturally prefer nonmetameric colors for matching their samples with the standard.

3.7.2 Color Space

The tristimulus values can be plotted to construct a three-dimensional color space or color solid. The XYZ values of a colored object pinpoint the color in the three-dimensional color space. An object that is a perfect nonmetameric match to the standard will be located in the same position; the worse the match, the further apart will be the object and the standard. However, XYZ space has never been used to quantify the closeness of a match because the range of distance ΔE values corresponding to equally close matches is almost 30 to 1. XYZ space is regarded as nonuniform color space.[46]

The color solid is projected as a two-dimensional color plane with CIE chromaticity coordinates x and y, equal to:

$$x = \frac{X}{X + Y + Z} \quad \text{and} \quad y = \frac{Y}{X + Y + Z}$$

Here, x represents the proportionate share of the red tristimulus value X in the sum total of tristimulus values $X + Y + Z$, while y represents the proportionate value of green. These chromaticity coordinates describe the chromaticity of a color by defining hue and saturation. The brightness (or luminosity) of a color does not affect the x and y values in any way since they are fractions of the total value. It is this separation of the chromaticity and luminosity that makes computation of the chromaticity coordinates useful.

Color, as described in the CIE system, is plotted in a chromaticity or tristimulus diagram, which is a plot of the chromaticity coordinates (or tristimulus coefficients) x and y (Figure 3.12). All possible colors are located within the area enclosed by the line connecting the points representing the characteristics of the spectral colors. The horseshoe-type-shaped line is known as the *spectrum locus*. The curve–spectrum locus is closed by the straight line of the purple colors. Near the center of the CIE tristimulus diagram, the white point (neutral point or CIE illuminant) represents the chromaticity of white, grey, or black achromatic colors. Colors of the same hue but of different saturation are situated along straight lines between the achromatic point and any point on the spectrum locus. The farther away the color point (chromatic point) is from the achromatic point, the greater the saturation of dyeing.

A measure of the brightness of a color is supplied by the tristimulus value Y because the CIE spectral distribution curve $\bar{y}$ coincides with the relative luminosity curve of the eye, as described earlier. The Y axis rises vertically on the plane-of-chromaticity diagram from the achromatic white point (C). The latter represents the chromaticity of a color that has the same reflectance for all wavelengths. For ideal white with 100% reflectance, $X = Y = Z = 100$ if illuminated by a standard light source with the same energy for every wavelength. For this point, $x = y = 0.3333$. If the illumination is affected by some other source of light, the Y tristimulus value of ideal white will still be 100, but X and Z will be different depending on the nature of the light.

The three standard light sources of the CIE system are tungsten-filament light (S_A), sunlight (S_B), and daylight (S_C). Using appropriate filters, S_A can be converted into the other two sources, except in the UV region which is already deficient in S_A. Filtered xenon-lamp light (S_{Xe}) is the best source of light in this respect, particularly, if the sample has a fluorescence that affects the color.[45] The tristimulus values of standard sources are given in Table 3.8.

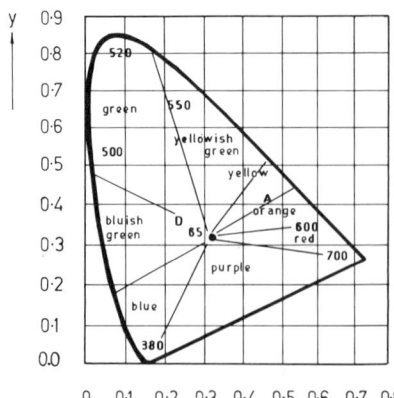

FIGURE 3.12. CIE chromaticity diagram showing achromatic points A and D 65.

TABLE 3.8   Tristimulus Values of Different Light Sources[44]

Standard Light Source	S_A	S_C	S_{Xe}
X	109.85	98.07	100.10
Y	100.00	100.00	100.00
Z	35.58	118.22	113.86
x	0.4476	0.3100	0.3188
y	0.4074	0.3162	0.3185

3.7.3   Dyers' Variables

The shade and strength of a dyeing are generally referred to during assessing and matching color with a standard. The dyer describes the difference in hue by any of the four hue terms—redder, yellower, greener, and bluer. The hues (red, yellow, green, and blue) are known as *unique or unitary hues* in the sense that it is possible to choose red, yellow, green, and blue hues that do not evoke two hue sensations. This cannot be done for other hues; for example, every orange hue evokes a simultaneous sensation of redness and yellowness; every violet, redness and blueness and so on. Colors are also called *primary* and *secondary* colors.

Various color-difference formulas that are based on *XYZ* values try to overcome the problem of nonuniform color space by transforming the tristimulus values to give uniform color space.[47] A large number of such formulas have been proposed and tested in order to determine the correlation between the color difference ΔE and % acceptance by visual assessment using a five-point scale. The uniform color space devised by Adams and Nickerson transforms the *XYZ* values into three values *L*, *A*, and *B*. The resulting color space is shown in Figure 3.13. This color space can also be expressed by cylindrical coordinates *L*, *C*, and *H°*, where

$$C = (A^2 + B^2)^{1/2}$$

and $H° = \arctan (B/A)$, expressed on a 0 to 360° scale, where $A^+ = 0°$, $B^+ = 90°$, $A^- = 180°$, and $B^- = 270°$. It is an angle between the line through the point and origin, and A^+ axis. The color difference ΔE is simply the distance in color space between the position of the object and the standard. ΔE is given by

$$\Delta E = (L^2 + A^2 + B^2)^{1/2}$$

The ΔE value is a measure of the perceived size of the color difference

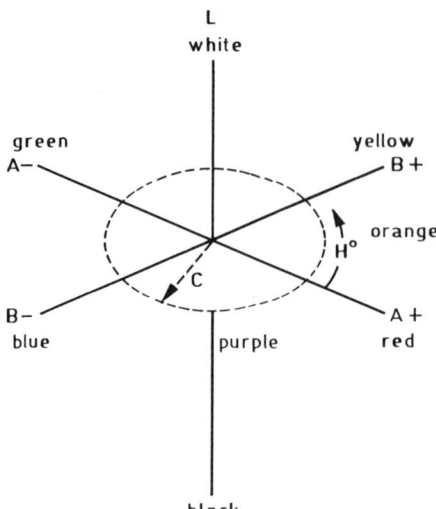

FIGURE 3.13. *ANLAB* color space diagram.

between the standard and the object. Additional information on the nature of the difference is acquired by the following two conventions:

1. The difference in *LAB* values is calculated by subtracting the *LAB* values of the standard from those of the object. If ΔL is positive, the object is lighter; if ΔL is negative, the object is darker than the standard. The signs of ΔA and ΔB indicate

$$+\Delta A: \text{redder}; \quad -\Delta A: \text{greener}.$$
$$+\Delta B: \text{yellower}; \quad -\Delta B: \text{bluer}.$$

2. Instead of *LAB* values, *L*, *C*, and *H°* are considered. $+\Delta C$ means chroma higher than the standard. *H°* indicates a difference in hue.

When *CIELAB* color space is viewed directly from above, a two-dimensional *AB* diagram is obtained in which the Cartesian coordinates *A* and *B* can be positive or negative, with achromatic colors at the center, $A = B = 0$. Another two-dimensional diagram is *L* versus *C* which describes the lightness level and the position of color in *CIELAB* color space.

When *LAB* values of a series of dyeings of increasing strength are located in *LAB* space, they fall on a smooth curve that starts at the position of undyed substrate having high *L* values and $A = B = 0$. These curves are best studied by considering two-dimensional *LC* and *AB* diagrams (Figure 3.14) rather than three-dimensional *LAB* space.[48] A typical relationship between dye concentration and $\Delta H°$, *C*, and *L* values is shown in Figure 3.15.

3.7.4 Dulling Effect

A neutral grey dye is added to the color in order to get dulled shades on fiber.[48] With increasing darkness, the *AB* values become smaller, even

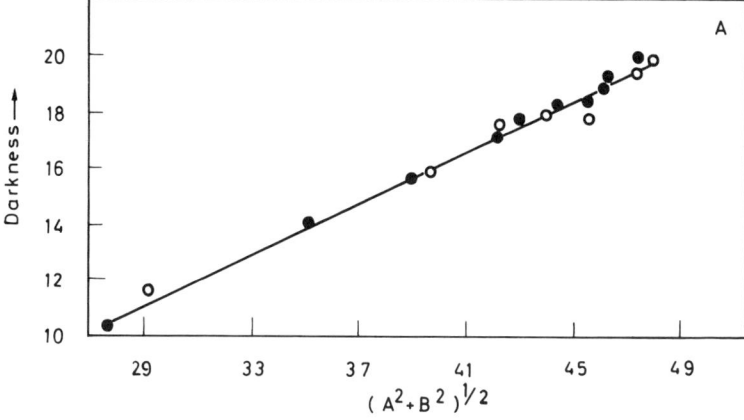

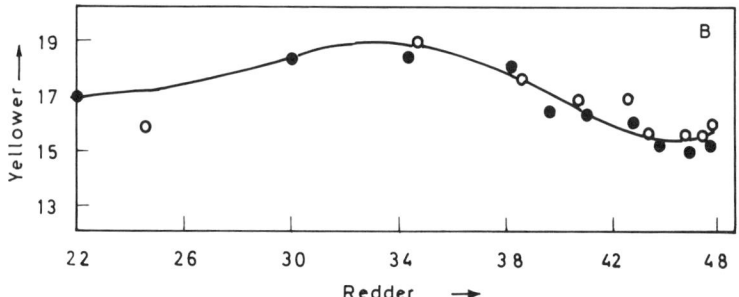

FIGURE 3.14. *LC* and *AB* diagram.[48] ●: Dyed from different dyebaths. ○: Dyed for different lengths of time.

though the dye concentration is same. If the diagrams are plotted on the same scale, the figures fall in an order similar to that of differently enlarged photographs of the same figure. Thus, the neutral dulling agent has no effect on the hue angle for almost all the dyes.[49] The position of a color in the *LAB* color space with increasing dullness traces a path that comes closer and closer to the achromatic lightness–darkness axis. In other words, the value of *C* becomes smaller and smaller with dullness until it becomes zero when a jet-black shade is produced. With the increase in dye concentration, the *C* value increases with some increase in dullness (decrease in *L* value).

3.7.5 Matching and Prediction of a Shade

A series of samples are dyed with systematic change in the depth of shade, and the hue and the color on these samples are compared with those on the standard. Based on the results of these comparisons, the process is repeated until an accepted match is obtained. This is an elaborate process and has to

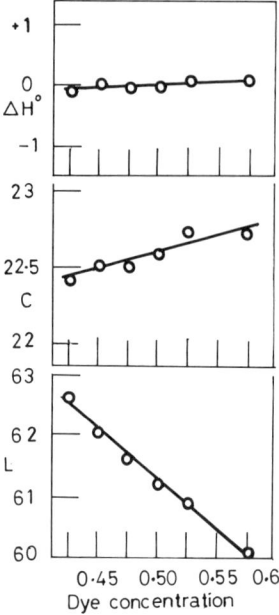

FIGURE 3.15. Relationship of dye concentration and ΔH^0, C, and L.

be carried out often. A catalogue of standard shades and their dye composition is maintained to minimize the number of dyeings.

The appearance of color varies with the direction of illumination and viewing due to surface gloss and texture, which causes reflection from the surface and uneven diffusion of light. Differences in illumination and viewing geometry are the biggest factors causing irreproducibility of results between different colorimetric measurements on the same sample.[50] CIE illumination and viewing conditions[45] are (a) 45°/Normal, (b) Normal/45°, (c) Diffuse/Normal, and (d) Normal/Diffuse.[51] Unless the sample is completely opaque, the nature of the backing will influence the evaluation. Datye and Mishra[48] have used white and black backgrounds to measure the color of nylon hose and the dulling effect due to titanium oxide. The results are shown in Figure 3.16.[48] By illuminating textured surfaces with three sources at 120°C azimuth intervals, the effect of texture on color can be suppressed to some extent.[52]

The colored textiles transmit, diffuse, scatter, and reflect light differently at different wavelengths of the spectrum. This light is measured by the spectrophotometer as a spectral reflectance. Only relative color measurements are required that form the basis of most of the colorimetric computations. For this purpose, reflectance standards and substandards are used. Smoked magnesium oxide is a standard white: purified barium sulfate powder or paint is also used as a standard white. Ceramic tiles with different shades are available as substandards. The internationally agreed physical standards and norms are used for colorimetry. To measure data throughout

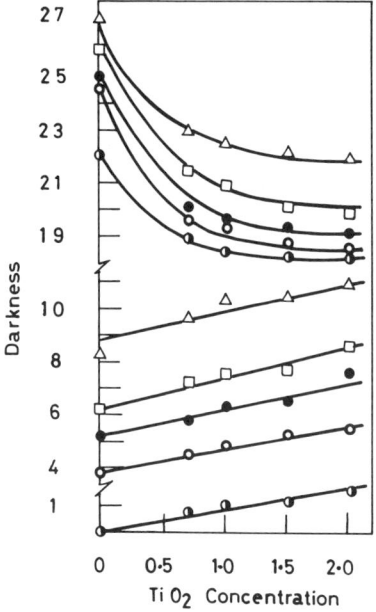

FIGURE 3.16. Darkness *vs*. TiO$_2$ concentration in nylon 6 as measured with white background (top section) and with black background (bottom section) with varying dye concentration.[48] Dye on wt. of fiber. △: 0.8%; □: 0.4%; ●: 0.2%; ○: 0.1%; ◑: 0%.

the spectrum and to sum it up simultaneously in order to obtain only three tristimulus values is the principle of physical colorimeter where spectral sensitivity of a single photodetector is modified so that when a sample is illuminated by a given light source, the response of the photodetector is proportional to one of the three tristimulus values. Subsequently, the other two tristimulus values are obtained by suitable modifications. The measurement of differences between samples of fairly similar color is done on a differential colorimeter. Units that directly give the differences in L, A, B, C and $H°$ are also available.

Dyeing recipe calculations begin with measurements of standard dyeings and the sample to be matched using a spectrophotometer or a special filter photometer. The dye concentration of each component dye is correlated with the reflectance density using the Kublka–Munk relationship

$$\frac{K}{S} = \frac{(1 - R)^2}{2R}$$

where K and S are, respectively, the absorption coefficient and the scattering coefficient of the dyed material. R is the reflectance density. The reflectance at 16 different wavelengths within the visible region are recorded for the sample to be matched and a number of standard dyed samples that are systematically prepared using different dyes and their concentrations. Using curve-fitting techniques, the correct combination of dyes to match can be predicted. The use of a digital computer makes the technique simpler.

The prediction is applicable to a given fabric–fiber–dye system. The Kubulka–Munk equation is not capable of predicting the color of the fabric from the optical and geometrical properties of the fabric–fiber–dye system, but does predict the effect of changes of dye concentration on the color. The indices of refraction of the system at appropriate dye concentrations, the geometry of the system, the fiber diameter (denier), the fiber cross-section, the fiber surface (roughness), the distribution of dye in fiber (ring-dyeing), the dulling agent that scatters the light, its concentration and distribution and so on are some other factors that decide the appearance of color.[53–58] This is illustrated by the following examples.

Sharp, triangular cross-sectional, glittering yarns show less color depth than the round cross-sectional (normal) yarns for the same concentration of a dye. The reflectance of a ring-dyed fiber for a given average dye concentration is higher than for a uniformly dyed fiber. This can account for the change in depth of shade of an incomplete dyed substrate, unpenetrated prints and so on when they are subjected to heat setting, steaming, or hot

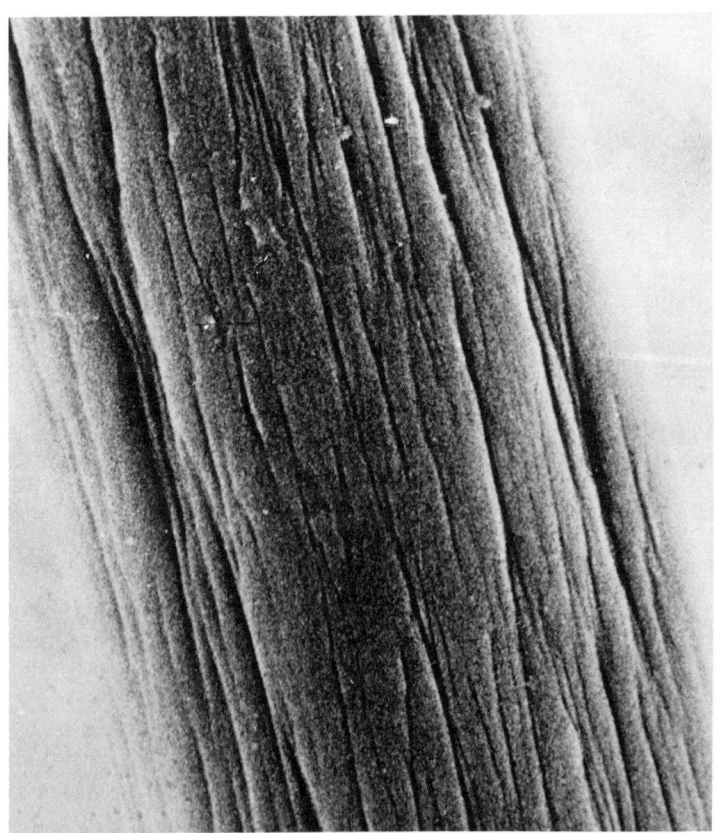

FIGURE 3.17.   Electron micrograph of pilled acrylic fiber.

washing treatments, which causes further dye penetration.[55,56] The color of the textile material changes on wetting. The weave of a fabric influences the color appearance.[57,58] The distribution of voids in wet-spun acrylic fibers (Figure 3.17) has a significant influence on the brightness of the color. Thus, it is not possible to predict the color for any unknown system; one has to try to match colors under specified conditions of fiber, dye, dyeing system, weave and so on.

3.7.6 Standard Depths

The depth of dyeing is how a dyer compares the strength of dyeings in different shades that appear to be of equal color strength if viewed by the eye. The International Organization for Standards has established 2/1, 1/1, 1/3, 1/6, 1/12, and 1/25 standard depths on both wool and rayon staple. CIE tristimulus values and standard depths cannot be correlated because of the nonuniform visual spacing of *CIEXYZ*, as described earlier.

It is often necessary to compare the strength of dyeings or to prepare dyeings of equal depth for testing fastness properties, since the accuracy of visual assessment of the change in depth during testing at equal depths of shade is the highest (Figure 3.18).[59] The comparison of fastness testing is made at equal depth of shade and, if possible, at a depth of shade equal to the reference standard. For this purpose, the percentage reflectance at the wavelength of maximum adsorption can be used to describe the depth of a dyeing, irrespective of hue. This method of describing the depth agrees well

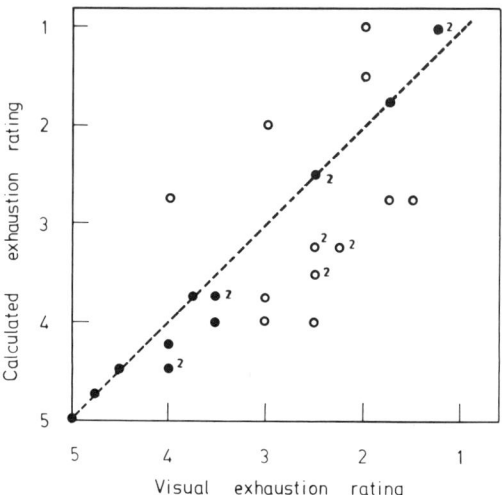

FIGURE 3.18. The exhaustion ratings by estimation of dye and visual assessment using 1% dye or dyeing 1/1 standard depth on PET fiber.[59] ○: 1% dyeing; ●: 1/1 standard depth. Visual assessment at standard depth is directly related to the calculated exhaustion ratings which is not the case when 1% dyeings were compared.

with the visual assessment.[60] For low standard depths, x, y, Y chromaticity coordinates may also be of use.

REFERENCES

1. McGregor, R., *JSDC,* **82** (1966) 450; **83** (1967) 52.

2. Datye, K. V., Unpublished work.

3. Datye, K. V., Kangle, P. J., and Milicevic, B., *Textilveredlung,* **2** (1967), 263.

4. Datye, K. V., Pitkar, S. C., and Purao, U. M., *Textilveredlung,* **8** (1973), 262.

5. Datye, K. V., and Pitkar, S. C., IVth Symposium, Centre of Advance Studies in Applied Chemistry, UDCT (Feb. 1969), 1.

6. Majury, T. G., *JSDC,* **72** (1956), 41.

7. Jones, F., and Seddon, R., *TRJ,* **34** (1964), 373.

8. Datye, K. V., Pitkar, S. C., Purao, U. M., *Textilveredlung,* **6** (1971), 593.

9. Peters, R. H., and Ingamells, W., *JSDC,* **89** (1973), 397.

10. Rosenbaum, R., *JPS,* **A3** (1965), 1949.

11. Williams, M. L., Landel, R. F., Ferry, J. D., *JACA,* **77** (1955), 3701.

12. Bell, J. P., *JAPS,* **12** (1968), 627.

13. Dumbleton, J. H., Bell, J. P., and Murayama, T., *JAPS,* **12** (1968), 2491.

14. McGregor, R., and Peters, R. H., *JSDC,* **84** (1968), 267.

15. Ingamells, W. C., Thornton, S. R., and Peters, R. H., *JAPS,* **17** (1973), 3733.

16. Donze, J. J., Bouchet, G., Freytag, R., Chabert, J., Schneider, R., and Viallier, P., *JSDC,* **91** (1975), 336.

17. Warwicker, J. O., *JSDC,* **88** (1972), 142.

18. Okijima, S., Nakayama, K., Kayama, K., and Kato, Y., *JAPS,* **14** (1970), 1069.

19. Nakayama, K., Okajima, S., and Kobayashi, Y., *JAPS,* **13** (1969), 659.

20. Patterson, D., and Ward, I. M., *Trans. Faraday Soc.,* **53** (1957), 1516.

21. Tsuruta, M., and Koshimo, A., *JAPS,* **9** (1965), 11.

22. Koshimo, A., and Kakishita, T., *JAPS,* **9** (1965), 91.

23. Hildebrand, J. H., and Scott, R. L., *The Solubility of Nonelectrolytes,* 3rd ed., Reinhold Publ. Corp., New York (1949).

24. Knox, B. H., Weigmann, H. D., and Scott, M. G., *TRJ,* **45** (1975), 203.

25. Ibe, C. E., *JAPS,* **14** (1970), 837.

26. Brown, A. H., and Peters, A. T., *ADR,* **57** (1968), 284.

27. Lemon, J. H., Kakar, S. K., and Cates, D. M., *ADR,* **55** (1966), 76.

28. Gerber, H., *JSDC,* **94** (1978), 298.

29. Ingamells, W. C., *JSDC,* **96** (1980), 466.

30. Harris, F. O., and Guion, T. H., *TRJ,* **42** (1972), 626.

31. Hopper *et al.*, *Diffusion in Polymers,* Crank, J., and Park, Eds., Academic, London (1968), p. 344.

32. Crank, J., *The Mathematics of Diffusion,* Clarendon Press, Oxford, (1956).

33. Crank, J., and Park, Eds. *Diffusion in Polymers,* Academic, London, (1968).

34. Peters, R. H., Petropoulous, J. H., and McGregor, R., *JSDC,* **77** (1961), 704.

35. Datye, K. V., Pitkar, S. C., and Rajendran, R., *Indian J. Tech.,* **4** (1966), 101, 202.

36. McGregor, R., and Peters, R. H., *JSDC,* **81** (1965), 393.

37. McGregor, R., Peters, R. H., and Varol, K., *JSDC*, **86** (1970) 437.

38. Datye, K. V., *Textilveredlung*, **4** (1969), 562; *Colourage*, **23**(2) (1976), 16.

39. Marshall, I., and Todd, A., *Trans. Faraday Soc.*, **49** (1953), 67.

40. Patterson, D., and Sheldon, R. P., *Trans. Faraday Soc.* **55** (1959), 1254.

41. Shibusawa, T., *JAPS*, **14** (1979), 1553.

42. Ito, I., Okajima, S., and Shibata, F., *JAPS*, **14** (1970), 551.

43. Okajima, S., Sato, N., and Tasaka, M., *JAPS*, **14** (1970), 1563.

44. *Color Measurement*, Bayer Farben Revue, Special Ed. No. 3E (Feb. 1964).

45. CIE Document on Colorimetry.

46. McLaren, K., *JSDC*, **92** (1976), 317.

47. McLaren, K., *Rev. Prog. Color*, **3** (1972), 3.

48. Datye, K. V., and Mishra, S. *Textilveredlung*, **18** (1983), 211.

49. Cooper, A. C., and McLaren, K., *JSDC*, **89** (1973), 41.

50. Adams, J. M., *J. Oil Col. Chem. Assoc.*, **50** (1967), 59.

51. Hunt, R. W. G., *Rev. Prog. Color.*, **2** (1971), 11.

52. Ward, J. W., *ADR*, **55** (1966), 1006.

53. Goldfinger, G., and Stafford, L., *Polym. Letters*, **11** (1973), 101.

54. Goldfinger, G., *Colourage*, **21**(19) (1974), 25.

55. Allen, E. H., and Goldfinger, G., *JAPS*, **16** (1972), 2973; **17** (1973), 1627.

56. Goldfinger, G., Lau, K. C., and McGregor, R., *Polym. Letters*, **11** (1973) 481; *JAPS*, **18** (1974), 1741.

57. Goldfinger, G., Goldfinger, H. S., Hersh, S. P., and Leonard, T. M., *JPS*, **C31** (1970), 25.

58. Allen, E. H., Faulkner, D. L., Goldfinger, G., and McGregor, R., *Polym. Letters*, **10** (1972), 203.

59. Datye, K. V., and Acharekar, J. Y., *JSDC*, **93** (1977) 413.

60. Taylor, M. E., *TCC*, **2**(1970), 149.

4 | MASS COLORATION AND TOW DYEING

The term *mass coloration* is used for the process of producing colored fibers and filaments by incorporating coloring matter during or after polymerization but before the polymer is converted into filaments by a spinning process. Natural and animal fibers cannot be mass colored. Mass coloration is possible with regenerated cellulosic and synthetic fibers. Polyolefin fibers are difficult to dye by known methods, but can be produced in various shades by mass-coloration techniques. Estimates of the production of mass-colored synthetic fibers for 1972 are shown in Table 4.1.[1] According to M. J. Wampetich,[2] mass coloration accounted for 0.5% polyester, 2–3% polyamide, 3–5% acrylic, and 40–50% polypropylene fiber of the total production of each fiber in 1976. It is estimated that in 1982, mass coloration will account for 3% and 4–5% of the total production of nylon and polyester fibers.[3,4] In the U.S.S.R., out of the total production in 1979–80, about 25% polyamide and 60% polyester were mass colored.[5]

4.1 ADVANTAGES AND LIMITATIONS

Mass coloration has many advantages over conventional dyeing and printing processes.[5,6] It gives even coloration without any problem of fastness properties. Any shade can be produced on filament material, including flat-filament yarns or filaments with very low deniers, otherwise a very tough problem for a dyer. Very high production rates are achieved with considerable savings in utilities and services and without any water pollution. The cost of coloring is very low. Colored yarns can be textured without any adverse effect on the shade. The processes of winding–rewinding (to get the yarn on

102

TABLE 4.1   Production of Mass-Colored Synthetic Fibers
(Year, 1972)[1]

	Tons of Fiber	Tons of Colorant Used
Melt-spinning nylon and polyester	50,000	500
Polyolefins	150,000	1200
Wet-spinning (acrylic)	40,000	400

a package into the proper form for dyeing) are completely eliminated by the mass-coloration process, thus giving savings in cost and reducing the wastage of yarn. Mass-colored fibers and filaments can be easily blended together to produce novel yarns and fabrics with multicolored effects. For example, heather mixtures of multicolored yarns and air-textured multicolored yarns are produced using mass-colored yarns.

However, there are many limitations and disadvantages associated with the mass-coloration processes. It is not always possible to economically color small lots in different shades. These processes also lack flexibility in production. Tonal adjustments by the conventional tinting process are not at all possible in mass-coloration techniques. Required variations in depth and tone of shades are difficult to manipulate. The aftertreatments given to dyed goods to improve their fastness properties are not possible in the mass-coloring process. For example, the tannic acid–tartaremetic treatment to improve the wet fastness properties of nylon dyed with acid dyes cannot be given to the dyed nylon chips that are to be spun later. There is a big time lag between coloration and the use of the colored material, which creates problems of high inventory cost, logistic disposition of colored material, a large variety of products and so on.

It is usually not possible to allot separate polymerization–drying–spinning units for individual shades except black. The same units are therefore used for different shades, one after the other. This involves the cleaning of the units which is a complicated, time-consuming process. The major problems is production loss due to a large amount of downtime so that polymerization, drying, spinning, and postspinning units can be cleaned after each shade. If the colorant has a vapor pressure at drying and spinning temperatures sufficient to contaminate these units, cleaning becomes a major time-consuming operation. Even in postspinning operations, for example, draw-twisting and texturing, the machine parts such as godets and guide rollers get a deposition of colorant or colored dust and need regular cleaning if they are to be used for white or any other delicate shade after any mass coloration.

Another complication is the recovery of waste materials. For economy in the manufacturing process, it is usual practice to recover and reuse the 5–10% polymer waste generated during the production of filaments. All the waste in mass-coloration is colored. The different colorants present in the mass-colored polymer waste may decompose during the recovery process and may contaminate the recovered product (usually the starting monomer). The carbon black pigment commonly used to produce black polymer may contain traces of sulfur, which may interfere during the recovery and reuse of recovered materials. Even though the mass-coloration process has many attractive features; it has equally serious limitations and drawbacks. The process is therefore used only when economy or ease-of-application demands mass coloration.

4.2   MASS-COLORATION METHODS

Five methods are available for the production of mass-colored fibers: (*a*) addition of a colorant during polymerization and spinning the colored polymer, (*b*) coating polymer chips with colorant and spinning the coated chips, (*c*) mixing a master batch of colorant with normal chips and spinning them together, (*d*) dyeing the chips with the colorant and spinning dyed chips, and (*e*) injecting the colorant formulations into the melt or dope of the polymer and spinning the colored dope or melt.

The polymerization process varies from fiber to fiber and the conditions during polymerization such as temperature, time, pH, pressure or vacuum, free-radical concentration, and catalysts are usually very drastic for the use of many colorants. Furthermore, the colorants should not interfere in the process of polymerization or adsorb or react with the ingredients. For example, carbon black slows down the transesterification and polycondensation or the polymerization process of PET.[7] The carbon black probably either gets deposited on the catalyst or adsorbs the transesterification and polycondensation catalyst. Carbon black is a cheap black colorant that is added during polymerization to many fiber-forming polymers. Carbon black is probably the only product that can be generally used as a colorant in any polymerization process.

Nylon melt has a powerful reducing action on colorants and very few colorants are stable under these conditions. The polymerization of acrylic-type polymers involves free radicals, which are likely to be blocked by the colorants, thus hampering the propagation of chain polymerization. It is much easier to add colorant to the dope during wet or dry-spinning than during polymerization. The chips-coating method is applicable to polymers that are melt-spun. Master batch preparation is the subject of many patents but can also be practiced by fiber manufacturers in certain cases. The chips-dyeing method is used for nylon but may be developed for polyester if a solvent-dyeing process for chips is standardized. For example, polyester can be rapidly dyed from chlorinated hydrocarbons without any problem of hydrolytic degradation. An injection of dye formulation during spinning has the advantage that the colorant is introduced only at the end of the preparation of a polymer for spinning. Thus, the equipment-cleaning problem is minimal in coloration by the injection process. However, the metering of the colorant has to be very accurate in order to get a fiber material without any shade variations. The colorant has to be specially formulated for injection purposes and the spinning unit has to be modified to take-up the injected colorant. These methods are suitably modified for individual fiber materials to get the best results.

4.3 MASS COLORATION OF POLYESTER

Mass-colored PET is available both in staple and continuous-filament forms, staple fibers being more common. Mass-coloration techniques may be increasingly used in the production of colored polyester, since the dyeing of polyester is a complicated, costly, and time-consuming process that needs specially formulated disperse dyes.[5] Furthermore, PET shrinks and loses strength during dyeing in aqueous baths. Textured material often gives uneven dyeing. Cationic dyeable polyester is colored preferentially by the dyeing processes and may not be mass colored.

Polyester is mass colored by the five methods stated above in the following manner:

4.3.1   Color Addition in Polymerization

Polyester melt does not have a powerful reducing action on colorants; hence, quite a few colorants are available to add to autoclave polymerization.[8-11] The colorant has to be stable under elevated temperatures ($\sim$260–290°C) and high vacuum (pressure less than 1 torr). Furthermore, it should not interfere with the transesterification and polycondensation reactions. This method is widely used for the production of black-colored PET with carbon black as the colorant. The cost of production of the jet-black shade by mass-coloration techniques is about 5% of dyeing fiber with disperse dyes or by a modified azoic process (See Chapter 7).

A carbon black (particle size 15–100 mμ) free of sulfur and volatile matter is ground in ethylene glycol in the presence of a sodium salt of polymerized alkyl naphthalene sulfonic acid-formaldehyde using an efficient ball mill or a colloid mill.[12] The slurry is filtered and added either at the beginning or after the transesterification reaction. Carbon black may adsorb the catalysts used in the transesterification and polycondensation and may lower the reaction rates. The fine particles of carbon black coagulate and form agglomerates because of the vigorous stirring and the high temperature of reaction. These agglomerates must be filtered off, either before the chips are made or before spinning. The life of the spin-pack is shortened by the presence of these agglomerates. If the size of the carbon black particles increases during the polymerization process, PET with a dull bronzy black color is obtained. The concentration of carbon black in the polymer to get a jet-black shade on the fiber depends on the type of pigment, particle size, and distribution of particles in the fiber, and is in the range of 1.5–2.5%. If the distribution of carbon black in the fiber is not uniform throughout the length and cross-section of the fiber but is localized, the color of fiber appears grey, even with 3% carbon black.[7] The presence of TiO_2 in the fiber adversely affects the black color.

Apart from the black shade, other shades can be produced on PET using special colorants, for example, those given in Table 4.2. These colorants dissolve in the polymer melt and need no severe dispersing and grinding prior to the addition of the reactor. Similarly, they create no coagulation problems during spinning. The colored fiber thus produced has excellent fastness properties. The colorant is dispersed in ethylene glycol and the slurry is introduced as in the case of carbon black. The late addition of colorant in polycondensation is preferred so that the time of contact of the colorant with the hot polyester melt is minimal. These colorants are usually anthraquinone derivatives or phthalocyanine-type colorants which contain carboxylic acid and/or hydroxyalkyl groups which may participate in the polymerization process and may give a copolymer containing the colorant residue. The colorant thus loses its capacity to sublime during drying and melt-spinning and does not contaminate the machinery.

It is not always easy to produce a jet-black shade with carbon black since carbon black coagulates and creates problems during filtration and melt-spinning. The agglomerated carbon black filters off and when the concentration falls significantly, the shade on the PET may not appear jet black. Very high concentrations of carbon black create unsurmountable problems. Appealing black shades can be conveniently produced at a considerably lower concentration of carbon black by adding a navy or blue colorant. Alternatively, the black fiber may be produced in two stages. A fiber of deep grey shade is produced without any serious problems of agglomeration and filter chocking when the concentration of carbon black is drastically reduced to less than about 0.5%. The grey fiber can be easily converted into a jet-black fiber by overdyeing with the conventional dyeing methods.[7] Heather mixtures with black-colored yarns are produced by doubling together this grey

TABLE 4.2 Typical Heat-Stable Colorants for PET

Manufacturer	Colorant
Sandoz	Estofil
Ciba-Geigy	Filester
ICI	Filomon
BASF	Basestren

yarn and a white yarn, and dyeing the doubled yarn with a disperse dye until the deep grey yarn turns black and the white yarn becomes colored. Novel color effects can be produced by combining white and various grey yarns containing different, though small, amounts of carbon black. Coloring such combined yarns or fabrics by the conventional dyeing and printing processes gives primary to tertiary dulled shades from a single dye.[7]

4.3.2 Chips Coating

The chips-coating method has the advantage that normal polymer chips can be used and the colorant does not have to have extremely high thermal stability as in the case of polymerization additives. The polymerization process is not disturbed and the reactors are not soiled by the colorants. The method involves tumbling dry polymer chips with the colorant, either alone or in the presence of an adhesive such as wax, silicone oil, or magnesium stearate. The dried, coated chips are melt-spun to get the colored filaments or fiber. The chips-coating method can be used for dyes that dissolve in the PET melt. The method fails with insoluble colorants. The loosely held colorant particles dust off and contaminate the drying unit. Better results are obtained on an extruder spinning unit than on a grid-type spinning unit.

4.3.3 Master-Batch Addition

In the master-batch addition method, colored chips containing a high percentage (10–50%) of a colorant are produced using a polymer such as polyethylene or low-melting polyester. Alternatively, the same PET may be used to prepare master batches of the colorant. The master batches of colored chips are blended together with the white chips in the desired proportion and are spun together to get a colored fiber. Such master-batch chips are also marketed by manufacturers of dyes and pigments.

The problems with the use of master batches are as follows: (a) the white and colored chips may not mix fully to form a homogeneous single phase in the fiber; (b) the separate domains of the two polymers cause breakages during the draw-twisting of the spun yarn. The incompatibility of the two polymers causes spinning problems such as unevenness of filaments and fisheyes and, at times, multifilaments of fine deniers cannot be spun. The colorant has to distribute itself evenly in the fiber; the shade reproduction may vary from machine to machine with spinning conditions and with the denier of the fiber depending on the distribution. The fiber formed by the mixing of the two polymers may not match in all properties with the fiber from the virgin polymer. The fastness properties of the colored fiber produced by this technique may be inferior to those produced by other coloring methods.

The biggest advantage of this method is that the dryer and other units are not soiled by the colorant. If proper care is taken to use master-batch chips produced under identical conditions as those for white chips or if their compatibility matched, many of the problems mentioned above would be eliminated.

4.3.4 Chips Dyeing

The chips-dyeing method consists of dyeing polyester chips with suitable disperse dyes followed by drying and melt-spinning. This method is not practiced commercially because complete removal of moisture from dyed PET chips is very difficult.[1] The residual moisture causes hydrolysis of PET during melting and spinning, which leads to a loss in strength. Another problem is the volatility of disperse dyes under the drying conditions of PET chips. The hot nitrogen or air used for drying carries away the dye vapor and contaminates the dryer and other units. Extruder and spinning blocks are also contaminated by the evaporated dye. This creates contamination problems during shade changes.

4.3.5 Injection During Melt-Spinning

The colorant is converted into a master batch in polyethylene or polyester in the form of powder or chips.[7] This master batch is then metered into the extruder spinning unit with a suitable device,[13] where it melts and gets blended with the PET melt. The mixture is spun into colored fibers. The method needs special injection equipment with a metering device for adding the colorant or its master batch into the extruder. Various attempts have been made with the conventional disperse-dye pigments and different carriers with limited success. The process has promise, since the dryer and other units are not contaminated by the colorant, and it may be possible to vary the composition of colorant and in turn the shades and tones of the colored fiber by manipulating the injection device. Disperse dyes with very low vapor pressures may be useful in this technique.

Except for the black shade with carbon black, PET is not colored by mass-coloration processes on a large scale. There is a wide scope to develop new techniques and colorants for this purpose. The field of polymer alloys is developing fast and new carrier polymers that will be compatible with PET polymer are being explored.

4.4 MASS COLORATION OF NYLON

The dyeing of nylon with direct, acid, metal-complex, and disperse dyes is not a complicated process. Mass-coloration processes therefore have to compete with conventional dyeing processes for economy, better aesthetic value, ease of production and so on. Thus, the mass-coloration method is of particular interest in producing the black shade with carbon black, colored textured yarns to avoid Barre' dyeing (see Chapter 9), and colored flat yarns and monofilaments in deniers such as 10, 20, and 40. Flat yarns and monofilament yarns cannot be easily dyed, either in the hank form or in the package form because their packages become very hard and hinder the circulation of the dye-liquor. On the other hand, slightly soft packages create tangles in the yarn because of slippage. The entanglement of flat yarns in the hank form can be as high as 10–12%. Such flat, colored yarns may be produced by mass-coloration techniques. The mass coloration of nylon can be done by the five methods described for PET.

4.4.1 Color Added in Polymerization

This promising process[14–21] has a number of shortcomings. Although considerable research effort has been made into colorants and processes for the mass coloring of nylon, an entirely satisfactory system has not yet been developed for producing various shades. The main difficulty is finding suitable dyes that are stable under the highly reducing character of the nylon melt and at high temperatures. The pigments have a tendency to flocculate during the polymerization reaction, which creates problems during the melt-spinning and draw-twisting of the colored yarn. The only satisfactory shade that is commercially produced is black using carbon black. The method consists of preparing a dispersion of carbon black in caprolactam solution in water with the help of a suitable dispersing agent. This dispersion is passed through a ball mill or a colloid mill to break down the aggregates of carbon black. The grinding–dispersion process is continued until a uniformly finely dispersed carbon black is obtained. The optical density of the dilute dispersion is checked every few hours until a constant optical density is obtained. The dispersion is then added to the autoclave containing caprolactam and other additives required for polymerization. The polymerization is carried out in the usual manner to get black-colored nylon polymer.

Production of neutral grey shades on nylon yarn is possible using a low concentration of carbon black. Very light grey to jet-black shades are produced using polymers containing different amounts of carbon black.[21] The typical color coordinates (see Chapter 3) of nylon 6 filaments containing carbon black are shown in Table 4.3. The carbon black in small quantities acts as a true neutral dulling agent as it does not influence the *AB* values or hue angle while increasing the darkness. The relationship of dullness and carbon black concentration in nylon 6 is shown in Figure 4.1[22] The logarithm concentration of carbon black exhibits a linear relationship with increase in darkness until the fiber becomes jet black. The light and wash fastness of all the shades including light grey are 7–8 and 5, respectively. These grey yarns can be further dyed to produce a variety of dulled shades.[19,22] For example, khaki, brown, and navy blue shades are produced by dyeing these grey yarns with yellow, red, and blue acid dyes. The dyeing is economical, easy to match, and gives colors with excellent fastness properties.[23] It is observed that the carbon black protects the dye during exposure to light thus giving

TABLE 4.3 Color Coordinates of Nylon 6 Filaments Containing Carbon Black[23]

Carbon Black in Fiber (%)	L^a	A^b	B^c
0^d	0	0	0
0.03	+8.0	+0.9	+1.4
0.075	+10.6	+1.2	+1.4
0.15	+13.1	+1.9	+1.1
0.30	+15.7	+2.3	+1.0
0.75	+16.1	+2.3	+1.0
1.50	+16.6	+2.3	+1.0

^{a} $+ L$ = Darker (see Section 3.7).
^{b} $+ A$ = Redder.
^{c} $+ B$ = Yellower.
^{d} Reference standard white fiber.

higher light fastness values[23] (Figure 4.2). If fabric is woven from the different grey yarns, tone-in-tone color effects are obtained from subsequent dyeing in a single bath. Thus, different color designs are produced from the same fabric by a simple piece-dyeing process.[21]

A few dyes are available for the production of shades other than black and grey. However, they are seldom added during polymerization. As an example, Nylofil dyes (Sandoz) and Filamide dyes (Ciba–Geigy) may be mentioned. In general, coloration of nylon by addition of the colorant during

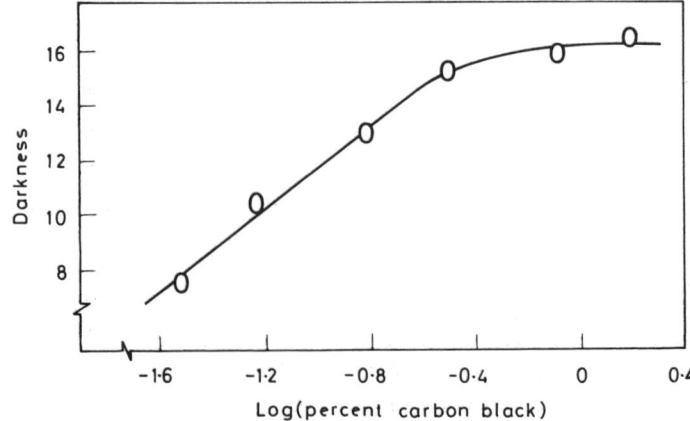

FIGURE 4.1 Relationship between log (percent carbon black) concentration in nylon 6 and darkness $(-\Delta L)$ value.[22]

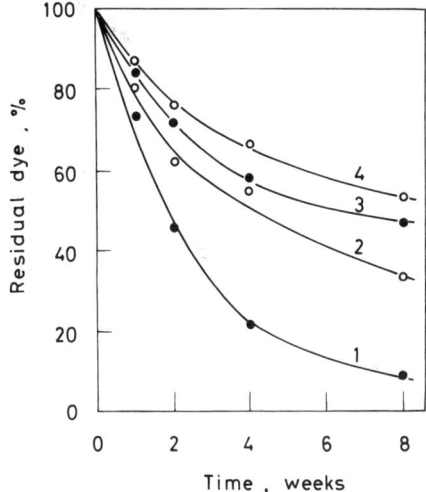

FIGURE 4.2. Fading of dyes on nylon 6 containing carbon black.[23] Carbon black (%) 1: 0.0%; 2: 0.15%; 3: 0.3%; 4: 0.75%.

polymerization is not practiced because of the difficulties in cleaning the equipment and the possible degradation of colorant and polymer.[24] The dyes mentioned above are more suitable for the chips-coating method than for addition to the polymerizer.

4.4.2   Chips Coating[21,25]

In the chips-coating method, the nylon chips are tumbled with colorants during drying. The adhesion of the colorant can be improved by the use of wax or silicone oil. The chips are thoroughly dried and melt-spun. Alternatively, coating is carried out by taking the colorant (as a dispersion) in a volatile organic solvent. This eliminates the problem of dusting of the colorant during tumble drying. A few typical colorants and their application methods are given in Table 4.4.[21] The fastness properties of a dye on nylon remain essentially the same for mass-colored goods and goods dyed by the conventional methods (Table 4.5).[21]

4.4.3   Master-Batch Addition

Suitable dyes are blended with a carrier polymer such as polyethylene or nylon 6 to get colored chips containing a high percentage of dye. Such master-batch chips are commercially available. They are tumbled with white chips and then spun to get colored fibers. When a master batch is prepared using nylon, the process of manufacture of the master batch and white chips must be one and the same in all respects. Otherwise, problems of compatibility of polymer chips may endanger the spinning process (see Section 4.4.6). The blending of the chips is done in a tumble dryer so that a statistically even

TABLE 4.4 Colorants for Nylon 6[21]

Colorant	Commercial Name	Mass Coloration by	
		Chips Coating	Chips Dyeing
—	Luramid Yellow R	$+^a$	$+^b$
C.I. Acid Yellow 12	Nylosol Yellow PA-10	−	+
—	Nylospin Orange 6R	−	+
—	Nylospin Red 2G	−	+
—	Luramid Red B	+	+
—	Luramid Blue RR	+	+
C.I. Acid Blue 227	Nylosol Blue PA 50	−	+
C.I. Pigment Blue 15	Nylofil Blue BLL	+	−
C.I. Acid Violet 48	Nylosol Violent PA 40	−	+
—	Nylospin Brown 5 BR	−	+
C.I. Pigment Black 7	Printex V (Degusa)	Added during polymerization	

^{a} + Suitable.
^{b} − Not suitable.

TABLE 4.5 Fastness Properties of Colored Nylon[21]

Shade[a]	Yarn-Dyeing Method		Mass-Coloration Method	
	Light Fastness	Wash Fastness	Light Fastness	Wash Fastness
Yellow	5	3–4	5–6	4
Orange	6	4	6	4
Red	5–6	4	6	4
Brown	6	4	6	4
Blue	5–6	1–2	5–6	1–2
Dark Blue	5–6	1–2	5–6	1–2
Bottle Green	5–6	3–4	6	3–4

^{a} Dyes from Table 4.4.

distribution is achieved that gives uniform shade. The proportion of the two chips should not be more than 1 : 20, beyond which uneven color may result.

4.4.4   Chips Dyeing[21,26–28]

The chips-dyeing method is in many respects the most suitable method for mass coloration of nylon fiber in shades other than black. The dyeing of nylon chips is very easy and does not involve any special equipment. Small batches can be dyed and processed economically. The contamination of the drying unit with the dyes is minimal. The contamination of the extruder spinning unit is very low since the anionic dyes used for nylon are not volatile and do not leave the nylon melt. The method is sometimes useful for covering yellowness of spinnable defective chips. The method is limited to heat-stable dyes and cannot be used for carbon black and other pigments. The chips-dyeing method is now adopted by many nylon producers for colored nylon 6. Since nylon 66 is melt-spun under more drastic conditions than nylon 6, the chips-dyeing method is not commonly used for the former. Nylon 6 chips are treated with boiling water to extract the monomer and oligomers. It is convenient to combine the dyeing of the chips with the extraction step. The dyed chips are then dried in the usual manner. The dyeing process does not interfere with the extraction step and vice versa is also true.

Many heat-stable acid dyes of metal-complex, azo, and anthraquinone-type are used for the chips-dyeing method (Table 4.4). However, their fastness properties are not always very high (Table 4.5). The conventional after-treatments such as tannic acid–tartaremetic for improving the wet-fastness of acid dyes on nylon are of no avail since the treatments lose their effectiveness during the melt-spinning process. Thus, there is a gap in the color range, for example, blue with excellent wash fastness is not available. Furthermore, it is observed[29] that there is a possibility of catalytic fading of dyes in a mixture, which must be checked by dyeing the chips, spinning the fiber, and exposing the fiber to light. Very heavy shades may not build up since the rate of the dyeing of the chips is not as high as that of the nylon fiber. The black shade with a mixture of heat-stable acid dyes invariably ends up a navy, deep brown, or olive shade. Bright, semidull or full dull chips can be used to get yarns of desired brightness. Trilobal colored yarns can be produced by using suitable spinnerets. Glittering, colored yarn of fine denier is easily produced by this method.

The dyeing may be carried out in a vessel in which the dye-liquor is circulated through a heat exchanger. Alternatively, the chips may be agitated with a paddle-type stirrer. The material-to-liquor ratio is kept as low as possible, for example, 1 : 2 to 1 : 5. Dyeing is started at 60°C and the temperature is raised to boiling in 30 min. Dyeing is carried out at boiling for usually 2–5 hours to exhaust the dyebath. The penetration of dye in the chips is never complete and the core portion of a chip remains white. However, this does not matter in getting a uniform shade on the filaments after melt spinning. The bath is then drained, and the chips are washed, soaped, washed, and dried. If dyeing is coupled with the extraction process when nylon chips are completely amorphous, the rate of dyeing is faster than if extracted chips are used. Unevenness in the chips does not influence the levelness of the shade, since during drying, the chips get mixed in the tumble dryer. In fact, colored chips can be mixed with white chips to produce a lighter shade. Any shade from a pale tint to a dark color can be produced with excellent levelness. This is almost impossible to get by any other method of dyeing with the acid dyes. The dyed yarn can be textured without any adverse effect on the shade. It may be advisable to give the fabric or garment prepared from mass-colored yarns an aftertreatment to improve the wet-fastness properties.

The dried chips are melt-spun, preferably on an extruder-type spin-head; the grid-type is not recommended because the thorough mixing of the melt of the colored polymer is essential for uniform level of shade on the filament. The color change from one shade to another on a spinning unit can be easily achieved by allowing the first lot of chips to drain completely into the extruder before adding the second lot of chips. Intermediate flushing with blank chips is seldom necessary during shade changes, if intermixing of chips in the holder tank is avoided.

4.4.5   Injection During Melt-Spinning

Unlike polyester, nylon is not mass colored by this method on an extensive scale. The method is similar to that described for polyester where a master batch as a powder or chips may be added during melt-spinning.

4.4.6   Properties of Nylon 6 Containing Carbon Black

Black nylon 6 is prepared using two different methods, namely, (a) by incorporating carbon black in disperse form during polymerization and melt-spinning, and (b) by the master-batch technique, that is, by blending chips with carbon black and chips without carbon black (blank) and melt-spinning. The other three methods are not used. The *CIELAB* darkness values of the fibers produced by methods (a) and (b) are similar (Table 4.6).[23] The darkness value is decided by the concentration of the carbon black pigment, its average particle size, and its size distribution. These factors are not significantly influenced by the production method of black nylon filaments. Nylon 6 containing carbon black gives filaments with marginally higher densities because the density of carbon black is 1.8–2.1 g/cm^3 while that of nylon 6 is 1.138 g/cm^3 (Table 4.7).[30] There is no increase in the X-ray crystallinity index when carbon black is added. Samples with carbon black show a lower modulus, lower tenacity, and higher elongation than blank samples.[21,30] This indicates that the small quantity of carbon black present in the sample does not act as

TABLE 4.6   Darkness Values ($-\Delta L$) (*CIELAB* Scale) of Black Nylon 6[23]

Carbon Black (%)	a	b
nil	0	0
0.015	5.6	5.0
0.030	11.0	10.9
0.075	15.0	15.2
0.15	18.8	18.8
0.30	20.3	19.4
0.75	20.4	—

a: Carbon black added during polymerization.
b: Master batch (1.5% carbon black) technique.

a reinforcing agent but appears to weaken the structure. With an increase in the carbon black content, the tenacity and modulus decrease. The carbon black particles act as fillers or voids because of a lack of interaction with the matrix and do not reinforce the matrix material. The electron micrographs of peeled nylon 6 are shown in Figure 4.3a and 4.3b. The figures clearly bring out the compact fibrillar morphological structure of nylon 6. In the samples

TABLE 4.7   Densities of Nylon 6 Containing Carbon Black[30]

Carbon black (%)	Density (g/cm³)	
	a	b
nil	1.138	1.138
0.015	1.138	1.138
0.030	1.139	1.138
0.075	1.139	1.141
0.15	1.138	1.143
0.30	1.141	1.143
0.75	1.140	1.140

a: Carbon black added during polymerization.
b: Master batch (1.5% carbon black) technique.

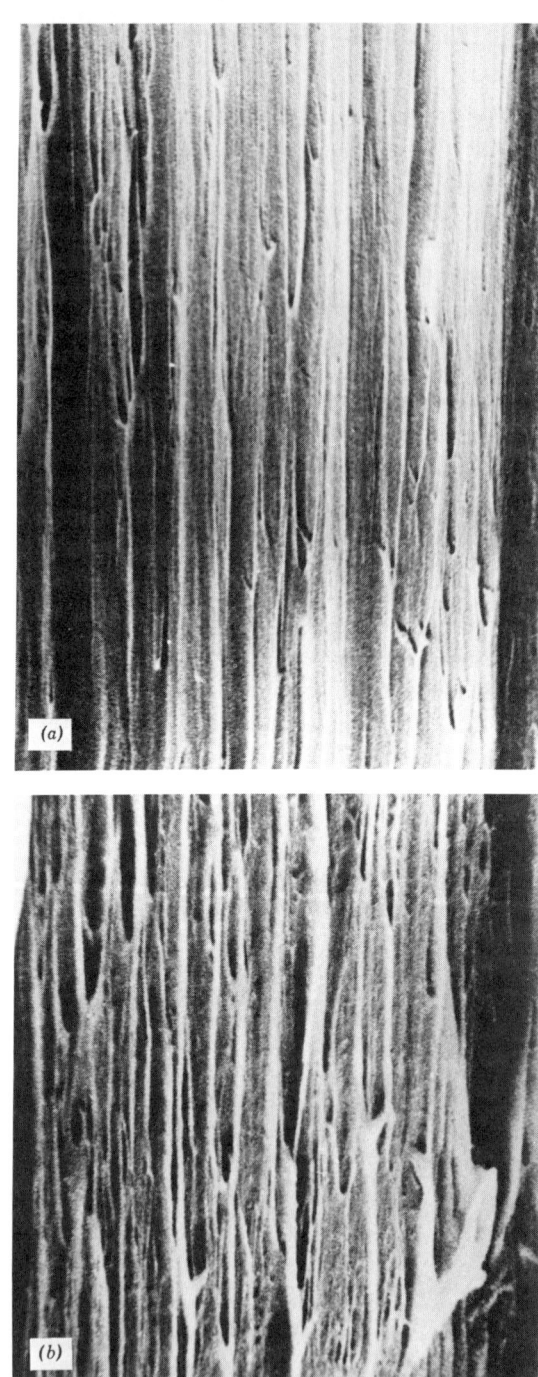

FIGURE 4.3. Electron micrographs of pilled nylon with and without carbon black (1.5%). (*a*): Without; (*b*): With 1.5% carbon black. Magnification: 650.

116

containing carbon black, the compactness is reduced and a slight porous structure becomes visible.

There are some remarkable differences in the properties of black nylons produced by the two methods. The DSC thermograms of black nylon samples are shown in Figure 4.4. The melting peak range of 220–223°C and the peak spread of over more than 20°C are common to the filaments produced by the two different methods. However, the composite nature of the peak is clearly visible in terms of the two peaks in the thermograms of samples from method *b*. This indicates that, in this set, the two phases are relatively better defined and melt at two distinctly different temperatures. Nonuniform mixing of the master batch with the blank polymer probably gives rise to these two phases, one close to the matrix material and the other close to the carbon-black-filled material. The electron micrographs of black nylon (0.75% carbon black) produced by the two methods are shown in Figure 4.5a and 4.5b. These results indicate that the heterogeneous structure of the carbon black containing nylon 6 is more pronounced in the samples produced by the master-batch technique than by the uniform black polymer. The moisture regain of the samples produced by the master-batch technique using different proportions of 1.5% carbon-black-containing chips is shown in Figure 4.6. It indicates that these samples absorb less moisture and that there is a minimum at 0.15% carbon black content (10% master batch chips: 90% blank chips). The weathering and exposure to light of these samples also show similar behavior (Figure 4.6). The viscosities of the original unexposed samples exhibit a slight decrease with an increase in carbon black

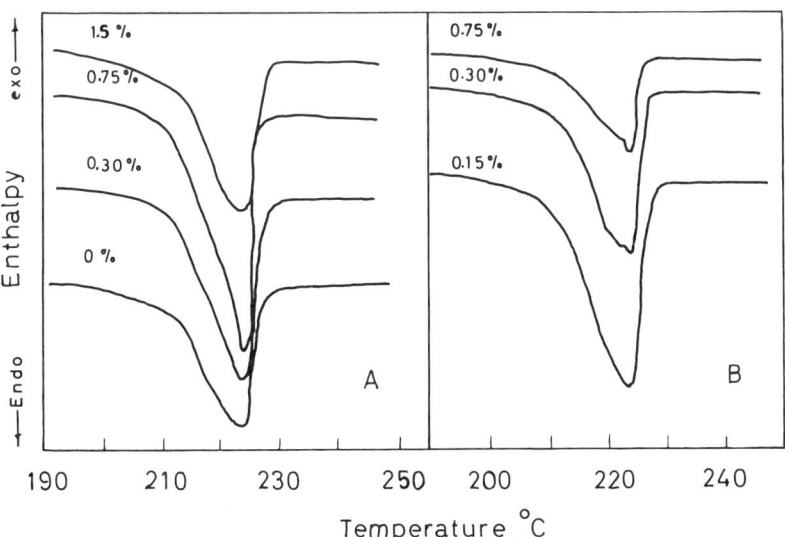

FIGURE 4.4. DSC thermograms of nylon samples containing carbon black produced by adding carbon black during polymerization (Sample A) and by the master batch technique (Sample B).

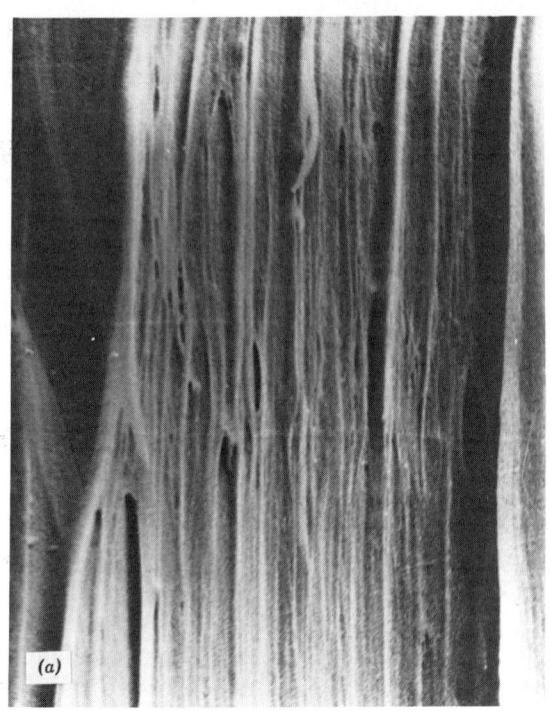

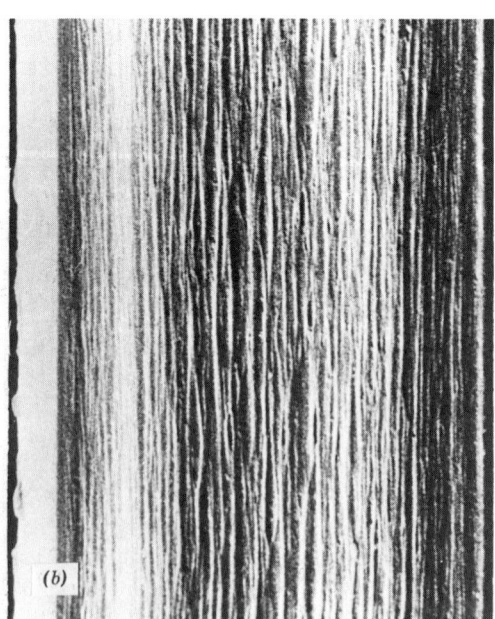

FIGURE 4.5. Electron micrographs of black nylon 6 (0.75% carbon black) produced by two different methods. (*a*) Addition during polymerization; (*b*) Master batch addition.

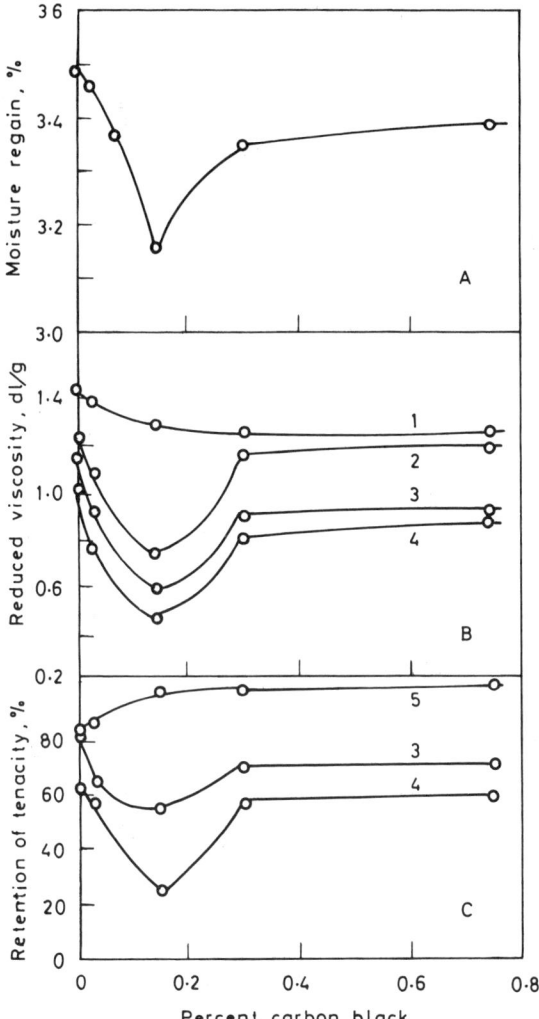

FIGURE 4.6. Moisture regain, percentage retention of tenacity, or reduced viscosity as a function of the carbon black content of nylon 6 fiber exposed to outdoor weathering. 1: Original sample; 2: Weathered samples 2 weeks; 3: Weathered samples 4 weeks; 4: Weathered samples 8 weeks; 5: Sample exposed in Fadeometer for 24 h.

content. The viscosity is reduced with an increase in time of weathering. The maximum reduction in viscosity at any time of weathering is around carbon black content of 0.15%. The results of tenacity on exposure show a similar trend. Carbon black is known to be a UV stabilizer.[31,32] It is expected that in the presence of carbon black, the fiber will not deteriorate rapidly. The nonuniformity in the fiber structure plays an important role in the weathering of black nylon fibers produced by method *b*.

4.5   MASS COLORATION OF ACRYLIC FIBER

Incorporation of colorant in the monomer prior or during polymerization is not feasible because most of the pigments and dyes interfere in the polymerization process. Since the acrylic fiber is spun by either dry or wet-process from a dope of the polymer, the mass coloration of acrylic fiber is best carried out by mixing the colorant with the polymer prior to the dissolution or after the polymer is dissolved in the solvent. Pigments are most widely used for mass coloration of the acrylic fiber.[33–36] The solubility of the pigment in the solvent–water mixture or in water should be negligible so that the colorant does not bleed into the coagulation bath used in the wet-spinning process. On the other hand, in the dry-spinning process, the colorant gets deposited during the evaporation of the solvent which causes unacceptable incrustations on the surface of the spinnerets. The selected pigments should have good stability in the thermal conditions encountered in the spinning and the stretching. They should exhibit good fastness properties, particularly to light and weathering, as mass-colored acrylic fibers are extensively used for outdoor purposes. Typical pigments suitable for the mass coloration are C.I. Pigment Yellow 3, C.I. Vat Yellow 4, C.I. Vat Orange 3 and 7, C.I. Pigment Brown 22, C.I. Pigment Red 2 and 7, C.I. Pigment Green 7, C.I. Vat Blue 1 and 15, and C.I. Pigment Black 7.

The pigments are finely dispersed in the spinning solution to achieve an optimum tinctorial strength, levelness of dyeing, and to avoid operational troubles such as the clogging of filters or spinnerets or the breaking of filaments by coarse agglomerates. Special cationic dyes free from diluents with very low aqueous solubility and high solubility in the solvent (dimethylformamide) are recommended for the addition to the dope.[37] These dyes get rapidly and quantitatively fixed on the polymer and are suitable for both wet and dry-spinning. Dyes in very high concentrations may coagulate the polymer and bleed in the stretching–washing bath. When inorganic salt or nitric acid is used as a solvent, the mass coloration of acrylic fiber cannot be carried out easily.

The fibers of the acrylic groups can be more conveniently colored by the gel-dyeing or the tow-dyeing process as is described later (Section 4.8). The addition of pigments and dyes to the dope is used more frequently to correct the whiteness of the fiber than to produce colored fibers.

4.6   MASS COLORATION OF OTHER FIBERS

Mass Coloration of Polypropylene.   Polypropylene is a fiber with no affinity for any class of dyes. Disperse dyes dissolve in the fiber but bleed out in a solvent that comes in contact with the dyed fiber. Thus, coloring polypropylene fiber by a dyeing process is difficult except when the fiber contains a metal that can form a complex with the dye. Because of these difficulties, polypropylene is mass colored using pigments.[1,38-44] The only requirement of the pigment is high thermal stability so that it can withstand the melt-spinning process. The fiber is essentially inert and does not attack the pigment. Both organic and inorganic pigments are employed for the mass coloration of polypropylene. The finely ground and dispersed pigment is incorporated in the polymer, either by the dry-blending method or by the use of master batches. In the dry-blending process, the polymer and a finely ground pigment are tumbled together before they are fed to the melt-spinning unit. However, this method cannot ensure fine dispersion of the pigment in the polymer.

The master batches of pigments are produced by mixing about 20–60% pigment with a carrier in a grinding mill, twin-screw extruders, or compounding extruders. These master batches and polypropylene are melt-spun together to get colored fiber. Carbon black is used to produce a black fiber.

The fastness properties of mass-colored polypropylene are superior to those of dyed fiber. Most of the fibers used in the carpet and upholstery industries are mass colored because of their excellent fastness properties and easy fabric-finishing methods.

Cellulose Acetates. Cellulose acetates—secondary acetate and cellulose triacetate—are mass colored by pigments dispersed in the solvent used to make the dope.[45-48] Master batches containing 30–50% colorants in acetate are now commercially available and can be used directly without any dispersion process. The fastness properties of mass-colored cellulose acetate fibers are satisfactory. Some solvent-soluble monoazo dyes are used to get yellow to red shades. Such dyes have to be free from dispersing agents and diluents.

4.7 TOW DYEING

Staples of synthetic fibers are produced by cutting the stretched continuous-filament tow. It is possible to color the tow in a continuous manner before, during, or after stretching, but before the tow is cut into staples. The steady rise in the consumption of PET and PAN staples is paralleled by a growing interest in continuous and semicontinuous methods of coloration of these staples, particularly in tow form.

In wet-spinning, the tow is dyed preferably before drying so that the open structure of the fiber can be exploited for rapid and complete penetration of the colorant. The undried tow is dyed within seconds, but the dried tow takes several minutes or hours to get fully dyed. Since tow dyeing is an integral part of fiber manufacture, there is a considerable savings in water, steam, electricity, and labor. The process gives level dyeing. All the unevenness in dyeing is covered up during the spinning of yarns from dyed staples. Multicolored fabrics can be prepared by blending fibers dyed in different shades.

The tow-dyeing process is of particular importance to acrylic fiber, since it is the only method of producing large quantities of uniformly dyed material in a continuous manner. (Continuous dyeing of acrylic fabrics is not practiced commercially.) The quality of tow-dyed acrylic fiber with respect to its bulking properties is better than that of fiber dyed in the conventional manner. Similarly, the tow dyeing of polyester is of importance because of the loss in mechanical properties and other difficulties under high temperature dyeing conditions.

The usual disadvantages of a continuous process are also present in tow dyeing. The process is unsuitable for dyeing smaller batches (less than 1000 kg) in a variety of shades. It is not possible to rectify the shade once it is produced on the tow. The dyes and colorants must have an affinity for the fiber in the tow and tailing effects must be carefully avoided by maintaining the concentration of color in the tow-dyeing unit. The temperature of the color bath and the functioning of the metering pump, delivery nozzle, agitator and so on are factors that have to be critically maintained and monitored. Dyes with a similar rate of pickup are selected so that there is no tonal variation with time. The dye on the filaments may exhaust almost completely so that the filaments do not soil the cutter and baling machines. The tow-dyeing installation does not interfere with the conventional spinning–stretching (drawing) operations, since the dyeing is completed within a few seconds and the flow of fiber material remains essentially the same as that without the tow-dyeing unit.

4.7.1 Dyeing of Polyester Tow

The tow-dyeing process can be carried out only in special machines because the filaments may become entangled during the tow-dyeing process. Several machines have been developed for the continuous dyeing of polyester tow based on high-temperature steaming or thermofixation by hot air. Owing to the small liquor ratio, high temperatures of fixation, and incomplete washing in tow dyeing, the oligomer content on the tow-dyed fiber is found to be high.[49] Typical PET tow-dyeing machines are described below.[50]

Segard Seracant Machine.[51] The machine consists of a padder, a steaming unit, and an aftertreatment section (Figure 4.7 and Figure 4.8). A horizontal two-bowl mangle gives good end-to-end levelness. The tunnel-type steaming unit consists of a perforated, stainless-steel tube, the cross-sectional area of which varies along its length and through which passes a stainless-steel conveyer chain that is used to transport the material through the tube. The padded tow is fed through a pair of rollers that overfeed the tow onto the chain and is then densely packed into the steaming tube. The time of steaming can be varied from 5–40 min by adjusting the speed of the machine. The production capacity of the machine is 100–300 kg/h.

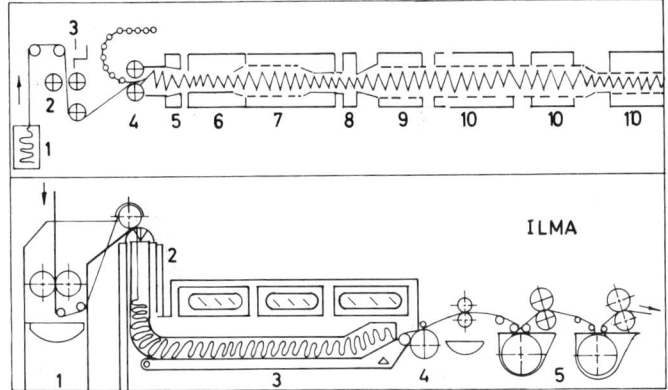

FIGURE 4.7. Seggard Serracant and ILMA tow-dyeing machines. Seggard Serracant Machine—1: Tow feed from box; 2: Pad; 3: Liquor feed; 4: Feed rollers and conveyor chain; 5: Presteamer; 6: Preheater; 7: Steaming zone; 8: Cooling zone (indirect cooling); 9: Rinsing zone; 10: Rinsing and aftertreatment compartments. ILMA Machine—1: Horizontal pad; 2: Heating-up zone (live steam); 3: Dwell chamber with conveyor belt (live steam); 4: Delivery roller; 5: Backwasher.

ILMA Machine.[52]   The fixation element of this machine consists of a rotating drum, around which are metal slats set parallel to the axis at 10–12 cm from each other (Figure 4.7). A cylindrical housing encloses the drum over three-quarters of its surface, leaving the top section open. The drum rotates at a speed of one revolution every 15 min. The padded tow enters through the

FIGURE 4.8. Photograph of Seggard Serracant tow-dyeing machine. (Courtesy of S. A. Serracant Sabadell.)

open section and is pressed tightly by slats into the space between the drum and the housing. The superheated steam at 145°C streams on to the tow through the perforated inner surface of the cylindrical housing. Average production is about 200 kg/h.

Thermosol Ranges. In the Fleissner or similar Conautex range (Figure 4.9), the disperse dye is fixed on the padded tow by hot air at 200–220°C. In another type, the dye is fixed with superheated steam at 190°C. The time of fixation is 2–4 min.

Dyeing Process.[53] The pad liquor is prepared consisting of

X g/kg	Disperse dye
1–2 g/kg	Wetting agent
5–10 g/kg	Thickener

at pH 5 (with acetic or formic acid) and a temperature of 30–50°C (see Chapter 7). Urea is added optionally to increase the rate of dye fixation.[54] The tow is passed through the pad bath and squeezed to 100–120% expression. The padded tow is steamed at 140–160°C for 5–40 min, reduction-cleared, washed, and dried. The dye fixation is around 90% for a steaming period of 5 min at a pressure of 2.1 kg/cm². Alternatively, the padded tow is conveyed into the thermofixation chamber without intermediate drying. The fixation time is 1–2 min at 200–220°C of which drying takes about 30 sec. The goods are then reduction-cleared, washed, and treated with a softener and an antistatic agent (each 2 g/liter).

Dyeing before Stretching. The process described above is used for coloring stretched/drawn tow. It is possible to dye unstretched tow on the polyester fiber line.[55] In a typical process, the unstretched tow is padded with the pad liquor of a disperse dye. The tow is then drawn to 3–4 times its original length by hot stretching at 160–180°C in 15–30 sec in the usual manner. The dye is fixed by thermofixation at 200–210°C in 15–30 sec. This is followed

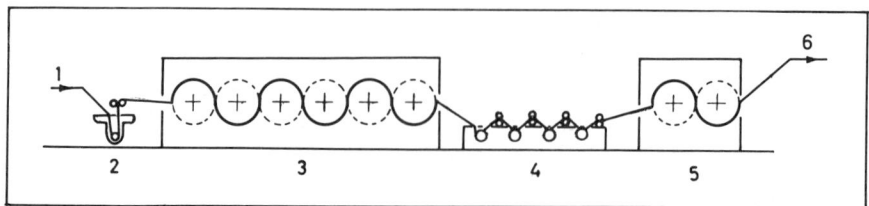

FIGURE 4.9. Continuous dyeing machine for polyester tow. 1: Tow from supply cans; 2: Dyeing pad; 3: Perforated-drum fixing unit; 4: Washing range; 5: Perforated drum dryer; 6: Take-up cans.

by the soaping–washing treatments before dyed tow is crimped and cut into staples. Very good penetration of the dye is achieved within a short thermofixation time because of the amorphous nature of the unstretched tow. Furthermore, the only machinery that needs to be added to the fiber line is a padding mangle and a small thermofixation unit.

4.7.2 Dyeing of Acrylic Tow

Acrylic tow cannot be dyed by the thermofixation process described above for PET tow. The method commonly employed is to pad the tow and steam it at 98–100°C for 10–45 min, depending on the dye and the depth of shade.[50,56–59] The dried tow is padded with a liquor consisting of

X g/liter	Cationic dyes
1–1.5 g/liter	Nonionic wetting agent
5–10 g/liter	Thickener
1–3 g/liter	Solvent

The tow is then steamed when 90–95% dye is fixed. The steamed tow is finally washed with dilute acetic acid (5–10 ml/liter) followed by water. During steaming, the acrylic fiber shrinks fully. The process is not suitable for fibers with high elongation. A finishing bath containing a softening and antistatic finish (5–20 g/liter) is used at the end of the washing treatment.

An interesting possibility of dyeing tows at low temperatures (20–75°C) without steaming exists in the case of fibers produced by the wet-spinning process. The polymer dope emerging from the spinnerets is coagulated into filaments that are drawn, washed, and subsequently dried in a continuous manner. The fiber structure after washing but prior to drying resembles a swollen gel with a minimum of intermolecular bonds between the polymer chain molecules. This open structure is lost during drying when intermolecular bonds are formed and a compact, collapsed chain-molecular structure develops. The dye molecules can penetrate in the compact, collapsed structure of the dried fiber only above the glass-transition temperature (see Chapters 2–3). On the other hand, the open gel structure of the undried fiber imparts little resistance to the diffusion of dye, even at low temperatures, and the cationic dyes can penetrate the fibers within seconds.[60–65] Thus, the dyeing of undried tow can become a part of the fiber manufacturing process without disturbing or prolonging the fiber-spinning operations. The extra unit needed for dyeing consists of two pairs of squeeze rollers, a dye-liquor trough with guide rollers, and airing rollers. The unit is placed after the washer and before the drying cylinders. The shade on the tow depends on the pH and the temperature of the dyebath. The dwelling time in the dye-liquor is 1–3 sec. The tow imbibed with dye-liquor is aired for about 15 sec before squeezing so that the dye-liquor penetrates the tow and the dye diffuses into the fiber before washing. Complete exhaustion of the dye-liquor is attempted, so that the dye is not lost during washing. Those cationic dyes that exhaust almost completely under these conditions are best suited for the gel–tow-dyeing process. Uniform and constant concentration of dye in the liquor is the important prerequisite for level dyeings. This can be achieved by continuous monitoring of the dye concentration by an on-line spectrophotometer coupled to a metering pump.[66] The fiber takes up the dye by adsorption and also by the absorption of the dye-liquor (similar to "tub-liquoring" of cotton yarns in the naphtholing in the azoic dyeing). The incoming water and outgoing dye-liquor on the tow must match. Squeeze rollers prior to the dye-liquor trough reduce the water content of the padded tow to less than 100%. The tow carries the dye-liquor with it during airing. It is squeezed again after airing (and not when it leaves the trough). The dyed tow is then washed, dried, crimped, and steamed, if necessary. The fastness properties of dyes staples depend on the dyes. Since the staples get mixed during the spinning of the yarn, minor variations in the levelness of shade pose no problem. The success of a gel–tow-dyeing process depends, apart from the selection of dyes, on the machinery used for dyeing, in particular, the metering pump, the control of the dye concentrations, the stability of the dye at pH 3–4.5 over a long period of time, the clarity of the dye-liquor, and the circulation of the dye-liquor. For optimum economy, a minimum lot for a shade has to be big, for example, more than 5 tons. A change of shade is very easy and there is no long downtime.

REFERENCES

1. Ackroyd, P., *Rev. Prog. Color.*, **5** (1974), 86.
2. Wampetich, M. J., *Chemiefasern/Textilindustrie*, **28/80** (1978), 1046.
3. Flex, S., and Wampetich, M. J., *MTB*, **61** (1980), 509.
4. Herbulot, D., *Chemiefasern/Textilindustrie*, **30/82** (1980), 941.
5. Yankov, V. I., Vikharev, S. A., Kavanov, Y. N., and Pleshanev, N. D., *Fiber Chem.*, **12** (1980), 267.
6. Gaunt, J. F., *Rev. Prog. Color.*, **2** (1971), 20.
7. SPRC Kota, India. (unpublished work).
8. DuPont, BP, 1,196,707 (1967).
9. Hoechst, A. G., BP, 1,276,213 (1969).
10. ICI BP, 1,185,030 (1966); BP, 1,326,941 (1970).
11. Ciba-Geigy, Swiss P., 567, 539 (1972); Ger. P., 2,334,064 (1972).
12. SPRC, Indian Patent application No. 898/DEL/78.
13. Andronova, A. P., Baramova, A. D., Timofeeva, G. F., and Aizenshteln, E. M., *Fiber Chem.*, **11** (1979), 368.
14. DuPont, USP, 2,868,757 (1959).
15. BASF, BP, 914,453 (1959).
16. Sandoz, BP, 984,014 (1960).

17. Ciba, BP, 984,853 (1962).

18. ICI, BP, 506,688 (1948); BP, 1,190,410 (1966).

19. Hoechst, BP, 1,501,370 (1978).

20. SPRC, Indian Patent application No. 878/DEL/78.

21. Datye, K. V., and Vaidya, A. A., Book of Symposium Papers, *Adaptation of Recent Innovations in Textile Technology to Indian Conditions*. IIT, Delhi, India, August 2–3, 1980 Also, *Man Made Text. in India,* **24** (1981), 361.

22. Datye, K. V., Mishra, S., and Gupta, V. B., *Indian J. Textile Res.* **7** (1983), 126.

23. Mishra, S., Ph.D. Thesis, IIT, Delhi, India, 1982.

24. Inam, E. R., *Rev. Prog. Color.,* **2** (1971), 62.

25. American Enka, USP, 3,035,003 (1962).

26. ICI, BP, 1,021,737 (1963); BP, 1,055,281 (1963).

27. Bayer, A. G., BP, 1,215,476 (1966).

28. Ciba, BP, 1,264,191 (1968); BP, 1,208,402 (1969).

29. SPRC unpublished work.

30. Datye, K. V., Mishra, S., and Gupta, V. B., Preprints, *International Symposium on Man Made Fibers,* Vol. 4, Kalinin, USSR, (1981), 246.

31. *Encyclopedia of Polymer Science and Technology,* Vol. 2, Interscience, New York (1965), p. 820.

32. Patten, T. C., *Pigment Handbook,* Vol. 1, Wiley-Interscience, New York, (1973), p. 709.

33. Courtaulds, BP, 1,284,891 (1972).

34. Mikheeva, G. M., Bolshakova, M. G., and Brisovk, E. P., *Fiber Chem.,* **8** (1976), 237.

35. Hoechst, A. G., USP, 3,996,192 (1976); USP, 4,020,037 (1977).

36. Bayer, A. G., BP, 1,514,558 (1978); USP, 4,087,494 (1978).

37. Hoechst, A. G., BP, 1,409,912 (1975); BP, 1,508,361 (1978); BP, 1,504,796 (1978).

38. BASF, BP, 1,220,795 (1968).

39. Maury, L. G., *TCC,* **4** (1972), 143.

40. Bash, D. P., *Fiber Prod.,* **6**(2) (1979), 24.

41. Biehler, B., *Chemiefasern/Textilindustrie,* **29/81** (1979), 848.

42. Ripke, C., *Chemiefasern/Textilindustrie,* **30/82** (1980), 30; **30/82** (1980), 110.

43. Ahmed, M., *Polypropylene Fibers–Science and Technology,* Elsevier Scientific Publishing Co., Amsterdam, 1982, p. 100.

44. Wishman, M., *Fiber Prod.,* **10**(2) (1982), 50.

45. British Celenase, BP, 687,481 (1949); BP, 866,329 (1956).

46. Courtaulds, USP, 2,661,299 (1949); BP, 1,274,207 (1968).

47. BASF, BP, 1,117,892 (1967).

48. Ciba-Geigy, BP, 1,287,491 (1969).

49. Keray, I., Decheva, R., and Duscheva, M., *Chemiefasern/Textilindustrie,* **28/80** (1978), 454.

50. Lemin, D. R., and Simpson, G. G., *JSDC,* **87** (1971), 257.

51. Cigarra, J., *JSDC,* **86** (1970), 26.

52. Sandoz Technical Inf. SB 006-68 *Continuous Dyeing Process for Polyester and Acrylic Tow and Tops.*

53. Boutler, H., and Ullrich, H., *MTB,* **53** (1972), 548.

54. Decheva, R., Karay, I., Duscheva, M., and Stoyanov, S., *Textilveredlung,* **12** (1977), 26.

55. ICI, BP, 921, 125 (1960).

56. Beal, W., *JSDC,* **83** (1967), 3.

57. Mayer, U., and Riechert, M. A., *ADR,* **57** (1978), 1104.

58. Mackinon, R. A., *JSDC,* **85** (1969), 661.

59. Shore, J., *Rev., Prog. Color,* **10** (1979), 33.

60. DuPont, BP, 1,416,851 (1972); USP, 3,932,571 (1973); USP, 3,944,386 (1973).

61. Bayer, A. G., BP, 1,412,231 (1973); BP 1,483,311 (1974); BP 1,508,025 (1974); USP 4,013,406 (1975).

62. Ciba-Geigy, BP, 1,498,069 (1974).

63. Lapple, A., and Schneider, A., *Textilveredlung,* **10** (1975), 63.

64. Chernetskii, E. K., Pakshver, E. A., and Kharkharov, A. A., *Fiber Chem.,* **10** (1978), 20.

65. Geller, A. A., *Fiber Chem.,* **11** (1979), 177.

66. Biedermann, W., Galafassi, P., and Ischi, A., *Textilveredlung,* **11** (1976), 417.

5 | PRETREATMENTS

The term *pretreatment* or *preparation* includes desizing, scouring, merceriz-ing, heat setting, bleaching and so on. Pretreatments are given to textiles to remove natural and adventitious impurities as well as spin-finishes, coning oils, and sizing agents. Pretreatments are also given to get yarns or fabrics with satisfactory whiteness and absorbency for subsequent dyeing, printing, and finishing. It is estimated that about 70% of all faults in finished goods can be attributed to inefficient pretreatments.[1,2] About 17% of the total energy consumed in the textile industry is used for fabric preparation.

Synthetic fibers contain spin-finishes, coning oils, sizing, and adventitious dirt as impurities. The removal of these impurities which may amount to 5–12% of the weight of a fabric can be carried out by a mild scouring treatment. Bleaching is usually not required, particularly for goods that will be dyed or printed. The original whiteness of synthetic fibers is usually sufficient unless extra-white material is required. Fabrics of synthetic fibers have a tendency to form permanent creases if they are processed in rope form at high temper-ature and pressures. The goods are therefore processed in open width if the processing conditions are drastic. Heat setting is often necessary as a first step to facilitate handling, to prevent wrinkling, and, at times, to insure the fabric width after processing. Heat setting is carried out as a separate opera-tion rather than as an in-line operation, since all fabrics do not require it.[1]

5.1 POLYESTER–COTTON (PET/CO) BLENDS

5.1.1 Desizing

In desizing, the sizing material is removed from the fabric as completely and uniformly as possible. The residual size can lead to uneven dyeing, stiff and

129

variable hand, variable responses to thermosetting resins, and reedy appearance of blend fabrics. Generally, the size for PET/CO blends consists of acrylic copolymer or polyvinyl alcohol together with carboxymethyl cellulose, starch, and lubricants such as fats, tallows, and waxes. Polyvinyl alcohol and acrylic copolymers are removed by a hot-water wash. It is essential to use enzymes for starch-based sizes.[3,4]

Desizing with Enzymes. Enzymes hydrolyze starch into water-soluble products without affecting cellulosic and PET fibers.[4] It is the safest process of desizing PET/CO blends. The desizing process consists of application of the enzyme solution, digestion of the starch, and removal of the digestion products.[5] The enzyme solution contains

5–10 g/liter	Enzymes
5–10 g/liter	Sodium chloride
1–2 g/liter	Wetting agent

at pH 7. The desizing process is carried out either by the pad-roll process, the continuous process in the *J*-box, or the pad-steam process.

Pad-Roll Process. The grey fabric is passed through the enzyme solution at 60–70°C, giving at least four dips and a 5-ton nip to get a pickup of about 100%. The wetting agent and the temperature help proper penetration and wetting of the grey fabric. The enzyme activity is maximum at pH 7 and increases further in the presence of sodium chloride. The padded fabric is wound on a roller that is then covered with polyethylene sheets to prevent drying of the fabric. The roll is slowly rotated at room temperature for 6–12 h. The fabric is then unwound and washed with hot (90–95°C) and cold water to remove the soluble degradation products of starch.

Continuous Desizing Process. In the continuous desizing process, the fabric padded with enzyme solution is passed through an open-width *J*-box maintained at 80–90°C. The speed of the fabric is adjusted to get a dwell time of 25–40 min. The fabric is then washed with hot and cold water in an open soaper. This method gives high production but requires costly equipment.

Pad-Steam Process. Enzymes that are stable at high temperatures are used in this method.[6] The padded goods are steamed for 1–5 min with wet steam. During this steaming, starch is hydrolyzed to soluble products and polyvinyl alcohol and other water-soluble sizes are hydrated and softened. A final washing operation with hot and cold water completes the process. This method gives very high production.

Desizing with Sodium Carbonate. When the size consists only of water-soluble materials such as polyvinyl alcohol, modified starches, or water-soluble cellulosic ethers, a treatment with a mild alkali such as sodium carbonate (4 g/liter) and a detergent (1 g/liter) at 70–80°C for 1 h is sufficient to desize the goods. The goods are then washed with hot and cold water.

Desizing with Sodium Bromite. Starch is rapidly oxidized into alkali-soluble products with sodium bromite.[7,8] The goods are padded with

1–2 g/liter	Available bromine (as bromite)
5 g/liter	Soda ash
1–2 g/liter	Wetting agent

from a short-bath padder with a V-shaped trough, stored for 5–20 min, and then treated with 5 g NaOH/liter at boiling in an open soaper. Washing with hot and cold water completes the desizing process. Conventional machinery is used at room temperature for bromite desizing. The process is rapid and gives uniformly clean fabric because bromite also destroys other impurities in cotton. Hot-alkali wash in an open width is necessary for thorough desizing. Even then fabric having a mixture of starch and tallow cannot be desized satisfactorily because of the poor penetration of the desizing liquor.[9] A continuous process of desizing consists of padding the fabric with

2–4 g/liter	Available bromine (as sodium bromite)
1–3 g/liter	Sodium hydroxide
1–2 g/liter	Wetting agent

steaming for 45 sec at 102°C, and washing with hot and cold water.[10]

5.1.2 Scouring

Scouring of blend fabric is the key to the future performance of the fabric. The use of good quality clean cotton and methods such as carding and combing eliminate the cotton seeds almost completely. Nevertheless, blend fabric must be scoured, that is, the oils, fats, and waxes on cotton fiber must be removed by saponification and emulsification processes, in order to get good whiteness and absorbency. The conventional method of kier boiling cannot be used for blends since scouring has to be carried out in open width. Polyester saponifies with caustic soda—the rate depends on the alkali concentration and the temperature.[11] Thus, PET loses about 3–5% in weight on treatment with 2% caustic soda at 100°C and the weight loss increases with temperature (Figure 5.1).[12] The blends on scouring with caustic soda lose weight and, in turn, tensile strength. It is therefore essential to optimize the scouring process with respect to the minimum loss in strength. According to Forrester and Caldwell,[13] 0.5% caustic soda at 93°C for 1 h does not lead to any strength loss; however, there is a considerable loss in strength if the time

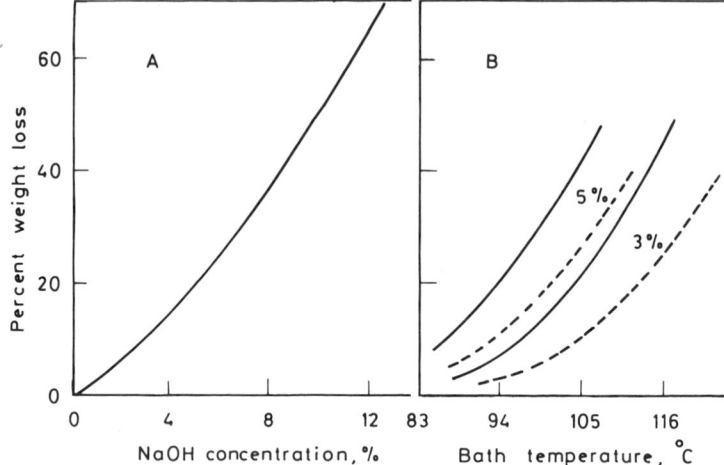

FIGURE 5.1. Loss in weight of polyester treated with sodium hydroxide solution.[12] (*a*): 2 denier Terylene for 2 h at 100°C; (*b*): 70/30 textured polyester. --- 30 min; —— 60 min.

of treatment is extended further. Sodium carbonate is a safer alkaline agent for scouring polyester-containing materials. After boiling with 2% sodium carbonate for 24 h, PET strength losses are negligible.

Batch Processes. PET/CO blend fabrics are scoured on a jigger using 0.2–0.5% detergent and 0.5–1% soda ash at boiling for 60–90 min. If sodium hydroxide (0.2–0.5%) is used in place of soda ash, the temperature is lowered to 75°C. The goods are washed off to complete the process.

Winch machines can be used for scouring when the fabric is in rope form. The conditions of scouring are milder than those used on the jigger in order to avoid the formation of permanent creases. The temperature is below 75°C and the concentrations of soda ash or caustic soda and detergent are half of those used on the jigger.

Semicontinuous Processes. The pad-roll process is a semicontinuous process in which the fabric is padded through a solution of soda ash (1–2%) and detergent (0.2–0.5%) using a double saturator to ensure an uniform and thorough exchange of liquor. The padded goods are passed through a chest for preheating and are wound on a roller inside a heating chamber. The roll is kept rotating in the chamber which is maintained at 90–95°C by passing open steam. Scouring is complete in 90–120 min. The fabric roll is then removed from the chamber and washed on an open soaper.

Continuous Processes. Continuous methods of scouring were developed to increase productivity. In a continuous process, PET/CO blend fabric is padded with a liquor having

10–20 g/liter	Sodium carbonate
1–2 g/liter	Detergent

It is then passed continuously through an open-width *J*-box at 93–99°C. The speed of the machine is adjusted to get a dwell period of 60–90 min. The fabric is then washed continuously at 90–95°C in an open soaper with a counter current flow.

The time of scouring is reduced to a couple of min and the quality of goods is improved by using high-pressure steam.[14-17] The fabric is padded as above or with

5–10 g/liter	Caustic soda
2–5 g/liter	Detergent

and is steamed about 2 kg/cm^2 pressure in a steam chamber (for an example of a vapor lock, see Figure 5.2). The temperature in the chamber is 120–125°C and the time is 1–2 min. After steaming, the goods are continuously washed on an open soaper. The pressure-steaming processes are rapid and therefore degradation of PET is insignificant. Very good and even absorbency is achieved by the fabric with these processes. The limitation of these processes is that they need expensive, special equipment.

With the advent of pressure scouring–bleaching methods, it is now possible to continuously produce blend fabric with adequate absorbency for dyeing or printing. Steaming under atmospheric pressure fails to give the desired absorbency and can damage polyester fiber if the treatment is extended to over an hour and if caustic soda is used in place of soda ash. After scouring, a final cold-water nip and a brief skying to reduce the temperature of the fabric are suggested.[1]

5.1.3 Singeing

A PET/CO blend fabric is carefully singed to remove loose protruding fiber ends.[18] It is also the best method of controlling pilling. The PET fiber ends melt during singeing. If the singeing machine is run too slowly, a portion of the fabric may be plasticized, resulting in a loss of tensile properties. (Figure 5.3) Thus, singeing is a critical operation and can spoil an otherwise satisfactory piece of fabric by uneven, deep plasticizing or melting of PET. The safety margins are larger in gas singeing than in plate singeing; gas singeing is preferred for PET/CO blend fabrics.

The molten beads formed during singeing of PET/CO blend fabric have higher dye pickup than the rest of the fabric. This is not apparent when dyeing is carried out by the thermofixation process. The singeing of goods to be dyed by the exhaust-dyeing method is carried out after rather than before dyeing. It is rarely carried out before scouring and cleaning since the re-

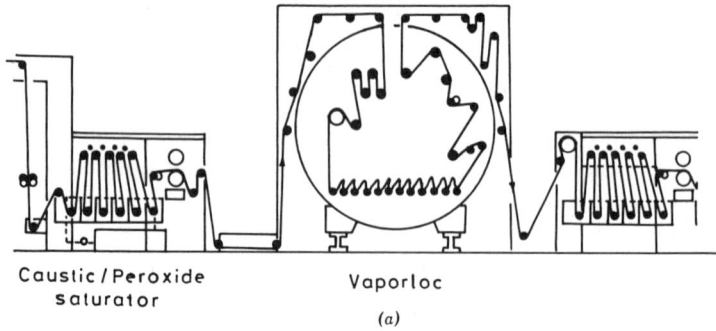

Caustic / Peroxide
saturator

Vaporloc

(a)

(b)

FIGURE 5.2.   Vapor lock unit. (a): Schematic diagram; (b): Mather and Platt unit. (Courtesy of M/s Mather and Platt Ltd.)

moval of dirt from PET after any heat treatment is difficult. The singeing operation is preceded by shear-cropping, that is, the removal of protruding fibers.[19]

Burners with short, even flames are used for blend fabrics as opposed to burners with long flames for cotton singeing. The fabric moves over the flame at a high speed followed by a water-filled roller (to prevent burn-through and to cool). This process may be repeated if necessary. Constant pressure governers and sensitive mixture controls are fitted to the gas and air supplies to get reproducible burner settings. For very fine and knitted goods, the fabric passes over a cooling cylinder while the flames impinge on its outer surface in order to avoid polymer fusion.

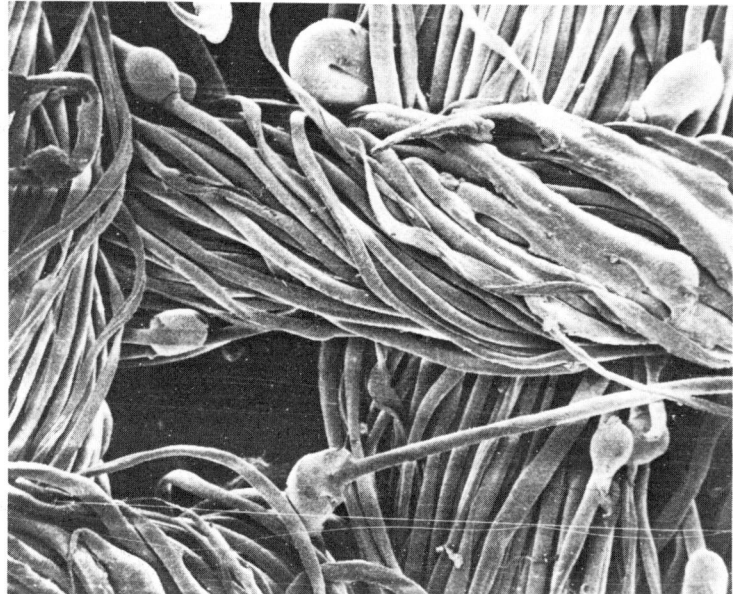

FIGURE 5.3. Electron micrograph (200×) of polyester fabric damages during singeing. (Courtesy of P. Neelakanthan, ATIRA, Ahmedabad, India.)

5.1.4 Mercerizing

PET/CO blend fabric is mercerized to improve the properties of the cotton component without significant damage to the PET component. Properties of cotton are luster, smoothness, dye affinity, dimensional stability, tensile strength, and chemical reactivity, especially with cross-linking agents. PET gets saponified by strong alkali under drastic conditions, and the fiber becomes finer in diameter and smoother in feel, which gives silklike handle. However, due to a decrease in the denier of PET, the yarn loses tensile strength. Since the mercerization treatment is carried out at low temperatures and within a short time, PET is not affected in spite of the concentrated alkali solution used for mercerizing.

PET/CO blend fabric is not subjected to severe scouring; therefore, the absorbency and wetability of the fabric is not very high. Special wetting agents are used during the mercerization of such blend fabrics. The concentration of sodium hydroxide is 18%. The other conditions are the same as those used for 100% cotton fabric. Thus, the fabric passes through a water mangle to permit uniform absorption of caustic soda of mercerizing strength. The fabric is saturated with the caustic liquor on a three-bowl mangle with 10-ton and 5-ton nips in order to force the caustic soda solution into the fabric and to give 100% pickup. The washing starts after about 40 sec. The fabric on the stenter is then stretched and washed with a 6–8 combination cascade and suction sections. An acid scour and final rinsing complete the process. The fabric is neutralized after mercerization until the pH is about 6. The use of ammonia for mercerization has the advantage that it can be removed from fabric by the "drying" process; thus, the wet-washing process may be eliminated. However, this process is still not used on a wide scale.

5.1.5   Heat Setting

Fabrics of synthetic fibers and their blends have a tendency to shrink and to form permanent creases and wrinkles in boiling water and hot air. A heat-setting process is carried out to eliminate this tendency. Besides this, heat setting improves dimensional stability, crease resistance, and resistance to pilling.[20] It also influences the dyeing and finishing process as well as the fabric's behavior during laundering. Polyester yarn shrinks about 7% in boiling water. Shrinkage in hot air increases with the air temperature in a linear fashion (Figure 5.4).[21] The degree of setting imparted to PET material depends on the time and temperature of setting, the previous treatments given to the fabric, and the tension on the fabric during setting. Heat setting at 180°C for 60 sec to 210°C for 20 sec gives sufficient dimensional stability. The fabric shrinks during heat setting. A relaxed heat setting gives better dimensional stability and improved pilling resistance to the set fabric

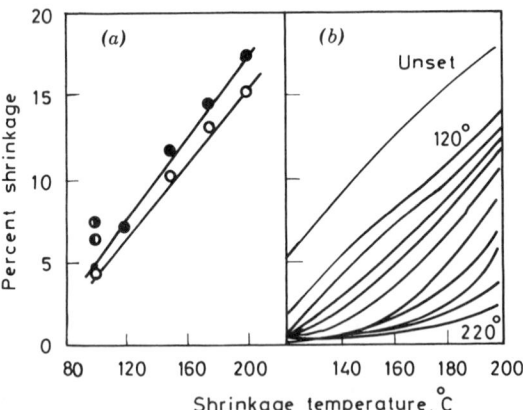

FIGURE 5.4.   Shrinkage of PET in hot air and boiling water and the effect of setting temperature on dimensional stability.[21]

(a):	Boiling water shrinkage	Hot-air shrinkage
High-tenacity polyester	◐	●
Medium-tenacity polyester	◑	○

(b): Setting temperature from top to bottom: 120°C, 140°C, 150°C, 160°C, . . . , 220°C.

than that given by setting under tension. Heat setting is carried out in hot air with IR radiations or by the thermal shock process. It is possible to set fabrics containing PET fibers in boiling water or by pressure steam. However, boiling water is rarely used since the setting effect is not as permanent and sufficient as the temperature of setting is very near to the temperatures applied during the subsequent processing and laundering. High-temperature steam gives good setting, but PET undergoes simultaneous hydrolysis and is of little use. Heat setting with hot air is the most widely used method for PET/CO blend fabrics. After heat setting, the fabric does not shrink more than 1% during boiling in water for 30 min. Besides the shrinkage in boiling water, the handle of the fabric before and after heat setting gives an idea of the extent of the setting.

Heat setting is carried out either immediately before any wet-processing (presetting), after scouring and mercerizing (intermediate setting), or after dyeing (post setting). Each method has its own advantages and disadvantages. Thus, the yellow color due to heat setting cannot be removed if setting is done after dyeing since there is no subsequent bleaching process. Similarly, dye migration is possible which lowers the fastness properties. Unset fabric is very susceptible to wrinkling and creasing during dyeing. Dyes fast to sublimation have to be used if setting is done after dyeing. On the other hand, faults in setting are shown as uneven dyeings if dyeing is carried out after setting. The residual carrier also remains in the fabric under these conditions while postsetting expels the carrier, thus, improving the light fastness of carrier-dyed goods. If setting is done prior to dyeing, many high-molecular-weight dyes may give lower build-up.

Hot-Air Heat Setting.   Pin-stenter setting machines, curing ovens, and cylinder setting machines are used for setting blend fabrics.[22] The advantage of a pin stenter is that the fabric is maintained during setting under complete dimensional control, both in warp and weft directions. The productivity of a pin stenter is, however, low and fluctuations of temperature across the width of the fabric must be kept within a narrow limit ($\pm1°C$), which is not always an easy process. A slight overfeed is given to allow shrinkage in the warp direction for better dimensional stability. High productivity can be achieved on curing ovens and contact cylinders. There is no control on fabric width in these machines. It is therefore essential to adjust the width of the fabric before heat setting, after considering the possible shrinkage during setting. A setting machine with a 50% perforated drum with or without thermochamber gives heat setting within 15–30 sec at a speed of 100 m/min.[23]

IR Heat Setting.   PET fabric is exposed to IR radiations of 3.35 μ wavelength using selective IR emitters at 80–120°C to heat set within 15–20 sec. Alternatively, heat setting is carried out within a second by using temperatures of 600–750°C when the fabric reaches temperatures up to 200–215°C. Very high production rates are possible by the latter process. However, serious limitations such as high energy consumption and severe fiber damage during sudden stoppages of the machine make the process less attractive.

Heat Setting and Properties.   Heat setting of PET and its blend fabrics imparts dimensional stability by bringing down the residual shrinkage to less than 1% if the setting temperature is 30–40° higher than that at which the dimensional stability is determined. Fabric set at a higher temperature exhibits lower shrinkage (Figure 5.4).[21] Since the handle of the fabric is impaired with increasing setting temperature with only a marginal advantage in stability, minimum best temperature of setting is preferred. The creases formed after setting are less permanent and are readily removed by ironing. In unset fabric, the wet treatment acts as a setting process and the creases are set in the shrunk fabric. The resistance to pilling of PET and PET/CO fabric improves with heat setting. The handle of the fabric becomes stiff with increasing setting temperatures which lowers the dry crease recovery. A scouring treatment restores the soft hand and, in turn, the loss in crease recovery.

Heat Setting and Dyeing.   Heat setting has a pronounced effect on the dyeing behavior of PET and blend fabrics.[24–26] The setting temperature is an important factor; very small variations in temperature can produce a very significant change in the dyeing rate. With an increase in setting temperature, the rate and extent of dye uptake first decreases and then increases.[24,27–28] Thus, the dye uptake of the set fabric is lower than that of the unset fabric up to the setting temperature of 210°C, the minimum being at 175°C (see Figure 2.18), after which it remains almost constant between 175°–200°C.[21] Above the setting temperature of 220°C, the dye uptake increases abruptly (because of softening prior to melting) with a drop in polymer chain orientation and crystallinity[26] until the PET fiber completely melts at about 265°C. The temperatures above 200°C are thus of no practical importance since small variations in temperature in this range produce large, uncontrollable variations in dyeability. The best setting temperature is in the range of 180–200°C and the best time of setting is less than 1 min.

The effect of heat setting on dyeability of disperse dyes is not the same for all dyes[27,28] or for all processes of dyeing.[29] The build-up on PET is lowered on setting by 10–40%, depending on the molecular weight of a dye.[28] The lower the molecular weight of the dye, the smaller is the effect of the setting temperature. The dyeability, that is, the rate of diffusion and saturation of the set fiber, is affected by the changes in setting temperature to a higher extent for the carrier dyeing at 100°C than that for HT dyeing at 130°C (Figure 5.5).[29,30] When the thermosol process (180–215°C) is used for dyeing, the effect of setting temperature on dyeability is very low.[31]

The tension on the fabric during heat setting affects the dyeability of the fabric.[28,32] It is essential to allow sufficient shrinkage during heat setting in

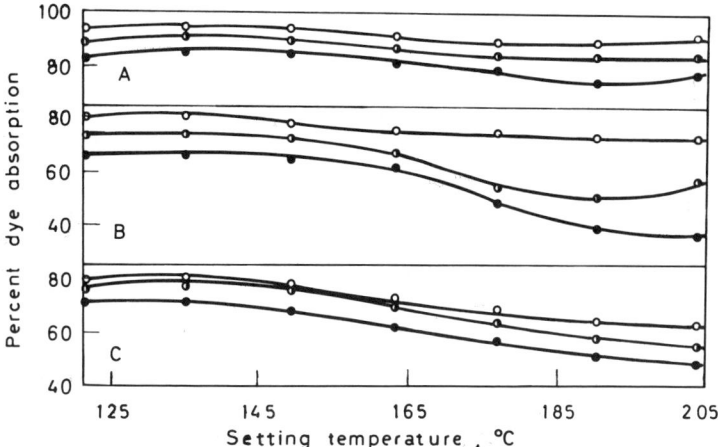

FIGURE 5.5. Effect of the heat-setting temperature on the absorption of disperse dyes by Dacron type 64 polyester fiber.[29,30] (*a*): 2% Latyl Cerise B; (*b*): 8% Latyl Blue 4R; (*c*): 2% Latyl Brilliant Blue 2G. Dyeing temperature: ○: 122°C; ◑: 98°C; ●: 93°C.

order to even out the structural differences caused by variations in tension during earlier mechanical or chemical processes.[33]

All these effects can be explained by considering the changes in the fine structure of PET fiber caused by the thermal treatments and tensions (see Chapters 1 and 2) and the diffusion and fixation processes of disperse dyes (see Chapter 7).

5.1.6 Bleaching

It is essential to bleach PET/CO blend fabrics to improve the whiteness of the cotton component. For white material, PET with an ivory color is also bleached to get a perfect white. The usual bleaching agents for cotton are used for bleaching blends. For goods that are to be dyed or printed in deep shades, bleaching is done with sodium hypochlorite or hydrogen peroxide. For the goods to be sold as white or lightly colored fabric, bleaching is done with either sodium hypochlorite or sodium chlorite followed by hydrogen peroxide.

Sodium Hypochloride Bleaching.   Blend fabric in rope form is treated with a solution of sodium hypochlorite (1–3 g/liter available chlorine; pH 10–11) in a rope-washing machine. The fabric is stored in pits for 1–2 h at room temperature and then thoroughly washed. Alternatively, the bleaching is carried out in a *J*-box in a continuous way, either in the rope form or in the open width. The fabric is padded with a hypochloride solution containing 2–4 g/liter available chlorine and is then passed through the *J*-box at room temperature. The speed of the machine is adjusted to get a dwell period in the *J*-box of 30–60 min. The fabric is then washed in an open soaper. There is no danger of setting creases during bleaching. The bleaching cost is also very low. Sodium hypochlorite does not produce good white on the polyester component of the blend. It is an efficient bleaching agent for cotton and is used where moderate whiteness of the blend fabric is required.

Hydrogen Peroxide Bleaching.   When PET/CO blend fabric is bleached with hydrogen peroxide, the PET does not get bleached. It is possible to get good whiteness by combining peroxide bleaching with optical bleaching with a fluorescent whitening agent. The ionized hydrogen peroxide gives the perhydroxyl ion (HO_2^-), which is an effective bleaching agent. Alkalinity in low concentrations favors the formation of perhydroxyl ion, while under highly alkaline conditions, hydrogen peroxide decomposes into oxygen and water. This wasteful reaction is also catalyzed by the presence of metal ions such as copper, chromium, iron, manganese, and nickel.[34] In order to minimize the decomposition of hydrogen peroxide under alkaline conditions and the high temperatures required for bleaching, sodium silicate is added as a stabilizer. The bleaching is carried out by a batch process, a semicontinuous pad-roll method, or a continuous process.[35]

Batch Process.   A small quantity of fabric is bleached on a jigger using a bath having

2–5 g/liter	Hydrogen peroxide (35%)
2–4 g/liter	Sodium silicate
0.5–1 g/liter	Sodium hydroxide
1–2 g/liter	Detergent

The fabric is run on a jigger at 90–95°C for 90–120 min. The goods are then washed with hot and cold water.

Pad-Roll Method.   In this semicontinuous method of bleaching, the blend fabric is padded with

10–20 g/liter	Hydrogen peroxide (35%)
10–15 g/liter	Sodium silicate
2–5 g/liter	Sodium hydroxide
2–5 g/liter	Detergent

and passed through a preheater before it is batched on a roller in a hot-batching chamber. The roller is rotated in the chamber at 90–95°C for 1–2 h before it is removed for washing on an open soaper. The hot-batching chamber is a separate unit on wheels and is moved away from the padding mangle and preheater after filling. Another load is wound in a fresh batching chamber. In this way, a number of chambers can be filled and stored for the dwell time to make the process semicontinuous.

Continuous Process. An open-width *J*-box is used for the continuous bleaching of blend fabric with peroxide so that no creases may set at bleaching temperatures.[36,37] The padded fabric is heated in the *J*-box at 90–95°C for 1 h instead of rolling it and storing in a hot chamber as in the pad-roll process. The bleached fabric is washed on an open soaper in a continuous manner. A continuous pressure scouring followed by bleaching with peroxide as described above gives fabric with satisfactory whiteness, absorbency, and strength.[36] Bleaching can be carried out in 1–2 min using high-pressure equipment. Thus, the padded fabric in open width is fed to the high pressure unit,[38] which is maintained at pressure of 2–3 kg/cm^2 using steam. The bleaching takes place at 120–135°C within 1–2 min to give fabric with good whiteness. The fabric is then washed on an open soaper.

A cold-bleaching process has been developed by ATIRA to conserve energy.[39] In this process, the desized goods are treated with sodium hypochlorite and then with hydrogen peroxide. A special catalyst is used in peroxide bleaching, which is carried out at room temperature.

Sodium Chlorite Bleaching. Sodium chlorite was commercially introduced in 1939 as a bleaching agent for cellulosic fibers and is the bleaching agent that produces satisfactory whiteness on all synthetic fibers. It bleaches both PET and cotton fibers in blends without any degradation and is effective under acidic conditions. This is of special interest in bleaching PET because PET is affected by hot alkaline liquors. However, acidic chlorite solutions are very corrosive and attack the materials of construction of the equipment. Special materials of construction such as glass, porcelain, earthenware, resistant plastics, or special steel containing molybdenum are required for bleaching with sodium chlorite. The chlorine dioxide fumes evolved during bleaching are very toxic and must be exhausted with proper ventilation.

Various chemicals and mixtures thereof protect the stainless steel from attack by sodium chlorite or chlorous acid (Table 5.1).[40] Sodium nitrate with or without hydrogen peroxide in the sodium-chlorite bleaching bath almost eliminates the corrosion of the machinery. A foaming agent that covers the surface of the bleaching bath with a thick foam decreases the obnoxious smell of chlorine dioxide. The trade products usually contain a mixture of sodium chlorite, sodium nitrate, and a foaming agent.[41]

Mechanism of Bleaching. Sodium chlorite under acidic conditions gives chlorine dioxide, chlorate, and chloride ions.

$$5ClO_2^- + 2H \rightarrow 4ClO_2^- + Cl^- + 2OH^-$$

$$3ClO_2^- \rightarrow 2ClO_3^- + Cl^-$$

TABLE 5.1 Corrosion of AISI 316 Stainless Steel
in Sodium Chlorite Liquors[40]

Additive to Chlorite Liquor	Iron Content of the Liquor (ppm)
None	807
NaH_2PO_4 (1 g/liter)	47
$NaNO_3$ (1 g/liter)	25
H_2O_2 (1 ml/liter 120 volume)	473
$NaNO_3$ (1 g/liter) + H_2O_2 (1 ml/liter 120 volume)	19

The chlorine dioxide is the active species for bleaching. The conditions of bleaching are such that the rate of formation of chlorine dioxide matches its rate of consumption by the impurities in the fibers. The maximum evolution of chlorine dioxide takes place at pH 2.5–3, and the control of pH and temperature decides the efficiency of the process. According to Hefti,[42] sodium chlorite decomposes to give active oxygen

$$ClO_2^- \rightarrow Cl^- + 2O$$

which is the effective bleaching agent. Sodium-chlorite bleaching is also carried out by a batch process, a pad-roll method, or a continuous process.

Batch Process. In an enclosed jigger, the bath is set with

1–5 g/liter	Sodium chlorite
1–2 g/liter	Sodium nitrate
1 g/liter	Foaming agent
2 g/liter	Formic acid (85%)
	(pH 3)

The blend fabric is put in the bath at 60°C and the temperature is raised to boiling in 20 min. After bleaching for 60–90 min, the goods are washed with hot water. An aftertreatment with anticolor 10–20 g/liter (sodium bisulfite) at 70°C for 10 min removes the last trace of active chlorine. Bleaching on a jigger gives very low productivity.

Pad-Roll Process. The blend fabric is padded with

10–30 g/liter	Sodium chlorite
2–3 g/liter	Sodium nitrate
2–3 g/liter	Wetting agent
3–5 g/liter	Sodium dihydrogen phosphate
0–5 g/liter	Formic acid (85%)

passed through a steam chest, and is then immediately rolled up in a hot-batching chamber.[41] After 1–2 h at 90–95°C in the batching chamber, the fabric is removed and washed.

Continuous Process.   The fabric is padded as in the pad-roll process. It is steamed in either a *J*-box or a reaction chamber in an open width for 30–90 min followed by washing on an open soaper. The materials of construction of the steaming unit have to withstand acidic fumes.

Full-Bleach Process.   For the best whiteness on PET/CO blends, it is essential to give a peroxide bleach after the sodium chlorite treatment. This process is used for producing whites and dyeing pastel shades. However, fluorescent whitening agents applied by the pad-bake technique are preferred to hazardous sodium chlorite for the same effect.[38]

5.2   POLYESTER–WOOL (PET/WO) BLENDS

The following pretreatments are given for polyester–wool blends:[42–48] (*a*) removal of heavy oil stains, (*b*) preset, (*c*) scour, (*d*) heat set, (*e*) brush and crop, and (*f*) singe.

Removal of Heavy Oil Stains.   Oil stains on PET/WO blend fabric are removed before the fabrics are given any heat treatment. The stains are removed by a spotting agent applied locally, followed immediately by a vigorous local hand scour before the solvent evaporates. Cold organic solvents such as trichloroethylene, perchloroethylene, trichlroethane, or white spirit are used as spotting agents.

Presetting.   PET/WO blend fabrics may be preset before scouring and drying to reduce distortion, cockling, rope marks, and shrinkage. Dry blowing or crabbings are two methods of presetting. In dry blowing, the fabric is wound on a perforated rotating roller and steamed for 10 min at a pressure of 1–3 kg/cm². Alternatively, a two-bowl crabbing treatment can be given in which the water in the first bowl is at 80–90°C and the water in the second bowl is at boiling. A period of 3–5 min in each bowl followed by a similar period of steaming, if necessary, gives satisfactory results.

Scouring.   PET/WO blend fabrics are scoured either in rope form or in an open width. A fabric scoured in rope form, particularly, on high-speed scouring machines, gives maximum softness of handle. The scouring is carried out with a solution of sodium carbonate (2–2.5%) and soap (3–4%) at 35–40°C for 10 min. This is followed by washing and a second treatment with soap (3–4%) at 30°C for 40 min. The goods are then thoroughly washed. Alternatively, the goods are scoured with sodium carbonate (2–2.5%) and nonionic detergent (0.2–0.5%) at 40°C for 10 min, washed, the above treatment is repeated, washed, treated with nonionic detergent (0.5%) at 40°C for 15 min, and washed.

Heat Setting. PET/WO blend fabric is heat set to impart dimensional stability, to reduce the cockling during pleating of skirting fabric or ironing of lightweight fabrics, to improve pilling resistance and drape, and to increase firmness of handle. PET/WO blend fabric is heat set at 170°C within 25–30 sec, allowing 1–5% relaxation shrinkage in both warp and weft directions of the fabric. Setting is carried out on a pin stenter fitted with an overfeed attachment and tapering chains. After heat setting, the fabric is brushed and cropped to cut the protruding fiber ends and to raise the fiber ends for removal by singeing.

Singeing.[49] This operation minimizes pilling. Both the face and the back of the fabric are singed on a gas-singeing machine to achieve maximum efficiency. The gas flame is directed at the cloth at an angle, so that globules of molten polymer are not formed in the body of the fabric where they are difficult to remove in subsequent treatments. This gives minimum harshening of the fabric. The singed fabric is given a light scour to remove degraded fiber and odor. After drying, the fabric is recropped to remove globules of melted polymer and to clear any fiber on the surface.

5.3 POLYESTER FILAMENT FABRIC

Fabric made from 100% PET flat-filament or textured yarns contains spin-finishes, coning oils, tinting colors, and adventitious dirt as the major impurities. These impurities are removed before dyeing.[50] The nonionic dyes exhibit solubility in coning oils; if a trace of coning oil remains in the dyed fabric, bleeding problems occur. Thus, the dye migrates to the surface, dissolves in coning oil, and bleeds easily during laundering.

Scouring. 100% PET fabric is scoured with a solution of sodium carbonate (1–5 g/liter) and a detergent (1–5 g/liter) at 60–70°C for 30–90 min, either in a winch, a jet dyeing machine, or on a jigger. The latter is used if the fabric is made of flat-filament yarns. This mild scouring easily removes all the impurities on the fabric.[51,52]

Heat Setting. As was described for PET/CO fabrics, flat-filament yarn fabric is heat set at 200–210°C for 30–60 sec on a pin stenter, ovens with rollers, or contact-cylinder machines (see Section 5.1.5). For textured yarn fabrics, setting is carried out at 160–180°C since there is a risk of losing the crimp of the yarn above 180°C.[53–55] The fabric is singed to minimize pilling (see Section 5.1.3).

Bleaching.   Bleaching of 100% PET fabric is rarely required except for full-white fabrics. The fabric is boiled for 60–90 min in a solution containing

1–5 g/liter	Sodium chlorite
1–2 g/liter	Sodium nitrate
1–2 g/liter	Formic acid (85%) (pH 4.5)

The goods are washed in hot and cold water and then dried.

5.4   POLYAMIDE FABRIC

Nylon fabric has spin-finishes, coning oils, tinting colors, and dirt that is easily removed by scouring with sodium carbonate and a nonionic detergent as described for 100% PET fabric (see Section 5.3). Anionic detergents have an affinity for nylon and thus are avoided,

Heat Setting.   Nylon fabric is heat set for 30–45 sec at 150–175°C if they are made from textured yarns, or at 190–200°C if they are made from flat yarns on pin stenters, roller-type ovens, or cylinder setting machines. Nylon stockings are heat set on aluminum boards by pressure steam at 120–125°C for 2 min. Similarly, nylon yarns in the form of cheeses are heat set by autoclave-steaming at 120–125°C for 30 min.

The density and crystallinity of nylon fiber increases markedly in steam-setting. The increase is smaller in dry-heat setting. The tensile strength drops slightly on steam setting while it increases slightly on dry-heat setting. However, there is no chemical degradation of the polymer and the end amino and carboxyl groups content remains the same on heat setting. There is an improvement in dimensional stability, shrinkage being very low on setting.[56] Dyeability of Nylon 6 is influenced by heat setting.[57-62] The rate of dyeing decreases with an increase in the dry-air setting temperature until it reaches a minimum at about 200°C (Figure 5.6).[62] The rate of dyeing of fabric set with saturated steam, on the other hand, increases with the temperature of steaming (Figure 5.6). Yarn set in steam at 2 kg/cm^2 pressure for 5 min exhibits a higher rate and extent of dye uptake than untreated yarn, while yarn set in air at 200°C for 5 min exhibits both a rate and extent of dye uptake lower than untreated yarn. (Figure 5.7). This effect varies from disperse dyes to anionic dyes. The extent of the increase is more pronounced at lower steam pressures for disperse dyes and at higher pressure for acid and metal-complex dyes.[62] The rate of dyeing of nylon 66 with metal-complex dyes, which are sensitive to variations in physical structure, first decreases with increasing setting temperature to a minimum of 230°C and then rises at a higher temperature. If the textured material is steamed (at 2 kg/cm^2 pressure), the dyeing rate decreases steadily from an initial high value as the

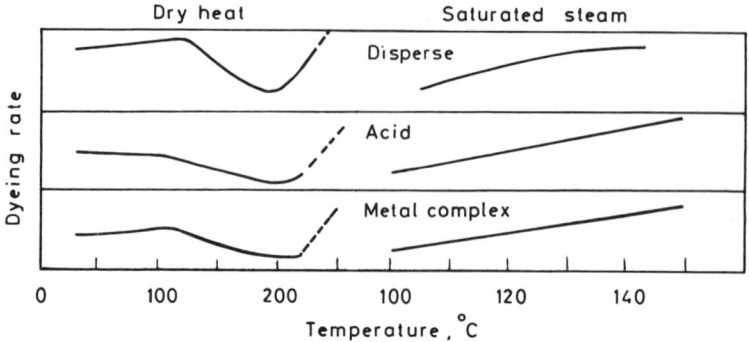

FIGURE 5.6. Effect of dry heat and saturated steam on rate of dyeing of PA 6.[62]

false-twist heater temperature increases (Figure 5.8). Setting in superheated (unsaturated) steam is similar to dry-heat setting, since superheated steam does not condense on the fiber and does not swell the fiber.

Bleaching. Nylon has a good natural whiteness and is not bleached except when a slight yellowing caused by the heat setting has to be removed in order to get full-white material.[63] Hypochlorite and hydrogen peroxide are not suitable for bleaching nylon since they damage the fiber. Sodium chlorite is the best bleaching agent for good whiteness without damaging the nylon material. The bleaching is carried out using

1–2.5 g/liter	Sodium chlorite
1–2 g/liter	Sodium nitrate
2 g/liter	Formic acid (85%), pH 4

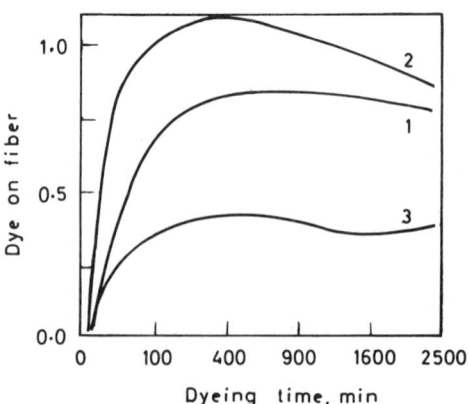

FIGURE 5.7. Effect of setting on the rate of dyeing of nylon with Ergalan Red 3G.[62] 1: Untreated (control); 2: Set in saturated steam (2 kg/cm² for 5 min); 3: Set in air (200°C for 5 min).

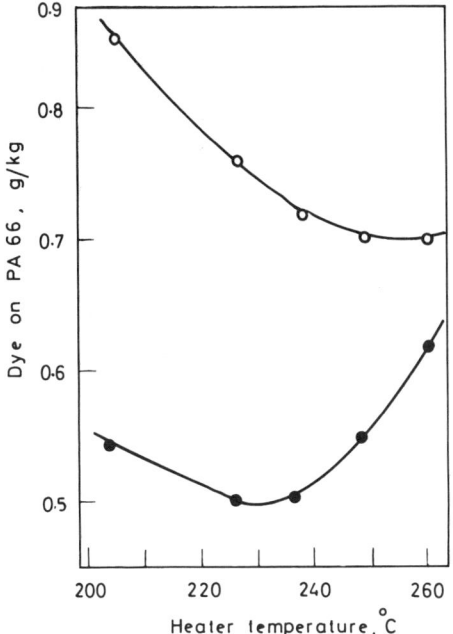

FIGURE 5.8. Effect of setting with dry heat and steam on the rate of dyeing of PA 66 textured at different temperatures.[62] Dye: Metal-complex type; ○: Heat set in steam (2 kg/cm²); ●: Controlled unset.

at 70°C for 60–90 min. The material is washed with hot and cold water and is treated with antichlor (sodium bisulphite, 1–2 g/liter) at 50°C for 30 min. A final wash with cold water completes the process.

Peracetic acid is another equally effective bleaching agent for nylon. Hydrogen peroxide and acetic acid in the presence of a strong mineral acid such as sulfuric acid gives peracetic acid.

$$CH_3COOH + H_2O_2 \xrightarrow{H^+} CH_3CO \cdot O \cdot OH + H_2O$$

It is marketed as a clear, colorless solution of 35–40% strength. Mixing of acetic anhydride and hydrogen peroxide at room temperature in the presence of a catalyst, for example, caustic soda or ethylene diamine tetraacetic acid (EDTA), gives peracetic acid.[34] The latter method is simple and safe and a solution of 3% strength is speedily made for daily use. Bleaching at pH 6–7 gives good white by removing the discoloration caused by heat setting and any other colored constituent without any degradation of nylon fiber.[34] The bath is set with

3 g/liter	Peracetic acid (35–40%)
0.25 g/liter	Sodium pyrophosphate
1 g/liter	Wetting agent

The goods are entered cold and the bath is heated to 80–85°C in 30 min. After 30 min at 80°C, the goods are washed. The bleaching equipment is made of ceramic, wood, or stainless steel; metallic impurities from copper, brass, or iron should be avoided.

5.5   ACRYLIC–COTTON BLENDS

Hot-alkali solutions are not used for treating materials containing acrylic fiber because the alkali attacks the nitrile groups of polyacrylonitrile and converts them into amide ($-CONH_2$) and carboxyl groups ($-COOH$).[64] The fiber becomes yellow by these and other concomitant chemical changes.

Desizing.   Yarn sized with starch is desized with enzymes as was described for PET/CO blends. Yarn sized with polyvinyl alcohol or acrylic copolymer is rinsed with hot water and then with cold water. The goods are scoured with a nonionic detergent solution (1%) at 80°C for 1–2 h. However, this treatment is not sufficient to get good absorbency for the cotton component of the blend.

Bleaching.   Sodium hypochlorite works under highly alkaline conditions and is never used for materials containing acrylic fiber. Hydrogen peroxide can be used for bleaching the cotton component, if proper control of the pH is maintained and sodium hydroxide is not used. The bleaching is carried out at 90°C for 1 h using

7.5 g/liter	Hydrogen peroxide (35%)
3.5 g/liter	Sodium silicate

The best results are obtained with sodium chlorite at 80–90°C for 60–90 min using a bath containing

1.5 g/liter	Sodium chlorite (80%)
2 g/liter	Formic acid
0.4 g/liter	Sodium pyrophosphate
1 g/liter	Detergent

After bleaching and washing, an antichlor treatment with sodium bisulfite (1.5 g/liter) is given at 60°C for 20 min. Bleaching may also be carried out using peracetic acid as was described earlier for nylon fiber (see Section 5.4).

5.6   ACRYLIC–WOOL BLENDS

The blend is scoured in a nonionic detergent (1–2 g/liter) at 80°C for 1–2 h and then washed with water.[65,66] A treatment with the fulling soap under conditions that are less severe than those used for 100% wool fabric is given as a fulling process. The blend is carbonized with sulfuric acid by drying at 80–85°C and baking at 100–110°C. The wool component is bleached with

2–5 g/liter	Hydrogen peroxide
2–3 g/liter	Sodium silicate

at 50–60°C for 1 h. The yellowing of acrylic fiber can be removed with an aftertreatment of formic acid and a detergent (see Section 5.2).

5.7 ACRYLIC FABRIC

Acrylic fiber is scoured in a bath of nonionic detergent (1–2 g/liter) at 80–85°C for 30–60 min to remove spin-finishes, tinting colors, and dirt, if any.[67] The goods are then washed. Bleaching is not necessary for acrylic fiber since it is sufficiently white. For extremely white goods, bleaching is carried out using

1–2 g/liter	Sodium chlorite
2–4 g/liter	Sodium nitrate
1–2 g/liter	Formic acid or oxalic acid (to get pH 4)

at 80–90°C for 30 min. The goods are washed, antichlored, and washed again. Alternatively, the following bath composition is used:

1–1.5 g/liter	Sodium chlorite
0.4 g/liter	Phosphoric acid (pH 2.5)
1–2 g/liter	Nonionic detergent.

The addition of a small amount of oxalic acid to the bleaching bath gives permanent whiteness on the fabric which does not turn yellow on exposure to light or heat.

5.8 POLYPROPYLENE FABRIC

Polypropylene fabric with a low softening point is singed with a low-temperature flame at a speed of 140–150 m/min so that the material exhibits minimum pilling tendency without any damage to the fiber.[68] The singeing is done twice on both the sides of the fabric. The size on the fabric is removed by either an enzyme or an nonionic detergent desizing process (see Section 5.1.1). The goods are scoured with

2–3 g/liter	Sodium carbonate
1–2 g/liter	Nonionic/anionic detergent
1–2 g/liter	Sequestering agent

at 80°C for 30–45 min to remove dirt, oil, and finishes. Bleaching is not necessary unless extra-white material is required. Sodium-chlorite bleaching (see Section 5.1.6) at 80–90°C and a treatment with an optical brightener give full-white material. The goods are dried on a stenter at a temperature below 125°C and heat set at 125–135°C for 1–3 min to minimize shrinkage and distortion of the fabric during subsequent dyeing.

5.9 MISCELLANEOUS SYNTHETIC FIBERS

Polyvinyl Alcohol Fiber.[69,70]   The PVA fabric is singed with a low-temperature broad flame at a very high speed. It is desized with enzymes or in a nonionic detergent solution at 60–70°C. Scouring is also carried out with a nonionic detergent solution (1–3 g/liter) at 80°C for 30 min. In order to safeguard the fiber against shrinkage, harsh handle, and yellowing, the wet treatments are carried out in a relaxed state below 95°C, even though the wet-softening temperature of PVA is 110–125°C. The fiber has a good natural whiteness. The goods can be bleached with sodium hypochlorite or sodium chlorite. The goods are soaked with a solution of hypochlorite (1–2 g/liter available chlorine) at 30–50°C for 15–30 min. After washing, the antichlor treatment is given with sodium thiosulfate. Sodium-chlorite bleaching is carried out at 60–70°C for 30 min. For wool blends, bleaching is done with hydrogen peroxide. The use of a fluorescent whitening agent gives good whiteness. The drying of PVA fabric is a critical step in deciding the handle of the fabric. The wet-softening temperature of PVA is low and the fabric easily stiffens on drying at high temperature on a cylinder range. The rate of removal of moisture is critical in the initial stages of drying, but once the moisture is substantially removed, high temperature has no adverse effect since dry-softening temperature is 200–220°C. Stiffness can be eliminated by a mechanical treatment or completely avoided by staggering the temperature of the drying cylinders or by wrapping the initial drying cans. PVA has thermoplastic behavior intermediate between cotton and cellulose acetate and does not need heat setting prior to dyeing.

Polyvinyl Chloride Fiber.   Polyvinyl chloride fiber has a very low-softening point. Above 70°C, these fibers retract in length, and in boiling water, the shrinkage is as high as 50%. All the pretreatments of polyvinyl chloride fiber are therefore carried out below 70°C. Since very few impurities are present on the fiber, only a mild scouring treatment with a nonionic detergent solution (1–2 g/liter) at 50–60°C is given. Polyvinyl chloride fiber does not require bleaching. For improving whiteness, treatment with an optical brightening agent may be given.

Polyurathane Fiber. The process of relaxation of polyurathane fibers decides the yield, width, and stretch properties. The relaxation is done by (*a*) solvent scour, (*b*) aqueous scour in a winch, (*c*) passing over a steaming table, or (*d*) stentering at 140°C using a presteaming and steam injection. The scouring is carried out with a solution containing

1–2 g/liter	Trisodium phosphate or tetrasodium pyrophosphate
1–2 g/liter	Nonionic detergent

at 60–70°C for 30 min. The goods are then washed. Incorporation of perchloroethylene (1–2 g/liter) in the scouring liquor helps to remove oil stains. Polyurathane fiber degrades under highly alkaline conditions and the pH of the scouring bath is, therefore, maintained below 11.5. Bleaching, if at all necessary, is carried out with

5–10 g/liter	Hydrogen peroxide (35%)
10–15 g/liter	Sodium silicate
1–2 g/liter	Nonionic detergent

at 80°C for 30 min. The goods are then washed. The whiteness is improved by treatment with an optical brightener of disperse or anionic type. The wet goods are dried on a stenter below 140°C and heat set at 180–185°C for 20–25 sec. Temperatures above 190°C cause thermal degradation of polyurathane fiber and should be avoided.

5.10 CELLULOSE ACETATE FIBERS

Cellulose acetate (CA) and triacetate (CTA) fibers are saponified with alkali. Severe alkaline conditions are therefore avoided during the wet-processing of these fibers. Desizing is carried out at 50–60°C, either with enzymes or a nonionic detergent. The goods are scoured in a bath containing

1–2 g/liter	Anionic/nonionic detergent
1–2 g/liter	Trisodium phosphate

at 60°C for 30 min and washed with cold water. Heat setting in hot air produces stiffness in CA fabrics due to slight filament cohesion.[71,72]

This can be avoided by a controlled mild saponification treatment before heat setting, so that a very thin film of cellulose is generated on the surface of each filament that blocks the filament cohesion. The total acetyl value of CA drops from 62% to 59%. The saponification treatment or 'S' finish is carried out with caustic soda (3 g/liter) at 80–90°C for 2 h.

Unlike secondary cellulose acetate, CTA resembles the synthetic fibers in its capacity for being heat set. As a result of heat setting, the resistance to dimensional changes (particularly, when wet) is enhanced. On setting, a CTA fabric becomes resistant to creasing during laundering. The fabric can be set in creased conditions as in pleated goods and the creases are retained during wet treatments. The setting of CTA is carried out using either a hydrosetting technique or steam or dry-heat setting process.[73,74] The hydro-setting is carried out using hot water in a pressure-dyeing machine. The fabric is set in steam at 1–1.2 kg/cm^2 pressure for 30 min. This steam setting is widely used for CTA garments. Dry-heat setting is carried out on a stenter at 190–220°C for 30–90 sec. Dry-heat setting, however, decreases the rate of dyeing.[75–77]

Acetate fibers do not usually need any bleaching. However, bleaching is sometimes carried out for full-white goods and to remove slight yellowing produced during heat setting. Cellulose acetate materials can be bleached with sodium hypochlorite, hydrogen peroxide, sodium chlorite, or peracetic acid.[78] The goods are treated with hypochlorite solution (0.5 g/liter available chlorine) at pH < 10 at room temperature for 45–90 min. This is followed by washing and antichlor treatment with sodium bisulfite (3 g/liter) at 70°C for 30 min. The goods are finally washed with cold water. Alternatively, the material is bleached with hydrogen peroxide (1.5–3 g/liter, 35%) and sodium silicate (2 g/liter) at pH 9.5 at 60–70°C for 1 h. Sodium chlorite bleaching is carried out using sodium chlorite (1–2 g/liter) and acetic acid (1–2 g/liter) at 80°C for 30 min as usual (see Section 5.1.6). A bleaching process using peracetic acid involves treatment in a bath consisting of

2–3 g/liter	Peracetic acid
5 g/liter	Sodium hexametaphosphate
1 g/liter	Wetting agent

at pH 5–6. The goods are bleached at 60–70°C for 1 h. The goods are then washed. The whiteness can be further improved by a treatment with a non-ionic optical brightener in the form of an aqueous dispersion.

REFERENCES

1. Turner, G. R., *TCC,* **13** (1981), 246.
2. Pragar, W., *ADR,* **67**(7) (1978), 24.
3. Davis, J. W., *JSDC,* **89** (1973), 77.

4. Seidel, M., *MTB,* **54** (1973), 533.

5. Casserly, J. J., *ADR,* **56** (1967), 477.

6. Kulkarni, G. G., and Trivedi, S. S., *Wet Processing of Polyester/Cotton,* ATIRA, Ahmedabad, India (1967), p. 30.

7. Freytag, R., *Bull. Inst. Text. France,* **17** (1963), 541.

8. Agster, A., Wirth, G., and Perchke, W., *MTB,* **47** (1966), 1279.

9. Gardner, H. S., and Kalinowski, S. E., *JSDC,* **81** (1965), 41.

10. Ehret, H., *MTB,* **49** (1968), 573.

11. Namboori, C. G. G., *TCC,* **1** (1969), 50.

12. Gorrafa, A. A. M., *TCC,* **12** (1980), 83.

13. Forrester, R. C., and Caldwell, M. A., *ADR,* **58**(2) (1969), 13.

14. Easten, B. K., and Gallagher, G. T., *ADR,* **53** (1964), 985.

15. Poser, S., *ADR,* **57** (1968), 668.

16. Duckworth, C., Horsley, J. V., and Thawaites, J. J., *JSDC,* **88** (1972), 281.

17. Elliott, E. J., and Silva, R. E., *TCC,* **13** (1981), 257.

18. Moore, H., *ADR,* **44** (1955), 153.

19. Anderson, J. H., and Swan, L. D., *ADR,* **44** (1955), 350.

20. Ritter, E., *Dyer,* **136** (1966), 263.

21. Marvin, D. N., *JSDC,* **70** (1954), 16.

22. Taylor, M. E., *TCC,* **2** (1970), 149.

23. Ulrich, H., *MTB,* **54** (1973), 1206.

24. Dumbleton, J. H., Bell, J. P., and Murayama, T., *JAPS,* **12** (1968), 2491.

25. Andriessen, J., and Van Soest, J., *Textilveredlung,* **3** (1968), 618.

26. Huisman, R., and Heuvel, H. M., *JAPS,* **22** (1978), 943.

27. Hallada, D. P., Keen, M. C., and Thomas, R. J., *ADR,* **50** (1961), 445.

28. Salvin, V. S., *ADR,* **54** (1965), 272.

29. Meunier, P. L., Thomas, R. J., and Hoscheit, J. S., *ADR,* **49** (1960), 153.

30. Merian, E., Carbonell, J., Lerch, U., and Sanahuja, V., *JSDC,* **79** (1963), 505.

31. Iannarone, J. J., Speakmens, P. L., Larson, O. S., Hurt, R. C., and Hinton, E. H., *ADR,* **46** (1957), 674.

32. Statton, K., *JAPS, Polym. Symposia,* **32** (1971), 219.

33. Donze, J. J., Bouchet, E., Freytag, R., Chabert, J., Schneider, R., and Viallier, P., *JSDC,* **91** (1975), 336.

34. Chesner, L., and Woodford, G. C., *JSDC,* **74** (1958), 531.

35. Dickson, K., and Heathcote, D., *JSDC,* **88** (1972), 137.

36. Rowe, M. H., *ADR,* **59**(5) (1970), 22; *TCC,* **3** (1971), 170.

37. Evans, B. A., *TCC,* **13** (1981), 254.

38. Butcher, J. V., *Rev. Prog. Color.,* **4** (1973), 90.

39. Mehta, H. U. and Mashruwala, M. N., *Colourage,* **29**(6) (1982) 9.

40. Hetherington, P. W., *JSDC,* **84** (1968), 359.

41. Skelly, J. K., *JSDC,* **76** (1960), 469.

42. Hefti, H., *Text. Rund.,* **11** (1956), 82.

43. Haden, I. E., *JTI,* **53** (1962), P820.

44. Houser, K. D., *ADR,* **52** (1964), 892.

45. Seddon, H. H., *Dyer,* **132** (1964), 894.

46. Blankenburg, G., *JTI,* **56** (1965), T145.

47. Jansen, H., *MTB,* **61** (1980), 357.

48. Sule, A. D., in *Blended Textiles* Gulrajani, M. L., Ed., The Textile Association, India (1981), p. 321.

49. Elsner, O., and Weintraub, J., *TCC,* **2** (1970), 358.

50. Haile, W. A., Willingham, W. L., Caldwell, M. A., and Forrester, R. C., *ADR,* **66**(3) (1977), 48.

51. Turner, G. R., Pratt, H. T., and Hug, G. T., *TCC,* **2** (1970), 235.

52. Klein, H., *Chemiefasern,* **20**(6) (1970), 472.

53. Pratt, H. T., *ADR,* **56** (1967), 671; **61**(4) (1972), 23; *TCC,* **2** (1970), 26; *Can. Text. J.,* **87**(4) (1970), 79.

54. Forrester, R. C., Schrum, F. F., and Caldwell, M. A., *ADR,* **60**(3) (1971), 21.

55. Dayvault, J. A., *TCC,* **4** (1972), 156.

56. Tsuruta, M., and Koshimo, A., *JAPS,* **9** (1965), 1.

57. Peters, H. W., and White, T. R., *JSDC,* **77** (1961), 601.

58. Davis, G. T., and Taylor, H. S., *TRJ,* **35** (1965), 405.

59. Warwicker, J. O., *JAPS,* **9** (1975), 1147.

60. Holfeld, W. T., and Shepard, M. S., *Can. Text. J.,* **94**(5) (1977), 72.

61. Holfeld, W. T., and Shepard, M. S., *TCC,* **10** (1978), 26.

62. Rush, J. L., and Miller, J. C., *ADR,* **58**(4) (1969), 37.

63. Shaw, S., and Wilson, W. S., *JSDC,* **71** (1955), 857.

64. Mehta, R. D., *TRJ,* **51** (1981), 57.

65. Rawicz, F. M., *ADR,* **50** (1961), 415.

66. Schuster, R. J., *ADR,* **50** (1961), 442.

67. Oyabu, N., *Japan Text. News,* **269** (1977), 33.

68. Metropolitan Section, *ADR,* **54** (1965), 107.

69. Nomura, S., and Tanabe, K., *JSDC,* **74** (1958), 359.

70. Hindle, W. H., *ADR,* **49** (1960), 463.

71. Mann, R. J., *JSDC,* **76** (1960), 665.

72. Murry, A., and Byrne, D. M., *Text. Mfr.,* **94** (1968), 342.

73. Meller, A., and Olpin, H. C., *JSDC,* **71** (1955), 817.

74. Ratcliffs, J. D., *JSDC,* **86** (1970), 482.

75. Stoll, R. G., *TRJ,* **25** (1955), 650.

76. Fester, W., and Liu, S. T., *JTI,* **57** (April 1966), A106.

77. Fester, W., and Liu, S. T., *Text. Prax.,* **21** (June 1966), 440.

78. Fortress, F., *ADR,* **44** (1955), 524.

6 | MECHANICAL AND ECONOMIC ASPECTS OF WET PROCESSING

Synthetic fibers are colored with suitable dyes usually in an aqueous medium with various auxiliaries and chemicals. The material can be dyed at any stage of its manufacture; colored fibers and filaments are produced by tow dyeing and the mass-coloration techniques described earlier (see Chapter 4). The material to be dyed may be fibers, yarns, warp sheets, knitted or woven fabrics, or garments. It may consist of one polymer or a mixture of two or three types of polymer. The material is dyed in the form of loose stock, hanks, packages, or as a continuous sheet. Thus, blends, unions, heather mixtures, monocomponent and bicomponent fibers, or differentially dyeable fibers are dyed using dyes that may differ in their application processes and programs. Furthermore, dyeing may be carried out in a batch process, a continuous process, or a combination of the two using one or more machines. The machines and dyeing techniques are described in later chapters (e.g., Chapter 20), although the machines will now be examined from the viewpoint of energy savings.

When machinery for dyeing is designed, two opposing factors have to be considered. The goal is the attainment of level dyeing by adequate movement of properly directed liquor without fabric distortion or other kinds of physical damage to the goods. However, this has to be achieved in the least possible time with a minimum liquor-to-goods ratio and the largest possible economy in steam, energy, dyes, and chemicals. At the same time, a volume of liquor sufficient to maintain these dyes and chemicals in solution must be taken into account. The layout of the machine must make the necessary preliminary handling of the material as simple as possible and the dyeing operation easy, facilitating the subsequent processes of liquor extraction, drying, and finishing.[1]

155

The earlier machines for dyeing synthetic fiber materials were developed from the machines used for natural and animal fibers. Soon, new and entirely novel machines were developed to cope with the newer techniques of dye application and the difficulties in dyeing polyester fiber and knitted goods. Machines for dyeing above 100°C, covered jiggers for carrier dyeing, high-pressure–high-temperature beam-package dyeing machines, vacuum impregnation padding mangles, thermosoling units, infrared predryers, high-pressure batch and continuous steamers, jet dyeing machines, and so on may be cited as examples of such machines. Similarly, a variety of machines have been developed for producing mass-colored yarn and tow dyeing. The dyeing machines work on the principle of liquor moving, material moving, and both liquor and material moving. The traditional processes of piece dyeing on the winch, in the jigger, or in padding sequences have tended to rely mainly on the movement of the fabric through the dye-liquor, but this movement, particularly, of synthetic fiber materials, may cause surface damage, stretching, and other kinds of distortion. The machinery manufacturers were encouraged to develop systems in which the fabric was kept stationary and the liquor circulated. For this purpose, the fiber material, loose stock, yarn, or fabric is prepared as a package through which the liquor flows. The liquor-to-material ratio varies from machine to machine and is as low as a fraction in foam dyeing or padding process to as high as fifty in hank dyeing machines for acrylic high-bulk yarns. Typical dyeing machines are described below.

6.1 LOOSE-STOCK DYEING MACHINE

Loose-stock dyeing or loose-fiber dyeing is used to produce multicolored yarns and fabrics with checks, stripes, and other color effects. In blends, one component may be stock dyed while the other may be dyed later to get various tones and color effects. PET staples are stock dyed before they are blended with cotton or wool and spun into blended yarn. The delicate natural or animal fiber thus will not be exposed to the high temperatures and pressures required for PET dyeing. The levelness of the dyed shade is not a very critical factor since after dyeing the fibers get thoroughly mixed when they are spun into yarn. The unevenness in shade is substantially concealed in the final yarn and fabric.

Loose-fiber dyeing machines are based on the principle of liquor moving through stationary material. Usually, the packing of loose fiber is not very uniform. A pump circulates the liquor through the package from the bottom to the top. The shape of the fiber-holder pan is slightly conical so that when the liquor enters from the bottom, the fiber material is forced up and pressed against the conical sides, thus assisting in getting uniform flow through the material. When the flow is reversed, the material is pressed downwards. Alternatively, the fiber is packed in a vessel with a perforated false-bottom plate, perforated cylindrical tube in the center, and a perforated lid. The liquor is circulated through the central tube, the material, and out of the perforated vessel or in the reverse direction with the help of a pump. The liquor ratio in such a machine is about 15 : 1. A typical conical-pan machine has been described by Jorder[2] and in the book edited by Bird.[3] The dyeing is carried out at 100°C or up to 130°C. After dyeing, the goods are washed and the vessel with the material is lifted out of the machine in which it was fitted for liquor circulation. It is possible to centrifuge the vessel with the fiber to remove the imbibed water.

Bale Dyeing Machine. Densely packed bales weighing 180–280 kg can be dyed without opening the bales.[4,5] The plastic bale wrappers are slit in places and up to four bales are loaded into a special carrier without removing the retaining bands. The dye-liquor is forced through the bales to the flow collector. Washing and hydroextraction are also carried out in the bale form.

6.2 YARN DYEING MACHINES

Yarns are dyed either in hank form or package form. Acrylic yarns are usually dyed in hank form. Textured yarns are dyed in a muff or a package form. Dyeing in hank form produces a finished yarn with good bulk and appearance.[6–8] This method is preferred for dyeing small lots efficiently. The drawbacks of the hank dyeing method are (*a*) high cost since larger amounts of dyes and chemicals are required because of the high liquor-to-goods ratio, (*b*) difficulty in matching shades from lot to lot and problems of unevenness because of temperature variations in the machine, (*c*) it is a labor-intensive costly process that involves hank reeling and after-dyeing rewinding onto cones, (*d*) entanglement of yarn during hank dyeing leads to the production of relatively higher waste, and (*e*) the textured filament yarns are almost impossible to handle in hank form. Multifilament (flat) yarns cannot be dyed by the hank (or package) dyeing process because of filament entanglement and breaks.

Hank Dyeing Machines. The machine essentially consists of a rectangular box into which the yarn carrier dips. The yarn is suspended from sticks fitted only at the top of the carrier (single-stick system) or attached to a further set of sticks fitted at the bottom of the carrier (double-stick system). The distance between the sticks is a little less than half the real length of the yarn. The dye-liquor is circulated by propeller and may flow either upwards or downwards through the yarn, depending on the direction of rotation of the propeller.

In a modified hank-dyeing machine, the hanks are loaded on a perforated arm. The dye-liquor is forced through the holes in the arm while the hanks are slowly rotated or rolled over the arms during dyeing. This allows for uniform and equal application and absorption of the dye by the yarn. Alternatively, series of perforated bars spray hot liquor under pressure on to the yarn. The arms holding the hank can be rotated and the entire hank-holding unit can be raised or lowered to control the bath dwell time and the amount of dye-liquor applied. The dyeing is rapid with a large savings (70%) in water and steam when the jet-spray system is used.

Because of the limitations of hank dyeing, the synthetic yarns are dyed in package form. Circular machines withstand high pressure better than rectangular ones, so that it is easier to construct a package dyeing machine capable of dyeing PET yarn at 130°C than a rectangular hank dyeing machine. Package dyeing machines can be easily fitted with interchangeable carriers for tops, loose stock, yarn packages and so on, thus, increasing the versatility of the dyehouse.

The yarn is wound on carriers,[9] such as cones, springs, perforated beams and so on.[10,11] The winding of the yarn on the carriers is strictly controlled for tension, humidity, package density, the geometry of package, angle of wind of cross-wound package and so on,[12] because this decides the flow of liquor through the package and, in turn, the evenness of the shade. In the package dyeing process, the shade of the dyed goods is likely to be uneven from the surface to center or in a random fashion if channeling of the liquor takes place. At times, disperse dyes filter off on dyed yarn apart from developing resistance to the liquor flow and high-pressure drop.[13] Hand-knitting yarns tend to flatten when dyed on conventionally wound packages. Perforated, extensible-spiral spring-tubes are used to wind such yarns under low tension. Sometimes the yarn is prepared in the form of muffs with no carrier and these muffs are packed in a basket that is introduced into the dyeing vessel. Pumps and propellers of sophisticated design are used for the liquor circulation. Cegarra[14] described a liquor-propulsion system consisting of a piston working vertically inside the pressure dyeing vessel. It allows 20–23 liquor flow reversals per min with a substantial reduction in dyeing time because of an increased rate of flow.

The diffusional boundary layer in package dyeing is said to be of importance.[15] However, Brooks and Nordon[16] found that the thickness of the diffusional boundary layer is too small to control the rate of dye uptake under conditions similar to those used in package dyeing (see Chapter 3). The role of the reversal of liquor flow on the degree of levelness is still a controversial issue. Level dyeing is optimized by ensuring uniform distribution from the beginning of the cycle and conditions favorable for rapid migration prior to fixation of the dye. Controlled absorption is facilitated by considering the rate of uptake in terms of the frequency of circulation of the dyebath rather than as merely a function of time.[17] If flow direction is reversed more frequently than once per dyebath circulation, the distribution of the dye is adversely affected. Dyers prefer to have a high rate of flow with one-way circulation only rather than reversal of direction of flow. Beckman and Hoffmann stressed the importance of local liquor flow rather than the number of times the liquor is circulated.[18] Flow reversal every four to eight circulations gives superior levelness during the exhaustion phase but is of little benefit during the migration phase.[19] On the other hand, maximum levelness is obtained by reversing the flow after each complete pass of the dye-liquor through the package. Optimum levelness is governed more by frequent renewal.[9] In fact, in new machine designs, rapid dyeing techniques involve high liquor reversal rate and high rate of heating.[20]

Package Dyeing Machines.   The development of package dyeing machines took place over a period of 100 years.[9] However, in the last decade or two, advances in theory led to new machinery designs and a new approach to dyeing techniques. It is now possible to obtain good yarn quality by package dyeing. (Formerly, only hank dyeing was considered suitable for obtaining good yarn quality.) The main line of development in package dyeing machines has been the ability to dye at high temperatures and pressures with greatly improved liquor flow [measured in terms of the frequency of the circulation of the total volume of liquor through the package (Figure 6.1)]. This results in improved uniformity of dyeings, in shorter dyeing cycles, and hence, in better yarn properties. The increased frequency of circulation is

FIGURE 6.1.   Photograph of package dyeing machine. (Courtesy of S. A. Serracant Sabadell.)

achieved by a combination of a reduction in the liquor-to-goods ratio and an increase in the capacity of the pump (see Chapter 8).

Apart from highly sophisticated control units, such machines have high-capacity heat-exchangers for rapid heating and cooling. Another feature of a modern package dyeing machine is the streamlining of the liquor path as compared to earlier designs. This avoids the adverse effect of sharp changes in the liquor path during changes in the direction of flow. The capillary flow theory was used to show that soft winding on large-diameter bowls augments the flow of liquor, as do low viscosity, high temperatures, and the use of wetting agents. An important factor in the success of package dyeing of yarn is the method of winding the yarn and the geometry of the package both before and during dyeing. The dyeing vessel can be vertical or horizontal in shape. A separate tank is provided for the addition of dye-liquor. The dye-liquor is circulated through the yarn packages with the help of a pump. The direction of the liquor flow can be altered from inside-out to outside-in depending on what is needed. The liquor-to-material ratio is about 10 : 1. The dye-liquor is heated by passing steam through a closed-coil heat exchanger situated inside the dyeing vessel. The temperature of the dye-liquor can be raised to as high as 140°C. The modern package dyeing machines have an automatic program controller so that the dyeing cycle is automatically regulated and very little manual work is involved in operating the machine.[21-23] A special sampling device is fitted for taking out the sample (for matching purposes) while the machine is running (Figure 6.1).

Rapid Dyeing Machines. Package dyeing in a normal machine is a long drawn out process with low production capacity that increases the total cost of dyeing. Rapid dyeing machines have been developed in which the liquor circulation and rise in temperature are very rapid.[17,24-29] These machines are described in Chapter 8.

6.2.1 Liquor Flow in Packages

The liquor flow through the material to be dyed is a complex process involving permeability of the liquor through the fibers, yarns, and layers of fabric, and depends on the spaces or pores that separate the individual components. The total flow of the liquor depends on intercomponent and intracomponent spaces in the material. These spaces are decided by the construction of yarns and fabrics and the packing of the material in the dyeing machine. In the material-moving machines, the intercomponent spaces may change their size, position, and distribution during movement. The flow that occurs between the components has a great influence on the levelness of dyeing while the intracomponent flow decides the penetration of the dye in the material. For example, in tightly woven material, the uniformity on the fabric surface may be good, but where warp and weft cross each other, there may not be any dyeing at all. Similarly, yarns that have high twist may exhibit uniform surface dyeing without any penetration of the dye inside the yarn. The package of the material decides interyarn flow. A tight package may not allow the liquor to flow through uniformly and may give uneven dyeing. Channeling because of uneven intercomponent flow is a major problem in dyeing packages. The liquor flow through a plain slab of material can be defined by Dacry's Law.[30,31]

$$Q = A \left(\frac{B}{\mu}\right)\left(\frac{\Delta p}{L}\right)$$

in which Q is the rate of flow of liquor normal to the surface of the slab through surface area A and thickness L, p is the pressure drop across the thickness, and μ is the viscosity of the liquor. B is the specific permeability of the material, which depends on the fractional volume of void space and the surface area per unit volume through which the liquor has to flow.

Fabrics with very open weave behave differently than tightly woven fabrics. The flow through open fabrics resembles a mechanism of free flow past the yarns, whereas the flow through more tightly woven fabrics corresponds more nearly to the capillary flow mechanism, particularly for a multilayer fabric.[21]

The maximum values for the mean flow velocity through yarns in cross-wound packages are low and changes in the rate of flow have a large effect on the rate of dyeing. The velocity within the yarn tends to determine the local rate of dye sorption, but the velocity of flow between the yarns will, in addition, influence the distribution and redistribution of dye within the package as a whole. Different textile materials and package constructions— cheeses, beams, bales and so on—present intrinsically different problems.

The dye solution flows rapidly through the interyarn spaces and around the outermost boundary of yarns, but penetration of the yarn itself is much slower. The velocity of the liquor decreases towards the center of the yarn, particularly, if the yarn has low porosity. Thus, penetration of the interyarn spaces is much more swift than penetration of the yarns. The yarns may, therefore, be ring-dyed, quite apart from any question of ring-dyeing of individual filaments in the yarn. Ring-dyed yarns may exhibit poor rubbing fastness since undyed filaments in the core get exposed during rubbing. Thus, openly constructed yarns or packages with a maximum rate of flow of liquor that is compatible with the avoidance of harmful stresses on the textile material give level dyeing with good penetration in the yarns. Uniformity of package construction is needed to avoid variations in flow rates and poor control of the rate of dyeing. Control of temperature, heating rate, conditions of dyebath pH, stability of dispersion and dyes and so on are other factors that are equally, if not more, important for uniform, well-penetrated dyeings. The tension on the yarn increases the compactness and thus decreases the intrayarn flow as shown in Figure 6.2.[3] (Figure 6.2 is obtained from a single yarn dyed under different tensions in a well-stirred dyebath.)

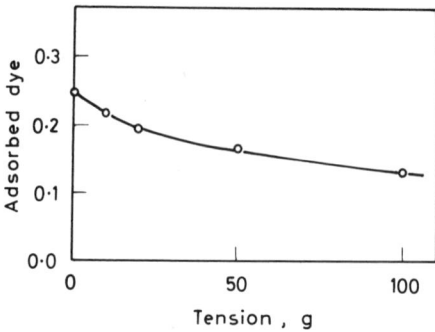

FIGURE 6.2. Effect of yarn tension on rate of interyarn flow and rate of dyeing.[31] Yarn: 100/34 nylon 66; Dye: Solvay Sky Blue B, 0.3 g/liter; pH: 3.1; Dyeing: 10 min at 75°C.

At higher tensions, the penetration of liquor in the yarn decreases to the extent that ring-dyed yarn is obtained.[31] Similar behavior is observed in package dyeing.[32]

6.2.2   Application of Vacuum in Package Dyeing

In package dyeing, a significant proportion of the processing time is required in preparing the substrate and in the removal of trapped air. Use is made of fiber-wetting agents, retarders, and the manipulation of temperatures along with the best possible form of liquor circulation through the goods. Removal of air by subatmospheric pressures is more rapid and efficient than the traditional degassing treatments. This method largely eliminates the need for surface active agents which in themselves can hinder the efficiency of flow.[33]

The package dyeing machine must have valves and closures to withstand both low (less than 10 cm Hg) and normal or high (3–4 kg/cm²) pressures without leakage. It is also necessary to provide an expansion tank that will be of sufficient capacity to allow the dyeing compartment to be completely filled before dyeing commences. The sequence of operation is: (a) load the dry material into the dyeing compartment—dry-beaming is possible since irregular selvages and tensions are permissible, (b) close and evacuate the dyeing compartment to the steady vacuum level required, (c) close the vacuum pump line and admit the prepared dye-liquor, and (d) continue with the normal dyeing cycle. Since impregnation is complete and uniform, diffusion into the fiber assembly and fixation of dye on fiber proceed unhampered by the usual hindrances that encourage differential dyeing conditions to be established at the start of dyeing. The time required in early stages of wet treatment for removing trapped air and overcoming the effect of nonuniform strike which occurs when the material is not uniformly saturated with the dye-liquor, are saved: considerable dyeing time can be cut from the normal procedure.[34] The Gaston County machine of a rapid dyeing system is based on an initial vacuum fill stage and a vacuum extraction stage at the end of the processing, thus reducing the processing time by 30–60%. However, so much dye goes on to the fiber within a few minutes that very uneven dyeing at the beginning may have to be converted into uniform dyeing by a long leveling time which may minimize the advantages of short fixation time.[35]

6.3 FABRIC DYEING MACHINES

Machines for dyeing continuous fabrics can be divided into two types, namely, a batch dyeing machine and a continuous dyeing machine. In these machines, the liquor or the fabric or both moves so that uniform and rapid dyeing can take place. The main types of fabric dyeing machines are winches, jiggers, beam dyeing machines, jet dyeing machines, and padding mangles.

Winch. A winch is the oldest type of dyeing machine. The fabric is dyed in rope form. The fabric, with ends stitched together to form an endless rope, is drawn continuously through the dye-liquor by passing it over a rotating fluted drum or winch that may be circular or elliptical in cross-section. Twelve or more pieces can be dyed side by side. They are kept apart by dividing rods situated below a small guide roller. In the front of the machine is a compartment known as the *stuffing or salting box,* that is formed by the insertion of a perforated baffle plate situated 15–25 cm from the front. This compartment houses the steam-heating pipes and provides a place for the addition of dyes and chemicals. The winches are enclosed and designed for use at above 100°C.[36] As the fabric leaves the winch, it falls into the liquor at the back of the machine. From there it is drawn forward through the main body of the dye vessel. An elliptical winch and a long, shallow dye vessel are preferred since the elliptical winch allows the fabric to plait down in folds (and not as a bunch) as it falls into the dye-liquor. The formation of creases in the thermoplastic synthetic fiber materials is thus avoided. Winch machines do not give a very efficient movement of goods through the liquor since for a large proportion of its time in the liquor, the piece lies in folds at the bottom of the dyeing vessel. They also have the disadvantage of using a long liquor ratio (20–40 : 1); a relatively large quantity of dye may be wasted (with systems which exhaust relatively poorly). In addition, it is difficult to obtain and maintain uniform temperatures in the winch dye vessel unless additional forced circulation of the liquor is provided.

Jiggers. In its simplest form, a jigger consists of two rollers and a dye trough. The material is wound in open width on a roller standing above the trough containing the dye-liquor. The rollers pull the fabric through the dye-liquor, first in one direction and then in the reverse direction. Complete immersion and almost tension-free movement with uniform speed are ensured by guide rollers in the trough, independent roller drives, same angular velocity regardless of the varying diameter of the roll of fabric and so on.[37]

Each passage of the fabric effects an exchange of liquor in the fabric with some of that in the trough, resulting in dye pickup; the immersion time of the fabric in the trough is short and dyeing takes place while the fabric is resident on the rollers.[37] In practice, an even number of ends (passages through the liquor) are used in order to avoid dyeing one end of the fabric a darker shade than the other. As a precaution, the dye is added in two portions, one before the first end and the other before the second end. At no stage is the fabric allowed to remain in one position for any length of time, as the liquor retained by the fabric will drain to the bottom of the roll. When this happens, air enters through the upper layers of the roll and cools the fabric. All the jiggers are covered to avoid cooling of the fabric rollers. High-pressure jiggers are used for dyeing polyester fabrics. The pressure jigger is closed in order to develop pressure and high temperature only after the goods have completed a few runs free of creases. This also helps in the dye distribution and leveling.

Jigger dyeing is a batch process suitable for both large and small batches. The $m:l$ ratio can be varied within wide limits. The unit requires small space and its operation is comparatively easy. The dyes in the liquor do not filter off on the fabric and there are very few problems about the dispersion stability of the dye. However, the jigger dyeing process has some limitations and drawbacks such as: (a) a low rate of production, (b) poor penetration of dye, (c) the high cost of the machine, (d) the possibility of formation and fixation of creases in the fabrics, and (e) jiggers exert a pronounced lengthwise tension on the fabric and delicate material can be distorted.

Mechanism of Jigger Dyeing.   Jigger dyeing is a system in which the fabric moves in and out of an essentially stationary liquor. For practical, low-volume ratios, end-to-end levelness is improved by adding the dye in lots, for example, half the dye at the beginning of the second run. In dyeing mixtures, the dyes are adsorbed in the ratio in which they exist in the dye-liquor, even when the dyes have distinctly different dyeing characteristics. The exhaustion is dependent on the number of runs of the fabric and not on the dyeing time in so far as this depends on the jigger roll. The jigger roll not only contains water of imbibition within the fiber, which is very small for synthetic fibers, but also interstitial liquor in an amount that is almost independent of temperature, fabric speed, and fabric tension, and is a characteristic of the fabric to be dyed. When the fabric is passed through the bath from one roll to the other, the composition of the interstitial liquor changes in an orderly manner with the result that a constant fraction of the liquid is replaced by the bath liquid. The latter is a characteristic of the fabric on a given jigger. The interstitial liquor pulsates vigorously at each revolution so that there is a rapid relative motion between the fiber and the liquor. The volume of the interstitial liquor is usually not more than 100% of the weight of the dry fabric. Dyeing takes place from this highly exhaustible bath. Thus, effectively instantaneous and complete dye absorption takes place in the low-volume ratio of the jigger roll.[38]

The major fault in jigger dyeing is the *ending,* where both sides of the batch are light or off-shade: ending gradually diminishes over a distance into the body of the fabric. Slow-dyeing dyes give more ending than fast-dyeing dyes. Similarly, slow-dyeing fabrics are more prone to ending; for example, triacetate and nylon fabrics adsorb dye at slower rates than cellulose acetate fabric and exhibit higher tendencies to ending, even at higher temperatures of dyeing. Compact fabrics are more prone to ending than more open fabrics. Enclosed jiggers are useful in reducing the incidence or extent of ending because of the higher temperature on the fabric roll. The dye uptake can be considered the result of two additive processes—bath strike and normal interchange of interstitial liquor. Significant bath strike will always increase the rate of bath exhaustion over that expected in its absence, especially in the earlier stages of dyeing, and it will cause considerable ending. This type of ending is diminished by adding the dye in two equal portions. If the dye sorption on the roll is not almost complete and instantaneous, the dye invariably gives ending. Since the sorption is slow, the dye uptake by the fabric in the second passage is considerably decreased by the unadsorbed dye already present in the interstitial liquid, which varies with the period that the portion of the fabric is on the roll before it goes back into the liquor. The end-portions stay on the roll alternatively for minimum and maximum time before they enter the bath. Depending on the half dyeing time of the dye, ending may be severe, minimum, or absent. The relation of dye properties to ending are shown in Table 6.1.[37] Leveling mechanisms play a negligible part in defining the final results of the jigger dyeing.

The jigger is also used for bleaching, washing of dyed goods, reduction-clearing and other treatments in which a nonsubstantive solute in the interstitial liquor in the fabric roll is removed. The mechanism of dyeing in jigger described above is also applicable to this process.[37] Washing in a single bath is clearly an inefficient process. Thorough rinsing necessitates many runs. Spray washing with water running to waste is a little more efficient in terms of time than washing in a fresh bath at each run. When the fabric is impregnated with a nonsubstantive solute such as a bleaching agent, two runs are sufficient to give 85% of equilibrium impregnation.

HT-Beam Dyeing Machines.   This is a liquor-moving machine in which the fabric is rolled or batched onto a perforated, hollow cylinder.[1,39,40] The latter is held horizontally in a chamber that is enclosed and capable of working at pressures of up to 5 kg/cm^2. The liquor is pumped radially through the beam and the flow may be reversed as in the typical package dyeing machine described earlier (see Section 6.2). A typical high-temperature beam machine is schematically shown in Figure 6.3.[1] For successful dyeing, the fabric is thermoset and wound on the beam so that the selvages are coincident and the tension is proper. Thus, the batch will not sag, the flow will not be uneven, and the liquor flow will not be too slow. In dealing with a fabric whose selvages are appreciably thicker than the fabric itself, it may be

TABLE 6.1 Relation of Dye Properties to Ending on Jigger[37]

Dyeing Properties		Rate of Exhaustion of Dyebath	Ending with Addition in		Hue Differences with Dye Mixture
Exhaustion	Strike		1 Portion	2 Portions	
Normal	Very rapid	Somewhat increased	I[a]	(I) or none	possible
Normal	Rapid	Normal	(I)	none	none
Normal	Moderately slow	Slightly decreased	II[b]	II	none
Normal	Extremely slow	Greatly decreased	none	none	none
Very slow	Rapid	Greatly increased	none	none	none
Very slow	Slow	Greatly decreased	none	none	none

[a] I: A end heavier or different in hue than Z end.
[b] II: A and Z ends lighter or different in hue from the uniformly dyed body of fabric.

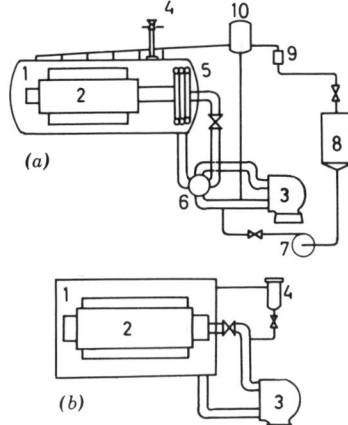

FIGURE 6.3. Longclose machines for dyeing at (*a*) high temperatures and (*b*) atmospheric pressure.[1] 1: Main tank; 2: Beam; 3: Main pump; 4: Sampling unit or tank; 5: Steam coil; 6: Reverse device; 7: Auxiliary pump; 8: Side tank; 9: Bleed cooler; 10: Buffer tank.

necessary to wind-on with a small traversing movement. Otherwise, the wound batch will have a greater diameter at the sides than in the middle; this can lead to poor dye penetration at the edges and an undesirable slackness in the rest of the fabric.[1] With knitted goods, the batch is limited to the size of the machine, but with woven fabrics, the batch size is determined by the porosity of the fabric.[41] HT beam machines with air-injection devices are particularly developed for dyeing delicate knitted goods of textured yarns. In these machines, air is introduced by means of a small jet mounted on the suction side of a circulation pump. It is claimed that this does not cause cavitation of the pump because the air is introduced into the dye-liquor in a very finely divided form. The air bubbles are distributed between the layers of the fabric, forming an air cushion that prevents the flattening of the surface effects on the fabrics. The bubbles do not grow in size because of high pressure.

The fabrics that are not sufficiently porous in the wet state, for example, PET/CO blend fabrics with tight fabric structures or sailcloth, are unsuitable for beam dyeing because of an inadequate rate of flow through the material. Pile or raised fabrics, ribbed fabrics, and those with other weaves, in which the effect is dependent on the peculiarities of the surface structure, may suffer from being treated in firmly wrapped batch form.[42]

Jet Dyeing Machines.[43–51]   A Gaston County machine is a batch-type, pressurized, modified winch unit in which both the liquor and the material move. Introduced in 1967, the Gaston County machine uses one of the few new principles in dyeing to appear in many years. The fabric is processed in endless, parallel loops and is essentially penetrated, conveyed, piled in the storage chamber, and dyed by precisely directed jets of the dye-liquor. The prime mover of the fabric around the system is the main jet, which might be described as an inside-out venturi. The cloth travels vertically through the center of the jet and is met at the throat by a circumferential converging jet of dye-liquor that emerges from a ring-shaped slot completely surrounding the fabric. The fluid simultaneously penetrates and propels the material through the fabric-guide tube to the rear of the machine. This produces a tremendous interchange of fluid with the fabric. The turbulence at the jet-throat and in the elbow immediately above the jet is quite intense, which creates a problem of foaming in partially filled jet-dyeing machines. In order to overcome this problem and to minimize lengthwise tension on the fabric, the jet machines are fully flooded with the liquors. The fully flooded machines rely solely on the jet action for fabric movement (without any driven reel). Although this may be suitable for lightweight fabrics, it fails to maintain a constant movement of heavier fabric. This results in the development of crush marks or crows' feet creases. Increased jet-pressure induces and inserts a twist in the rope, resulting in both wrap way and lateral creasing.

To overcome these drawbacks, new horizontal jet-dyeing machines based on "soft flow" systems have been designed, thus ensuring that the fabric follows a path similar to that in the conventional winch. To minimize tension and creasing, the rope of fabric is supported in a transportation tube for the period it is out of the main dyeing vessel. For a more gentle action than that given by previous designs of jet machines, a driven reel is frequently installed to assist the rope movement and to allow a reduction in the jet pressure. The liquor-to-goods ratio in newer jet machines is as low as 5 : 1, and with the use of foam, it can be brought down still further.[51]

Jet dyeing machines are used for double-jersey polyester, knitted and woven fabrics of 100% acrylic, and polyester and their blends, but mainly for material from textured filament yarn or spun yarn (see Chapter 8). Jet machines are not specially suitable for materials from flat-filament yarns.

Padding Mangles.   The dye-liquors or finishing agents are applied to the fabric on a padding mangle in almost all the semicontinuous and continuous processes. In principle, the fabric passes through a small-dimension trough to pickup the liquor and then through the nip formed by two or more squeeze rollers, where excess solution is squeezed out and runs back into the trough. The pickup or expression of the dye-liquor is decided by the nature of liquor and fabric,[52] the bowl hardness, and the nip pressure. In order to get uniform pickup, it is essential that the pressure in the nip is uniform along the length and both sides of the fabric. The bowls are ground with a slight camber to offset higher pressures at the ends where force is applied, which causes the bowls to bend compared to the center. The processes after padding are concerned not with leveling but with fixing the dye or finishing agent on the fabric without any migration.

For a pad-dry sequence, the real bath-immersion time for the fabric is often only a fraction of a second. After leaving the nip of the mangle, the liquor in and on the fabric is evaporated by drying. Rapid removal of surface liquor, that is, before the fabric is fully saturated, is not conductive to the diffusion process required for deep penetration. As the medium that carries the dye (or a finishing agent) is vaporized, the dye tends to migrate and to aggregate on the surface.[52] Under normal padding conditions without an efficient wetting agent, the fabric remains an *aerogel* and water-logging is not achieved. The expulsion of air at the nip of the mangle occurs at a late stage in the impregnation process. For this reason, use of nips before the fabric enters the impregnation bath is suggested to reduce the air content. However, the use of preentry air-expulsion nips, although providing improvement over the system of direct dip followed by nip, does not give air-free saturated fabric. This is because air is not readily displaced from capillaries that are closed at one end, for example, by constriction in tightly crossed or twisted yarns or fibers, as in densely packed fabric assemblies or where local or irregular swelling of the fiber has already formed a barrier to the displacement of air. If the fiber material has no opportunity to relax, deaeration of closed capillaries is not readily achieved by high-pressure nips. The air itself can be compressed by mechanically applied pressure, thus presenting further resistance to the entry of water.[53,54]

Removal of atmospheric air can increase the speed and efficiency of processing. The sequence—(a) remove air by vacuum, (b) pass directly into impregnating liquid, and (c) return to atmospheric pressure—leads to almost complete water-logging of the material, provided that the subatmospheric pressure at step (a) is near zero. This is explained by the structure of a fiber assembly such as a fabric. The fiber assembly consists essentially of fibers in atmospheric air and can be thought of as an aerogel. In pretreatments, dyeing, and finishing processes, the air, which exists either as discrete pockets or as a continuous phase in the more open parts of the fiber assembly, must be fully replaced by liquid to promote maximum efficiency of treatment. Thus, the aerogel must become a *hydrogel* so that solutes or dispersions can be uniformly transferred from the water to the fiber without hindrance from the enveloped air. Even a fine film of air presents a barrier to the adsorption of dyes on the fiber surface. Air is present in the fissures, natural cavities, and void spaces in the fiber structure and may retard the diffusion in the fiber phase. Impregnation of difficultly wettable substrates by exposing the dry substrate to a sufficiently low atmosphere to remove occluded air, followed by immersion in aqueous bath, gives thoroughly saturated substrate in a fraction of the normal wetting time.

Exploitation of the principle of vacuum impregnation can be seen in several times of padding equipment.[55] In each case, the dry material is exposed to a high vacuum (less than 100 mm Hg) to remove the air rapidly. The principle of Kleinwefer's Vacupad Unit is shown in Figure 6.4.[54]

6.4 DRYING

The removal of entrained water from the package after the dyeing sequence but prior to complete drying can be carried out by (a) centrifugal extraction, (b) vacuum extraction, and (c) partial drying with hot air. After dyeing,

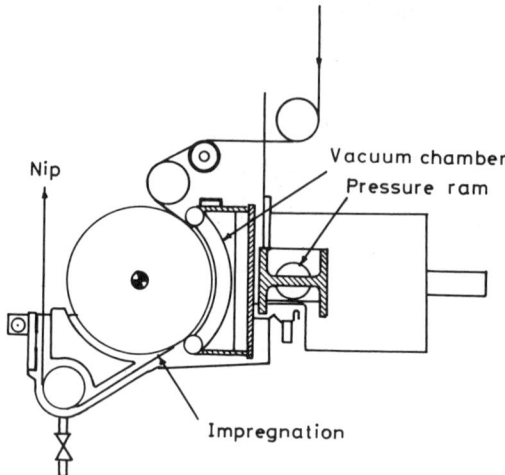

FIGURE 6.4.   Kleinwefer's vacupad unit.[54]

while passing from the batch (beam) to the stenter for final drying and set-ting, the fabric passes over a slit-suction extractor. After the yarn is dyed in package form, the wet material is dried in a separate dryer. Alternatively, it can be dried in the dyeing unit itself after draining the dye-liquor. The en-trained liquor is forced out by applying compressed air or a vacuum. The yarn is finally dried by hot air or steam.[56] Various other drying methods are chamber drying, low and high-pressure drying, and vacuum drying.[9,57] Re-cently, dielectric drying employing radio-frequency (rf) systems has been used for drying textile goods.[58–63] These machines are discussed in Chap-ter 20.

Mechanism of Drying.   In the course of textile processing, it is often neces-sary to apply heat, either simply to raise the temperature (as in dye fixation or heat setting) or to dry products. There are three ways to transfer the heat to the material: radiation, conduction, and convection, used either sepa-rately or in combination. Irrespective of the heat source and the method of heat transfer, heat must pass through the surface of the material. The surface varies with the type of material and the method of heat transfer. Since textile materials are poor conductors of heat, in order to achieve a certain tempera-ture in the center of the block, the temperature on the outside must be somewhat higher, that is, there must be a temperature gradient. As long as the surface is wet, the heat reaching it will cause evaporation which in turn will prevent the material from exceeding the wet-bulb temperature. When there is no water present or drying has reached the falling rate, the surface can reach the temperature of the heating medium. The heating temperature must not be excessively high or the material must move through the heating zone rapidly to avoid permanent damage.

In the first phase of drying, the material (or at least the outer layer) is brought up to the evaporation temperature. In the constant rate phase, a fixed amount of water evaporates for a given quantity of heat when evaporation takes place at the physical surface. Eventually, the capillary movement of water ceases, which results in the liquid surface retreating from the physical surface. This is the beginning of the falling-rate phase. The drying process is hindered by two aspects—increasingly effective thermal insulation created by the outer dry layers and the fact that the movement of water to the atmosphere has to be by vapor-phase diffusion.

6.5   DIELECTRIC HEATING[58-63]

An alternative method of heating that does not depend on heat transfer through a surface is to use high-frequency dielectric energy. In simple terms, the heating effect of dielectric energy is caused by the reaction of the molecules in a mass to a rapidly alternating electric field. Water molecules with dipoles will attempt to rotate to a position of minimum stress. If the direction of the electromagnetic field is reversed, the position of minimum stress is also reversed. In high-frequency heating, these reversals take place several million times a second: typically at 13 and 27 MHz (rf) and 896 and 2450 MHz (microwave frequency). The effect of the rapid oscillation of a dipole is to generate heat within the molecule itself which leads to evaporation of water. There is little generation of heat in the polymers that are electrically symmetrical. For example, tap water heats up about 100 times more easily than many textile fibers. The temperature of the polymer will rise because of the hot water vapor. Thus, dielectric heating has the ability to heat water selectively and to do so volumetrically without heat passing through the surface. No mass being dried can exceed the evaporation temperature of 100°C as long as residual moisture is present. This selective effect is used to good effect in dielectric drying. Wet areas in a product will be selectively heated and dried, but dry areas will absorb little or no energy. This is most effective in making products of uniform moisture content and ensuring good energy economy.

In dielectric heating, the heat generated is given by:

$$P = 2\pi f \, \epsilon_0 \epsilon_r'' \, \epsilon^2 \quad (W/m^3)$$

where f = frequency of the supply in Hertz (Hz); ϵ_0 = constant (dielectric permittivity of free space, 8.85×10^{-12} farads/m); ϵ'' = loss factor that is a measure of the degree to which a material will react to an alternating electric field; ϵ = strength of the electric field in V/m; and ϵ_r = dielectric constant of the material.

The loss factors of common materials are listed in Table 6.2. Under given installations, the frequency is fixed, and if $\epsilon_r'' (\epsilon_r'' = \epsilon_r \times \tan \delta; \tan \delta$ = loss

TABLE 6.2 Dielectric Constants and Loss Factors (Approximate Values at Room Temperature)[63]

Material	at	Dielectric Constant (ϵ_r)		Loss Factor (ϵ_r'')	
		10 MHz	3000 MHz	10 MHz	3000 MHz
Cotton					
(210 kg/m³, 7% regain)		1.5 ⎫		0.03	
Wool		⎬ at 27 MHz			
(68 kg/m³, 20% regain)		1.2 ⎭		0.01	
Cellulose acetate				0.07	0.09
Nylon 66		3.2	3.0	0.09	0.04
Polyester		4.0	4.0	0.04	0.04
Polyethylene		2.25	2.25	0.0004	0.001
PVC (Pure)		2.9	2.8	0.03	0.02
+40% Plasticizer		3.7	2.9	0.4	0.1
Water					
Ice ($-12°C$)		3.7	3.2	0.07	0.003
Water (25°C)		78.0	76.0	0.36	12.0
+0.1 M NaCl		80.0	75.5	100.0	18.0

tangent) is assumed to be independent of frequency, for a given heat generated within the material, the electric field strength in microwave systems is less than that in rf systems. The higher the electric field strength, the greater is the risk of an electrical discharge and, for this reason, microwave energy is preferred. These cases occur when the material being heated does not have any water present. In most drying and heating processes where water is present, the loss is high enough to ensure that the electric field does not exceed that at which a discharge would take place.

The rf and microwave heating equipment is safe for operators provided it meets the standards and is operated per manufacturers' instructions. The present recommended maximum leakage or emission level from rf and microwave equipment is 10 mW/cm² at 50 mm or 0.01 mW/cm² at 500 mm from the radiation source. Microwave power is most commonly generated for industrial purposes with magnetron electronic valves which require a high-tension DC electric supply and low-voltage heaters. They are usually water-cooled.

The rf and microwave heating techniques have been established successfully in the textile industry (see Chapter 20). There is a considerable savings in energy because of the efficient conversion of electricity into heat; for example, 70% conversion by 900 MHz microwaves and 50–60% by 2450 MHz microwaves. The equipment is very compact. The wastage is low and continuous pad microwave drying of hank may eliminate hanking and rewinding operations.

6.6 QUALITY CONTROL

A variety of products and processes are used in dyehouses to produce textile materials with desired properties. Analysis of dyes and chemicals, checking of processing lots, identification of merge numbers, assessing physical and chemical processing parameters, and testing the finished products for their properties have to be routinely carried out in a wet-processing house. It is necessary to develop representative sampling techniques, testing specifications, and reporting–analyzing systems. The aim is to eradicate the causes of faults rather than only correcting the faults.

Water is a major raw material in the wet-processing industry. Water must be tested for pH,[64] hardness, dissolved solids and so on before it is used in any of the standard dyeing methods. Demineralized water is easily obtained using a battery of anion and cation-exchange resins. The pH of the demineralized water may not be neutral: a check of the pH is essential. Trace elements in water alter the tone of dyed goods. It is sometimes advisable to add sodium hexametaphosphate or ethylenediamine tetracetate to water to bind the metal ions. Recovered water may contain dissolved solids that may influence wet processing. Inorganic salts and alkalies are usually analyzed for purity and the nature of any impurities. For example, iron salts may autooxidize the fiber in a bleaching process and should not be present in any additive. Auxiliary products are marketed either in pure form or as formulations. These products are tested for efficiency or effectiveness in use. For example, a wetting agent is tested for the time required to wet a fabric under standard conditions. If the auxiliary product remains on the finished fabric, its influence on the properties of the goods, for example, light or wash fastness, is evaluated. Biodegradable products are preferred.

Dyes (and pigments) are tested for their tinctorial strength, fastness properties on dyed goods, color-yield in different dyeing methods, build-up of depth in relation to the dye concentration, staining of fibers in blends, rate of dyeing, compatibility, storage stability and so on. They must have consistent strength, hue, tone, dyeing characteristics, and fastness properties. This is particularly important when dyeing is controlled by computer matching systems or the process is automated.

In process control, it is important to clearly identify the incoming grey goods. Synthetic fibers are produced with a variety of properties and forms. Any mistake in identifying the grey goods can ruin the fiber material in the wet-processing. Similarly, the process parameters of desizing–singeing to finishing for synthetic fiber materials have to be strictly controlled. At the same time, the industrial engineering department has to synchronize and to optimize the utilization of machinery, materials, and energy so that wet-processing becomes economical.

In general, faulty dyeings can be grouped into four technical types (a) weaver's faults which are usually tagged at make-up, (b) quality of goods from dyeing to finishing processes (e.g., two-sided, listed, spots, creases, roller marks), (c) color and strength beyond tolerance levels, and (d) fastness properties below the specifications. The faults usually invoke strong and sometimes panic measures to put them right. At a technical and commercial level, dyed goods need a postmortem on the most recent production. Random spot checks can be introduced to advantage in well-organized testing and quality control systems. A control strategy can be developed by identifying critical areas, defining operations and targets, collecting information on important variables, quantifying errors, introducing controls, and implementing and monitoring the control measures.[65]

Some of the quality control jobs have been taken over by microprocessers fitted onto automated units. Verification of on-line instrumentation for reliability is done for quality control. The probability of faulty functioning of microcomputers is very low and the probable sources of faults and errors lie outside the computers, that is, in the actuators, sensors, and acknowledgment hardware that communicates with the computer's input/output unit. It is essential to keep strict control and to check on temperatures of ovens and stenters where synthetic fibers are dried and heat-set. The wet-processing behavior and properties of textile products are significantly affected by variations in temperature (see Chapter 2).

When a fault is noticed in the finished goods, the nature of the fault or the problem and its likely cause or source have to be identified so that corrective measures can be suggested. This is not always easy since the problem may have a variety of causes. A typical example of problems in polyester printing is given in Table 6.3 to illustrate the point.[66] The filtration of a disperse dye

TABLE 6.3 Problems in Polyester Printing[66]

Problem	Cause
1. Yellowing and uneven prints	Faulty fabric preparation
2. Specky prints	Poor dye dispersion, foaming of print-paste
3. Change in shade	Chelation of dye with metal contaminants in paste, change in pH
4. Lighter prints	Foaming of print-paste, alkaline pH during steaming (volatile acid for pH adjustment), inefficient steaming or thermofixation.
5. Splashing on white ground	Low print-paste viscosity, tacky print-paste
6. Bleeding during steaming	Poor ability to hold water by thickener, improper print-paste additives.
7. Moire effect during rotary screen printing	Too viscous print-paste, absence of fabric wetting, impression of black-grey structure
8. Staining of white ground	Poor sublimation fastness of dyes, excessive steaming, hot and improper washing before reduction-clearing.

on a package during HT dyeing may be caused by poor dye-dispersion stability; tight packages; breaking of dispersion because of ionic, electrical, or mechanical shock; or because of vibrations in the pump.

6.7 AUTOMATION AND COST EFFECTIVENESS

The automatic controllers are identical for different sized machines. The cost differences among the different machines are related to the capital cost for pneumatic valves and other ancillary equipment required for automation. The cost of valves increases rapidly with increasing dimensions which increase the capital cost of machines of bigger size.[67]

Out of the operating costs, only those directly affected by automation contribute to the cost reductions. Fixed costs and costs for dyes and chemicals are not lowered by automation. The cost of labor rapidly decreases with increasing machine capacity and does not dominate operating costs. For big machines, the cost of services (power, steam, and water) becomes increasing important. With automation, the labor cost is very much reduced, particularly, if the machine is small. For larger machines, the influence of automation on the other variable costs is quite significant. Costs of both water and energy consumption can be substantially reduced by automation. Furthermore, there is a 10–30% rise in production.

The economic trend in costs before and after automation of typical dyeing machines is given in Table 6.4.[67] The cost increases in three years (1978–1980) are assumed to be

30%	Labor
30%	Machines
5–20%	Automation
100%	Services
30%	Variable costs
25%	Production value

The cost of automation has increased much more slowly than the cost of machines, the reason being the increased use of relatively cheap and, at the same time, more complex microelectronic components. The new technology has some very important added values. One can obtain information or process status, productivity, and quality directly from the machines in a way previously not economically possible. The increased reproducibility, the better utilization of machines, and the control of energy and water consumption are added advantages of automation of dyeing and finishing processes. Dyehouses have, in general, not been able to entirely cover the increased costs by increasing the sales price for the finished product. Control engineering in dyeing and finishing, that is, dyehouse automation, gives increased

TABLE 6.4 Comparison of Costs (1978–1980) with and without Automation (sewing-thread dyeing machine)[67]

Machine Capacity (kg):	50		425	
Year:	1978	1980	1978	1980
1. Manual m/c (£ × 10^3)	27.0	34.5	55.4	71.0
2. Cost of automation (£ × 10^3)	8.0	8.8	11.0	12.6
3. Operating costs (£/kg)				
Without automation				
Labor	0.195	2.54	0.024	0.032
Services	0.059	0.118	0.055	0.110
Other variables	0.055	0.072	0.034	0.044
Total	0.309	0.444	0.113	0.186
With automation				
Labor	0.145	0.189	0.022	0.029
Services	0.059	0.118	0.055	0.110
Other variables	0.046	0.060	0.029	0.038
Total	0.250	0.367	0.106	0.177
4. Savings due to 10% increased production (£ × 10^3)	2.70	3.38	5.54	6.93
5. New cost of automation [(2) − (4)] (£ × 10^3)	5.30	5.42	5.46	5.67
6. Operations—20 dyeings/week (£ × 10^3/year)	2.77	3.62	2.80	3.67
7. Payback period (months)	23	18	23.5	19

reproducibility which has led to shorter cycle times, a reduced number of dye additions (for shade matching), and reduced consumption of water, dyes, chemicals, energy, and labor.

Microcomputers in the Dyehouse.[67,68] Microprocessors have opened up the possibilities for making small, specialized, discrete systems. The well-defined functions in a dyehouse are: (*a*) weighing and registration of the dyes and goods; (*b*) weighing, registration, and stock control of dyes and chemicals; (*c*) recipe estimation from color measurements; (*d*) calculation of recipes; (*e*) automatic transfer of dyes and chemicals from stores to machines; (*f*) production planning; (*g*) production control; (*h*) production supervision; (*i*) data acquisition. Different functions make very different demands on the microprocessor, for example, calculatory capacity, memory capacity, ease of programming, and combinations of the above. Software and hardware are now available for specific uses in a dyehouse. A complete flowchart of the machine or system and a functional description is required that specifies which values and components are activated when a function is in operation. The availability of a wide range of synthetic fibers and blends processed by numerous different processes required relatively sophisticated controls.

6.8 ECONOMICS OF DYEING PROCESSES

The economic scene of the world is changing fast and increasing complexity in the balance of supply and demand have made the economic survival of a textile enterprise a difficult and unpredictable challenge. Increasing relative cost of energy (see Chapter 20) and concern about environmental pollution so that more money is allocated to hazard testing of dyes and chemicals and to the treatment of water and effluents are some of the reasons for this change.[69,70] External economic pressures have forced wet processors to consider in more detail how the cost distribution depends on the products and processes selected for a specific wet treatment.

Dyeing of Synthetic Yarn and Tow. A coloration process such as mass coloration or tow dyeing (see Chapter 4) is much cheaper than dyeing yarn packages. Thus, the cost of tow-dyed fiber with respect to package-dyed fiber (100%) is:[71]

31–41%	Acrylic
47–56%	Polyester
80%	Wool

This has resulted in less yarn dyeing. In 1979, yarn dyeing for acrylic was only 30% of that in 1973. A typical distribution of costs for yarn dyeing is given in Table 6.5.[72,73] It is not easy to significantly lower the contribution of dyes and chemicals to the overall cost distribution. Instrument color matching and automation in handling are used for the best results. The dyes should give reproducible recipes so that multiple shading and reprocessing are avoided. It is worth paying more for products that enable an overall saving of process cost.[74]

TABLE 6.5 Cost Distribution for Yarns[73]

% Cost Distribution	Acrylic	Nylon	Polyester	Wool
Dyes and chemicals	29–44	39–45	32–45	55
Overheads	27–32	24–27	25–28	20–22
Labor	18–22	16–18	16–19	13–14
Services[a]	11–17	15–17	13–21	10–12
Electricity[a]	15–23	13–14	9–17	15
Steam	53–59	46–53	65–77	49
Water	23–27	34–41	6–22	36

[a] Services (100%) are subdivided into electricity, steam, and water.

Dyeing and Printing of Fabrics.[69]   Many of the trends noted for the dyeing of yarn are applicable to the piece-dyeing sector. Relative dyeing cost-indices for HT-beam dyeing (as 100) and three different designs of Thies Jet Machines operating at liquor ratios in the range 6:1 to 20:1 are given in Table 6.6. The major influences of batch size and liquor ratio on dyeing cost are clearly seen from the data in Table 6.6.[75]

Technoeconomic assessment of the application of disperse dyes to textured polyester fabrics (*a*) by vapor-phase transfer from printed transfer paper and (*b*) in direct printing by rotary screen has been made by Gibson.[76] Each system has its advantages and limitations which cannot be easily expressed in monetary terms. The transfer-printing method is particularly effective for short runs. The two methods become closely competitive for

TABLE 6.6   Batch Size and Liquor Ratio in Textured-PET Fabric Dyeing[75]

| | | Cost Indices | | |
| | | One Machine (55–90 kg/h) | | Several Machines (55 kg/h) |
Machine Type	Liquor Ratio	Cost Index	Sizes	Cost Index
R jet 140	6:1	55	3 × 600 kg	58
	8:1	61		64
Soft stream	8:1	75	9 × 300 kg	70
	12:1	87		83
Jumbo jet	12:1	68	5 × 300, 1 × 150	76
	20:1	92		100
HT Beam	12:1	100	14 × 200 kg	84

TABLE 6.7   Effect of Productivity on Carpet Printing Cost[77]

Production Level (m³ × 10⁶ *pa*)	10	5	2	1	0.5
Relative printing cost	20	26	36	58	100
Cost distribution (%)					
Print-paste	66	52	36	23	13
Services	9	7	5	3	2
Depreciation	9	16	22	28	32
Interest on capital	6	10	20	25	29
Labor	7	11	10	12	14
Sampling	2	3	5	6	7
Maintenance	1	1	2	3	3

lengths around 5000 m, after which rotary screen printing becomes cheaper for long runs. However, low capital outlay, wide availability of designs, quick delivery and turnaround times, and the convenience and versatility of holding stocks of printed paper awaiting transfer rather than fabric already printed and so on make the transfer-printing process more attractive (see Chapter 15). The details of cost distribution varies with the production index (Table 6.7).[77]

The possibility of replacing expensive fiber with less expensive fiber to bring down the cost has been explored. Thus, nylon is replaced by polyester in uses such as floor coverings and automotive fabrics. Relative cost indices are shown in Table 6.8.[78]

TABLE 6.8   Cost for Nylon and PET in Carpet Coloration[78]

Relative Cost Indices	Cost of Polymer Production	Cost of Undyed Yarn	Dyeing Cost	Printing Cost
Nylon	100	100	100	100
Cationic dyeable polyester	73	86	163	160
Normal polyester	67	80	227	171–247

TABLE 6.9   Relative Dyeing Cost for PET-Blend Fabric[79]

Machine Type	Liquor Ratio	Dyeing Time (min)	Mixture of Dyes (%)	Dyes and Chemicals	Process
				Relative Cost Indices	
Blend with Wool					
R Jet 95	6:1	185	1.70	85	45
R Jet 140	6:1	180	1.70	85	46
R Soft-Stream Jet	16:1	240	1.90	95	70
HT Winch	25:1	315	1.95	98	88
Open width	25:1	310	2.00	100	100
Blend with Cotton					
R Jet 140	6:1	255	4.0	80	64
R Soft-Stream Jet	11:1	300	4.25	85	75
Jumbo Jet	15:1	265	4.50	90	84
Soft Stream	16:1	355	4.50	90	85
HT Winch	25:1	400	5.00	100	100

Dyeing of Polyester Blends. A comparison of the actual performance of six types of batchwise machines for dyeing a 55 : 45 PET–WO blend and 70 : 30 PET–cotton fabric is shown in Table 6.9.[79] Dyeing of a PET/CO blend fabric with disperse and reactive dyes is a very lengthy process when the two classes of dyes are applied in separate baths. A one-bath process offers considerable savings (see Chapter 7) and there is a considerable scope for modifying the methods of application of reactive and disperse dyes in order to reduce costs.

REFERENCES

1. Limbert, K., *JSDC,* **82** (1966), 97.
2. Jorder, H., *MTB,* **50** (1969), 267.
3. Bird, C. L., *The Theory and Practice of Wool Dyeing,* 3rd Ed., The Society of Dyers and Colourists, Bradford (1963).
4. Anon., *Dyer,* **64**(3) (1980), 16.
5. Chaplin, C. H., Mason, J. S., and Park, J., *JSDC,* **96** (1980), 103.
6. Robinson, G., and Jagger, D., *JSDC,* **67** (1951), 557.
7. Derbyshire, A. N., and Lemin, D. R., *JSDC,* **80** (1964), 363.
8. Burley, R. W., Flower, J. P., and Rattee, I. D., *JSDC,* **85** (1969), 187; **87** (1971), 278.
9. Fleming, R., and Gaunt, G. F., *Rev. Prog. Color,* **8** (1977), 47.
10. Park, J., *Dyer,* **155** (1976), 70.
11. Mason, J. S., Park, J., and Thompson, T. M., *JSDC,* **96** (1980), 246; **96** (1980), 580.
12. Neuhaus, L., *Text. Prax.,* **26** (1971), 230.
13. Siegrist, G., and Liddiard, A. G., *JSDC,* **89** (1973), 523.
14. Cegara, J., *Teintex,* **37** (1972), 9; *MTB,* **54** (1973), 394; **54** (1973), 503.
15. Etters, J. N., *JSDC,* **97** (1981), 170.
16. Brooks, J. H., and Nordon, P., *JSDC,* **89** (1971), 12.
17. Carbonell, J., Hasler, R., Walliser, R., and Knobel, W., *MTB,* **54** (1972), 68.
18. Beckmann, W., and Hoffman, F., Book of Papers, AATCC Intl. Tech. Conf. Montreal, Canada, (1976), p. 14.
19. Carbonell, J., Walliser, R., and Hasler, R., *Teintex,* **38** (1973), 73; *MTB,* **55** (1974), 149.
20. Ulrich, H., and Reuther, A., *Text. Prax.,* **29** (1974), 1565.
21. Camp, J. G., *ADR,* **62**(11) (1973), 25.
22. Walters, J. E., *ADR,* **63**(3) (1974), 52; **63**(3) (1974), 66.
23. Gailey, I., *JSDC,* **91** (1975), 165.
24. Reither, A., Waasmuth, J., and Ulrich, H., *MTB,* **55** (1974), 549.
25. Ulrich, H., *Can. Text. J.,* **92**(4) (1975), 83.
26. Schoeupflug, E., and Richter, P., *TCC,* **7** (1975), 136.
27. Nakayuma, T., *Japan Text. News,* **256** (March 1976), 73.
28. Siegrist, G., *Rev. Prog. Color.,* **8** (1977), 25.
29. Vaidya, A. A., and Sunder, R., *Colourage,* **24**(16) (1977), 35.

30. Scheidegger, A. E., *The Physics of Flow of Porous Media*, U. of Toronto Press, Toronto, Canada, (1957).

31. McGregor, R., *JSDC*, **81** (1965), 429.

32. Fox, M. R., *JSDC*, **78** (1962), 393.

33. Hadfield, H. R., and Lemin D. R., *JSDC*, **77** (1961), 198.

34. Ameling, B., *MTB*, **54** (1973), 403.

35. Richter, P., *Text. Prax.*, **28** (1973), 12.

36. Ordway, C. B., *ADR*, **51** (1962), 22; **51** (1962), 791.

37. Morton, T. H., *JSDC*, **81** (1965), 52; **81** (1965), 93; **81** (1965), 150.

38. Kilby, W. F., *JSDC*, **76** (1960), 479.

39. Hollis, N. M., *ADR*, **55** (1966), 502.

40. Davis, W., *JSDC*, **89** (1973), 77.

41. Garrett, D. A., Haden, I. E., and Smith, F. R., *Dyer*, **124** (1960), 405.

42. Charlesworth, A., *JSDC*, **82** (1966), 89.

43. Carpenter, W. T., *TCC*, **1** (1969), 265.

44. Eikoetter, W. H., *TCC*, **1** (1969), 268.

45. Olley, M. J., *JSDC*, **88** (1972), 321.

46. Limbrt, K., *JSDC*, **88** (1972), 385.

47. Patterson, M., *Rev. Prog. Color.*, **4** (1973), 80.

48. Ratcliff, J. D., and Birtwistle, F., *JSDC*, **90** (1974), 313.

49. Ratcliffe, J. D., *Rev. Prog. Color.*, **9** (1978), 58.

50. Ratcliffe, J. D., *JSDC*, **96** (1980), 94.

51. Simborowski, V., *JSDC*, **96** (1980), 111.

52. Datye, K. V., and Pitkar, S. C., Seminar, UDCT, Bombay, India (1967), p. 47.

53. Fox, M. R., Marshall, W. J., and Stewart, N. D., *JSDC*, **83** (1967), 493.

54. Fox, M. R., *ADR*, **61**(3) (1972), 48; *JSDC*, **89** (1973), 46.

55. ICI, BP, 1,158,284 (1966).

56. Karrer, F. W. J., BP, 1,242,316 (1971).

57. Ulrich, H., *Text. Prax.*, **31** (1976), 416; **32** (1977), 64.

58. Puschner, *Heating with Microwaves*, Eindhoven, The Netherlands (1966).

59. Jones, P. L., *JSDC*, **98** (1982), 248.

60. Hull, P. J., *JSDC*, **98** (1982), 250.

61. Henderson, K., McAulay, T., Young, R., and Smith, G., *JSDC*, **98** (1982), 303.

62. Harrison, D. H., *JSDC*, **98** (1982) 305.

63. Electricity Council (UK), Technical Information Sheet, *IND*, 16 (1979).

64. Dowson, T. L., *JSDC*, **97** (1981), 115.

65. Harrison, V. M., Roberts, J. E., and Ward, A., *JSDC*, **97** (1981), 106.

66. Patel, I. M., in *Advances in Textile Chemical Processing*, Chavan, R. B., Ed., IIT, Delhi, India (1981), p. 599.

67. Svenson, R., *JSDC*, **97** (1981), 327.

68. Bailik, Z., Pack, J., and Walker, D. C., *Rev. Prog. Color.*, **10** (1979), 55.

69. Shore, J., *Rev. Prog. Color.*, **11** (1981), 58.

70. Dwek, J. C., *JSDC*, **97** (1981), 390.

71. Chaplin, H., Park, J., and Thompson, T. M., *JSDC*, **96** (1980), 580.

72. Carbonell, J., Hasler, R., and Walliser, R., *JSDC*, **92** (1976), 100.

73. Boyd, W., Park, J., Thompson, T. M., and Warbis, T., *JSDC,* **96** (1980), 497.

74. Hansford, J. A., Seaman, H., and Shore, J., *JSDC,* **98** (1982), 225.

75. Ulrich, H. U., *Textilveredlung,* **11** (1976), 169.

76. Gibson, L. J., *JSDC,* **93** (1977), 164.

77. Mitter, M., *Text. Prax.,* **29** (1974), 949.

78. Dowson, T. L., and Robert, B. P., *JSDC,* **93** (1977), 83.

79. Nikko, W. D., *Textilveredlung,* **10** (1975), 338.

7 | DYEING OF POLYESTER AND ITS BLENDS

Three factors are responsible for making PET fiber difficult to dye: (*a*) high-fiber crystallinity, (*b*) a marked hydrophobic character, and (*c*) an absence of chemically reactive groups in the polymer. Owing to these factors, PET cannot be dyed with the same dyes that are generally employed for cellulosic, protein, nylon, or acrylic fiber. Since the ester groups content of cellulose acetate and polyester fiber is nearly the same (40–45%), attempts have been made to dye polyester fiber with disperse dyes by the same method used for cellulose acetate. However, it was observed that PET was not dyed at 80–100°C. This was due to a very slow rate of diffusion of disperse dyes into the compact polyester fiber.

In the early years attention was directed to finding a means of improving dyeability. The yield of a disperse dye on PET is limited and is vastly inferior to the yield on nylon and cellulose acetate because of the low rate of dyeing rather than the low substantivity of early disperse dyes for PET.[1] The problem is solved by using different approaches to increase the rate of dyeing.

1. Building up dye molecules inside PET (azoic dyeing).
2. Opening up the fiber structure to bring down the T_g (carrier dyeing).
3. Using temperatures above 100°C [high-temperature (HT) dyeing].
4. Heating the dye and the PET in the dry state together near the softening temperature of the fiber (thermosol or thermofixation dyeing).
5. Replacing water with an organic solvent as a dyeing medium (solvent dyeing).

Apart from the above approaches, chemical modification of PET (to im-

183

part affinity for dyes other than nonionic dyes) is commercially practiced in order to get cationic dyeable PET (see Chapter 12). Similarly, the transfer-printing process (see Chapter 15) is used to color polyester in solid shades.

The use of solvents for dyeing PET was intensively investigated in the early 1970s. Even though PET can be dyed to any depth of shade using solvents, none of the solvent-dyeing methods ever reached a state of a commercial feasability. Solvent dyeing is therefore not discussed in detail. The azoic dyeing process was once used to color PET, but with the development of disperse dyes and various dyeing methods, it has now lost its importance. It is now used mainly to produce black shades.

7.1 DYES FOR POLYESTER (PET)

PET is now dyed with nonionic dyes specially synthesized to suit the dyeing processes. Nonionic dyes with low aqueous solubility at dyeing temperatures (100–130°C) are the best dyes for PET. The solubility of nonionic dyes in water is so low that these dyes are considered water-insoluble. It is essential, however, for the dyes to have some solubility in the dyebath to get dyeing in the aqueous bath. These dyes are applied in the form of an aqueous dispersion; the size of the dye particles in dispersion is on the order of 0.5–2.0 μm. The small aqueous solubility and the particle size of a disperse dye play vital roles in the rate of dissolution and the rate of adsorption of dye by PET. A dye that has different crystal forms gives a different rate and extent of dyeing depending on the presence of a particular crystal form in the dispersion of the dye.[2] Fine particles in a dye dispersion oversaturate the dyebath so that big dye crystals build up. These stable dye crystals have low aqueous solubility and a poor rate of dissolution. Such unstable dye dispersions create problems in dyeing. Apart from this, the particles in a dispersion coagulate due to thermal, hydrodynamic, mechanical and ionic shocks. The vibrations introduced by a pump may break the dispersion. The state of dispersion and dispersion stability on storage and during dyeing are of critical importance. Dispersing agents play a vital role in deciding these characteristics. Laboratory equipment has been developed recently to study the dispersion properties of dyes under high-temperature dyeing-conditions.[3,4] Some disperse dyes are sensitive to heavy metals and form chelated compounds with calcium ions giving tonal variations. Soft water is therefore used for dyeing.

The usual way of intensifying the dyeing is by raising the rate of heating of the dyebath. High rates of heating and circulation require considerable stability of the disperse dye in the bath to prevent faulty dyeings. In package dyeing (discussed later), this can result in the filtration of agglomerated dye on the material, thus building up resistance to the liquor flow and giving uneven dyeing. The solubility of the dye in the bath influences the colloidal state of dispersion. The latter is also affected by the temperature. When a hot dyebath is cooled, a proportion of the dissolved dye may precipitate out as a solid that may grow crystals and large particles.

The dyeing properties of disperse dyes on synthetic fibers can be determined by tests such as the critical temperature test, the migration test, the build-up test, and the diffusion test reported by the Disperse Dyes Committee.[5] The pH stability, fastness properties by different dyeing methods, level dyeing property on textured yarn, exhaustion rating,[6] build-up in binary mixtures and so on are examined as and when required before the disperse dyes are used.

Disperse dyes are available in two forms—microdisperse granules or powder and liquid dyes. The dispersion of a dye is spray dried to get solid granules and powders. The amount of dispersing agent required to get stable dispersions can be 40–90% and usually 60% of the dried disperse-dye powder. This large proportion of a dispersing agent in the granules and powders of disperse dyes creates problems such as increasing aqueous solubility, inducing migration during drying of padded goods, lowering the exhaustion of dyebath and so on. The properties expected of microdisperse-dye granules include stability, dryness, uniformity, free flowing, nondusting, and nonhygroscopic nature, good bulk density (~0.5 or more), and ready dispersibility. Liquid dyes are dispersions with a low concentration of a dispersing agent. Dispersion stability, easy miscibility, proper pH, and free-flowing nature are some of the prerequisites for liquid dyes. Since metering pumps can be used for liquid dyes, the additions, weighing and so on pose no problems. Liquid dyes are easy to dissolve and to use. They pose none of the other problems that are associated with the granular dyes. However, liquid dyes are likely to dry up, to settle, and to alter in concentration during storage. Special precautions are required to store and handle liquid dyes. Many times, disperse dyes have poor storage stability, particularly, if they are exposed to a humid atmosphere. Under these conditions, the dispersion breaks into lumps. Such a dye is likely to give uneven, specky dyeing. The state of the dye dispersion can be easily checked by dispersing the dye in water and dropping it on filter paper. If the dispersion is good, no particle will be visible on the filter paper. Improvements in the physical form of the dyes improve the final color results.

Chemically, the disperse dyes come from various classes such as azo, anthraquinone, methine, and diphenylamine. The dyes usually have NO_2, CN, OH, halogen, and primary, secondary, and tertiary amines groups, but they never have any polar groups, which easily ionize in an aqueous bath. Some of the dyes have a free COOH group. Such dyes are usually applied by printing techniques under acidic pH so that this group does not ionize substantially. Free aliphatic hydroxyl groups that impart high aqueous solubility are esterified with acetic acid or a mixture of acids. These dyes generally have low molecular weight which facilitates their entry and diffusion into the highly crystalline polyester fiber. The higher the molecular weight of the dyes, the slower is the diffusion in the fiber. They have significant, though low, vapor pressure, particularly, at elevated temperatures. Disperse dyes are sensitive to pH. Methine dyes hydrolyze or dimerize under alkaline conditions. The pH of the dyebath for dyeing PET is therefore maintained on the acid side. A redox buffer is usually also added to the dyebath to avoid reduction of disperse dyes.[7]

Disperse dyes can be divided into groups based on their suitability for the dyeing process and fastness properties. Dyes with poor fastness to heat but good aqueous dyeing properties are used for dyeing cellulose acetate, triacetate, and polyamide fibers. They are seldom used for the thermofixation process. Dyes with moderate to good sublimation fastness are used for dyeing by the carrier or HT method. Occasionally, such dyes are used to get pale shades on PET by the thermofixation process. Similarly, in the end uses where high sublimation fastness is not required, they are applied by all the methods including the thermofixation method. Dyes with very high sublimation fastness are selected for the thermofixation process. Such dyes may have poor aqueous dyeing properties and may need high-energy, temperature, and time and optimum control during their application by the thermofixation process.

The fastness properties and dyeing characteristics of disperse dyes are considered with particular reference to the subsequent treatments. In the case of yarn dyeing and, to a lesser extent, piece dyeing, wet fastness after heat setting is important since the knitting or coning oils on dyed goods can lead to the migration of the dye into the oil. Besides the usual light and wash fastness, the sublimation fastness of disperse dyes is very important since dyes of low sublimation fastness give problems during subsequent treatments such as resin finishing. A similar high standard of fastness is required for dry and wet rubbing. These fastness properties are assessed before and after heat setting to simulate the performance of the fabric in processing and end use conditions. Migration of dye to the surface of the fiber during the heat-setting process frequently results in failure in the above tests.[3] Dyes with high sublimation fastness are therefore used for the dyeing of yarns. Similarly, dyes, auxiliaries, and dyeing conditions are selected to give optimum coverage of small variations in dye affinity of textured yarns.[3] Thus, dyes used for yarn dyeing must meet with the following specifications:

1. Good dispersion properties, so that the dye is not filtered on a package of yarn that constitutes an effective filter. Paste brands of disperse dyes are usually preferred for yarn dyeing.
2. Good stability in HT bath during dyeing (130°C/2 h).
3. Good leveling properties, at least with the addition of a surface active agent. Use of certain carriers helps in getting level dyeing of a yarn package.
4. Good sublimation fastness.

7.2 MECHANISM OF DYEING

The mechanism of dyeing PET with nonionic dyes under different conditions of dyeing have some common features and some significant differences. HT dyeing and carrier dyeing involve dye transfer from aqueous baths, while in thermofixation dyeing, the water in the pad liquor is completely expelled by a drying process before the dye is fixed on PET.

The contribution of the PET structure to the dyeing mechanism remains the same for the three processes because the fiber does not absorb any significant amount of water and the presence or absence of water on the fiber does not play any significant role in the sorption of dye by PET. The dye is adsorbed only in the amorphous regions of PET, that is, it does not enter the crystalline regions. Thus, if calculated on the basis of the amorphous content of the PET materials, the fiber saturation values of a dye on different PET materials are similar (Figure 7.1).[8] The percentage composition of the crystalline and noncrystalline regions in the fiber may vary from fiber to fiber and the fiber may exhibit apparent differences in its dyeing behavior.

The penetration of dyes in the PET structure is explained by the free-volume theory for the low-molecular-weight compounds in an amorphous polymer (see Chapter 3). The energy effects in dyeing show abrupt changes over a very short range of temperatures at T_g. The concerted movements of

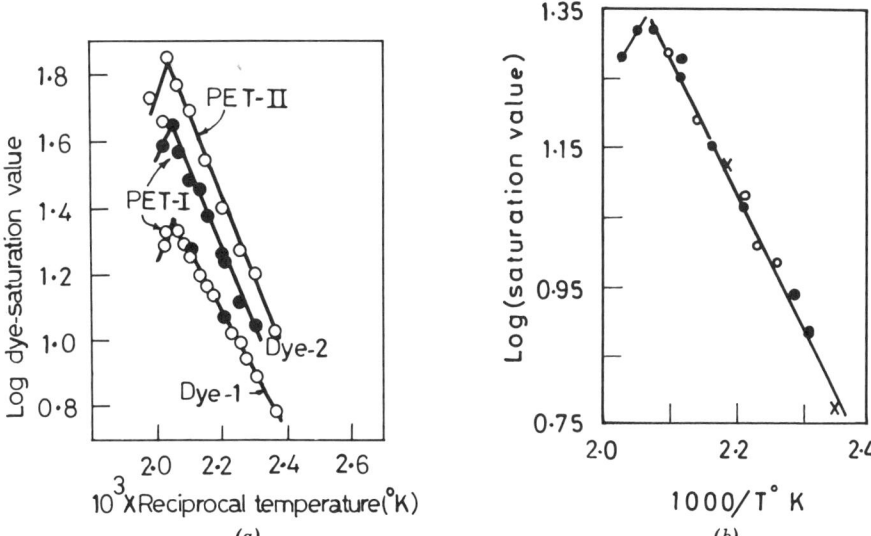

FIGURE 7.1. Temperature dependence of true saturation values of dyes on polyester material.[8] (a) ●: Fixed in air oven; ○: Fixed in metal press; (b). Saturation values were calculated for three polyesters on the basis of their amorphous content when the data on all polyesters lie on the same plot.

chain segments of polymer molecules are started at T_g. An increase in temperature to above T_g raises the frequency and amplitude of the movement of chain segments. This facilitates diffusion of dye and the rate of diffusion increases with the temperature (Figure 7.2). In the thermofixation process, however, as the thermofixation temperature increases and approaches the softening temperature of PET, there is a sudden drop in the fiber saturation value (Figure 7.1). This is attributed to the increased crystallization of PET chains during the premelting stage, which lowers the amorphous content of the fiber.[8,9]

Under dyeing conditions, the rate of adsorption of dye molecules on the fiber surface is always higher than the rate of diffusion into the fiber. Therefore, the former does not exhibit any influence on the overall rate of dyeing and the diffusion of dyes within the fiber is the rate-determining step. Disperse dye has a tendency to deposit on the fiber surface. In the course of dyeing, this deposited dye has to be desorbed to migrate to some other part of the fiber material to get the level uniform dyeing.[10] This prolongs the dyeing process. Significant surface deposition takes place only from oversaturated dyebath[8] (Figure 7.3). This is because the surface of the polyester fiber is full of C—O—C (ether) linkages that are hydrophobic, while the C=O (ester) linkages that are hydrophilic face towards the interior of the fiber.[11]

Since dye has to diffuse through the holes and spaces formed by the vibrations of the chain segments of PET molecules,[12] the shape and size of the dye molecule influences the rate of dyeing. The higher the molecular size of a dye, the higher is the space required for the dye to diffuse. Because of this, as the temperature increases, the effect of the size of a dye molecule on the rate of dyeing decreases; that is, the activation energy of diffusion increases with the molecular weight of the dye[8,13] (see Figure 3.10). The rate and extent of adsorption of a dye are decided by the fiber structure, time, and temperature of aqueous dyeing or thermofixation.[13]

Disperse dyes are combined to produce mixed shades. Neither the rate

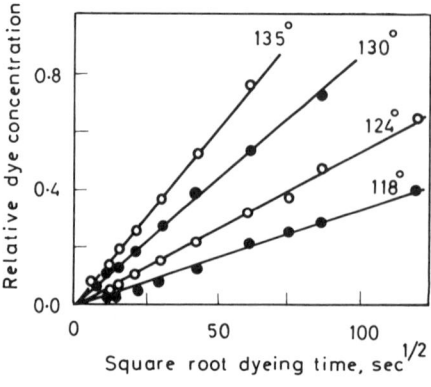

FIGURE 7.2. Rate of dyeing of PET with C.I. Disperse Brown 1 from an infinite saturated bath in perchloroethylene.[8]

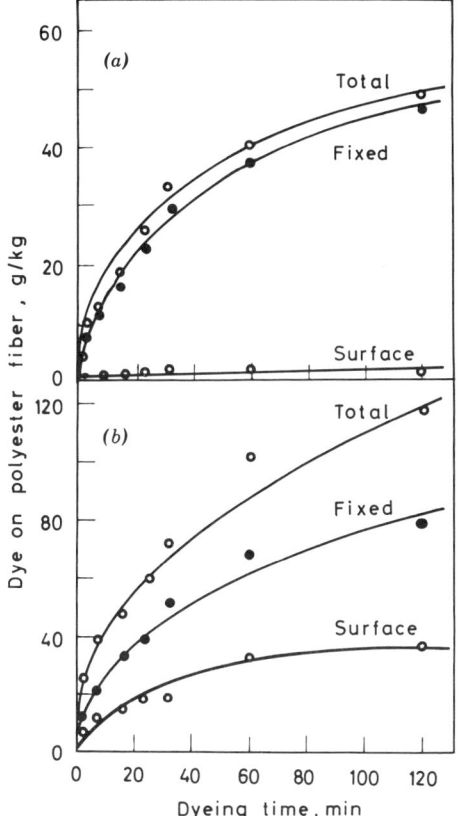

FIGURE 7.3. Dye on PET in HT dyeing (130°C/1 h).[8] Dye: C.I. Disperse Brown 1 (Microdisperse). Concentration: (a) 0.8 g/liter (unsaturated bath); (b) 1.6 g/liter (oversaturated bath); $M:L$ ratio 1:4000.

nor the equilibrium adsorption of dyes in mixture is influenced by the presence of the other dye.[13] The dyes build up on PET, independent of each other, up to their saturation values. This is also the case with dyeing from an organic solvent.[8] We will now consider the individual dyeing processes.

7.2.1 HT Dyeing

The mechanism of dyeing PET with nonionic dyes in an aqueous dispersion has been investigated by many workers.[14-26] Earlier investigators believed that dyeing involved the attraction of positively charged particles of suspended dye to negatively charged fiber surfaces to build up a surface layer of dye particles. Subsequently, the solid dye dissolves in the fiber to form a solid solution. This mechanism, which was first suggested for dyeing cellulose acetate with disperse dyes by Kartaschoff,[27] is now rejected. It is now established that dyeing takes place in a saturated solution of dye in an aqueous bath; the suspended particles in dispersion form a reservoir of dye that replenishes the solution as the dye molecules are removed from the dyebath by the fiber. The dye in solution is assumed to be in a monomeric form, even though experimental difficulties prevent any conclusive proof from being obtained on the monomolecular state of dye in solution. Disperse dyes have a definite water solubility.[28] The solubility of a dye in the bath increases with temperature.

Dyeing takes place in three simultaneous steps: (*a*) dissolution of dye particles in the bath to give a dye solution, (*b*) adsorption of the dissolved dye from solution on to the fiber surface, and (*c*) diffusion of adsorbed dye from the fiber surface to the interior of the fiber substance.

The rate-of-dyeing curves, that is, plots of dye concentration on fiber vs. square-root time, are linear in shape, at least in the initial stages of dyeing. The rate of dyeing increases with temperature (Figure 7.2).

The dyes show an approximately linear equilibrium partition between the dyebath and PET[29] (Figure 7.4). The partition coefficient is dependent on the chemical nature of the dye, the history of PET, the dyebath composition, and the temperature.

7.2.2 Carrier Dyeing

A large variety of organic compounds act as carriers, accelerating the rate of dyeing. Almost all the carriers are aromatic compounds[14] with rigid structures. They swell PET[30] and increase the solubility of the dye in the bath.[31] Based on these two observations, various theories were proposed.[32,33] A carrier causes fiber swelling, which facilitates the diffusion of dye molecules in the fiber, or increases the water of imbitation in fiber, which produces a more hydrophilic environment into which dye moves readily.[32] The improved solubility of the dyes in the bath in the presence of a carrier does not

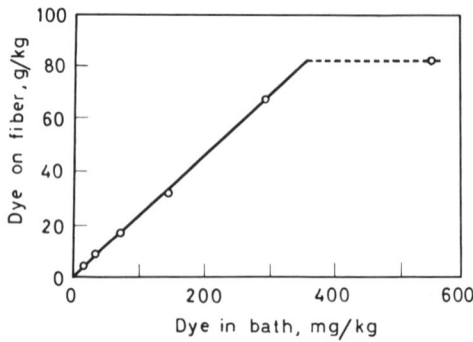

FIGURE 7.4. The relationship between the dye on PET and that in the bath at 130°C at equilibrium.[8] Dye: C.I. Disperse Brown 1 from infinite bath.

always improve the rate of dyeing, for example, diethyl phthalate has virtually no carrier action, even though it has a good solubilizing power for disperse dyes. Similarly, a good carrier like trichlorobenzene causes less swelling than water and does not have any hydrophilic groups to increase the water of imbitation of PET.

It was suggested that a water-insoluble carrier surrounds the fiber as a film in which the dye dissolves. Dyeing takes place from the dye solution in this carrier film.[34] In another theory, the carrier adsorbed within the fiber dissolves the dye. However, the solubilizing property is not correlated with the carrier action. In still another theory, the carrier replaces the intermolecular fiber–fiber bonds with weaker fiber–carrier bonds which helps to loosen the fiber structure and, in turn, to increase the rate of dyeing. Based on the available evidence,[35-45] all of the above-mentioned theories are now discarded in preference to the lowering of the glass-transition-temperature theory.

The carriers lower the T_g of the PET. The drop in T_g is a function of the carrier concentration in the fiber.[43] The carrier causes a disruption of intermolecular cohesive forces between polyester chain molecules, thereby allowing increased mobility of chain segments. The solubility parameter of PET is 10.7 $(cal/cm^3)^{1/2}$ and PET polymer exhibits strong interaction with liquids having similar solubility parameters (see Chapter 3). When the glass-transition temperature is lowered, the effective dyeing temperature, that is, the difference in the actual dyeing temperature and the glass-transition temperature, increases. The diffusion of dye molecules into the fiber increases either when the actual dyeing temperature is raised or when the effective temperature is raised by lowering the T_g by the carrier action[41] (See Figure 3.5).

7.2.3 Thermofixation Dyeing

PET Material.   The stages of padding the liquor and drying in the thermosol process are aimed at depositing the dye evenly on the fabric. At this stage, the dye can be readily stripped from the fabric surface by rinsing with cold water, soaping at boiling, and/or extracting with cold acetone. The dye-liquor is uniformly and evenly deposited in the channels formed by fibers and, on drying, the dye deposits as a solid on the fiber surface. The dye transfer takes place only in the thermofixation step when a part of the solid dye on the fiber surface is transferred to the fiber phase in the absence of water or any other liquid. The dye particles are too big to enter the fiber which has a very compact structure. The dye forms a solid solution in the PET fiber.[46] The relationship between the fixed dye and the remaining unfixed dye on the PET surface at different thermofixation temperatures is shown in Figure 7.5. Though small, the disperse dyes have a measurable vapor pressure at the thermofixation temperature. Datye et al.[9,13,47-49] have shown that the dye solubility in PET is related to this vapor pressure. The

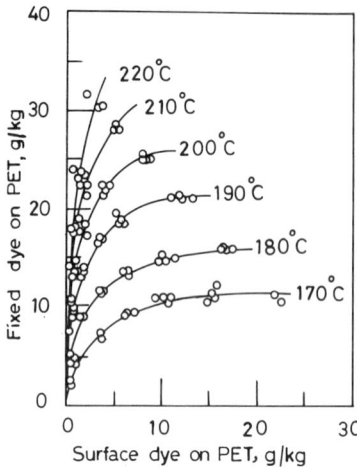

FIGURE 7.5. The fixed dye on PET in relation to residual surface dye after thermofixation at different temperatures in an oven.[49] Temperature (°C)/time: 170°/30 min, 180°/15 min, 190°/10 min, 200°/5 min, 210°/3 min and 220°/1.5 min.

kinetics of dyeing at elevated temperatures (200°C) is explained by the adsorption of dye from its vapor phase.[49] Apparent enthalpy changes associated with dye fixation in the thermofixation process are related to the sublimation properties of a dye (see Chapter 3). It is therefore generally accepted that the dye on the fiber surface vaporizes before it enters the fiber phase. The mechanism of dye transfer, thus, involves formation of dye vapor, its rapid adsorption on the fiber surface, and slow diffusion into the fiber phase.

The interyarn migration of a dye is an absolutely essential step for dyeing a fabric to level shades. The microdisperse dye deposited on the PET surface will have to migrate during thermofixation through the fibers and the air surrounding the fibers to reach the undyed portions of the fabric. This leveling out of the uneven distribution of the dye involves diffusion of dye in the vapor form. This diffusion process is studied by the cylindrical roll method. A typical relationship between the diffusion coefficient and the thermofixation temperature is shown in Figure 7.6.[13]

PET–Cotton Blend.[47]   The blend fabric consists of hydrophobic polyester fiber and hydrophilic cellulose fiber. The latter helps in wetting and increasing the pickup of the aqueous dye-liquor. The padded dye-liquor is uniformly distributed within the channels formed by the parallel fibers in the yarn apart from that imbibed in the cellulosic fiber. During drying, the liquor in the channels starts losing water by evaporation until the channels dry up completely. The dye is deposited at the place where the last traces of water dry up. The liquor has a tendency to remain attached to cellulose and to leave hydrophobic PET fiber if it can migrate to cellulose. This creates migration of dye to cellulose during drying (Table 7.1). The volume of water on padded blend fabric therefore exhibits an adverse effect on the dyeing of polyester by disperse dyes[47] (Figure 7.7).

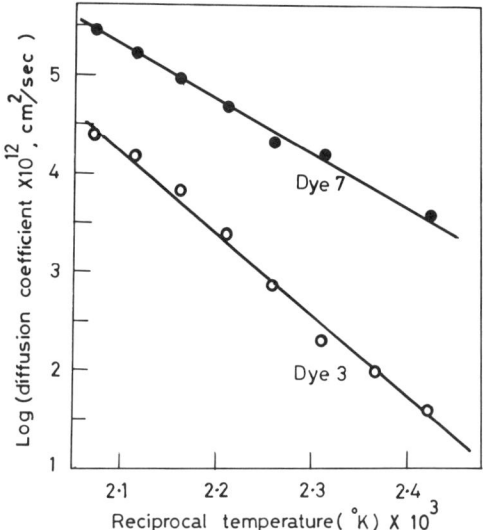

FIGURE 7.6. Relationship between the mean apparent-diffusion coefficient (D*) by pad-roll method[13] and temperature of thermodiffusion.[48] (For the formulas of the two dyes, see Ref. 13.)

TABLE 7.1 Distribution of a Disperse Dye in a Blend Fabric after Drying[47]

Fiber[a]	Dye on Fiber	
	(g/kg)	(%)
Blend	13.57	100
Polyester	9.48	50
Cotton	19.45	50
Concentration ratio		
Cotton : Polyester	2.05	

[a] Fibers are separated by the floatation method after padding 40 g/kg Terasil Brilliant Pink 2GL (microdisperse, Ciba-Geigy) and drying the blend fabric containing PET : CO : 66 : 33

The dye on cotton migrates to PET during thermofixation. This leaves cotton almost reserved giving efficiency of dyeing PET in blend fabric as 75–90%. The high percentage of fixed dye is doubly significant when one considers that more than half of the total dye that is present on cotton (Table 7.1) has to migrate to polyester.[47–53] The remigration of dye on cotton to PET in the thermofixation is thus an important step in dyeing blend fabrics. This

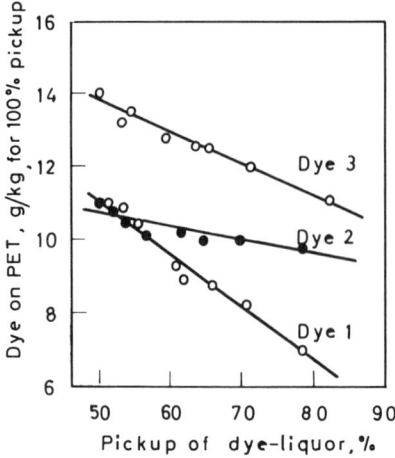

FIGURE 7.7. Relationship between % pickup of dye-liquor on padding mangle and fixed dye (%) on PET in blend fabrics.[13] Pad-bath: Dye 40 g/kg, alginate thickening (4%): 200 g/kg (all values are calculated on the basis of 100% expression). Dyes: 1, 2, 3 are disperse dyes.

remigration is related to the rate of sublimation of dye on cotton[49] and the intimacy of PET and cotton fibers.[48] The dye migrates from cotton to polyester, even when the two are not in physical contact with each other. The earlier theories of dye remigration,[54] which assumed contact transfer from cotton to PET across the area of contact or through a layer of molten additive surrounding the fiber acting as a solvent, are now discarded. It is now accepted that the migration of dye from cellulosic fiber to PET has the following stages.[48]

1. Sublimation of dye on cellulosic fiber.
2. Diffusion of dye vapor in the air to reach the PET surface.
3. Adsorption of dye on the PET surface followed by diffusion into the PET core.

The first stage depends on the vapor pressure of the dye and the fineness of the dye particles. The nature of cellulosic fiber plays no role in the migration step. Dye vapor in the air may diffuse over short distances because of the intimate mixture of the two fibers in the blend fabrics. The third stage of the adsorption of dye on the PET surface depends on the fiber saturation value and the rate of diffusion of dyes in the fiber.[13,47]

7.3 CARRIER DYEING

Waters[14] observed that the rate of dyeing of PET is considerably increased by the presence of many organic substances in the dyebath. These substances are termed *carriers* or *accerlants* and the process of dyeing using carriers is called *carrier dyeing*. PET is usually dyed in the presence of a carrier at boiling in order to get deep shades within a reasonable dyeing time. The effect of the carrier on the dyeing of PET is shown in Figure 7.8.[30]

7.3.1 Carriers

The desirable characteristics of a carrier[33] are (a) ready availability, (b) price, (c) effective at a low concentration, (d) nontoxic and no unpleasant odor, (e) easy removal from dyed PET, (f) inert to the dye and fiber during and after dyeing, (g) nonvolatile, (h) biodegradable, and (i) compatible with the dyebath. No carrier possesses all the desirable characteristics. The most efficient carriers are relatively insoluble in water and are supplied as self-emulsifiable liquids. A large variety of organic compounds act as carriers but only a few exhibit sufficiently attractive properties to justify their commercial use.[55–57] Such carriers are described below.

2-Hydroxydiphenyl. This carrier is insoluble in water and is used as a self-emulsifiable liquid but has no carrier action. It is necessary to regenerate the original phenolic compound during dyeing by changing the pH to the acidic side with ammonium phosphate. This carrier promotes rapid and good build-up of many disperse dyes and does not have carrier-spotting problems since the carrier is not very volatile in steam. It has a low level of odor. However, the residual carrier in the dyed fiber adversely affects the light fastness of the dyeing. Simple scouring treatments do not remove the carrier completely. A heat treatment at 150–160°C is needed for its complete removal. The carrier is not biodegradable.

Monochloro-2-hydroxydiphenyl. This carrier is the most powerful and outstanding carrier suitable for producing heavy shades with high-molecular-weight dyes with good sublimation fastness. The carrier is, however, unsuit-

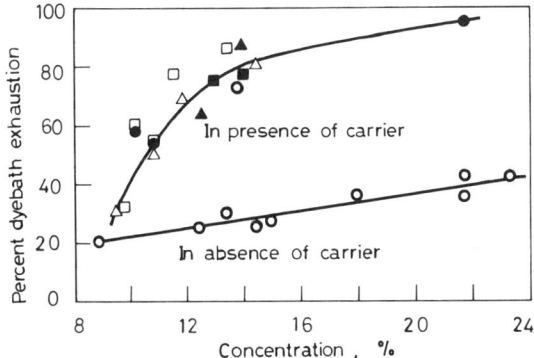

FIGURE 7.8. Exhaustion (%) of dyebath as a function of carrier action on PET fiber.[30] Dye: Latyl Violet 2R (3%); $M:L$ ratio 1 : 200. Carriers: ●: *O*-phenyl phenol; ■: Methyl benzoate; ▲: Methyl, salicylate; □: Benzoic acid; △: Phenol; ○: Carrier is stripped before dyeing.

able for rapid-dyeing low-molecular-weight dyes as their rate of dyeing may be too high to control, giving rise to unlevel results. Other properties are similar to 2-hydroxydiphenyl. The carrier is nontoxic. It also needs a heat treatment at 150–160°C for its complete removal from the dyed fiber.

p-Hydroxydiphenyl.   This carrier is a quite efficient and cheap carrier, has very little odor, and is nonvolatile in steam. However, the removal of this carrier is very difficult and the residual carrier lowers the light fastness of the dyed fiber.

Diphenyl.   This carrier is a moderately effective cheap carrier. The removal of carrier from dyed fiber is easy. The light fastness of dyeing is also not affected by this carrier. However, the carrier has a powerful unpleasant odor and problems of carrier spotting because of its volatile nature. Diphenyl has a low toxicity, but is not biodegradable.

Methylnaphthalene.   This carrier has a weak carrier action and is suitable for dyeing with rapidly diffusing disperse dyes. It creates the problem of carrier spotting and adversely affects the light fastness of the dyed fiber. Methylnaphthalene is an effective solubilizing agent for disperse dyes and lowers the exhaustion, if it is used in large quantities. In small quantities, it acts as a leveling agent in the high-pressure (HT) dyeing process. It is readily biodegradable.

Methyldichlorophenoxyacetate, Ethers of 2,4-Dichlorophenol, and Dimethylterephthalate.   These effective carriers possess little odor, easy removability from dyed fiber, no serious adverse effect on light fastness of dyed fiber, and easy biodegradability. However, they are very expensive.

Methyl cresotinate, Methyl Salicylate, and Butyl Benzoate.   These carriers have many of the good properties described above. They exhibit very low toxicity. However, they have a distinctive, moderately sharp, though not unduly unpleasant odor. They are expensive and volatile in nature. They can be used as a leveling agent in the HT dyeing process.

N-**Alkylnaphthalimide Derivative.**   This efficient carrier has little odor, low toxicity, and is easily removed from dyed PET. It is biodegradable and is especially useful for garment dyeing.

7.3.2   Carrier Dyeing Process

PET fabric is heat set before dyeing in a closed automatic jigger with direct and indirect steam-heating arrangements. The bath is set in the jigger with

0.1–0.5%	Anionic wetting agent
0.1–0.5%	Dispersing agent
$X\%$	Carrier
0.1–0.5%	Acetic acid (for pH 5.5–6.5).

The fabric is run at 60°C for half an hour in the blank bath. The dye is dispersed in water at 40–50°C. This dispersion is filtered through a muslin cloth and is added to the jigger during two ends at 60°C. The temperature is slowly raised to boiling in half an hour and the dyeing is continued at boiling for 90 min. A sample is taken out after dyeing for about an hour to match the shade. Before adding the shading dyes, the temperature is lowered to 80–85°C. The temperature is then raised to boiling and boiling is continued for half an hour to complete the dyeing. The dyed fabric is then soaped (1 g/liter soap, 1 g/liter soda) at boiling to remove the carrier from the PET. It is then given a reduction-clear treatment (2 g/liter sodium hydrosulfite, 2 g/liter caustic soda, 60°C) for half an hour to destroy the unfixed dye on the PET. The fabric is given a hot and cold wash and is then dried.[58] The cotton, viscose, and wool in PET blend fabrics are then dyed with suitable dyes as is described later.

The pH of the dyebath has to be 5.5–6.5 so that the disperse dye or wool fiber in the blends are not hydrolyzed during dyeing. Under highly acidic conditions, the cellulosic fibers in the blends are damaged. The *tailing* effect can be minimized by operating the jigger at the maximum possible speed and simultaneously blowing steam in the upper part of the jigger (see Chapter 6).

It was predicted in the early 1960s that the interest in carrier dyeing would decrease with time when new dyes and, particularly, new pressure-dyeing equipment became available. The carrier-dyeing process, however, has retained its importance since it permits PET dyeing at boiling under atmospheric pressure using conventional equipment such as liquor-circulating machines, covered jiggers, beaks, and winches. In fact, carrier dyeing is the method for polyester–wool blends and, to some extent, texturized fiber because of the limitations of temperature imposed by the presence of wool and temperature-sensitive textured and knitted PET. The addition of small amounts of a carrier is found to be beneficial, even in the HT dyeing process, because the carrier improves the leveling and migration properties of a dye.

The carrier dyeing has some limitations such as the high cost of carrier, difficulties in complete removal from the dyed fabric, spotting problems, and pollution problems. Some carriers adversely affect the light fastness of the dyed goods if they are present even in small quantities. Many disperse dyes with excellent fastness properties are not fully exhausted from the bath to build up a heavy shade. Because of these limitations, HT and thermosol dyeing processes are preferred.

7.4 HIGH-TEMPERATURE (HT) DYEING

The influence of temperature on the rate of dyeing[14] was exploited as soon as machinery was developed that permitted dyeing above the atmospheric pressure at temperatures up to 140°C. The main advantages of HT dyeing are reduction in dyeing time, no carrier cost, complete penetration of PET even with high-molecular-weight dyes, and a higher build-up of shade than by carrier dyeing. As the temperature of dyeing increases, the rate of dyeing increases remarkably, and the extent of exhaustion of the dyebath during the short span of dyeing time increases to a satisfactory level. The rapid growth of HT dyeing for both yarn and piece goods, apart from loose stock and sliver, in the last two decades has been one of the most significant developments in polyester processing, particularly, because of the rapid growth throughout the world of knitted, textured PET fabric. The introduction of HT rapid-dyeing processes in recent years has changed the picture of HT dyeing.

7.4.1 HT Dyeing Process

The process of HT dyeing is schematically depicted in Figure 7.9.[8] The disperse dye dissolves in the dyebath and the dye in solution is taken up by the fiber after it reaches the fiber surface by diffusion through the liquid phase. Vigorous movement of the PET material and/or the liquid minimizes or even eliminates the influence of this transport process. Various machines for HT dyeing have been designed so that the liquor circulation around the fabric is uniform and very rapid. However, the boundary layer on the fiber surface remains a stationary liquid layer through which the dye has to diffuse

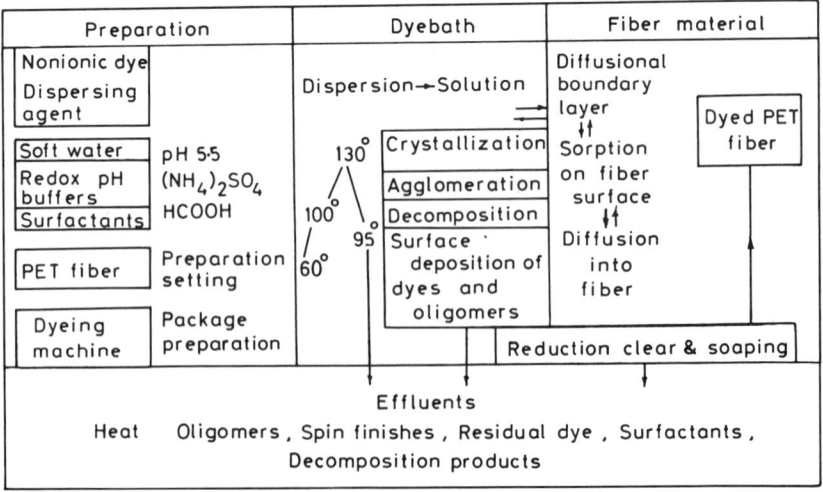

FIGURE 7.9. Process of HT dyeing.[8]

in order to reach the fiber surface (see Chapter 3). Therefore, in principle, the dye is adsorbed on the fiber surface from the diffusional boundary layer, which, in turn, receives the dye from the bulk of the agitated liquor. The thickness of the boundary layer under circulation conditions in different machines may vary, which, in turn, may vary the rate of dyeing from machine to machine.

The migration of dye has been regarded as the prime factor in level dyeing because it usually prolongs the dyeing process. Attempts have been made to dispense with the migration phase by manipulating the temperature, the heating rate, and the rate of addition of dye to the bath in order to avoid surface deposition. For example, Figure 7.3 shows the uptake of dye on polyester in HT dyeing at 130°C. When the bath is unsaturated (0.8 g/liter), there is very little surface dye on the PET fiber. When the dyebath is oversaturated (1.6 g/liter), a large portion of the dye is deposited on the fiber surface that must migrate in order to give level dyeing. If dye is present in the solution without any dispersed dye in the bath, the migration phase can be avoided.[8]

Only microfine disperse dyes with good dispersion stability under drastic HT dyeing conditions are used for HT dyeing. The dye crystals and aggregates filter off on the fabric, giving poor fastness properties. Rubbing fastness of heavy shades is very poor at times. There is a danger of oligomer deposition on the fiber and surface of the HT unit. The residual dye in the bath crystallizes out and deposits on the fiber surface during cooling. To avoid these problems, the exhaust bath is dropped at as high a temperature as possible. A reduction-clear treatment also improves the rubbing fastness.

7.4.2 HT Dyeing Equipment

The newer HT units have been fitted with very efficient high-capacity pumps and heat exchangers. Arrangements are made to drop the exhaust liquor at the dyeing temperature under pressure. There are units in which air is injected under high pressure. HT dyeing machines are expensive and energy-consuming. The machines for HT dyeing can be classified into two groups: (*a*) pressure jiggers and (*b*) package dyeing machines.

Pressure Jiggers. The new pressure jiggers exhibit the same advantages and limitations as the normal jiggers[59] (see Chapter 6). Jigger dyeing is a batch process suitable for both large and small batches. These machines exert a pronounced lengthwise tension on the fabric and hence, a delicate material such as a knitted fabric is likely to get distorted. Pressure jiggers are, therefore, best suited for heavy fabrics. New tensionless jiggers are best suited for delicate materials.

HT Package-Dyeing Machines.[60-62] In package dyeing machines for PET and its blends, the yarn or fabric to be dyed is wound on a perforated holder or a beam. The package is then mounted and moved into the pressure vessel. The dye-liquor is forced through the wound package with pumps. A special device is provided to take out a sample for matching while the machine is running under pressure. These units are usually used under the extra pressure given by the compressed air so that the liquor does not boil and generate pockets in the package. The pressure in the machine is more than the steam pressure of water at the dyeing temperature (130°C). This process is called the *high-pressure HT process*. Careful package preparation is an essential prerequisite to the success of the package dyeing operation. The package dyeing of a yarn is a costly process and much of the extra cost must be associated with transferring the yarn from one form to another. (This adds no intrinsic value to the end product.[63]) HT package dyeing machines are developed for dyeing knitted goods made from textured polyester yarns with an air-injection device[64] (see Chapter 6).

Irregular dyeing may occur in HT machines if the individual yarns within the load differ in their dyeing properties. Both the rate and equilibrium dye uptake may differ in individual yarns. Dyes with good migration properties can cover yarn variations. However, such dyes enhance the problem of variations in equilibrium dye uptakes. The color yield in heavy depth is generally poor due to the incomplete exhaustion of the dyebath. Furthermore, there is a tendency for a disperse dye to deposit on the fiber surface when heavy depths are dyed. Thus, goods in heavy depths need repeated reduction-clearing treatments. Individual fibers in the package should be sufficiently well-penetrated with dye to avoid lower fastness properties, change in color on subsequent heat treatment, and matching problems from batch to batch.

Other HT Units.   Coloring of PET materials can be carried out in HT winches and HT steamers. HT winches are usually used for dyeing carpet materials. Batchwise pressure steamers have been developed to deal with PET materials, particularly, to fix the prints of disperse dyes. They are rarely used for the uniform application of a dye. These two methods are described elsewhere.

7.4.3   Preparation of Fabric for Dyeing

A fabric to be dyed must be desized, scoured, bleached, if necessary, and heat set. In order to avoid wrinkles, creases, and shrinkage of the fabric during dyeing, heat setting is an essential pretreatment.[65-71] The temperature during heat setting is controlled within narrow limits because the dye uptake varies with the temperature of heat setting[45,65,72] (see Chapters 2 and 3). Correct beaming up of the fabric is essential to ensure satisfactory results. If any air is trapped, it will gradually move to the top of the batch during dyeing and may leave undyed areas. Irregular tensioning of the fabric causes channelling or deflection of the flow and therefore unlevelness. In order to produce evenly dyed selvages, the edge of the batch should slightly overlap the end of the perforations. If the overlap is too small, the liquor tends to flow rapidly at the selvages, resulting in deeper selvages. If the overlap is too large, pale selvages can result. Depending on the quality of the fabric and the thickness of the roll, one must determine the correct conditions. The best results are obtained if the overlap is equal to half the thickness of the batch or 2.5 cm, whichever is greater.[73] The quantity of the fabric per batch depends on the quality of the fabric, size of the dye vessel, and the capacity of the pump. Beams of large diameters are preferred since a considerable amount of material can be rolled on them, while the thickness of the roll remains comparatively small. Typical thin fabrics can have a batch size of 1200–2000 m; shirting fabrics, 700–1000 m; and suiting fabrics, 250–500 m on a beam of diameter 50–60 cm.

Cotton cloth (10 m) of not-too-tight construction is stitched to the ends of the fabric. These end-pieces prevent the marking off of the perforations on the fabric and act as filters for agglomerated dye particles, incompletely dissolved chemicals, and oligomers. The batching of the fabric is done in a wet state to minimize the moiré effect on the dyed fabric.

7.4.4 Selection of Auxiliaries[3,74]

Dispersing Agent.[3,75] Although the solid disperse dye contains 50–90% dispersing agent, this amount is usually not sufficient to maintain the dye in finely dispersed form in the large volume of dye-liquor. This is the case particularly when pale and medium shades are to be dyed, where the amount of dye in the liquor is very small. The addition of a small amount of dispersing agent helps in avoiding agglomeration of dyes during dyeing. The anionic dispersing agents give the best results. If a nonionic dispersing agent is to be used, it is essential that its cloud point is high. Very severe aggregation of dyes can take place in the presence of a nonionic dispersing agent of relatively low cloud point, when the temperature of the dyebath is raised above the cloud point. Some of the nonionic dispersing agents also lower the rubbing fastness of dyeings because of the higher deposition of dye on the surface of the fabric.[3] Addition of a dispersing agent influences the rate of exhaustion of dyes. However, it does not influence the rate of diffusion of dyes in the fiber. The dispersing agent lowers the extent of exhaustion.[76] This is because the solubility of disperse dye in the bath increases with the addition of a dispersing agent (Table 7.2).[3] According to Carbonell,[75] for better fastness properties, particularly, at saturation point, a small amount of dispersing agent on the weight of the goods may be added. It also gives better exhaustion and better crock fastness with long dyeing cycles. However, for very long dyeing cycles involving shading, a very small amount of dispersing agent is preferred.

A dispersing agent remains in the exhaust liquor and holds back some dye.[8] Many dispersing agents are not biodegradable and create pollution

TABLE 7.2   Solubility of C.I. Disperse Red 151 at 130°C[3]

Additive	Solubility (mg/liter)
None	20.8
Irgasol DA (1 g/liter)[a]	38
Neolan Salt P (1.5 g/liter)[a]	48.8
Methylsalicylate (1.75 g/liter)	32
Experimental product (5 g/liter)[a]	26

[a] Ciba-Geigy Product.

problems. Due to their surface activity, many dispersing agents interfere with the effluent treatment.

Leveling Agent.[74,77]   The leveling agent for HT dyeing is either a noncarrier or a carrier type. Furthermore, the noncarrier agent is nonionic or anionic in nature. The nonionic leveling agents retard the rate of dyeing initially and thus, produce uniform dyeings. However, some of these products also desorb the dye and thus, there is a loss in color yield. The anionic agents do not produce any restraining (desorption) effect at the end of the dyeing process and seem to be more suitable.[78] The carrier type of leveling agents are increasingly used in HT dyeing to get uniform dyeing. These leveling agents increase the migration of dye during dyeing, as shown in Table 7.3.[78] The carrier type of leveling agent is of particular interest for dyeing with dyes with poor migration properties. These leveling agents are added to the HT bath only after the maximum dyeing temperature is reached. This avoids an excessive strike of the dye during the heating period. The usual effect of a carrier on the fastness properties of dyed goods should be taken into account during the selection of a carrier and dyes. These agents are exhausted onto the fiber.[3,39] The carrier-type leveling agents avoid the deposition of oligomers on the fiber surface.

Antifoaming Agents.   The HT dye-liquors have a tendency to foam under the influence of dispersing and wetting agents and the pumping action. Excessive foam gives specky dyeings and may impair the pump action. It is therefore a practice to add a silicone-type defoamer to the dyebath. Diluted defoamer is added to the cold dyebath so that it is well dispersed before the temperature is raised. The addition of the defoamer to the hot dyebath can cause a breakdown of the silicone emulsion, which results in spotting.

Wetting Agents.   Polyester is an extremely hydrophobic fiber and addition of a wetting agent to the HT dyebath is a usual practice. Wetting agents usually tend to foam and must not be used in large quantities.

TABLE 7.3   Migration(%) of Disperse Dyes at 120°C in the
Presence of Carrier-Type Leveling Agents[78]

	Migration(%)[a]			
C.I. Disperse Dye	No Agent	Palanil Carrier B[b]	Remol LE[c]	Tumescol D[d]
Yellow 42	42	84	92	96
Orange 60	10	70	32	22
Orange 54	32	68	96	50
Orange 13	18	78	80	38
Red 131	48	70	74	76
Red 82	48	80	72	92
Violet 33	52	82	82	76
Blue 122	34	64	62	58
Red 60	44	70	100	78
Red 11	56	100	98	98
Blue 72	34	72	72	38
Blue 83	32	62	68	54

[a] Dyed and initially undyed samples match at 100% migration.
[b] BASF product.
[c] Hoechst product.
[d] ICI product.

pH and Redox Buffer.   A mixture of ammonium sulfate and formic acid is the best redox buffer to protect the dyes from degradation in the HT bath. The pH of the dyebath is adjusted to 5.5 with formic or acetic acid. At the end of the dyeing, the pH should not rise above 6.5.

Other Dye Formulations.[8]   Many of the problems in the HT dyeing process can be traced back to the metastable dispersion of nonionic dyes. Attempts are made to replace the metastable dispersions with stable solutions or to use an inert adsorbent on which the dye is mounted. A saturated solution of the dye in an organic solvent is gradually injected into an aqueous dyebath of desired pH and temperature (130°C). The dye gets dispersed and, depending on its solubility in the dyebath at 130°C, dissolves in the bath and is taken up by the fiber. The solvent may be recovered by an azetropic distillation. The rate of injection of the dye solution has to match the rate of exhaustion for the best results. Nonionic dyes dissolve in many polymers to give a "solid solution" of desired size, shape, and composition. Alternatively, the dye is adsorbed or deposited on an inert surface, for example, wood or paddy husk, after a suitable surface modification of the support of desired shape and size. These preparations are suspended in the liquor of the HT machine in either a bag or a cage and the liquor is allowed to flow through the dye formulation in a holder. The dye gradually dissolves in the liquor and is transported to the PET, where it is adsorbed. The residual preparation is fished out from the dyebath. The exhaust liquor remains free from the preparation as dyeing takes place in the saturated solution of the dye in bath. In these two techniques, the dye-migration step during dyeing is almost eliminated.[8]

The temporary solubilization of nonionic dyes, for example, ionamines for cellulose acetate fiber, is, in theory, a convenient way to avoid the dispersion process. The rate of regeneration of the insoluble dye from its soluble derivative must match the rate of its adsorption on PET. Such modified dyes or formulations, as are described above, are not available in the market.

Laboratory Scale Dyeing.   Laboratory-scale dyeing is carried out to find out the dyes, their proportions, and their ability to develop the desired shade on the PET material to be dyed, before dyeing on a HT-machine is tried. Nonionic dyes are adsorbed on the walls of dyepots, which can introduce matching errors in delicate shades. The pots are cleaned with acetone before use. Error may be introduced in the final composition because of this factor.[29] Shades are matched as precisely as possible to minimize the shading during final dyeing on the commercial unit. The latter is a time and energy-consuming process and adds to the cost of dyeing.

7.4.5   Dyeing on HT Machines

Two-Bath Process.   The dyebath is set with

1 g/liter	Dispersing agent
0.5 g/liter	Wetting agent
0.5–1 g/liter	Leveling agent (optional)

The redox buffer is added and the pH is adjusted to 5.5. The liquor is circulated through the package at 50°C for 15 min. The dyes are dispersed in 10 to 20 times their weight of water at 45–50°C. The dispersion is filtered through a thin cloth into the dyebath. The liquor is circulated and the temperature raised to 130°C in 30 min. The pump pressure is slightly higher than the static pressure. Dyeing is continued for 40–60 min. The dye-liquor is circulated inside-out throughout the whole dyeing cycle so that the position of the fabric is not disturbed. Or the flow of liquor may be from inside-out for 7 min to outside-in for 3 min in every 10 min to get very uniform results. After about 40 min, a sample is taken out for the shade matching. The liquor is cooled to 80–90°C before the shading disperse-dye liquor is added to avoid uneven dyeing.

On completion of dyeing, the dyebath is drained at the highest possible temperature. Otherwise, the shade appears bronzy because of the deposited ashlike oligomer powder that may dust off and create problems in finishing. The precipitated dyes are difficult to remove from the PET surface and they lower the rubbing fastness of dyed goods. Thus, the bath must be dropped at the dyeing temperature or as near that temperature as possible. However, there are difficulties in dropping the bath (which is under pressure), since a lot of steam is generated when the hot liquor comes to atmospheric pressure. Special arrangements are usually made to take care of this problem. After the bath is dropped, the material is washed, reduction-cleared, and soaped. In the reduction-clear treatment, the material is treated with sodium hydrosulfite (1 g/liter) and caustic soda (3 ml/liter, 30%) below 60°C. (Hydrosulfite decomposes rapidly above this temperature.) Hydrosulfite cannot destroy the dye unless the dye dissolves in the bath. It is therefore absolutely essential to bring the dye on fiber into the solution in order to destroy it. As the dissolved dye is destroyed, more dye dissolves and the process continues until all the surface dye is destroyed. If the solubility of dye in an alkaline bath at 60°C is very very low, the dye on fiber remains as is and is not destroyed in the reduction-clear treatment. Addition of a wetting agent in the reduction-clear bath helps the dissolution process. For very heavy shades, it may be necessary to repeat the treatment. The concentration of the surface unfixed dye can be checked by extracting the sample with acetone at 0–5°C. The soaping is carried out in the same bath or in a fresh bath at temperatures up to boiling in order to get excellent wet and rubbing fastness. After soaping, the material is rinsed with hot and cold water. If PET/CO blend fabrics are to be dyed, the polyester is usually dyed first and the cotton is dyed later with direct, reactive, or vat dyes. The reduction-clear treatment and the vat dyeing can be combined. For all other dyes, the reduction-clear treatment is essential since disperse dyes have a tendency to stain cotton and other cellulosic fibers in the blend, which impairs the shade and fastness properties of the goods. However, for a number of pale and medium shades and for deep shades with dyes that do not stain the cellulosic portion heavily, the reduction-clear treatment is superfluous. For such shades, the reverse dyeing method is suggested.

Reverse Dyeing Method.   In the reverse dyeing method, the cellulosic portion is dyed first followed by the dyeing of the PET portion,[79–81] so that one of the soaping operations is eliminated. Because of their good stability under mild acidic conditions employed during disperse dyeing, the vinyl sulfone type of reactive dyes are suited for the reverse dyeing process. Remazol reactive dyes (Hoechst)[79] or highly reactive Procion MX dyes (ICI)[80,81] are applied in the first step. The goods are soaped and rinsed to remove the unreacted dye before the PET portion is dyed with disperse dyes. After dyeing, the material is rinsed to remove the surface dye. Besides the considerable savings in energy, water, and chemicals, the process claims to reduce the pollution problem.[82]

One-Bath Process.   Dyeing PET and cotton or other cellulosic fiber in a single bath has the advantages of high production and low cost of energy, water, chemicals, and labor.[83] However, the matching of shade is difficult in the single-bath dyeing process. Furthermore, the time required on the HT machine is longer since cotton, which does not need high pressure for its dyeing, is also dyed in the HT machine. In two-bath processes, cotton is dyed in open machines such as jiggers and only PET is dyed in the HT machine. The combination dyes for the single-bath process are marketed as a single product by a special brand name. For this purpose, dyes from two chemical classes are selected and mixed. The following combinations of dyes can be used for the single-bath HT dyeing process.

Direct and Disperse Dyes.[83–85]   Dyeings in pale to medium shades of moderate fastness properties are produced by this combination of dyes. The direct dye should be stable under HT dyeing conditions of high temperature and the acidic dyebath. Dyeing conditions are the same as those used for disperse dyes alone, when direct dyes also give excellent level dyeings on cotton. The dyebath is cooled to 85–90°C and is maintained for the exhaustion of the direct dye. Salt may be added to improve the exhaustion. Direct dyes that are not stable at higher temperatures can also be added at this stage. This is the shortest single-bath dyeing process for PET–cotton blend fabrics.

Reactive and Disperse Dyes.   Disperse dyes are applied to the PET portion under HT dyeing conditions. The bath is cooled to 90–95°C when reactive dye and salt are added. The fixation of reactive dye is accomplished by adding soda ash at 80°C. A full range of colors with high fastness properties can be produced by this technique. The unfixed disperse dye on the fiber surface is removed while the reactive dye is fixed by the soda ash solution at 80°C. The main problems with this combination of dyes are difficulties in matching and the addition of a large quantity of the salt during dyeing.

It is possible to dye the cotton portion with the reactive dyes first. The bath is then adjusted to pH 5.5 and the polyester is dyed under HT conditions. The second stage serves as a very efficient washing-off treatment for the unfixed reactive dye. However, the large amount of salt in the bath can break the dispersion of the disperse dye causing aggregation, surface deposition, and problems in the migration–leveling phase of HT dyeing. Furthermore, the matching of the shade is difficult, since the shade on cotton has to be matched first because this component is dyed first. No method is available to remove the PET component from the blend in order to judge the shade of the cotton portion.

Vat and Disperse Dyes.[86]   The dispersion of both dyes is added together after running the blank bath with a dispersing agent and acetic acid at 60°C for 10 min. The temperature is raised rapidly to 95°C and then slowly (1°/

min) up to 130°C to dye the PET within 45–90 min as described earlier. The bath is cooled to 85°C when caustic soda and hydrosulfite are added to reduce the vat dye within 30 min. The bath is slowly cooled (1°/min) and maintained at 60°C for 10 min. Alkali and hydrosulfite are added, if necessary (phenolphthalein and vat papers are used for testing). After dropping the bath, the goods are rinsed in water for 10 min, oxidized, soured with dilute acid, rinsed, soaped, and washed. Although this method gives very high production, there are certain problems such as difficulty in matching, stability of vat dyes at high temperatures and so on. Vat dyes may stain PET fiber. This method requires strict parameter controls and is not very commonly used.

Yarn Dyeing.[87–89]   It is not possible to dye single or multifilament flat yarns in the hank form. Textured yarns can be dyed in the muff form only if the yarn is doubled after texturing. PET yarn dyeing is therefore carried out only in the package form. The yarn is wound on a perforated package that is then mounted on a holder through which the dye-liquor can flow. The package winding is a tricky business and extreme care has to be taken to wind the yarn under uniform tension. The package should not be so tight that it will hinder the flow of liquor. It should not be too loose (soft), otherwise, the yarn may slip and entangle. The package hardness has to be uniform in all the cheeses, since a number of cheeses are packed together. The process of dyeing is the same as that for the piece goods in the beam dyeing machine described earlier. After dyeing, the yarn is wound on cones after a finishing-oil preparation is applied to avoid static electricity generation and surface friction. The dyed yarn is used for weaving or knitting into fancy fabrics. The fabric receives the usual treatments such as heat setting, scouring, finishing and so on. The dyes on the yarn have to withstand all these treatments. The requirements of fastness properties for effect threads are more stringent than those for dyed fabrics (See Section 7.1). The PET portion of a blend yarn is dyed with disperse dyes followed by reduction-clearing, washing, and soaping. The cotton portion of the yarn is then dyed with direct, reactive, or vat dyes.

Matching of Shades.   The correct matching of a shade on PET yarn or fabric usually poses no major problem. The same is not the case with the PET/CO or cellulosic blend yarn or fabric. The most widely used method for the shade matching on PET in a blend is to dissolve the cotton portion in sulfuric acid. Seventy percent sulfuric acid at 40–50°C is used for this purpose. The shade on the residual PET fiber is then matched. The shade on cotton cannot be checked or matched by any simple method. If the cotton portion is dyed first, matching becomes exceedingly difficult. A rapid method of separation of components of blend fabric or yarn is to powder the sample and to wet the powder with water and hydrophobic water-immiscible solvent (e.g., white spirit).[54] The cotton portion and the PET portion are separated into two liquid phases, namely, water and white spirit, and can be further separated by decantation and isolation. The matching of powdered fibers needs some practice, otherwise, the technique is very simple. The contribution of staining of cotton with disperse dyes always remains unevaluated.

Stripping Faulty Dyeings.   The correction of offshades on PET is very difficult and troublesome. The stripping of the disperse dye from PET is done either by (*a*) sodium sulfoxylate formaldehyde and a carrier at 100°C and pH 4–5 or (*b*) sodium chlorite and formic acid at 100°C and pH. 3.5. The best results are obtained by giving treatment (*a*) followed by treatment (*b*). Redyeing the stripped material may pose problems of even dyeing. The best way to handle faulty dyeings of PET is to overdye them into the jet-black shade. Since the capacity of PET to adsorb disperse dyes is almost unlimited for practical purposes, a jet-black shade can be produced on any PET material without tonal variations.

7.5   MODIFIED AZOIC PROCESS

For many years there was a severe restriction on the choice of disperse dyes and the attainable depth of shade was dictated by the earliest PET dyeing processes. Therefore, considerable attention was directed to the modified azoic dyeing process, which forms the dye *in situ.*[8,14,55,68,90] Simple aromatic amines and hydroxy compounds possess good affinity to PET and diffuse rapidly to the center of the fiber. The amines and the naphthols are applied at boiling as their dispersions, either together or sequentially in the case of those products that tend to interact in the bath to form a tar. The diazotization and coupling are carried out hot. The rate of development is governed by the diffusion of nitrous acid into the PET fiber. This method requires careful selection of azoic components with respect to their uptake rate, resulting fastness properties, and their compatibility with each other. The use of a closed (pressure) vessel eliminates the loss of nitrous fumes and improves the color yield. There should not be any residual free azoic base on the dyed fiber, that is, there has to be an excess of coupling component over the base for good sublimation fastness. Azoic combinations on PET exhibit variations in light and sublimation fastness. The range of coupling components that could be used increased as carrier and high-temperature dyeing processes were introduced and C.I. Disperse Black 1 became very important for the production of blacks. The azoic base is marketed in a disperse form as a diazo black base.

   The use of azoic combinations was common for a number of years for the production of heavy reds, browns, navys, and blacks, particularly on polyester slubbings. The method of application of azoic combinations is complicated, and fastness properties, particularly, rubbing and sublimation fastness, are poor at times. As the range of disperse dyes was extended and HT equipment came into general use, azoic combinations declined in importance except for the production of blacks. Azoic combinations are applied to the PET component of the blend for the economic production of black, navy, brown, and maroon shades. They give dyeings of good fastness properties if special precautions are taken. The large dye molecule synthesized *in situ* in the fiber from small molecules of satisfactory diffusion properties gets trapped in the fiber. It is not easily removed by subsequent heat and wet treatments. Thus, many azoic combinations exhibit excellent heat and wet fastness properties compared to the low-molecular-weight disperse dyes.

Under the special dyeing conditions of azoic combinations for PET, the cotton portion of the blend remains undyed. It is then dyed with vat, sulfur, direct, or reactive dyes. The vat dyes are preferred because of their satisfactory fastness properties. The PET–cotton blend is dyed to get a black shade by the following methods:

Method 1.   The dispersed diazo base is applied by either the carrier process or the HT dyeing process. The azoic coupling component is then applied followed by diazotization and coupling. The bath is set with

3–5%	Disperse diazo black base (owf)
3.5 g/liter	Carrier (for dyeing at 100°C)
1.0 g/liter	Wetting agent

The dyeing is started at 60°C, the temperature is raised to boiling in 30 min, and is maintained for 45 min. In the HT process, dyeing is carried out at 110–130°C without a carrier. The bath is dropped at the end of the period. A new bath is set with

5%	Dispersed coupling component (owf)
1 g/liter	Wetting agent

The dyeing is started at 70°C, the temperature is raised to boiling, and is maintained for 30 min. After a reduction-clear treatment, diazotization is carried out with

2–8%	Sodium nitrite
11–14%	Hydrochloric acid

at 85°C for 20–30 min. After rinsing, the goods are soaped at 70–80°C for 15 min with soap (1 g/liter) and soda ash (2 g/liter), washed, and dried.

Method 2.   The azoic base and coupling component are applied in the same bath by the carrier method. The sodium salt of the coupling agent (1 part) is prepared by pasting with methylated spirit (0.75 part) and dissolving in caustic soda (0.5 part, 28%) and water (1.5 parts). The dyebath is set as in Method 1 with the addition of the sodium salt of the azoic coupling agent. The temperature is raised to boiling and maintained for 15 min. At this stage, acetic acid (30%) is added after dilution to bring the pH of the bath to 5–6 (about 5.5 parts of 30% acid). Boiling is continued for 90 min after the coupling component is adsorbed. The bath is drained and the diazotization is carried out as in Method 1.

Method 3.   The azoic base and the coupling component are applied in the same bath by the HT dyeing method. The bath is set at 85°C with

3–4%	Disperse diazo black base
2.5%	Disperse coupling component
1 g/liter	Wetting agent

The azoic base and coupling component are dispersed separately with 10 to 20 times their weight of warm water, allowing the dispersions to stand for 10 min with occasional stirring before sieving individually into the dyebath.

After dyeing at 85°C for 15 min, the temperature is raised to 130°C in 15–20 min, and is maintained for 45 min. The pH is between 7.00 and 7.5 during this period. The bath is drained and the goods are rinsed with cold water, followed by reduction-clear, washing, and development of color by diazotization and coupling, as described in Method 1. Final soaping and washing complete the process.

7.6   THERMOFIXATION PROCESS

One of the most important discoveries in the field of dye application to textiles is the continuous process of coloring PET fabrics using a dry-heat fixation technique patented by Gibson.[91] This DuPont Thermosol process comprises impregnating PET fabric with an aqueous dispersion of a nonionic dye, drying, and exposing the dried fabric to heat at a temperature of 180°–230°C for a few seconds to effect the fixation and the uniform distribution of the dye within the fiber. The process is widely used in dyeing and printing of PET and PET–cotton blends with disperse dyes as such or in a mixture with reactive or vat dyes. The cotton portion in the blends is not dyed under thermofixation conditions unless the process is further modified and special dyes are used. The latter process is known as Dybln Process (DuPont).[92] In this process, the nonionic dye is used for coloring both PET and cotton. Another outcome of the studies on the thermofixation process (and of great technoeconomical importance) is the sublimation transfer-print process.[8,93] The latter process is still being developed into newer textile-coloring techniques (see Chapter 15).

Thermofixation processes such as the DuPont Thermosol process are dry-heat dyeing processes in the sense that water is absent during the fixation of the dye. The actual dye transfer takes place from the solid dye deposited on the fiber surface by padding and drying. The dyeing time is very short, a few seconds. Thus, the thermofixation process differs from all the known aqueous dyeing and printing processes. The process consists of four steps (Figure 7.10):

1. Padding with dye-liquor.
2. Drying at 100–120°C.
3. Heating at 180–230°C for 60–90 sec.
4. Washing.

In spite of it being a sound technical and economical method, the thermofixation process was commercially accepted only in the 1960s. PET is extremely hydrophobic, having very low water of imbition. A PET fabric exhibits a low capacity for retention of aqueous dye-liquors and it is difficult to deposit sufficient liquor on PET fabric to produce deep shades. This problem is overcome to a certain extent by increasing the viscosity of the pad-liquor (by incorporating a thickener). PET fabric is dyed in small lots for which a continuous thermofixation process is not ideally suited. Both these problems disappeared with the introduction of PET–cotton blend fabrics with good water-retention capacity because of the hydrophilic nature of cotton. Another problem of dyeing by the thermofixation process is the migration of dye during drying. The evaporation of water on the fabric starts on the fabric surface which comes in contact with the hot-air or hot-metal surface. This changes the water content through the thickness of the fabric and the dye-liquor migrates continuously towards the dry surface. The dye is deposited where all the water evaporates.[47,48,94] If the temperature along the width or on the two sides of a fabric during drying is not uniform, the rate of water evaporation will not be uniform, which results in the migration of the disperse dye, and, in turn, in uneven dyeing, shade variations, or two-sided coloring. This problem will be discussed later. The migration of a disperse dye is retarded by thickeners; for example, sodium alginate acts as a migration inhibiter by stopping the movement of dye particles when the water

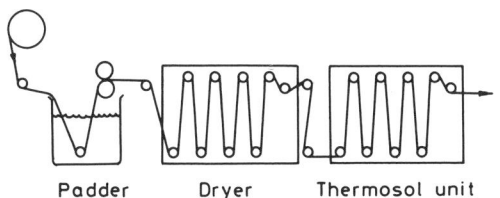

Padder   Dryer   Thermosol unit

FIGURE 7.10. Thermofixation process for PET fabrics.[91]

moves during drying. On the other hand, the dispersing agent enhances the migration by keeping the dye particles dispersed in the moving water. Drying by IR radiations gives minimum migration as these rays are absorbed by the fiber and water, and the fabric is heated from within and not only at the surface.[50] Better padding mangles with shallow troughs or rollers acting as a trough bring down the volume of liquor to be dried and, in turn, the migration of dyes. Liquid disperse dyes with a low concentration of a dispersing agent exhibit less migration than the disperse dyes in granular form with 50–90% dispersing agent.

There are many advantages to the thermofixation process such as high production in a continuous open-width process which eliminates the use of a carrier and a separate heat-setting operation, high color fixation efficiency (Table 7.4) because of good penetration and fixation of the dye, and an one-bath process for PET–cotton blends. As in any other continuous process, this process is economical only when large quantities of a fabric are dyed in a given shade. The fabric is free from wrinkles and rope marks but the feel of the thermofixed fabric is rather harsh and metallic. Unlike aqueous dyeing, the staining of the cotton portion in blends is heavy.

7.6.1 Thermofixation Units

The heating units for the thermofixation of disperse dyes on PET and its blend fabrics vary in construction, the mechanism of heat transfer to the fabric, and the medium of transfer, namely, hot-metal surface, hot air, and superheated dry steam.[96] Typical units are described below.

Hot-Flue Ovens. The fabric passes over rollers in a hot-flue oven where it is heated to the desired temperature. The productivity of these ovens is high, maintenance is easy, and the space requirements are not excessive. How-

TABLE 7.4 Dye Fixation (%) Efficiency[95]

Dye (C.I. Disperse)	% Shade	Fixation (%)	
		Carrier Dyeing	Thermosol Dyeing
Yellow 42	1.5	62	85
Yellow 42	3.0	50	85
Yellow 39	1.5	84	90
Yellow 39	3.0	75	85
Violet 33	1.5	64	90
Violet 33	3.0	54	90
Blue 35	3.0	60	85
Blue 35	6.0	58	85

ever, there is no control on the width of the fabric and there is a danger of the creases being set permanently. These problems can be overcome by the addition of a short stenter after the hot-flue unit. The possibility of cooling the fabric before stentering should be avoided.

Hot-Air Stenter.   The dimensions of the fabric are controlled in this machine by holding the fabric along its selvages while the fabric moves through the heating zone of the machine. There is no danger of crease formation, but plate and pin marks due to the cooling effect are common. The pin stenter simplifies the mechanism of overfeed and avoids the pin or plate markings. The productivity of a stenter is however quite low, the maintenance costs are high, and space requirements are large.

Perforated Drum Machines.[95]   The padded and dried fabric is passed over a series of 4–8 hot drums with a 50% perforated surface. The fabric is in contact with the hot-metal surface as well as the hot-air circulated around it. Because of its contact with the metal surface, the fabric is rapidly heated and within a few sec reaches 210°C. The time of thermofixation is thus brought down to about 20 sec so that the speed of the fabric through the machine is 25–30 m/min. The space requirement of the machine is low. A combination of hot-flue ovens and perforated drums is also used.

Cylinder Machines.[97]   A typical machine consists of eight contact-heating cylinders arranged in two vertical rows in a stand. The fabric in contact with the hot-metal surface is heated to the thermofixation temperature within 2–4 sec. A fabric speed of 35–40 m/min is possible. (For heat setting, a high speed of 100 m/min is used.)

Stenters with Superheated Steam.   Several stenter units use superheated steam as a heat-transfer medium for the thermofixation of dyes. Superheated steam has a high heat capacity and superior heat transfer properties compared to air at the same temperature. It means that dyes may be fixed at lower temperatures, for example, 190°C instead of 210°C for the same color yield. Dyes with very high sublimation fastness are fixed in hot air at very close to the softening point of polyester. Superheated steam gives full fixation of these dyes at lower temperatures.

7.6.2   Preparation of Fabric for Dyeing[98]

The fabric is carefully desized, washed, scoured, rinsed, and evenly dried so that it wets out instantly and uniformly during the padding process. Any residue of soap, desizing agent, or surface active agent may agglomerate the disperse dye or help in the frothing of the dye-liquor during padding, giving specky dyeing. The pH of the fabric is 6–7 before padding. The fabric need not be heat set or wetted out before padding.

7.6.3   Preparation of Pad-Liquor

The padding liquor consists of dyes, dispersing agent, thickener, wetting agent, acetic acid, urea (optional), and fixation accelerant (optional) in deionized water.

Dyes.   The disperse dyes used for padding must be in very fine form with particle sizes <2 μm. They must disperse well and should not agglomerate in the pad bath so that the dye is uniformly distributed on the fabric during padding. The particle diameter of the disperse dyes is <2 μm while the fiber diameters are on the order of 7–14 μm. The cross-section of the dyed fiber shows that the penetration occurs uniformly around the surface of PET fiber,[99] even though the area of contact between the two is too small to give uniform surface adsorption on PET unless the particle is vaporized and the vapor is adsorbed on the fiber surface. Since the dye has to sublime to reach the fiber surface, it must have sufficient vapor pressure and rate of vaporization to be efficient in dyeing during the short heating period of the thermofixation.[100] The vapor pressure of nonionic dyes has been determined by various techniques.[9,101,102] The relationship of the vapor pressure (P, mm Hg) of typical nonionic dyes and temperature is shown in Figure 7.11.[9] Dye with very high vapor pressure, however, sublimes away from the fiber, which entails a loss of color yield and a shade change apart from the contamination of the thermofixation unit.[46] Under such conditions, the batch to follow in the same thermofixation unit may be stained by the earlier evaporated dye.

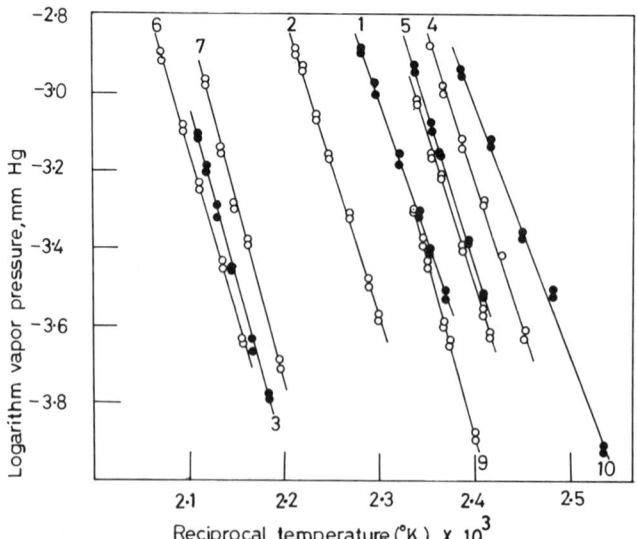

FIGURE 7.11.   The temperature dependence of the vapor pressure of dyes.[9] (For dye formulas, see Ref. 9.)

The rate of vaporization of a dye is decided by its particle size; finely divided disperse dyes have a very high rate of vaporization.

The disperse dye in PET forms a virtually ideal solution with little or no specific interaction between the dye and the PET.[102] The fastness to sublimation is, therefore, largely a function of the vapor pressure of a dye. The introduction of certain polar groups ($CONH_2$, SO_2NH_2) decreases the vapor pressure of the dye and increases the fastness to sublimation. In other words, dyeing will not change the sublimation behavior of the dye. The higher the vapor pressure of the dye, the lower is the sublimation fastness, as shown in Figure 7.12.[9]

Thus, on the one hand, the dye must sublime efficiently to give excellent dyeings under thermofixation conditions and, on the other hand, it must have minimum vapor pressure so that the fastness to sublimation of the dyeing is very high for the reasons stated above. The contradictory demands from the disperse dyes are met by building dye molecules with the required chemical structure for sublimation fastness and grinding the dye to the lowest possible size to get the maximum sublimation rate of dye particles. Alternatively, if dyes with low sublimation fastness are to be used, the padded and dried fabric may be coated with a thickener to form a film on the fabric that is impermeable to the dye vapor. A starch or sodium alginate film was found to stop the loss of dye vapor to the ambient air, thus, improving the color yield and functioning of the thermofixation unit.[103] Dyes with low sublimation fastness are not used for printing or for dyeing goods that are to be pleated or resin-finished.[54,104,105]

In aqueous dyeing, disperse dyes generally do not stain cotton heavily. This is not the case in thermofixation dyeing of PET–cotton blends. It is impossible to avoid staining cotton since the disperse dyes penetrate cellulosic fiber at elevated temperatures (Table 7.5).[48] The stain of a disperse dye

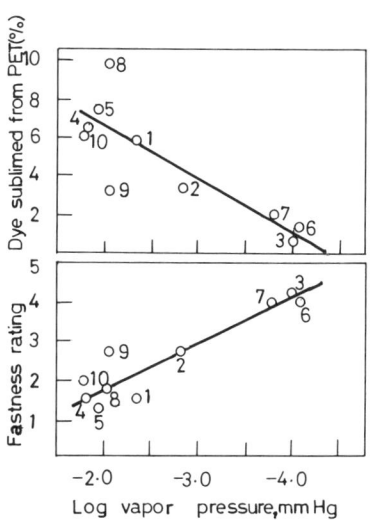

FIGURE 7.12. Sublimation fastness and vapor pressure (p) of disperse dyes.[9] (For dye formulas, see Ref. 9.)

TABLE 7.5 Diffusion of Disperse Dyes in Cotton at Elevated Temperatures[48]

Temperature (°C)	Dye 1	Dye 2	Dye 3
Apparent Diffusion Coefficient[a] *(cm²/sec) × 10^{15}*			
170		0.008	2.07
180		0.05	16.9
190		0.37	19.6
200	3.10	2.68	79
210	16.0	18.0	186
Apparent Activation Energy (kcal/mole)			
	61	86	49

[a] For the formulas of dyes, see Ref. 48.

on cotton is dull and detrimental to the achievement of bright shades on blends.[72] Additionally, the stain exhibits poor light fastness which is reflected in the light fastness of the final dyeing. (This is not the case if the cotton portion in the blend is dissolved to get 100% PET material.) Thus, for the success of thermofixation dyeing, a disperse dye with excellent dispersion properties, good sublimation properties, and minimum staining of cotton is selected.

Dispersing Agent.[89,99,106] The dye particles are kept in a finely dispersed state without any aggregation or agglomeration during padding by a dispersing agent. However, dispersing agents have a tendency to hold back the dye, which gives significantly lower color yield. For example, Figure 7.13 shows the relationship between the dye retained by a dispersing agent after thermofixation and the concentration of the dispersing agent in the pad bath.[49] The relationship between the fixed dye on PET at equilibrium and the concentration of the dispersing agent in the pad bath (Figure 7.13) clearly indicates that the dye molecules that are retained by the dispersing agent are not available for dyeing, thus, lowering the equilibrium fixed dye on the fiber. The addition of a small amount of dispersing agent may be beneficial but an excess of dispersing agent in the pad bath definitely lowers the color yield on PET in the thermofixation process.[49] Another problem with the dispersing agent is that it increases the migration of dye on a padded fabric during drying. This creates problems of unlevel dyeing.

Thickener.[106–108] The hydrophilic fibers in a blend fabric help to pickup the dye-liquor, although a thickener has to be added to the pad-liquor to improve

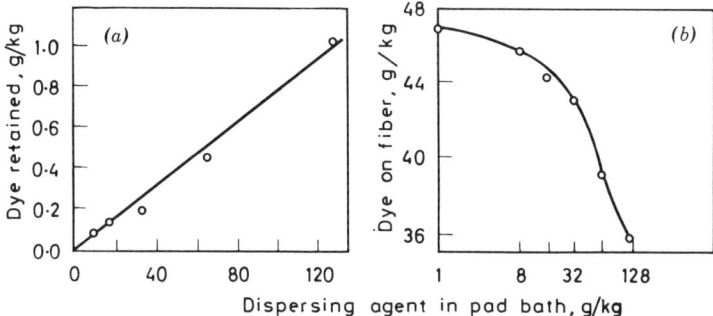

FIGURE 7.13. The influence of dispersing agent concentration on dye transfer in thermofixation dyeing of PET. (*a*) Dye not stripped by dry boiling chlorobenzene (dye retained) from padded, dried PET in relation to the dispersing agent concentration.[106] (*b*) Equilibrium fixed dye on PET with increasing concentration of dispersing agent in pad bath.[106]

the pickup of the liquor by the 100% PET fabric. The thickener prevents the migration of disperse dyes during drying and its presence on the fiber avoids uneven results.

The essential requirements of a thickener are: (*a*) it should have the capacity to give high viscosity at low concentration; (*b*) it should be inert to the dye and its dispersion; (*c*) while preventing migration, it should not hinder the transfer of dye to the fiber; (*d*) it should be stable under padding, drying, and thermofixation conditions; (*e*) it should be easily removable from the fabric. Sodium alginate meets these requirements and is the best thickening agent. Synthetic polymeric materials are also used as thickeners. They form a thin film on the fiber that inhibits the migration of dye during drying. At the same time, they do not interfere with the transfer of dye to the fiber surface during the thermofixation. The added advantage of these products is that they prevent the growth of dye particles by agglomeration. The dye remains in the state of fine dispersion during drying and the rate of sublimation and dye transfer during the thermofixation remains very high. The net result is that the color yield on PET by a short thermofixation process is higher in the presence of thickeners that prevent agglomeration.[108]

The concentration of a thickener in the pad-liquor influences the efficiency of dyeing PET and blend fabrics to a very great extent (Figure 7.14). The decrease in the efficiency of the dye fixation with an increase in the thickener concentration is attributed to the impermeability of the thickener film to the dye vapors.[49]

Wetting Agent and Other Products.[109] PET and PET–cotton blend fabrics are not as absorbent as 100% cotton fabric. Therefore, it is essential to incorporate a small amount of wetting agent in the pad-liquor, which facilitates quick and uniform wetting of the fabric during the short time of its

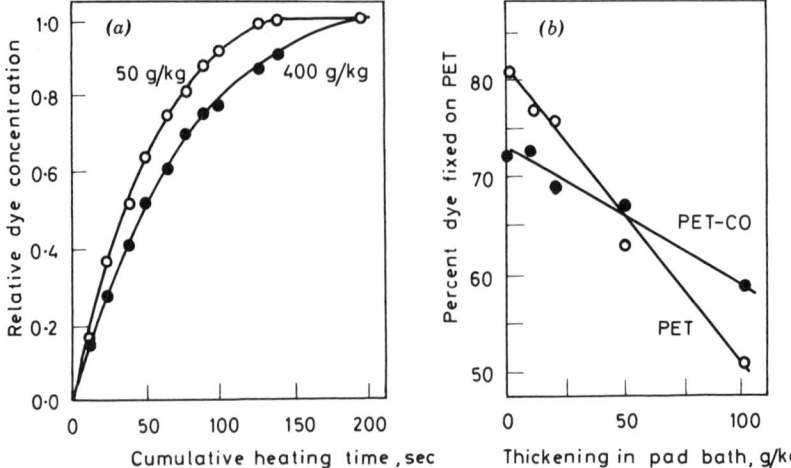

FIGURE 7.14. Influence of sodium alginate thickening on rate of dyeing PET at 210°C. (*a*) Sodium alginate thickening (4%) in pad bath;[9] ○: 50 g/kg (fixation cycles of 13 sec at 210°C); ●: 400 g/kg; (*b*) Fixed dye (%) on PET with alginate thickening (4%) in bath[13] (fixation: 210°C/60 sec). ○: PET fabric. ●: PET/CO (65:35) blend fabric.

contact with the pad-liquor. Cleaning of the fabric after thermofixation is also facilitated if a wetting agent is present on the fabric.

Urea in the pad-liquor increases the rate of fixation of disperse dyes, which allows thermofixation under milder conditions of temperature and time. Urea prevents the discoloration of the cellulosic portion during thermofixation. However, urea decomposes near its melting point, giving ammonia and guanidine and at times, degrades the disperse dyes.

Addition of high boiling solvents such as fatty alcohols and polyglycol ethers to the pad-liquor has been suggested as away to increase the rate of fixation of disperse dyes.[110] However, this increases the staining of cotton to a very great extent and thus, is of very little practical use.

Pad-Liquor. Many disperse dyes are sensitive to metal ions, calcium salts, and ionic bodies, and give tonal difference in their presence. The dispersion is also sensitive to ionic shocks. The pad-liquors are, therefore, prepared in deionized water. The thickening agent is first dissolved in water to get a smooth paste. The required quantity of this paste is mixed with warm water before the dye is strewn in it under vigorous stirring. The dye must immediately disperse into fine particles. The pH of the pad-liquor will usually be on the acidic side. The acidifying agent is carefully added in dilute form so that the thickening is not precipitated locally. The liquor is filtered through a muslin cloth before it is added to the pad bath. The concentration of the pad-liquor is kept as high as possible considering the padding mangle expression. Dilute pad-liquors will entail higher expression on the padding mangle which increases the migration of dye during drying. In the case of blends, the efficiency of dyeing is lowered if dilute liquors are used (Figure 7.7).[48]

7.6.4 Padding and Drying

Padding.   A fabric is padded on a two-bowl or three-bowl padding mangle. For a two-bowl padding mangle, it is preferable to have soft rollers with 60–70° Shore hardness.[111] A combination of one soft and one hard roller leads to uneven squeezing effects on the face and back of the fabric and consequently, to two-sided dyeings. A pickup of 60% is quite satisfactory. An even squeeze is absolutely essential because irregularities in padding will not level out during subsequent stages of the process. The speed of the fabric is maintained constant to get an even squeeze. The expression of the padding mangle influences the migration of dye in drying and the efficiency of color yield on PET in blend fabrics (Figure 7.7).[9]

Drying.   This is the most critical step in ensuring good final results. Any uneven drying gives rise to migration, listing, two-sidedness, and general poor appearance. The drying is usually done either on a hot flue or a pin stenter. It is preferable to predry the fabric on infrared equipment which brings down the water content of the fabric below 40%. This minimizes the migration of dyes during further drying. However, infrared predrying equipment is expensive. Satisfactory results are also obtained using a migration inhibitor and controlling the conditions during drying. The temperature of the drying unit is maintained at 100–110°C. It is preferable to run the fans in the machine at low speeds. The dried goods are thoroughly checked because it is possible at this stage to wash out the dye if uneven drying is observed. For deep shades, double padding and drying is required. In this case, the face of the fabric is changed during the second padding and drying stage to minimize the two-sided effect, if any.

The migration of dyes during drying can be minimized by replacing a portion of the water in the pad-liquor with methanol and burning out the methanol. Thus, in the Ramaflame Process (Hoechst), the fabric is padded with a dispersion of dyes in a 50:50 mixture of water and methanol. The padded fabric is then passed over a flame so that the methanol is ignited and burnt off. The fabric dries quickly without any migration of dyes.[95,112]

The padding and drying of blend fabrics need more attention since the dye migrates during drying, disturbing the distribution between component fibers of a uniformly padded fabric.[47] A large fraction of the total dye is found on the cellulosic fiber after drying. The fraction depends, apart from the expression of the mangle, on the composition of the blend fabric (Figure 7.15).[8] The interfiber migration during drying ceases when the volume of liquor drops below that required to imbibe the individual fibers.[48] Up to this stage drying is the most critical step. Thus, a two-stage drying involving an infrared predryer to bring down the volume of liquor below the level of imbibed liquor in the fiber ($\sim$ <25% for 66:33 blend) followed by a hot-air stenter to

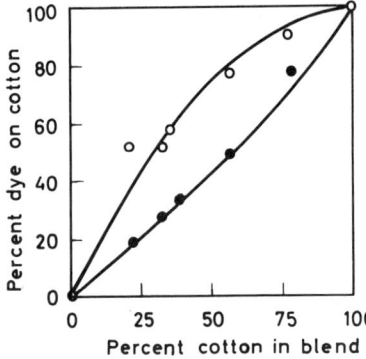

FIGURE 7.15. Distribution of dye in PET/CO blends after padding and drying.[8] ○: Aqueous pad bath (preferential deposition on cotton). ●: Perchloroethylene pad bath (almost uniform deposition). (The continuous lines represent theoretical values.)

dry it completely, gives efficient uniform drying with a minimum migration of dye on blend fabrics.

7.6.5 Thermofixation

A padded and dried fabric is maintained at a temperature between 180–230°C for 5–120 sec when the dye is fixed on PET. The exact conditions of the thermofixation depend on the nature and concentration of the disperse dye, type of equipment, the construction of the fabric and so on. The actual temperature of a fabric decides the rate and efficiency of dye transfer to PET. Drums and cylinder machines heat the fabric very rapidly because of the direct contact between the surfaces of the hot metal and the fabric. The temperature of the fabric rises rapidly to reach the thermofixing level within a few seconds. Hot-air ovens and stenters take a longer time to bring the fabric to the desired temperature. Timings of typical machines to get the satisfactory color yield during thermofixation at 210°C are shown in Table 7.6.[97]

The rate and extent of the fixation of a dye are decided by the fiber structure and the time and temperature of the thermofixation.[13] Fibers differing in crystallinity (i.e., the amorphous content) exhibit different rates and

TABLE 7.6 Thermofixation Timings and Machines[97]

	Thermofixation Time (sec)		
Machine	Initial Heating	Residence at 210°C	Total for Full Fixation
Hot-air oven	30–60	20–30	50–90
Stenter	5–20	20–25	30–40
Drum and cylinder	3–5	7–20	10–25

extents of dye fixation. Inadequate thermal treatment will lead to ring-dyeing with a large amount of unfixed dye on the surface of the fabric.[99] The latter will not only give poor color yield but will impair the fastness properties of dyed fabric. It is not necessary to give extra time for the penetration of dye through the PET since good fastness properties are achieved once the dye is transferred to the PET phase. Thus, Anselrode[113] has reported good fastness properties on PET after a dye transfer of a second, although the fiber may remain ring-dyed, even up to 30 sec at 180–210°C.[114] The dye may sublime out into the thermofixation unit if the thermal treatment is overdone.[46] The feel of the fabric is also affected. Thus, it is essential for each dye and depth of shade to standardize the conditions of thermofixation for getting a satisfactory color yield and feel of the fabric under local dyeing conditions. The time of thermofixation is the time when the fabric reaches and continues to remain at the thermofixation temperature. Different heating methods may have different times of thermofixation. The method of heating does not alter the color yield but influences the feel of the thermofixed fabric.

After the thermofixation, the fabric is washed with hot and cold water to remove the thickener, residual dye, dispersing agent and so on. Sodium alginate is easily removed by the washing treatment. It is essential to remove the dye superficially adsorbed on PET and cotton materials by the reduction-clear and soaping treatments described earlier. The cotton portion is then dyed by a separate dyeing treatment.

7.6.6 Continuous-Blend Dyeing Processes

Although the thermofixation process is a continuous process of dyeing polyester, it can be a batch process for dyeing blend fabrics since the cotton component can be dyed separately. In the continuous process, both the component fibers are dyed, either simultaneously or one after the other using special dye combinations. The advantages of a continuous process, such as high production and savings in labor, energy, and materials can be fully achieved in dyeing blends. Some such dyeing systems are given below:

Disperse and Reactive Dyes.   Bright shades are produced with combinations of disperse and reactive dyes using either a one-bath process or a two-bath process.[115–119]

One-Bath Process.   The blend fabric is padded through a liquor consisting of

X g/liter	Disperse dye
Y g/liter	Reactive dye
10–15 g/liter	Sodium bicarbonate or carbonate
0–200 g/liter	Urea
1–2 g/liter	Anionic wetting agent
2–10 g/liter	Migration inhibitor

The fabric is carefully dried, thermofixed, washed, soaped at boiling with soap and soda ash, washed, and dried. The disperse dye is fixed on PET and the reactive dye is fixed on cotton during the thermofixation at 190–210°C in 20–90 sec. Reactive dyes are sensitive to the reduction-clear treatment and hence, the latter is avoided. Thorough soaping is usually sufficient to give good fastness properties. Disperse dyes for this combination must withstand alkaline conditions and give the expected color yield in the presence of the reactive dye. All reactive dyes are not suitable for this process, since only a few give good fixation under the dry conditions of thermofixation. The fixation is increased by giving it a short steaming treatment (1–2 min at 100°C) before thermofixation. By this pad-dry–short-steam–thermofix technique, it is possible to get a good fixation of many reactive dyes. The process is useful for the production of deep shades even with dyes with poor reactivity. The fixation of disperse dyes under one-bath conditions is, however, generally low and the process is unsuitable for very deep shades.

Two-Bath Process.   The blend fabric is padded through a liquor consisting of

X g/liter	Disperse dye
Y g/liter	Reactive dye
1–2 g/liter	Anionic wetting agent
2–10 g/liter	Migration inhibitor
(5.5 pH with acetic acid)	

The padded fabric is dried and thermofixed as usual. The goods are again padded at room temperature with

5–20 g/liter	Sodium hydroxide or carbonate
50–150 g/liter	Sodium chloride

and steamed 1–2 min at 100°C. Washing, soaping and so on are the same as in the one-bath process described above.

The fixation of both disperse and reactive dyes in the two-bath process is very good, which eliminates the necessity of critical selection of the dyes. The pad bath stability is unlimited under the acidic conditions used in the initial padding. Deep shades can be produced by this process. However, this is a longer process with a higher consumption of materials and energy than the one-bath process.

A development in disperse–reactive dye combinations was the introduction of reactive dyes that are fixed on cellulose under mild acidic conditions. Thus, Procion T(ICI) dyes exhibit the identical responses to the pH of fixation as the selected disperse dyes (e.g., Dispersol T of ICI) and can be applied under acidic pH.[120–122] The pad bath consists of:

X g/liter	Dispersol T
Y g/liter	Procion T
1–2 g/liter	Ammonium dihydrogen phosphate (pH 6–6.5)
1–2 g/liter	Wetting agent
1–2 g/liter	Antimigrating agent

After padding, drying, and thermofixing in the usual manner, the fabric is padded with caustic soda at 80°C followed by soaping and washing. The process has the advantages of good pad bath stability, compatibility of dyes, easy washing off without reduction-clear treatment with excellent fixation of dyes, and good build up, even on unmercerized goods.

Disperse and Vat Dyes.   A full range of hues in deep shades with excellent all around fastness properties on both polyester and cotton can be produced with these combinations.[123–129] However, heavy shades, at times, exhibit poor crock fastness and a few bright shades are not available in these combinations. Disperse dyes with good sublimation fastness and minimal staining of cotton are required for the best results. Special brands of selected disperse and vat dyes in blended form are marketed for this purpose. A blend fabric is padded with a liquor consisting of

X g/liter	Disperse dye
Y g/liter	Vat dye
1–2 g/liter	Anionic wetting agent
5–10 g/liter	Antimigrating agent
(Acetic acid to pH 5.5–6.0)	

The padded fabric is predried on an IR drier, fully dried at 100–110°C in a hot flue, and thermofixed at 190–210°C for 30–90 sec to fix the disperse dye. The fabric is then padded through

40–175 ml/liter	Caustic soda 28%
15–45 g/liter	Hydrosulfite
1–2 g/liter	Wetting agent
1–2 g/liter	Dispersing agent

steamed at 100°C for 30–60 sec, oxidized with 2–3 ml/liter hydrogen peroxide (35%) at pH 8–9 at 50°C (or with sodium dicromate solution), soaped, washed, and dried. A separate reduction-clear treatment after the fixation of disperse dyes is not required since the vat dyebath acts as a reducing medium.

Vat Dyes.   Since polyester can be dyed with water-insoluble nonionic dyes, it is expected that vat dyes will dye PET fibers.[116,130] These dyes have larger molecular sizes with more rigid molecular ring-structures than disperse dyes. Therefore, they are difficult to apply to polyester fiber at the temperatures of HT and carrier dyeing. During the thermofixation, however, many vat dyes diffuse into polyester, which gives dyed polyester with excellent fastness properties. The light fastness of such dyeings is very good, except for a few blue dyes, and the shades are brighter than those on cotton. The high fastness and comprehensive shade range have made vat dyes one of the most important groups of dyes for cellulosic fibers. If the vat dye is properly selected, it gives good dyeings on both components of a blend fabric. The addition of a small amount of polyethylene glycol (molecular weight of around 600) improves the uptake of vat dyes on polyester. Thus, vat dyes can be used to produce pale to medium shades on blend fabrics by the same method as that for disperse and vat combinations described earlier. The cost of dyeing is higher than for combination dyes since vat dyes are generally more expensive than disperse dyes.

Solubilized Vat Dyes.   Pale shades on blend fabrics are easily produced by solubilized vat dyes. Even though solubilized vat dyes do not penetrate PET fiber, a reasonable solidity of shade is obtained on the blend fabric.
    The method consists of padding the fabric with

X g/liter	Solubilized vat dye
0.5 g/liter	Soda ash
5.0 g/liter	Sodium nitrite
0.5 g/liter	Wetting agent

followed by repadding with 10–20 g/liter sulfuric acid solution at 40–60°C. The fabric is then washed, soaped, and dried. In order to stabilize the shade to subsequent heat treatments, the fabric is heat set at 200–210°C for 10–30 sec.

Disperse and Direct Dyes.   This dye combination has a low cost of dyeing and a simple method of application.[116,131] However, the fastness properties of the dyed material are not very good. The fabric is padded with

X g/liter	Disperse dye
Y g/liter	Direct dye
1–2 g/liter	Wetting agent
20–50 g/liter	Sodium chloride
1 g/liter	Antimigrating agent

dried carefully, and thermofixed, as usual. The fabric is then aftertreated to improve the fastness of the direct dyes.

Disperse and Sulfur Dyes.   Heavy shades for both solid and cross-dye effects and a wide choice of colors are possible with this combination.[132] However, brightness of shade and fastness to chlorine are generally very poor. This combination lowers the absorbency of blend fabrics.

The method of application consists of fixing disperse dye by the usual pad-dry-thermofix process, followed by passing the fabric through the chemical pad-liquor of reduced sulfur dye. The fabric is steamed at 100°C for 25–40 sec, washed, oxidized, soaped, and washed on an open soaper.

Resin-Bonded Pigments.   The advantages of this system are low cost of dyeing, ease of application, good bright appearance, and exceptionally good light fastness.[116] However, with the pigments, it is difficult to produce deep shades, the fastness to rubbing is not very good, and the handle of the fabric is harsh. In order to get deep shades, a combination of resin-bonded pigment with disperse dye is used.

The fabric is padded with

X g/liter	Pigment
Y g/liter	Binder
1–2 g/liter	Antimigrating agent
1–2 g/liter	Wetting agent

dried, and thermofixed as usual. A soaping operation completes the process.

Dybln Process.[92]   This is a modified thermofixation process in which special disperse dyes are fixed on both cotton and polyester. The pad-liquor contains a nonvolatile water-miscible compound such as polyglycol ether which swells the cellulose and makes it receptive to a nonionic disperse dye during the thermofixation. After padding, the fabric is dried and thermofixed when the dye is fixed on both component fibers. The fabric is then washed with cold water to remove the swelling agent and soaped to improve the fastness properties. The dye has to be trapped in the cellulosic fiber during the thermofixation and washing treatments. The molecular size of the Dybln dye is bigger than that of the conventional disperse dye. By this technique, a blend is dyed in a solid shade with nonionic dyes in a single step. Datye[103] tried to improve the fastness properties using reactive disperse dyes and an acid binding compound in the pad-liquor. Many dyes are fixed on cotton under these modified conditions. However, the stability of reactive disperse dyes under alkaline conditions is not very high. Fixing such dyes on cotton in the washing treatment rather than during thermofixation is a possibility that has yet to be explored. The fastness properties of commercial Dybln dyes on PET are excellent.

7.7 DYEING OF POLYESTER–WOOL BLENDS

The dyeing and finishing of PET–wool blends is difficult compared to PET–cotton blends.[50,55,133–143] The problems are dimensional stability, pilling, avoiding crease marks and stains, improving disperse dye uptake and its fastness, and reducing wool damage while obtaining excellent drape and easy care properties at the same time. It is not possible to expose wool to HT dyeing conditions as wool is severely damaged at temperatures over 106°C. On the other hand, when the dyeing temperature is below 106°C, the exhaustion of disperse dyes on PET is not sufficient unless a carrier is used. In the absence of a carrier and at temperatures above 100°C, disperse dyes stain the wool heavily, leading to poor wet fastness. Furthermore, wool cannot stand the strong alkaline reduction-clear treatment. It is, therefore, necessary to take precautions so that the staining of wool is minimal. Disperse dyes that stain wool minimally are selected. Dyeing conditions such as time and temperature are only as severe as necessary so that the disperse dye exhausts on polyester without any damage to the wool component. PET is dyed first, then the wool is dyed in a fresh bath after an intermediate scouring to remove the stain on wool.

Two methods are commonly employed in applying disperse dyes on PET–wool blends.

1. Dyeing at boiling with carriers.
2. Dyeing at slightly above 100°C using a small quantity of a carrier.

The disperse dye is adsorbed at the beginning of the dyeing cycle by the wool. During the course of dyeing at boiling, as the dye diffuses into PET, the dye on wool migrates to PET. The process continues until an equilibrium is reached. At this stage, not all the dye on wool is lost and the fiber still has some disperse dye. This equilibrium is disturbed if the dye on wool is destroyed by the reduction-clear treatment. When wool is dyed with an acid dye, the disperse dye on PET partially remigrates to the wool until new equilibrium concentrations are established. Thus, the reduction-clear treatment can be dropped if the fastness properties and the final shade on the dyed blend are not improved by the treatment.[133]

Disperse dyes are hydrolyzed or reduced rapidly in the dyeing of PET–wool blends compared to the dyeing of PET–cotton blends. The hydrolysis of many disperse dyes is pronounced when the pH of the dyebath rises to neutral or to the alkaline side (which happens when the bath is not properly buffered). The reduction of disperse dyes is due to the wool. Above pH 5, wool has a tendency to hydrolyze and to develop functional groups that can reduce the dye. This process may continue throughout the dyeing cycle when the absorbed dye on PET remigrates to the wool and gets reduced. When prolonged dyeing cycles are encountered, the shade on PET may become lighter and lighter due to the migration and reduction of the dye on wool. Nonionic surfactants in the dyebath enhance the migration of the dye, and, in turn, its reduction on wool. Thus, it is necessary to use a buffer system to stabilize the pH and to retard the reduction of disperse dyes. The buffer system functions as an inhibitor in the acid medium. Ammonium sulfate/formic acid is the most suitable buffer and 2 g/liter ammonium sulfate and formic acid to pH 5.5 are used for optimum color value on PET–wool blends. The formic acid is added to the dyebath only after adding the carrier, dye, ammonium sulfate, and running the material for 10 min to adjust the pH to 5.5. This way, pH changes due to the carrier, dye, and fabric are prevented.[133]

The PET–wool blend fabric is preset before or after scouring to remove cockles caused by setting the wool. The scouring is carried out using two baths at about 50°C with a nonionic or anionic detergent. The first scour is a short one to rinse out the surface impurities. The second scour is for 30–45 min. A thorough rinsing, a sour-wash with dilute acetic acid, and a final rinsing with water completes the scouring process. The fabric is dried below 110°C on a stenter 2–4 cm above the wet width. The dried fabric is heat set to improve the dimensional stability of the PET component. Full overfeed in the warp and weft directions is given during the heat setting. The temperature of heat setting is 180 ± 10°C and the time is about 30 sec. Wool is significantly damaged above 190°C and under alkaline conditions (Table 7.7).[143] Blend fabrics containing 40% or more PET are heat set to impart dimensional stability and resistance to creasing and shrinkage.

7.7.1 Carrier Dyeing

PET–wool blend fabric is dyed using the carrier dyeing method described for PET–cotton at temperatures close to 100°C. In the carrier dyeing at boiling, the time of dyeing is long, the penetration of dye in the PET fiber is poor, and

TABLE 7.7 Effect of Heat-Setting Conditions on Wool[143]

Setting Temperature (°C)	pH of Wool	Yellowness Index	Handle
Untreated	7.5	1.1	soft
150	5.5	1.23	soft
170	5.5	1.49	soft
170	9.0	2.37	crisp
190	7.5	1.98	crisp
190	9.0	3.75	harsh
200	7.5	2.33	crisp
200	9.5	4.66	harsh

the staining of wool is very heavy. To overcome these limitations, the quantity of carrier is reduced and the dyeing is carried out under pressure at 103–106°C.[140–142] This method produces dyeings with good wash and rubbing fastness.

The material is treated with

0.5–1 g/liter	Anionic wetting agent
0.5–1 g/liter	Dispersing agent
X g/liter	Carrier
2 g/liter	Ammonium sulfate

at 60°C for 15 min. The dye dispersion is then added and dyeing is continued for another 15 min until the pH is finally adjusted to 5.5 with formic acid. The temperature is raised to 105–106°C in 20 min and dyeing is carried out for 1–2 h. The bath is then drained, the goods are washed, and, if necessary, are given a reduction-clear treatment in a fresh bath consisting of

4 g/liter	Ammonia (sp. gr. 0.95)
3 g/liter	Sodium hydrosulfite
0.5–1 g/liter	Dispersing agent

at 50–60°C for 15 min. A final wash and a rinse with 1 g/liter acetic acid (30%) complete the process. The wool component of the blend is then dyed with suitable acid or metal-complex dyes. The reduction-clear treatment may be dropped for the reasons described above.

Simultaneous dyeing of polyester and wool components is possible with selected combinations of disperse and acid or metal-complex dyes. The bath is set with

0.5 g/liter	Anionic wetting agent
0.5–1 g/liter	Dispersing agent
X g/liter	Carrier
50–100 g/liter	Glauber's salt
2 g/liter	Ammonium sulfate
(pH 5 with Formic acid)	

and the dyeing is carried out as above, except that the acid or the metal-complex dye is added after the disperse dye. Dyeing of wool at 105°C takes place very rapidly while PET dyes exhaust very slowly. After dyeing, the bath is drained and the goods are washed with 1–2 g/liter nonionic wetting agent at boiling for 15 min. A mixture of disperse and acid or metal-complex dyes is used to produce solid shades on blend fabric with good fastness properties.

The staining of the wool component in blend fabrics with disperse dyes is a major disadvantage of dyeing PET–wool blends with disperse dyes.[136,144-146] The surface deposition of the disperse dye on wool results in the lowering of fastness properties and the loss of color yield on PET component. Wool rapidly adsorbs the disperse dye at low temperatures and this dye is transferred to PET as the temperature is raised. The dye taken up by the wool is evenly distributed through the fiber cross-section.[145] Thus, the uptake of dye by the wool component reaches a maximum at 90°C which rapidly drops at higher temperatures with the desorption of the dye into the bath from which it is adsorbed by the PET component.[136] Nevertheless, in certain instances, the dye can penetrate wool fiber to give acceptable fastness properties, particularly, if the dyes are from the reactive disperse class.[147] The range of five colors (Procinyl® ICI) are effective as print dyes on wool and wool-blend fabrics if the goods are steamed at 100°C for 30 min before thermofixation at 190°C for 1 min. It is observed that the range of Procinyl reactive dyes have a significant affinity for wool.[146] By designing the structure of such dyes, approximately equal uptake of dye by wool and polyester from the same dyebath can be achieved. Using such dyes, it appears possible to dye PET–wool blends with a single disperse dye.

7.7.2 Thermofixation Process

Wool is damaged and yellowed and the handle of the blend fabric becomes harsh under dry-heat treatment[50] (Table 7.7). Because of this, the thermofixation process has not received wide acceptance in coloring PET–wool blend fabrics. The thermofixation process is the same as that for PET–cotton blend fabrics; it is carried out on a pin stenter because the feel of fabric is adversely affected in contact-type (roller) thermosoling units. The staining of wool with disperse dyes under thermofixation conditions is very low and a mild reduction-clear treatment removes the disperse dye deposited on the surface of wool. Final washing treatment completes the process of dyeing PET.

The wool component is then dyed with acid or metal-complex dyes. Special dyes are marketed for the one-bath thermofixation process for polyester–wool blend fabrics.[148] For example, Lanestren® dyes (BASF) are recommended for such processes. After the thermofixation process, the fabric is treated in a winch with

0.5–1 g/liter	Anionic wetting agent
0.5–1 g/liter	Dispersing agent
50–100 g/liter	Glauber's salt
(pH 5.5 with acetic acid (30%))	

at 50°C for 10 min and then at boiling for 1 h. The goods are then washed. It is claimed that the process gives good shades of medium depth. For deep solid shades, it is recommended that extra disperse dye be added to the pad-liquor. Alternatively, reactive disperse dyes may be used as described above.

REFERENCES

1. Derbyshire, A. N., *JSDC*, **90** (1974), 273.
2. Biedermann, W., *JSDC*, **87** (1971), 105; **88** (1972), 329.
3. Skelly, J. K., and Evans, D. C., *Textilveredlung*, **8** (1973), 102.
4. Jones, F., and Patterson, K., *JSDC*, **92** (1976), 442.
5. Disperse Dyes Committee, SDC, **JSDC, 80** (1964), 237, **81** (1965), 209; **88** (1972), 296; and **93** (1977), 228.
6. Datye, K. V., and Acharekar, J. Y., *JSDC*, **93** (1977), 413.
7. Schnider, E. F., *ADR*, **52** (1963), 370.
8. Datye, K. V., *Colourage*, **24**(4) (1977), 27.
9. Datye, K. V., Kangle, P. J., and Milicevic, B., *Textilveredlung*, **2** (1967), 263.
10. McGregor, R., and Peters, R. H., *JSDC*, **81** (1965), 393; 429.
11. Ellison, A. H., and Zisman, W. A., *J. Phys. Chem.*, **58** (1954), 503.
12. Peters, R. H., and Ingamells, W., *JSDC*, **89** (1973), 395.
13. Datye, K. V., Pitkar, S. C., and Rajendran, R., *Indian J. Tech.*, **4** (1966), 101; **8** (1970), 6.
14. Waters, E., *JSDC*, **66** (1950), 609.
15. Fern, A. S., *JSDC*, **71** (1955), 502.
16. Remington, W. R., and Schroeder, H. E., *TRJ*, **27** (1957), 177.
17. Glenz, O., Beckmann, W., and Wunder, W., *JSDC*, **75** (1959), 141.
18. Piedmont Section, AATCC, *ADR*, **48**(22) (1959), 23; **48**(23) (1959), 37.
19. Patterson, D., and Scheldon, R. P., *Trans. Faraday Soc.*, **55** (1959), 1254.
20. Salvin, V. S., *ADR*, **49** (1960), 600.
21. Merian, E., *TRJ*, **36** (1966), 612.
22. Jones, F., *Rev. Prog. Color*, **1** (1969), 15.
23. McDowell, W., and Weingarten, R., *MTB*, **50** (1969), 59; **50** (1969), 814; **50** (1969), 1340; *JSDC*, **85** (1969), 589.
24. McDowell, W., *TCC*, **10** (1978), 131; *MTB*, **61** (1980), 946.
25. Dowson, T. L., and Todd, J. C., *JSDC*, **95** (1979), 417.
26. McGregor, R., and Etters, J. N., *TCC*, **11** (1979), 202.
27. Kartaschoff, V., *Helv. Chem. Acta*, **8** (1925), 928.
28. Patterson, D., and Sheldon, R. P., *JSDC*, **76** (1960), 178.
29. Datye, K. V., Pitkar, S. C., and Purao, U. M., Proceedings of Symposium on Physical Chem. of Dyeing Processes, UDCT (February 1976), p. 10.
30. Rawicz, F. M., Cates, D. M., and Rutherford, H. A., *ADR*, **50** (1961), 320; **50** (1961), 354.
31. Zimmermann, C. L., Mecco, J. M., and Carlino, A. J., *ADR*, **44** (1955), 296.
32. Vickerstaff, T., *MTB*, **35** (1954), 765; *Hexagon Digest*, **20** (1954), 7.
33. Murray, A., and Mortimer, K., *Rev. Prog. Color*, **2** (1971), 67.
34. Millson, H. E., *ADR*, **44** (1955), 417.
35. Fortress, F., and Salvin, V. S., *TRJ*, **28** (1958), 1009.
36. Lemons, J. K., Kakar, S. K., and Cates, D. M., *ADR*, **55** (1966), 76.

37. Balmforth, D., Bowers, C. A., Bullington, J. W., Guion, T. H., and Roberts, T. S., *JSDC*, **82** (1966), 405.

38. Ibe, E. C., *JAPS*, **14** (1970), 837.

39. Weckler, G., *Text. Prax.*, **27**(2) (1972), 117.

40. Ribnick, A. S., Weigmann, H. D., and Rebenfeld, L., *TRJ*, **42** (1972), 720; **43** (1973), 176.

41. Ingamell, W., Peters, R. H., and Thornton, S. R., *JAPS*, **17** (1973), 3733.

42. Hemming, D. F., and Datyner, A., *TRJ*, **45** (1975), 235.

43. Olsen, E. S., and Mendozo-Vergara, C., Book of Papers, AATCC Natl. Tech. Conf., Chicago, (1975), p. 239.

44. Keller, K. H., *Textilveredlung*, **13** (1978), 140.

45. Roberts, G. A. F., and Solanki, R. K., *JSDC*, **95** (1979), 226; **97** (1981), 220.

46. Bent, C. J., Flynn, D. P., and Sumner, H. H., *JSDC*, **85** (1969), 606.

47. Datye, K. V., and Pitkar, S. C., Seminar on *Contribution to Chemistry of Synthetic Dyes and Mechanism of Dyeing*, UDCT, Bombay, (1967), p. 47.

48. Datye, K. V., *Textilveredlung*, **4** (1969), 562.

49. Datye, K. V., and Pitkar, S. C., IVth Symposium UDCT, Bombay, (February 1969), p. 6.

50. Keaton, J., and Preston, D. T., *JSDC*, **80** (1964), 312.

51. Landerl, H. P., *ADR*, **51** (1962), 552.

52. Beckmann, W., and Kuth, R., *MTB*, **48** (1967), 1441.

53. Gerber, H., and Somm, F., *Textilveredlung*, **6** (1971), 372; *ADR*, **64**(9) (1975), 44.

54. Fox, M. R., *TCC*, **1** (1969), 566.

55. Fern, A. S., and Hadfield, H. R., *JSDC*, **71** (1955), 840.

56. Hallada, D. P., Keen, M. C., and Thomas, R. J., *ADR*, **50** (1961), 445.

57. Vaidya, A. A., and Trivedi, S. S., *Textile Auxiliaries and Finishing Chemicals*, ATIRA, Ahmedabad, 1975, p. 35.

58. Vaidya, A. A., *Textile Industry and Trade J. Annual*, India (1975), p. 3.

59. Duckworth, C., *Rev. Prog. Color*, **1** (1969), 3.

60. Newcomb, W. J., and Ward, G. C., *ADR*, **49** (1960), 86.

61. Hollis, N. M., *ADR*, **55** (1966), 502.

62. Preston, D. T., Skelly, J. K., and Wilson, P., *JSDC*, **82** (1966), 176.

63. Mason, J. S., Park, J., Thomson, T. M., and Shore, J., *JSDC*, **96** (1980), 246.

64. Patterson, M., *Rev. Prog. Color*, **4** (1973), 80.

65. Marvin, D. N., *JSDC*, **70** (1954), 16.

66. Dumbleton, J. H., Bell, B. P., and Murayama, T., *JAPS*, **12** (1968), 2491.

67. Forrester, R. C., and Caldwell, M. A., *ADR*, **58**(2) (1969), 13.

68. Kirner, H., *Textilveredlung*, **4** (1969), 3.

69. Taylor, A. R., and Fries, R. E., *TCC*, **2** (1970), 147.

70. Warwicker, J. O., *JSDC*, **88** (1972), 142.

71. Vaidya, A. A., and Shah, R. C., *Text. Dyer Printer*, **7**(2) (1973), 49.

72. Salvin, V. S., *ADR*, **54** (1965), 272; **55** (1966), 490; **56** (1967), 421.

73. Koshti, R. S., *Colourage*, **21**(15) (1974), 29.

74. Leube, H., and Ruttiger, W., *MTB*, **59** (1978), 797.

75. Carbonell, J., *ADR*, **51** (1962), 83.

76. Merian, E., Carbonell, J., and Sanahuja, V., *MTB*, **42** (1961), 238.

77. Schonpflug, E., *TCC*, **11** (1979), 166; *MTB*, **60** (1979), 229.

78. Derbyshire, A. N., Miles, W. P., and Shore, J., *JSDC*, **88** (1972), 389.

79. Kenyon, G. H., *ADR*, **68**(3) (1979), 19.

80. ICI, Technical Bulletin 655, Rapid Inverse Dyeing of Polyester/Cotton Blends.

81. Stetson, G. R., and Thompson, C. W., *ADR*, **68**(3) (1979), 28.

82. Southeastern Section, AATCC, *TCC*, **11** (1979), 246.

83. Norris, R. T., and Ward, A., *JSDC*, **89** (1973), 197.

84. White, M., and Houser, N. E., *ADR*, **68**(3) (1979), 22.

85. Ordway, C. B., *ADR*, **54** (1965), 279.

86. Carter, S., *Can. Text. J.*, **87** (1970), 67.

87. Iannarone, J., and Thackrah, J., *ADR*, **52** (1963), 357.

88. Fleming, R., and Gaunt, J. F., *Rev. Prog. Color*, **8** (1977), 47.

89. Siegrist, G., *Rev. Prog. Color*, **8** (1977), 24.

90. Mayston, R. S., *ADR*, **53** (1964), 142.

91. Gibson, J. W., USP, 2, 663, 612 (1953); also *TCC*, **11** (1979), 241.

92. USP, 3,794,463 (8 Aug 1972); 3,706,525 (8 March 1971); *ADR*, **60**(3) (1971), 25.

93. de Plasse, M. N., French P. 1,223,330 (1960), as described by Datye, K. V., in *Textile Printing*, Chavan, R. B., Ed., IIT, Delhi, India (Sept. 14, 1979), p. 16.

94. Urbanik, A., and Etters, J. N., *TRJ*, **43** (1973), 657.

95. Ulrich, H., *MTB*, **54** (1973), 1206.

96. Liddiard, A., *Rev. Prog. Color*, **1** (1969), 64.

97. Houben, H., and Pabst, M., *MTB*, **54** (1973), 153.

98. Iannarone, J. J., Meunier, P. L., and Wygand, W. J., *ADR*, **52** (1963), 1014.

99. Tullio, V., *ADR*, **55** (1966), 412.

100. Rhode Island Section, AATCC, *ADR*, **54** (1965), 13.

101. Bradley, R. S., Bird, C. L., and Jones, F., *Trans. Faraday Soc.*, **56** (1960), 23; also Green, H. S., and Jones, F., *Trans. Faraday Soc.*, **63** (1967), 1612.

102. McDowell, W., *JSDC*, **89** (1973), 177.

103. Datye, K. V. (unpublished work).

104. Hawkyard, C. J., *JSDC*, **97** (1981), 213.

105. Somers, H. W., *TCC*, **3** (1971), 259.

106. Fox, M. R., Glover, B., and Hughes, A. C., *JSDC*, **85** (1969), 614.

107. Etters, J. N., *TCC*, **4** (1972), 160.

108. Etters, J. N., and Urbanik, A., *TCC*, **9** (1977), 102.

109. Murray, A., and Mortimer, K., JSDC, **87** (1971) 173.

110. N. Piedoment Section, AATCC, *TCC*, **4** (1972), 260.

111. Hoverath, A., *TCC*, **13** (1981), 41.

112. Birke, W., Ulrich, H., and Schoen, F., *Textilveredlung*, **9** (1974), 3.

113. Anselrode, L., *JSDC*, **93** (1977), 201.

114. Datye, K. V., *JSDC*, **94** (1978), 415.

115. Stetson, G. R., *ADR*, **53** (1964), 229.

116. King, J. C., *ADR*, **53** (1964), 132.

117. Harrison, V. W., Norris, R. T., and Ward, A., *JSDC*, **93** (1977), 8.

118. Buser, R., Capponi, M., and Somm, F., *Colourage, Annual* (1978), 78.

119. Wallenberger, F. T., Holfeld, W. T., and Turner, G. R., *TCC*, **13** (1981), 173.

120. Graham, C. A., and Suratt, L. A., *ADR*, **57**(7) (1978), 36.

121. Stewart, N. D., *Text. Prax.*, **33** (1978), 1377.

122. Shore, J., *Rev. Prog. Color,* **10** (1979), 47.

123. Wilcoxson, W. C., *ADR,* **53** (1964), 298.

124. Ellis, J. R., *ADR,* **54** (1965), 234.

125. Goorhuis, M., *Teintex,* **31** (1966), 327.

126. Davis, J. W., *JSDC,* **89** (1973), 77.

127. Kochling, G., *MTB,* **57** (1976), 825.

128. Hughey, C. S., *ADR,* **68**(9) (1979), 42.

129. Baumgrate, U. and Schluter, H., *MTB,* **61** (1980), 434.

130. Ulrich, H., *Chemiefasern/Textilindustrie,* **29/81** (1979), 474.

131. Hollis, N. M., *ADR,* **65**(3) (1976), 19.

132. Malton, O. E., *ADR,* **65**(10) (1976), 59.

133. Carbonell, J., and Sanahuja, V., *Textil-Rundschau,* **10** (1962), 1.

134. Schonpflug, E., *Chemiefasern,* **16** (1966), 208.

135. Bihn, G., *Text. Prax.,* **21** (1966), 291, **21** (1966), 337.

136. Cheetham, R. C., *JSDC,* **83** (1967), 320.

137. Thiel, G., *TCC,* **2** (1970), 196.

138. Konigs, R., *MTB,* **53** (1972), 1375.

139. Sherrill, W. T., *TCC,* **10** (1978), 210.

140. Konard, H. H., and Turschmann, K., *Text. Prax. Inter.,* **33** (1978), 932.

141. Romar, S., *Textilveredlung,* **14** (1978), 332.

142. Beckmann, W., *Chemiefasern/Textilindustrie,* **29/81** (1979), 339.

143. Sule, A. D., in *Blended Textiles,* Gulrajani, M. L., Ed., The Textile Association, India (1981), p. 321.

144. Lacy, R. E., Salvin, V. S., and Schoenberg, W. A., *ADR,* **50** (1961), 978.

145. Baumgarte, U., *MTB,* **45** (1964), 1267.

146. Stapleton, I. W., and Waters, P. J., *JSDC,* **96** (1980), 301; **97** (1981), 56.

147. Brady, P. R., and Cookson, P. G., *JSDC,* **95** (1979), 302.

148. Buckholz, S. H. W., *ADR,* **54** (1965), 545.

8 | DYEING OF TEXTURED POLYESTER

A large volume of PET filament yarn is textured and is used in weaving and knitting. Textiles made from textured yarns have many advantages, but they also pose various problems during dyeing that are unique to textured materials.[1–3] When these materials are dyed they show a type of unevenness that is known as *barré effect,* that is, light and deep lines of colored yarns at regular intervals across the fabric width. Higher dyeing temperatures and prolonged dyeing times reduce the barré effect. But these very measures endanger the desirable properties that are typical of textured materials—hand, bulk, and elasticity. Thus, special dyeing processes and machines are developed to dye textured materials.

The dyeing problems are created because of defects in textured yarns (see Section 1.9.4) or defective fabric manufacture or heat-setting conditions. The textured yarn has a regularly deformed structure (see Chapter 2) and any variations in the deformations or regularity in the deformations over long lengths give defective fabric. Slight variations in texturing parameters such as twist, tension, temperature, and surface friction can lead to the barré effect. Thus, barréness can usually be attributed to the physical irregularities in the textured yarns. Barréness can be minimized by avoiding the use of faulty yarns. However, the fabric manufacturing process can also introduce unevenness leading to the barré effect. Some of the defects such as variations in yarn density, denier of filaments and yarns, tight spots, and melted or undrawn portions, once introduced in the fabric, cannot be corrected later.

Slight variations in unwinding tensions give faulty knitted goods that give barré dyeing. Special attention is given to uniformity of tension on the yarn during knitting, pern winding, and weaving at each stage so that defective

234

fabric is not produced.[4] It is essential to examine the fabric, preferably against light, before dyeing. A small piece may be dyed and thoroughly examined, since practically none of the irregularities that arise during the production of textured yarns can usually be detected until after dyeing. If it is suspected that the fabric does not have absolutely uniform dyeability, the material is checked further to identify the causes of nonuniform dyeing behavior. A replica on a plastic sheet (polystyrene film) is produced by pressing the sheet on the dyed fabric in a hot press.[5] If the replica shows unevenness, it is likely that the fabric is defective. Such a fabric is then dyed under special dyeing conditions using selected disperse dyes as described later. If the faults are very serious, it may be better to print the fabric than to dye it uniformly. For the dyeing trial, the piece should be long enough to have two or three pern changes so that color variations on dyeing are clearly seen. Whether it is barréness, weft bars in irregular patterns, ladder effect, specky dyeing and so on, they must be carefully identified and evaluated for the source of the defect. These defects can be minimized by the proper selection of dyes, the use of dispersing and leveling agents, optimizing dyeing conditions, use of rapid or jet dyeing machines, or by dyeing by continuous processes. Solvent dyeing techniques were attempted to get even dyeing.[6] However, they are not used commercially.

8.1 DYES

For dyeing textured PET goods, the dyes, apart from the other properties described in Section 7.1, must have the ability to cover yarn irregularities.[7–9] The disperse dyes can be classified into groups according to this ability:

Group 1   Dyes that have excellent covering power but poor sublimation fastness.

Group 2   Dyes that have good covering power and moderate sublimation fastness.

Group 3   Dyes that have moderate covering power and good sublimation fastness.

Group 4   Dyes that have no ability to cover irregularities but excellent sublimation fastness.

The dyes generally appear to produce a lower barré effect as their diffusion coefficient in the textured PET increases. But the dyes with the highest diffusion coefficients are also those with the lowest fastness to sublimation. Thus, there are no dyes with excellent covering power and good sublimation fastness. The dyes are therefore selected depending on the quality and the end-use requirements of the material.[7] The temperature of heat setting for textured PET fabric is 170°C and dyes from Group 2 or Group 3 are selected to get satisfactory level dyeings. If the heat setting is carried out at higher than 170°C, it is essential to select dyes from Group 3 or Group 4. In such a case, the choice for dyes becomes limited since these dyes cannot cover yarn irregularities under normal dyeing conditions. It may be essential to adjust the optimum dyeing conditions.

Sublimation fastness and wash fastness of disperse dyes on PET are not interrelated; dyes with high sublimation fastness need not necessarily have good wash fastness. The sublimation fastness of a dye increases with the presence of groups with ionic character, for example, OH and $CONH_2$, respectively, in place of acetylated derivatives and CN groups. Such groups increase aqueous solubility of dyes and, in turn, may lower the wet fastness. The exhaustion of the dyebath may also decrease with the presence of such groups. The fastness to washing is therefore checked whenever the end use demands high wet fastness.

The light fastness for a given dye on textured yarns could be lower than that on flat yarns. The textured yarn is bulked and therefore, a larger area of the material is exposed to light than that of the material made from flat yarn under similar conditions.[10] When selecting the dyes, the light fastness on textured material has to be considered.

The dyes are selected to suit the method of dyeing. Thermofixation is carried out at 170–180°C and dyes should fix under these conditions. Tailor-made dyes for each type of dyeing process and equipment are marketed by manufacturers.[11]

8.2 AUXILIARIES

A variety of auxiliaries have been described to increase and decrease the rate of dyeing to get barré-free dyeings on textured PET materials.[7–19] The addition of nonionic products has a retarding effect on HT dyeing of PET materials. Thus, the dyeing phase at low temperatures (at which dye goes slowly onto the fiber to give a level dyeing from the very beginning in the presence of the nonionic product) may allow one to dispense with the subsequent leveling operation. Even though use of such auxiliaries help in getting uniform dye transfer in the critical temperature range, these auxiliaries may lower the final exhaustion of the dyebath. They may also cause precipitation of dyes because of their low cloud point.[19]

The addition of a carrier-type auxiliary increases the rate of dyeing at low temperatures when the heating rate may be very high and may offset the desired retarding effect of the nonionic auxiliaries in the heating-up phase. A commercial product, for example, Palegal A (BASF), is a combination of a retarder during the heating-up phase of the dyeing and an accelerant in the final high-temperature phase.[15] Alternatively, the nonionic auxiliaries are added first and the carrier is added at 130°C in the last stages of the leveling–dye-fixing phase of HT dyeing. The dispersing agents, leveling agents, accelerants, and other auxiliaries used for dyeing PET have been described in Section 7.4.4.

The carriers act as a leveling agent that gives barré-free dyeings on textured PET material.[13,20–22] This is illustrated in Figure 8.1, in which the yarns differing in their physical structure because of different texturing temperatures are dyed at 120°C without and with a carrier.[21] The carrier increases the rate of diffusion of a dye in the fiber by lowering the glass-transition temperature of the fibers and thus, increasing the effective dyeing temperature (see Chapter 3). The internal variations in the fine structure of textured yarns, which influence the diffusion of dye in the fiber, become less and less significant with higher diffusion rates and thus, do not decide the rate of dyeing of textured yarns. The dye rapidly diffuses in and out of the fiber in the presence of a carrier. This facilitates the migration of dye. The contribution of the carrier in the migration process can be demonstrated by carrying out a migration test in which a dyed sample along with an equal weight of undyed sample is treated in a blank bath with and without a carrier. The dyeings will match (100% migration) if 50% of the initial dye migrates to the undyed sample. Typical results of a migration test are given in Table 8.1.

Apart from the migration-leveling action, a carrier influences the build-up of a dye on texturized PET, which need not be in harmony with the former (Table 8.2).[1] The use of a carrier is of particular importance for dyes from Groups 3 and 4. The concentration of a carrier is critical and excessive use of a carrier leads to uneven dyeing. After dyeing, the carrier is completely removed from the dyed fabric to achieve optimum light fastness.

8.3 CARRIER DYEING

Textured polyester fiber holds knitting lubricants more tenaciously than other fibers, especially if the goods have been stored for some time. Therefore, a good scour, preferably a solvent scour, is desirable and can be mandatory in some cases for good results. The knitted, textured yarn fabric is

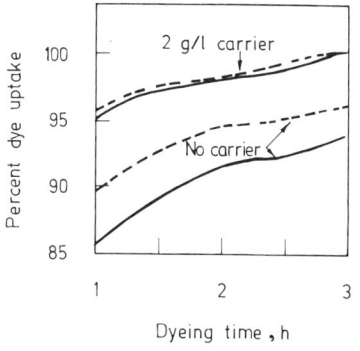

FIGURE 8.1. Effect of carrier and texturing temperature on dye uptake of polyester.[21] Dyeing temperature: 120°C. Dye: Eastman Polyester Brilliant Red B-SLW. Texturing temperature: ——— 205°C; ——— 220°C.

TABLE 8.1   Migration of Dyes on Textured PET in
the Presence of a Carrier[8]

Dye A: C.I. Disperse Red 151
Dye B: C.I. Disperse Violet 57
Shade: 4%
Migration Test: 120°C/1 h
(100% migration when originally dyed and
originally undyed samples match.)

	Migration (%)	
Carrier (2 g/liter)	Dye A	Dye B
None	32	30
2-Hydroxydiphenyl	58	54
Butylbenzoate	66	66
o-Dichlorobenzene	50	62
Dichlorophenoxy ethylacetate	82	85
Irga Carrier HTP (Ciba-Geigy)	40	43

TABLE 8.2   Influence of Carriers on Build-up and Leveling of Disperse Dyes
on PET[2]

Shade Build-up	Leveling
1. Methyl-2-hydroxy-3-methyl benzoate	Trichlorobenzene
2. 2-Hydroxybiphenyl	Butylbenzoate
3. 2-Methyl naphthalene	Methyl-2-hydroxy-3-methyl benzoate
4. Diphenyl	2-Methyl naphthalene
5. Trichlorobenzene	2-Hydroxydiphenyl
6. Butyl benzoate	Diphenyl

Note: Carriers arranged according to decreasing influence.

heat set (150–180°C/15 sec) on a pin stenter and cooled before dyeing. The
temperature employed for the heat setting of the fiber has a tremendous
influence on the subsequent uptake of disperse dyes. The dyes reflect differences within the material and disclose irregularities in the heat-setting conditions. The delicate knitted fabric is dyed in tension-free conditions, preferably in a covered winch. The bath is set with diammonium phosphate and the
goods are treated for 10 min at 50°C before the carrier (2-hydroxydiphenyl
sodium) is added. After 10 min, the dye dispersion is filtered into the dyebath
and the temperature is gradually raised to boiling within 45 min. Dyeing is
continued for 1–1.5 h during which the carrier is slowly converted from its

sodium salt to free phenol by the loss of ammonia and the formation of sodium phosphate. The goods are then thoroughly rinsed with warm and cold water. After dyeing, the traces of the carrier in the fiber are removed either by treating the goods at 155°C for 3 min or at higher temperatures for a shorter time, or by treating the goods with caustic soda solution and a wetting agent. The former method is preferred because it also acts as a setting operation.

8.4 HT DYEING

Dyeing conditions are selected according to the group of disperse dyes. The dyes in Groups 1 and 2 (with good-to-excellent covering power) are dyed below 125°C. For dyes in Groups 3 and 4 (with poor covering power), the dyeing temperature must be 130°C and the addition of a carrier is recommended for better coverage of the yarn irregularities.

Sometimes barré-free dyeings are obtained by giving a relaxation treatment in a blank bath at 120–130°C. After the treatment, the bath is set at 60°C at pH 5.5 and the HT dyeing is carried out in the usual manner, if necessary, with the addition of a carrier. If the relaxation treatment is carried out under alkaline conditions, the oligomer problem is minimized (see Section 8.7).

Rationalization of the HT Dyeing Process. Under practical HT dyeing conditions, the temperature increases from the start up to a time when it reaches the maximum where it remains constant.[23] Typical rate-of-dyeing curves under such a condition are shown in Figure 8.2.[15] There are three distinct phases where the dye gets transferred from the bath to the fiber. In the initial stage, the dye transfer is very slow and small and there is no diffusion of dye in the fiber phase. Then, over a temperature range of 30°C, about 80% of the dye is taken up by the fiber when a small fraction of the total dye on PET diffuses into the fiber. In the last stage, the dye in the boundary layer on the surface of the fiber diffuses into the fiber or remigrates into the dyebath. The latter process is more predominant in heavily dyed portions than in weakly dyed portions. This migration process is accelerated when the dyebath is depleted of the dye. Attempts have been made to get even dyeing without the migration process.[1,15,24–27] The dyeing cycle is organized in such a way that the dye goes onto the fiber evenly from the very beginning so that no time is wasted in the subsequent migration-leveling process and the high-temperature phase can be restricted only to the diffusion of dye into the fiber.

The characteristic temperature range in which most of the dye is transferred from the dyebath to the fiber surface is not the same for different dyes. The dyes can be grouped into 3–4 sets with nearly similar characteristic temperature ranges.[23–25,27] It is possible to manipulate this range with the addition of auxiliaries (Figure 8.2).[15] The dyebath is rapidly heated before

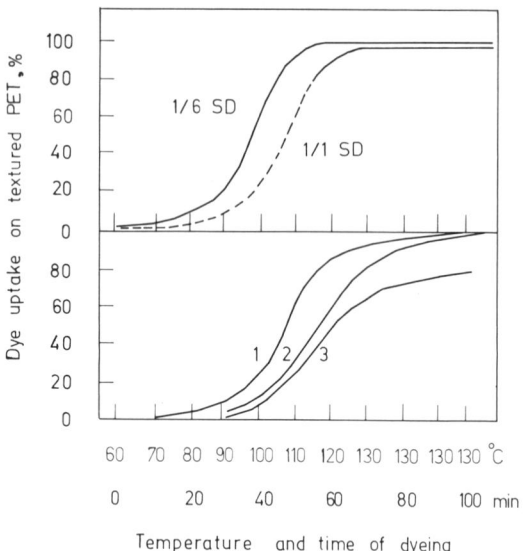

FIGURE 8.2. Rate-of-dyeing curve of disperse dyes on textured polyester at two standard depths (SD).[15] 1: Without auxiliary. 2: With 2% Besosoft SG. 3: With 1.5% Palegal A.

and after the characteristic temperature range (when 80% dye is transferred to the fiber), thereby, reducing the dyeing cycle time. The temperature control in the heating-up phase (when 80% dye is transferred to the fiber) should be accurate so that the dye goes on to the fiber uniformly. Initial and final heating can be very rapid while in the critical temperature range of about 30°C, the heating rate should be low (~1°/min). This method is suitable for dyeing textured PET yarn, fabric, and knitwear in HT and jet dyeing machines.

Rapid Dyeing Machines.[14,27–33]   Although it is possible to cut down the dyeing time considerably by using the rationalized dyeing method, the total dyeing time is still quite long. In the characteristic temperature range, the dyebath temperature cannot be increased rapidly with the existing dyeing machines, which have a liquor circulation rate of 1–2 cycles/min and give uneven dyeing. To obtain level dyeings, the rate of rise-of-temperature must be decided from knowledge of the exhaustion of the dye and the time required to circulate the total volume of liquor.[28] To obtain acceptable dyeings, a linear rate of exhaustion must be achieved using a rate of rise-of-temperature that gives 1–2% exhaustion of the dye for each circulation cycle of the dye-liquor. Thus, the more rapidly the liquor circulates, the more rapidly the temperature can be raised without any risk of unevenness and the more rapidly the liquor will exhaust.

Rapid dyeing machines were developed (Table 8.3)[14] based on the above contentions. In these machines, the dye-liquor is circulated at the very high

TABLE 8.3 Typical Rapid Dyeing Machine[14]

Manufacturer/ Machine Name	Type of Pump	Liquor-to-Goods Ratio	Flow Rate (liter/kg/min)	Bath Circulation Rate (cycles/min)	Maximum Heating Rate (°C/min)
Argelich, Termes	Centrifugal	10 : 1	70–80	7–8	10–12
Bruckner, Tupulsar	Reversible axial pump	8 : 1	56–60	6–7	10–12
Callebaut de Blicquy, TR	Reversible axial pump	10 : 1	up to 200	up to 20	12–16
Frauchiger, Favorit	Regulated Centrifugal	7 : 1	70	10	10–12
Gaston County Rapid Reversal	Centrifugal	11 : 1	40–70	6	6–10
Pegg, HTU	Mixed flow pump	11 : 1	80–100	7–11	10
Thies, Burl-Vac	Regulated Centrifugal	10 : 1	40–60	4–6	Above 2

rate of 6–12 cycles/min. Two methods have been used to get rapid circulation. In the first method, high-capacity pumps were used. The second method depends on the shortening of the liquor-to-goods ratio so that the smaller volume of the liquor is pumped to complete the liquor circulation cycle. Such machines may have pumps of a somewhat larger capacity than those of conventional machines but they are not as large as the pumps required for the first method.

The rapid dyeing machines are equipped with powerful heat exchangers so that the liquor can be heated at a rate of 6–10°/min. Special arrangements are made for the quick filling and draining of the liquor. The rate of heating is adjusted according to the rate of circulation of liquor in the machine (Table 8.4).[30] Below and above the characteristic temperature range, the heating is very rapid, at a rate of up to 6–10°/min. Dyeing at maximum temperature is carried out for 15–45 min, depending on the depth of shade and the rate of diffusion of the dyes. A typical time–temperature heating cycle program for dyeing textured PET in medium depth is shown in Figure 8.3. The rapid dyeing machine can dye 18–22 batches of textured PET yarn/24 h.[14] Furthermore, the dyeings are very uniform and therefore these machines are widely used for dyeing textured PET yarn. However, the initial cost of these machines is higher than that of the conventional HT machines.[16,27,29,31]

Jet Dyeing Machines.[34-43]   These HT dyeing machines were developed for handling very delicate knitted fabrics of textured yarns. The speciality of these machines is that both the fabric and the dye-liquor move during the dyeing process. In this rope dyeing machine, the goods are kept moving rapidly and shifting constantly so that a wrinkle or fold is worked out. The fabric is processed in endless parallel loops and is essentially penetrated, conveyed, piled in the storage chamber, and dyed by precisely directed jets

TABLE 8.4   Rapid Dyeing Machine Variables
($M:L$ Ratio 1 :10)[30]

Flow Rate (liter/min/kg)	Bath Circulation (cycles/min)	Critical Heating Rate (°C/min)
15	1.5	0.9
20	2	1.2
25	2.5	1.5
30	3	1.8
40	4	2.5
60	6	3.7
80	8	5.0
100	10	6.2
120	12	7.5

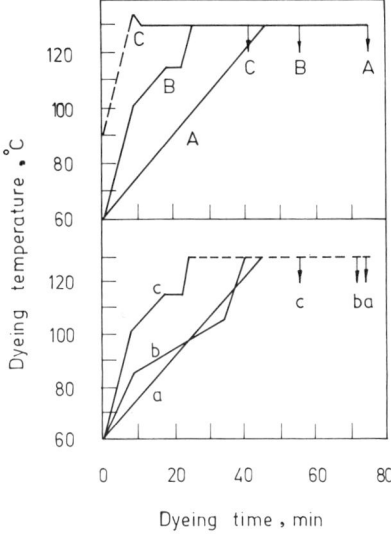

FIGURE 8.3. Time and temperature heating cycle for medium depths on textured polyester.[14] (A) Linear heating (75 min with diffusion time), (B) Rapid dyeing (56 min with diffusion time), (C) Injection method (42 min with diffusion time), (a) Linear heating (45 min without diffusion time), (b) Optimum heating (42 min without diffusion time), (c) Rapid dyeing (26 min without diffusion time).

of dye-liquor. The prime mover of the fabric in the machine is the main jet (Figure 8.4).[34] The liquor simultaneously penetrates and propels the material at a speed of 90–130 m/min through the cloth-guide tube to the rear of the machine and produces a tremendous interchange of fluid with the fabric. Various types of jet dyeing machines have been marketed (see Chapter 6) that can be broadly classified into two types: (*a*) partially flooded and (*b*) fully flooded. The latter is more suitable for textured yarn goods since it minimizes the lengthwise tension. The problem of foaming is absent in the latter type of machines. These machines are used for dyeing double-jersey polyester, knitted goods, and woven fabrics. These machines give improved levelness and quality of dyeing and have no problems of shading, ending, listing, patchiness and so on. The feel of the fabric remains unaffected and there are no distortions because of tension on the fabric. Dyeing is economical in time and chemicals (short liquor ratio of 1 : 12) and gives minimum barréness. However, there are some inherent problems in jet dyeing machines such as blockage of jets, entanglement of fabric, crease marks and so on. Apart from these problems, the initial cost is also very high.

Double-Knits of Textured Set Polyester. Two major type of defects have been attributed to the dyeing of double-knits of textured set polyester. The first is the variation in dyeability of a single end, which can be a few mm long or can run for several meters. The shade might be light or heavy. This type of defect is attributed to the variations in texturing temperature or other thermal history of the yarn. The second type of defect consists of light and dark bands, 1–1.5-cm wide, called *shade barré* or *tapered barré*. The shaded barré are caused in knitting machines by a slight, but immeasurable variation

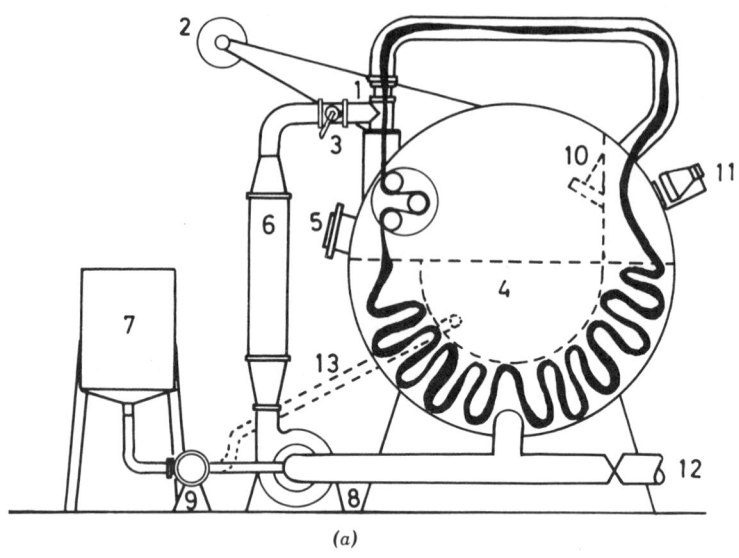

(a)

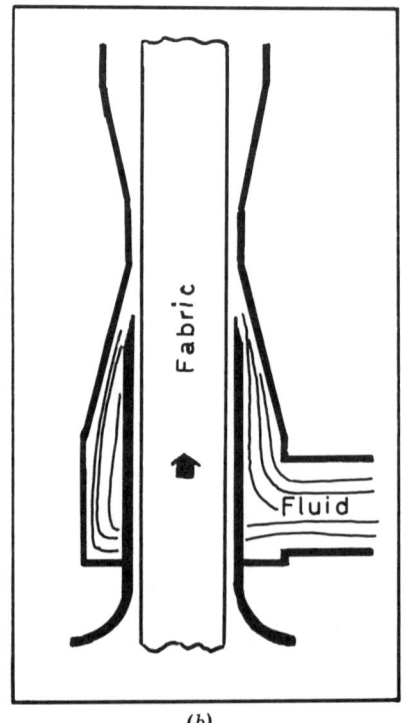

(b)

FIGURE 8.4. Jet dyeing machine (a) and cross-section of the main jet (b).[34] (A) 1: Jet; 2: Unloading reel; 3: Throttle valve; 4: Main vessel; 5: Loading port, 6: Heat exchanger; 7: Add dye tank; 8: Centrifugal pump; 9: Add dye pump; 10: Doffing jet; 11: Light; 12: Drain; 13: Alternate connection.

244

in stitch formation, specifically by a variation in the gap between the dial and the cylinder of the double-knit machine. The amount of barré can vary from machine to machine and from day to day, depending on how well the machine is maintained. Dark or light-dyeing ends in the fabric can magnify or reduce machine barré depending on how they fall in the band. Machine barré is not peculiar to textured-set polyester yarn. But the nylon, acetate, or wool yarns in double-knits give more cover so that the barré pattern is not as noticeable as in textured PET yarns.

Machine barré can be reduced or entirely eliminated during dyeing if the stitches are allowed to equalize in size and shape. Thus, carriers that give high fabric shrinkage reduce machine barré, particularly, under HT dyeing conditions.

Rope marks, wrinkles, and cracks in the fabric can be prevented by keeping the fabric moving rapidly and shifting constantly in the beck or jet machine. Lubricants (0.5–1%) in the bath that allow the yarn to slip freely and cool slowly (1–2°C/min) are helpful in avoiding wrinkles and cracks. The wrinkles get hydroset if the rate of cooling is very high and cannot be removed by later heat setting. Once the fabric is cooled below the glass-transition temperature, the cooling rate may be high.

Heat setting after dyeing removes many soft wrinkles and creases, but cracks and sharp creases caused by the slow movement of the goods during a rapid temperature rise or drop cannot be removed. The higher the setting temperature, the more wrinkles are removed.[44] However, the heat-setting temperature and the carrier level influence the hand, appearance, and performance of the fabric. Fabric hand can also vary with the depth of shade. Heat setting above 165°C gives the best fabric performance and evens out differences between dark and light shades. For menswear, crisper, more worsted-type hand and less than 15% stretch are specified.[45] Better stabilization is also needed because most menswear is subjected to more steam pressing than womenswear. Dyes with sufficient sublimation fastness to withstand the higher temperatures of heat setting are therefore selected for menswear. This is likely to give more of a barré effect. For example, the following dyes may be used for the dyeing of menswear.[46]

C.I. Disperse Yellow 42, 44, 58, and 63
C.I. Disperse Red 53
C.I. Disperse Blue 60, 61, and 73.

8.5 THERMOFIXATION DYEING

Thermofixation dyeing is a continuous process of coloring textured PET fabrics.[47,48] Apart from a high production rate, it gives minimum barréness, uniform consistent shade over long lengths without rope marks, creases and so on, and firm hand and dimensional stability. It is economical in utilities, energy, and labor. However, it has some limitations for textured PET goods. It imparts stiffness to the material with a loss of bulk in textured yarns.[49] Thermofixation conditions, time, and temperature have to be mild to protect the textured yarns from flattening. Many dyes do not fix on PET, which gives poor penetration into the fiber, under these conditions. The usual problems of migration of dye during drying, uneven heating from selvages to center or side to side, poor reproducibility of shades and so on which are associated with the thermofixation process of coloring PET, are also present in the dyeing of textured goods. The textured PET fabric is scoured and heat set. The dyeing operation consists of padding, drying, thermofixation, and aftertreatments.

Padding.   The fabric is padded with

X g/liter	Disperse dye
1 g/liter	Anionic wetting agent
1 g/liter	Dispersing agent
2–10 g/liter	Antimigrating agent
2 g/liter	Acetic acid (to pH 5)

A fixation accelerant is incorporated in the pad-liquor to facilitate thermofixation under mild conditions. This agent should not interfere with the transfer of dye to the fiber in any other way and should be easily removable. Ethylene glycol monophenyl ether (3–4 g/liter) acts as a good fixation accelerant and thermofixation can be carried out at 160°C in 4 min or at 180°C in 2 min using superheated steam.[50]

Drying.   Because of its bulk, the wet pickup on textured PET fabric is 90–100% compared to 60% on PET/CO blend fabric. This induces the migration of dye during drying (see Figure 7.7) and, in turn, uneven distribution of dye on the fabric after drying. Thus, drying of padded textured fabric becomes a critical step in getting uniform dyeing. Predrying with IR radiations gives minimum migration. The air circulation during drying in a hot flue or a pin stenter is kept to a low level without losing the uniformity of temperature (100–110°C).

Thermofixation.   When the textured PET fabric is exposed to a dry-heat treatment at 190–210°C, the texture of the fabric is adversely affected and the handle and appearance of the fabric are no longer comparable with those of goods dyed in a long liquor. It is, therefore, essential to thermofix the padded dried goods under conditions that cause the least harm. Thus, fixation temperature is limited to 160–180°C and the mechanical stresses on the material at elevated temperatures must be as low as possible. At the same time, a satisfactory color yield with deep shades has to be achieved. For this purpose, fast diffusing dyes are used. The dwell time in the fixation chamber is increased and superheated steam is used for fixation.[50] The heat capacity of steam is much higher than that of air at the same temperature. Considerably greater color yield is obtained with disperse dyes at 160–170°C by superheated steam than by hot air. Besides giving a high fixation at a lower temperature, it does not adversely affect the feel of the textured fabric.

Aftertreatments. After thermofixation, the fabric is given the usual reduction-clear, washing, and soaping treatments. (see Chapter 7). The thermofixation process is suitable for woven fabrics in which the filling consists of textured PET yarn. Such fabrics behave in many respects like staple fabrics. The process is used for dyeing flat knitted fabrics and is never used for very bulky fabrics. The process is particularly useful where a firm full hand is required. Knitted fabrics, whose characteristics are soft handle, full appearance, and high bulk should not be dyed by the thermofixation process.

8.6 TEXTURED POLYESTER BLENDS

Woven fabrics containing textured PET filament yarns of 100–400 denier as weft and PET staple–cotton spun yarn of 65 : 35 or 50 : 50 PET : CO ratio as warp offer many advantages,[51] such as:

1. The fabric contains about 80% PET and thus, the cost is lower than 100% spun-yarn material.
2. Special characteristics such as drape, luster, and brightness of shade are achieved.
3. Fewer filling breaks increase the efficiency of weaving.
4. Same coverage with fewer picks per inch.
5. Release of spinning capacity in composite mills blocked in spinning the weft yarns.

The delicate nature of this blend fabric necessitates careful handling throughout the wet-processing sequence. Built-in, material-induced barréness of textured weft yarns has to be avoided by special care in dyeing or printing. Marking off is a potential problem during the resin-finishing of these fabrics. Dyeability of textured filament yarns and polyester staples in the blend spun-yarn is not the same, which can create difficulties in getting a solid shade unless special precautions are taken.

8.6.1 Preparation for Dyeing

The preparation of these blends involve the usual steps—singeing, desizing, scouring, relaxing, bleaching, mercerizing, and heat setting. Some of these steps may be eliminated or combined into one step. Singeing is required for minimizing pilling. In order to avoid the fusing of filaments in the textured yarn, singeing should be done at a high speed with a low flame. For twill constructions, the burners are sometimes used only on the warp-face side. If the singeing operation cannot be well-controlled or there are major equipment limitations, it is advisable to omit singeing, whether it is done before or after dyeing. Desizing is usually done with enzymes (see Chapter 5). These blended fabrics are scoured using an alkali such as sodium carbonate or caustic soda. The bulk and crimp of textured yarns are constantly subjected to adverse conditions during the mechanical and chemical-processing operations. Relaxation and reconditioning treatments are required to restore the bulk and crimp and to confer the original handle. Special-purpose open-width relaxation machines are developed to relieve stress and strain caused by weaving and preparatory processes. Conventional open-width becks, continuous washing ranges, and drum washers are also used. The relaxation and scouring treatment can be combined. Hydrogen peroxide is generally employed for the bleaching of blended fabrics because it produces higher cotton whiteness and removes motes. Mercerization helps to ensure coverage of the immature cotton during dyeing and improves the dyeability of the cotton. Mercerization is needed more for carded cotton than for combed cotton and for an improved dyed appearance of heavy shades. The purpose of heat setting, which is carried out before dyeing, is (a) to pull out wrinkles, creases, and crack marks and (b) to improve the width stability of the fabric during the dyeing operation. Before heat setting, the prepared fabric should be free of caustic soda and should have a pH of less than 7. Heat setting of blended fabric is carried out at 170–180°C for 30–60 sec in order to preserve the bulk and handle of the textured yarn.

The drying of the fabric containing textured polyester-filament yarns is carried out carefully.[44,51] When such a fabric is passed over a large number of hot cans, the shine of the fabric increases. Overdrying can effect the absorbency of the fabric. Low-cotton–polyester blends dry very fast and it is possible to reduce steam in the cans or to bypass some hot cans and still obtain satisfactory drying at production speeds. The fabric is not allowed to slip on the hot cans as this will have a polishing or ironing effect and will give a metallic shine.

8.6.2 Dyeing

The polyester staple fiber has a higher rate of dyeing than textured filament yarns (Figure 8.5).[52] This creates a problem in getting a solid shade on the blend fabric. The staple fiber is blended with cotton and this makes it possible to overcome the problem. The cotton is dyed lighter than the spun polyester which allows the overall shade of the warp and filling yarns to almost match. For improved dyed appearance, the depth of shade on cotton

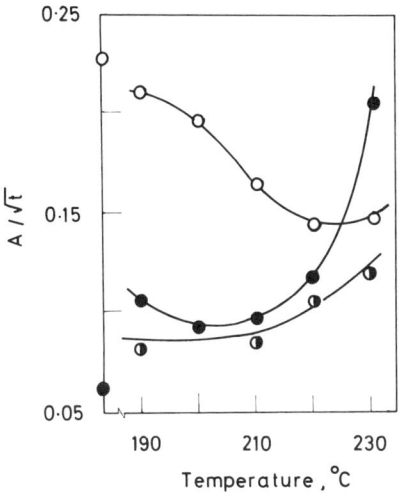

FIGURE 8.5. Dyeability of textured and flat PET yarn dyed without and under tension.[52] ●: Flat PET dyeing without tension. ○: Textured PET dyeing without tension. ◑: Textured PET dyeing with 10 g tension. $A/\sqrt{t}$: dyeing rate.

is between the darker shade on spun polyester and the lighter shade on textured filament filling, preferably closer to the latter.

Blended fabrics can be dyed either in a HT-beam dyeing machine or in a jet dyeing machine at a temperature of 120–130°C. The addition of a small amount of carrier gives more uniform results because it helps the migration of disperse dyes. The carrier is also useful in minimizing the difference in depth between warp and weft. After dyeing the polyester component, the usual reduction-clearing treatment is given and the cotton portion is then dyed with the appropriate dyes. It is possible to dye the PET in these blend fabrics by the conventional thermosol process for textured materials described in Section 8.5.

8.7 OLIGOMER PROBLEM

Polymer chains of high-molecular-weight are built up from monomers, such as ethylene glycol and terephthalic acid. Many different types of compounds are also formed simultaneously. Cyclic trimer is one of the compounds of polymerization formed by the intramolecular or intermolecular cyclization process.[8] Among the various low-molecular-weight polymers (oligomers) in PET, the cyclic oligomer occurs in the largest amount. Oligomers in the polyester are not removed before melt-spinning because they cannot be extracted easily and do not interfere with the spinning process. Furthermore, if extracted from the PET chips by solvent, the extracted chips reproduce the cyclic oligomers on remelting. The process can be repeated which demonstrates the equilibrium nature of the formation of these oligomers. All the commercial polyester fibers contain the cyclic oligomer, the amount of which is 1.3–1.7%.[8,53] There is no known way to prevent its formation, although the possibilities of production of PET with low trimer content have been explored.[54] The oligomers can be extracted with 1,4-dioxan at boiling in several hours. Use of cold dioxan for extraction removes only the surface oligomers.

PET cyclic trimer consists of three molecules of ethylene terephthalate which are reacted to form the cyclic compound (see Figure 1.9). It has a molecular weight of 576 by mass-spectrograph and a melting point of about 320–322°C. It is extremely difficult to dissolve the crystals in water because of poor water solubility (1–3 mg/liter in a HT bath at 130°C). Under the influence of heat, the molecular structure of PET fiber contracts and the cyclic trimer is exuded on the fiber surface. The exact mechanism of its emergence on the fiber surface as crystal is not yet fully established. The diffusion rate of cyclic oligomer in the fiber phase has to be very very low considering its molecular weight and cyclic structure.[55] It probably separates out by the phase-separation mechanism for the two immiscible polymers.[56] Pratt described it to be a process analogous to water being squeezed from a sponge.[57] The more severe the hot–wet conditions, the more trimer is exuded.[58,59] The total amount of oligomer removed during heating and dyeing processes may thus be only a small portion of the total present in the fiber. The amount of trimer exuded by PET textured yarn during autoclaving before dyeing increases with temperature (Figure 8.6).[57] The thermal history of the fiber and the dyeing conditions are the two most important factors that decide the gravity of the oligomer problems. Carriers in dyebath promote[57] while auxiliaries do not prevent the oligomer emergence.[59] Cationic dyeable polyester exudes more trimer than the disperse dyeable polyester. Texturing PET yarns creates acute oligomer problems since the oligomers migrate to the surface because of heat treatment during the crimp setting in the texturing process.[60] PET yarns textured under different conditions were analyzed by Skelly[8] before and after a treatment in a blank dyebath for 60 min at 130°C for the surface and total oligomer content. His results are given in Table 8.5.

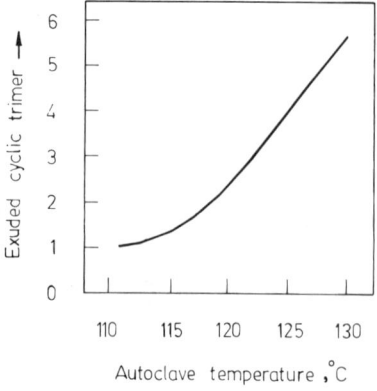

FIGURE 8.6. Effect of autoclave temperature on the cyclic trimer exudation by a textured polyester yarn.[55]

TABLE 8.5 Surface and Total Oligomer Content of PET
Textured Yards after Blank Dyeing at 130°C for h;
% Surface Oligomer Content Before Treatment Is Less than
0.005% in All the Samples

| | Heater Temperature | | % Oligomer Content | | |
| | | | Surface Content | | Total |
	First	Second	A	B	Average
0	Untreated		0.13	0.14	0.72
1	210	190	0.15	0.22	0.70
2	200	190	0.18	0.20	0.60
3	210	180	0.12	0.15	0.62
4	200	180	0.09	0.16	0.74
5	190	180	0.11	0.17	0.77
6[a]	210	190	0.12	0.18	0.65
7[a]	210	190	0.14	0.21	0.71
	Average		0.13	0.18	0.69

[a] Sample 6 and sample 7 have −1% and +2% overfeed.
Auxiliary in dyebath:
A: 0.5 g/liter Irgasol DA (Ciba-Geigy)
B: 1.0 g/liter Irgacarrier HTP (Ciba-Geigy)

The total oligomer content is not significantly changed by texturing or high-temperature dyeing conditions, although the surface oligomer has varied.

The oligomers create a number of problems.[57,58] They show up on the yarn surface or on the package dyeing machines. In extreme cases, the oligomers show up as a white crystalline deposit (see Figure 1.10). They commonly show up as a white dust or gummy deposits on guides and other contact surfaces on coning machines, knitting machines, looms and so on. Gummy deposits may also contain moisture, dye, fiber, and finish. Noticeable deposits on walls of dyeing apparatus can clog the pumps and heat exchangers, resulting in reduced pump and heat-transfer efficiency. At times, deposits on heavily dyed yarn packages and textiles cause dusting, loss of brilliance, and uneven appearance. The moving yarn gives abrasion deposits at yarn guides and the frictional properties are altered which, in turn, affect the knitting and weaving because of high and nonuniform yarn tensions. The reeds and knitting needles wear rapidly since the oligomer crystals embedded on the fiber surface have sharp edges (Figure 8.7).

During wet-processing and, particularly, under HT dyeing conditions, a small fraction of the oligomers diffuse into the dye-liquor, crystallizes, and is deposited in the form of loose dust (see Figure 1.10) when the bath is cooled before draining. The dyebath is drained at maximum temperature without any cooling so that the liquor carries away the dispersed and dissolved

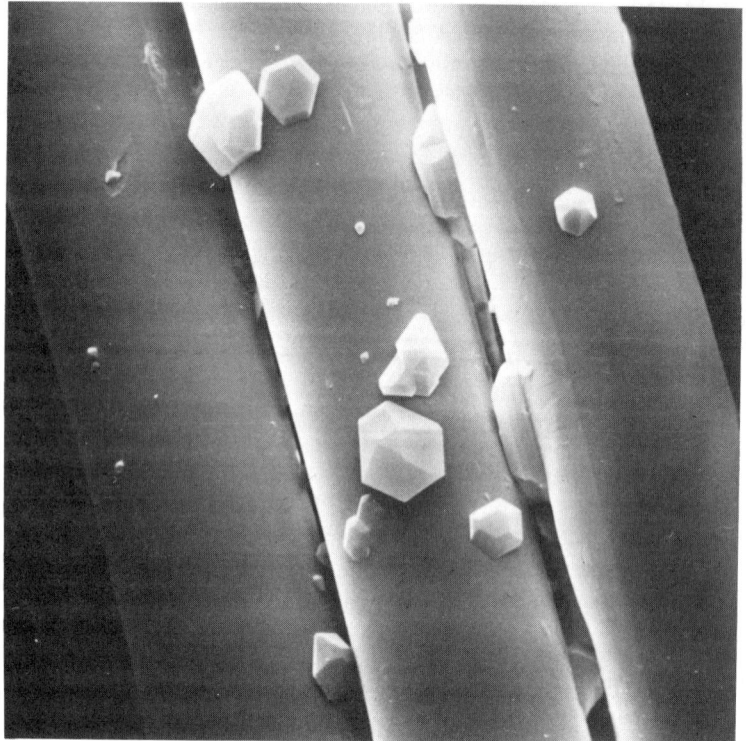

FIGURE 8.7.   Electron micrograph of embedded oligomer crystals on PET fiber surface.

oligomers before they precipitate. This, however, does not solve the problem of crystals on the fiber surface since the oligomer fraction dissolved in the liquor is negligibly small. Setting the yarn in the dyebath without autoclaving appears to reduce the amount of trimer on the fiber.[57] The trend towards high dyeing temperatures to cover dye differences partially accounts for more trimers on the fiber. Yarn packages are ideal filters for cyclic trimer crystals so the problem is critical in yarn dyeing in package form. The oligomer is flushed off the fabric rather easily during rinsing. The presence of a carrier in the dyebath helps to keep the oligomers in solution since the latter are soluble in the carriers.[61] However, since carriers also promote oligomer emergence,[58] the full advantage of the solubilizing action cannot be exploited for maximum efficiency of oligomer removal. Some of the auxiliaries[57,59] keep the oligomer bound to the fiber surface to prevent dusting. However, they may interfere in dye absorption and may affect the fastness properties of the dye. Dispersing agents are ineffective in keeping cyclic trimers in the solution. Certain auxiliaries promote the washing off of the oligomer dust. Reducing agents have no effect on the oligomer content on the fiber surface. An intensive alkali treatment may partially peel off the fiber surface together with oligomers. A solvent pretreatment does not prevent emergence of the oligomers. It merely influences the surface content, probably by removal of crystallization nuclei. Boiling out the dyeing machine periodically removes deposited oligomers and minimizes build-up. It reduces filtration of deposits left by previous lots. The dyeing machine is cleaned by circulating a solution containing caustic soda (2–5%) and ethylene glycol (2%) or trichloroethylene (1%) under HT dyeing conditions for 30–60 min.

The how-to-eliminate-trimer-in-dyeing-PET-materials problem has been discussed by many workers.[57-62] The best solution is to give a pretreatment in an alkaline HT bath with or without a carrier prior to dyeing.[58] No machine cleaning or aftertreatments are necessary. Furthermore, the structure leveling during the pretreatment improves uniformity in dyeing. It improves brilliancy and pilling resistance without affecting the force–elongation properties and crimp indices of the dyed fiber. Some of the fastness properties are also improved. This pretreatment may be applied to textile goods in any make-up. The dyeing may be carried out in the same or different bath after adjusting the pH.

REFERENCES

1. Beckmann, W., *TCC,* **2** (1970), 350.
2. Vaidya, A. A., *Text. Ind. and Trade J.,* India, (March–April 1976), 11.
3. Elliot, A., *Text. Dyer and Printer,* **11**(4) (1977), 29.
4. Kirjanov, A. S., *ADR,* **67**(3) (1975), 28.
5. Mahal, K., *Textilveredlung,* **15** (1980), 373.
6. Vaidya, A. A., and Sunder, R., Book of Papers presented at Conference on Texturing, Organized by IIT, Delhi, India, Sept. 1977, p. 107.
7. Hemphill, J. E., *ADR,* **58**(4) (1969), 28.
8. Skelly, J. K., *JSDC,* **89** (1973), 349.
9. Derbyshire, A. N., *JSDC,* **90** (1974), 273.
10. Schonrock, G. F., and Weigold, S., *TCC,* **1** (1969), 579.
11. Hasler, R., Jakob, H., and Palacin, F., *MTB,* **62** (1981), 671.
12. Murray, A., and Mortimer, K., *JSDC,* **87** (1971), 173.
13. Beckmann, W., and Brieden, H. H., *TCC,* **5** (1973), 118.
14. Siegrist, G., and Liddiard, A. G., *JSDC,* **89** (1973), 523.
15. Richter, P., *Textil-Praxis,* **28** (1973), 12; Shirley Institute Publication, **S-31** (Nov. 1977).
16. Schoenpflug, E., and Richter, P., *TCC,* **7** (1975), 136.
17. Leube, H., *TCC,* **10** (1978), 39.
18. Slack, I., *Can. Text. J.,* **96**(7) (1979), 46.
19. Derbyshire, A. N., Mills, W. P., and Shore, J., *JSDC,* **88** (1972), 389.
20. Forrester, R. C., Schrum, F. F., and Caldwell, M. A., *ADR,* **60**(3) (1971), 21.
21. Haile, W. A., Willingham., W. L., Forrester, R. C., and Caldwell, M. A., *ADR,* **66**(4) (1977), 32.

22. Vaidya, A. A., Sunder, R., and Nigam, J. K., Proceedings of National Textile Seminar, Delhi, India, April 1978.

23. Carbonell, J., and Lerch, U., *Textilveredlung,* **4** (1969), 229; *Colourage* (Jan. 1972), 23.

24. Beckmann, W., and Brieden, H. H., *Chemiefasern,* **20** (1970), 553.

25. Beckmann, W., Brieden, H. H., Leckebusch, H., and Weigner, H. D., *Textilveredlung,* **7** (1972), 499.

26. Krieg, U., *ADR,* **70**(3) (1981), 17.

27. Ullrich, H., *ADR,* **70**(4) (1981), 30.

28. Carbonell, J., Hasler, R., Walliser, R., and Knobel, W., *MTB,* **54** (1973), 68.

29. Ullrich, H., Reither, A., and Waasmuth, H. B., *MTB,* **55** (1974), 549.

30. Ullrich, H., and Reuther, A., *Can. Text. J.,* **92**(4) (1975), 83.

31. Nakayama, T., *Japan Text. News,* **256** (March 1976), 73.

32. Siegist, G., *Rev. Prog. Color.,* **8** (1977), 24.

33. Meisa Chemical Works Ltd., *Japan Text. News* (April 1979), 75.

34. Carpenter, W. T., *TCC,* **1** (1969), 265.

35. Haigh, D., *Rev. Prog. Color,* **2** (1971), 27.

36. Ullrich, H., and Reither, A., *MTB,* **52** (1971), 216, 318.

37. Proceedings of the Society of Dyers and Colorists, *JSDC,* **88** (1972), 321.

38. Patterson, M., *Rev. Prog. Color.,* **4** (1973), 80.

39. Koshti, R. S., *Colourage,* **21**(4) (1974), 25.

40. Ratcliffs, J. D., and Birtwistle, F., *JSDC,* **90** (1974), 313.

41. Elliot, A., *TCC,* **10** (1978), 147.

42. Haile, W. A., and Somers, H. W., *TCC,* **10** (1978), 203.

43. Ratcliff, J. D., *Rev. Prog. Color,* **9** (1978), 58.

44. Pratt, H. T., *ADR,* **56** (1967), 671.

45. Dayvault, J. A., *TCC,* **4** (1972), 156.

46. Celanese Tech. Bull. TBP-21, Celanese Co., USA.

47. Leube, H., and Richter, P., *TCC,* **5** (1973), 29.

48. Rau, R. O., Sello, S. B., and Stevens, C. V., *TCC,* **7** (1975), 31.

49. Dugal, S., and Heidemann, G., *MTB,* **61** (1980), 742.

50. Thurner, K., and Wurz, A., *MTB,* **49** (1968), 76.

51. Haile, W. A., *ADR,* **71**(3) (1982), 19.

52. Warwicker, J. O., Shirley Inst. Publication, **S25** (1976), 49.

53. Goodman, I., and Nesbitt, B. F., *J. Polymer Sci.,* **48** (1960), 423.

54. Humbrecht, R., *MTB,* **61** (1980), 450.

55. Datye, K. V., *Colourage,* **24**(4) (1977), 27.

56. Datye, K. V. (unpublished work).

57. Pratt, H., *Knitting Times,* **39**(35) (1970), 1.

58. Valk, G., *MTB,* **59** (1978), 843; *Text. Industries* (Dec. 1970), 147.

59. Keller, K. H., Furer, L., and Wegner, K., *Textilveredlung,* **16** (1981), 72.

60. Ullrich, H., Kloss, E., and Kunze, W., *MTB,* **8** (1969), 953.

61. DeMaria, A., and Gajendragadkar, P., *ADR,* **68**(9) (1979), 32.

62. Calhoun, E. S., Brafford, K. C., and Holden, W. H., *TCC,* **7** (1975), 141.

9 DYEING OF NYLON

Nylon, a more polar molecule than polyester, contains carboxyl groups and amide nitrogen atoms that form strong hydrogen bonds between chains. These bonds are much stronger individually than the van der Waals and other small forces that bind polyester chains. Nevertheless, nylon can be dyed more readily than most other synthetic fibers. This is because the hydrogen bonds in nylon can be easily disrupted by the polar molecules of water. Water increases the segmental mobility of the polymer chains and lowers the glass-transition temperature T_g. Thus, nylon 66 and polyester have T_g of 80°C and 110°C in dry state and −20°C and 80°C at 100% R.H., respectively; that is, on wetting, the T_g of nylon is lowered by 100°C while the T_g of polyester is lowered by only 30°C. The dyeing of nylon in water is analogous to dyeing PET in a 100% carrier (solvent) bath.[1] The effect of the water is a lowering of the temperature range needed for dyeing nylon which enables it to be dyed at or below 100°C (see Chapter 3). Water causes both reversible and irreversible changes in the nylon fiber structure. Reversible changes occur at temperatures up to 71°C in water. Under these conditions, the polymer chains remain in approximately the same positions and take the same configurations when the water is removed. At or above the hydrosetting temperature (around 80°C), irreversible changes occur in the fiber structure that fix the configuration of the polymer chains. The hydrosetting temperature depends on the thermomechanical history of the fiber.[1]

Nylon is commercially dyed using disperse dyes, nonmetallized anionic (acid) dyes, and neutral dyeing metallized anionic (metal-complex) dyes. Several other dye classes that can be used in special situations are azoic, cationic, chrome, direct, developed, fiber reactive, and vat dyes.[2,3] The disperse dyes are used to give light to medium shades with moderate fastness properties. These dyes are soluble in nylon. During the dyeing process, a very small amount of disperse dye dissolves in water and is then extracted from the water by the stronger solvent, that is, the nylon. The disperse dyes are held on nylon by the relatively weak forces of solution and are easily stripped from the fiber back into hot water. If the initial dye strike is nonuniform, the disperse dye will migrate and level with ease. The problems of streakiness or barré dyeing (see Section 9.10) are seldom serious with disperse dyes. Thus, ladies' hosiery dyed in paddle machines or on dye-boarding equipment is dyed exclusively with disperse dyes to minimize rings or streaks. Similarly, the piece dyeing of nylon carpets is carried out with disperse dyes.

Apart from the poor wet fastness of disperse dyes, the shades are not bright and the light fastness is not very high. All the disperse dyes are not suitable for nylon because some of the dyes are easily reduced by nylon. Acid dyes are preferred for bright shades with excellent fastness properties. The dye anions react with the primary amino groups at the end of polyamide molecules to form a salt.[4-7] The salt linkages between the end amino group and the sulfonic acid group in the dye molecule are strong enough to withstand aqueous treatments, particularly, when the dye forms other weak linkages with the polymer and dyeings with good fastness properties are obtained.[7]

9.1 MECHANISM OF DYEING WITH ACID DYES

All the polyamide fibers have terminal amino groups (—NH$_2$), terminal carboxyl groups (—COOH), and amido groups along the chain (—CONH—). The polyamide can be titrated with an acid or an acid dye to get a titration curve as shown in Figure 9.1.[4,5] The curve is divided into three parts:

Part *a*. In the range of pH 6–10, acid or the dye is taken up and the fiber

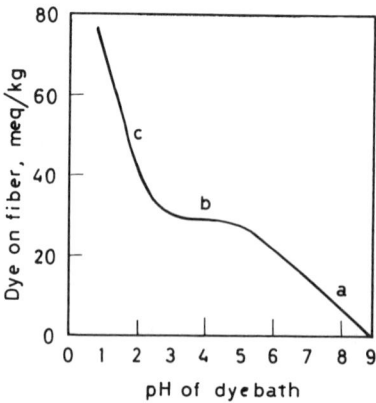

FIGURE 9.1. Titration curve of nylon with Solvay Blue B.[5]

accepts protons that are attached to the terminal amino end-groups, as if acid is titrating a free amine. The titration is complete at about pH 6.

Part *b*. The absorption of the acid or dye in the pH range of 2.5–6 remains constant and unaltered. Here, the addition of acid only lowers the pH of the liquor and the fiber accepts no more protons.

Part *c*. Below pH 2.5, the fiber again takes up the acid or the acid dye and the protons are probably attached to the amido groups to form $-CO\overset{+}{N}H_{-2}$ groups. Thus, nylon acquires a positive charge below pH 2.7.[8] This is called *overdyeing* of nylon.

The mechanism of dyeing at pH below 2.7 is rather complex. According to Peters,[4] the attachment of anions of acid dyes to terminal amino groups in nylon takes place at all pH values above pH 2.7, the isoelectric point. The increased adsorption at lower pH is entirely on protonated and ionized amido linkages. This is reflected in reduced tenacity and extensibility of the fiber. Remington and Gladding[9] and Bhatt and Daruwala[10] showed that acid dyeing reaches an equilibrium, the fiber molecules are depolymerized at low pH, and more amino groups are formed which, in turn, are responsible for the increased dye uptake. However, at any given time of dyeing, the amount of acid dye adsorbed is appreciably in excess of that needed to saturate the total end amino groups.[10] When adsorption of acid dye occurs in excess of amine groups, the dye is adsorbed in a nondissociated form at any pH.[11] Thus, overdyeing may also be the consequence of the high substantivity for the fiber of a dye anion with a low degree of sulfonation.[12,13] The large extent to which overdyeing may occur is common to both nylon 6 and nylon 66[14] and depends on the substantivity of the dye for the polymer and on the number of sulfonic acid groups that it contains (Figure 9.2).[13]

The carboxyl end groups of polyamides have some effect on the sorption of acid dye anions, especially at higher pH values.[13] Thus, the equilibrium dye sorption depends on end amino and carboxyl groups, pH, ionic charge on dye anions, and affinity of the dye.

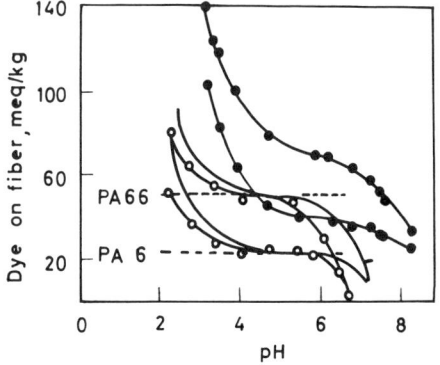

FIGURE 9.2. Titration curve for the sorption of acid dyes by PA 6 and PA 66.[13]

	C.I. Acid Red	Dye concentration (g/liter)	Number of sulfo groups
-●-	8	0.500	1
—	13	0.502	2
-○-	18	0.446	3
---	Amino group content of PA 6 and PA 66		

According to Brody,[11] dyeing of acid and disperse dyes occurs by filling vacancies caused primarily by chain folding (see Chapter 2). During dyeing with acid dyes, hydrolysis is assumed to take place at these folds. Thus, maximum dye capacity is determined by packing requirements for a fixed free volume within the fiber. In the case of acid dyes, additional dye can be accommodated by the further hydrolysis of chains bordering the vacancies.

Suganuma[15] found that the nylon filament yarn dyed with acid dyes has greater yield stress than that dyed in the blank dyebath (Figure 9.3). Acid dyes exhibit this effect while disperse dyes of similar molecular structure (but without sulfonic acid groups) show little effect. The molecular forces that affect the yield stress of undyed nylon are hydrogen bonds between the amido groups. It is considered that the molecular force between disperse dye and the amido group is nearly equal to that between amido groups. On the other hand, the bond between the acid dye and the amido group is an ionic interaction[11,16] so that the yield stress of nylon dyed with acid dye becomes larger than that of the undyed one.

Dyeing of nylon with acid dyes is influenced by the number of sulfonic acid groups in dyes.[17–20] The equilibrium dye uptake decreases with an increasing number of sulfonic acid groups in the dye molecule. The ionic interaction between dye and the amino groups in nylon increases because of the addition of sulfonic acid groups.[18,19] The pH at which overdyeing occurs varies with the dye and is dependent on the nature and size of the dye molecule and the degree of sulfonation; the addition of sulfonic acid groups to the dye molecule tends to decrease overdyeing.[17] The overdyeing may be due to attachment of dye ions of high affinity to the fiber at nonionic sites.[17,21] This affinity decreases with increasing solubility of the dye in water or with

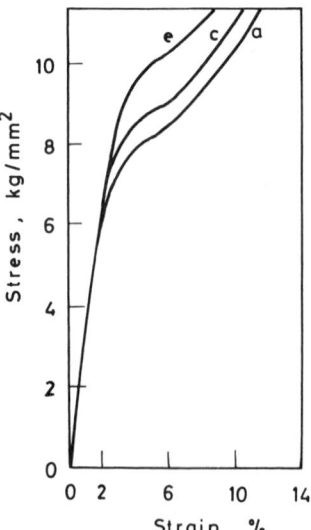

FIGURE 9.3. Typical stress–strain curves of nylon filament dyed at pH 1.4.[15] C.I. Acid Red 88 on filament. (*a*) Nil (Blank dyed). (*c*) 0.0537 mole/kg. (*e*) 0.1316 mole/kg.

increasing sulfonation. The yield stress of dyed nylon, which is larger than that of undyed nylon, increases with the dye concentration in the filament (Figure 9.3).[15] The effect is much smaller when dyeing takes place on the terminal amine groups than when overdyeing. The concentration of dye in the filament in which the amine groups become saturated decreases with increasing sulfonation, so that for the same concentration of dye in nylon, the yield stress increases with increasing sulfonation and the number of sulfonic acid groups have an effect on the yield stress (Figure 9.4).[15] The net effect of the dyes on the yield stress is independent of the number of sulfonic acid groups in dyes (Figure 9.5). Thus, in the region of overdyeing, the number of sulfonic acid groups plays no role.[11,15]

9.2 RATE OF ACID DYEING

In dyeing practice, the pH of the dyebath is never below 2.7 because of the degradation of the fiber and many commercial dyes. For practical dyeing, the dye sorption is limited essentially to the terminal amino groups. The number of free amino groups in nylon is limited to about 10% of those in wool.[22] Each available amino group can be neutralized with one sulfonic acid group in the dye molecule. Once the group is blocked, it is no longer available for further adsorption. Once all the available amino groups are blocked, the fiber gets saturated with the dye and no more dye is added to the fiber on continuing the dyeing. Because of this, the build-up of a shade of a mixture of acid dyes can pose a problem. The higher the number of sulfonic acid groups in a dye molecule, the higher will be its capacity to block the amino groups per molecule and, in turn, the lower will be the saturation value. The dyes with three sulfonic acid groups in the molecules (equivalent wt: $\frac{1}{3}$) thus

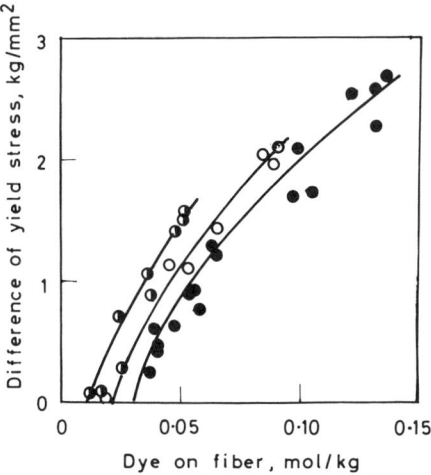

FIGURE 9.4. Effect of number of sulfonic groups in dyes on yield stress of dyed nylon filament.[15] ●: C.I. Acid Red 88 (mono-sulfonic dye). ○: C.I. Acid Red 13 (disulfonic dye). ◑: C.I. Acid Red 18 (trisulfonic dye).

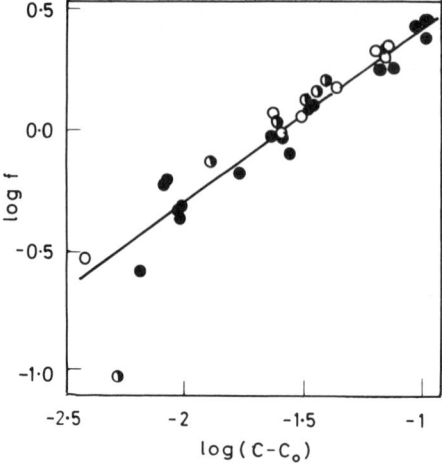

FIGURE 9.5. The relation between logarithm of the increment of the yield stress (log f) and the logarithm of the dye concentration on amide groups.[15] C: Total dye concentration. C_O: Intercepts in Figure 9.4. (Signs are the same as in Figure 9.4.)

have a lower saturation value on nylon fibers than those with only one sulfonic acid group (equivalent wt: 1). Typical equilibrium sorption behavior of nylon 66 and nylon 6 with acid dyes is shown in Figure 9.2.[13] When a mixture of dyes is used, there is competition among the dyes for the free amino groups, and the dyes that have more affinity for the fiber will be the ones to block the available free amino groups. This may, in turn, block the adsorption of the dye of lower affinity. In order to avoid the blocking effect, it is essential to use dyes with similar affinity for the nylon fiber when they are applied in a mixture. If a mixture of monosulfonic acid dye and trisulfonic acid dye is used, colorless monosulfonic acid is incorporated in the dyebath as a leveling agent. This agent competes with the dyes for sites in the fiber and thus minimizes the blocking effect of the trisulfonic acid dye. During dyeing, the leveling agent is replaced by the monosulfonic acid dye with the higher affinity for the fiber.

The variation in amine end-group content causes variation in the rate of dyeing of anionic dyes[20,23] (Figure 9.6). It appears from the figure that when free amino groups are absent, no dye enters the fiber; the curves pass through origin. The high rate-of-dyeing, though desirable for production efficiency, is not conducive to uniform dyeing. It is undesirable to increase the dyeing rate by incorporating a large number of free amino groups in the polymer since the dye strikes the surface of such a fiber and builds up the concentration at the surface without increasing the rate of penetration in the fiber, giving ring-dyeing.[20] The dye may diffuse towards the center of the fiber during further processing or washing and the apparent shade will become lighter. Furthermore, migration of dye is difficult under these conditions, giving uneven dyeings.[9,20] Thus, the end amino groups in the fiber must be just sufficient to act as a driving force for the diffusion of dye anions into the fiber and to adsorb sufficient dye to build up a full shade before

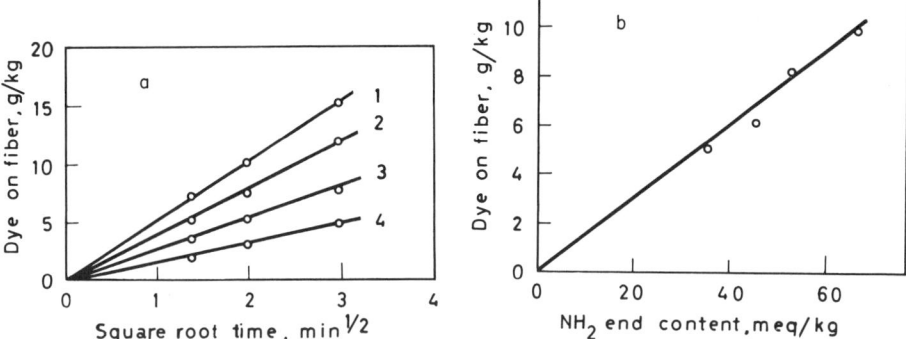

FIGURE 9.6. The effect of end-amine group concentration on rate of dyeing.[20] (*a*) C.I. Acid Red 1, 0.75 g/liter, pH 3.2, 70°C (a) meq NH_2 group/kg 1: 66.4; 2: 52.9; 3: 45.1; 4: 35.2. (*b*) Dyeing time: 4 min.

reaching saturation. The amino end group concentration is kept constant by the addition of a terminator such as acetic acid during polymerization to block some of these groups. Thus, the variation in the concentration of these groups will usually not be in a given merge lot of a fiber. However, any variation in these groups will invariably produce variations in the depth of shade[5] and it is extremely difficult to minimize this unevenness by manipulating the dyeing process. As long as the amino group concentration remains constant, variations in the molecular weight of the fiber do not contribute significantly to the variations in dyeing rate for dyeing uniformity.

The rate of dyeing is usually controlled by the diffusion in the fiber. The big-size dye anions diffusing into the fiber and the small-size anions diffusing out have very different mobilities. This difference gives rise to an electrical potential gradient that slows the fast ions and speeds up the slow ones so that the fluxes of the two anions become the same, as required by electroneutrality.[24] The presence of a salt in the dyeing system affects both the rate and fiber saturation of the dye (Table 9.1).[20] The salt anions compete with the

TABLE 9.1 Effect of Added Salt on Dyeing Rate and Dye Sorption Equilibrium[20]

Salt (owf)	Diffusivity (cm²/sec)	Equilibrium Sorption (Equiv. dye/g fiber)
None	1.81×10^{-9}	39.5
20% NaCl	1.27×10^{-9}	26.0
100% NaCl	7.65×10^{-10}	13.9
20% Na_2SO_4	1.46×10^{-9}	33.8
100% Na_2SO_4	9.86×10^{-10}	32.9

dye anions for the dye sites. The presence of salt greatly enhances leveling of nonuniform dyeing caused by variations in the physical structure of the fiber. However, the exhaustion of the dyebath is incomplete depending on the amount and type of salt. Although the dyeing rates are lowered in the presence of salt, this is more than compensated for by the improved leveling.

The rate of dyeing nylon with acid dyes is significantly affected by the physical variations in the fiber such as denier per filament, cross-section, orientation, and crystallinity,[23,25] and by the thermal history of the fiber. Different dyes are affected to different extents by the same degree of variation in the physical structure. For example, a few dyes which are very sensitive to chemical variations in the fiber exhibit different sensitivity to physical variations (Table 9.2). The variations in the fine structure of the fiber lead to uneven dyeing and barréness.[23] However, physical variations do not affect the dye uptake at equilibrium. By adjusting the rate of dyeing, it is possible to cover these variations. The rate of dyeing is influenced by the filament denier. The surface area per unit weight is inversely proportional to the square root of the denier. The finer the fiber, that is, the smaller the denier, the larger the surface area per unit weight of a fiber through which the dye has to diffuse and the quicker the dye uptake[20,26] (Figure 9.7).[20] Since the migration of an adsorbed acid dye during dyeing is difficult, denier variations in a given lot invariably result in light and heavy dyeings. The dyeing of PA 6 is easier than that of PA 66 (Figure 9.8).[5]

TABLE 9.2   Sensitivity of Acid Dyes to Physical Variations in Nylon Fiber[23]

C.I. Acid	Sensitivity
Red 37	Relatively insensitive
Orange 10	Slightly sensitive
Red 58	Moderately sensitive
Red 57	Sensitive
Blue 127	Very sensitive

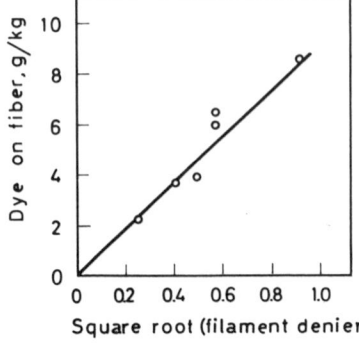

FIGURE 9.7.   The relation between the fiber denier and dyeing rate.[20] Dyeing conditions are the same as in Figure 9.6. Values corrected for variation in NH_2 content.

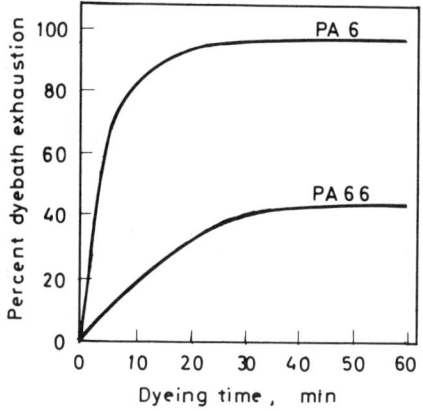

FIGURE 9.8. Rate of dyeing of nylon 6 and nylon 66 with acid dyes.[5]

9.3   SELECTION OF DYES

Dyes can be grouped on the basis of their ability to cover or accentuate the physical nonuniformity of the fiber.[27,28] Dyes with one sulfonic acid group and small molecular size exhibit good coverage of physical irregularities. However, this leveling type of dye has poor affinity to the fiber and therefore exhibits poor wash fastness. The metal-complex neutral dyeing acid dyes form a coordinate complex in the fiber involving groups in the polyamide molecules. They differ from the conventional acid dyes in that they need not have a sulfonic acid group to confer the anionic character. The charge is the consequence of the trivalent ($+3$) metal chelated in a tetravalent (-4) ligand. They are usually very fast to washing and light. Once the dye is adsorbed on the fiber, it cannot migrate from heavily dyed portions to lighter portions during dyeing. Hence, metal-complex dyes invariably give barré dyeing on a fiber with physical variations. These dyes are large molecules, a necessary consequence of two dye moities bonded to a single metal atom (2 : 1 premetalized dyes). The large size interferes with the dye transfer. In fact, the metal-complex dyes accentuate the irregularities in nylon and are never used for dyeing pastel and light shades. Barré-free uniform dyeing of nylon thus involves a compromise with the wet fastness of dyed goods. Acid dyes are used to dye medium and deep shades and metal-complex dyes are used for very heavy shades. If the barréness is due to physical variations and not to differences in free amino groups, the selection of dyes based on the above considerations will usually give satisfactory dyeings. Light and pastel shades are dyed with dyes from other classes.

9.4   LEVELING AGENTS

The inability of a dye to migrate from a heavily dyed portion of the fiber material to a less heavily dyed one is the cause of uneven dyeing. If the migration step is avoided by controlling the initial stages of dye adsorption, uneven dyeings can be avoided. It is necessary to reduce the initial "strike" of the acid dyes in order to overcome the migration problem. The leveling agents function by lowering the initial rate of dyeing, thus producing even dyeings free from the barré effect. A common feature of all the leveling agents is that besides controlling the initial rate of dyeing, they invariably improve the migration properties of a dye. This is easily demonstrated by redyeing a dyed sample and an undyed sample of equal weight with and without a leveling agent in a blank bath. Double the concentration of dye on undyed sample as a percentage of that on the initially dyed sample indicates the extent of migration (Table 9.3). At 100% migration, both initially undyed and dyed samples match.[29]

The following types of leveling agents and their mixtures are used.

1. Anionic agents: Sulfonated fatty alcohols, alkyl aryl sulfonates.
2. Cationic agents: Quaternary ammonium salts.
3. Nonionic agents: Ethylene oxide condensates of fatty alcohol or alkylphenol.

Anionic Leveling Agents.[30–32]   These agents have a sulfonic acid group in their molecule and, like acid dyes, they combine with amino groups in nylon. The addition of an aromatic sulfonic acid sodium salt to the dyebath increases the migration of the dye on nylon. The influence of the leveling agent varies from dye to dye depending on the affinity of the dye for the nylon fiber.[32] The anionic agent may block the end amino groups by forming a salt

TABLE 9.3   Migration of Acid and Direct Dyes[29]

Dye	No Agent	2% Anionic Agent	2% Cationic Agent	2% Anionic + 2% Cationic Agent
		Migration %		
C.I. Acid				
Yellow 135	75	90	90	90
Red 266	81	87	91	93
Blue 25	93	96	100	97
Orange 116	22	54	67	74
Red 299	34	66	56	65
Blue 128	27	69	69	76
C.I. Direct				
Yellow 12	4	46	29	38
Red 1	5	38	47	42

if nylon is treated with them.[6] The initial rate of dyeing of an acid dye therefore slows down. However, the affinity of an acid dye to nylon is higher than that of the leveling agent. With time and a rise in temperature, the dye replaces the leveling agent almost completely. Thus, though the initial rate of dyeing is lowered significantly, the exhaustion is not significantly affected by a leveling agent (Table 9.4).[31] Since the acid dye does not strike the nylon fiber treated with a leveling agent, the problem of unevenness and migration of dye is eliminated.

Dyeing Method.   The bath is set with

| 10–30 g/liter | Anionic leveling agent |
| 2 g/liter | Acetic acid (30%) to pH 6.5 |

The material is run at 40°C for 10 min before adding the solution of an acid dye and dyeing for 10 min more. The temperature is raised to boiling in 30 min and maintained for 45 min. The goods are then washed, soaped, washed, and finished with a softener.

Cationic Leveling Agents.[33–35]   These agents have virtually no affinity for nylon fiber. Their first effect lies in slowing down the rate of adsorption of dye anions by the nylon. The action may be said to be the reverse of blocking. In blocking, the adsorption of the dye is slowed down by partially satisfying the sites (end amino groups) with an anionic leveling agent, a substitute for the acid dye. This makes the site less ready to react with the

TABLE 9.4   Dyeing of Nylon 6 at 95°C[31]

1% Dye, $M:L$ ratio: 1:50.
A: Blank (No leveling agent).
B: 2% anionic leveling agent.

| Dyeing Time (min) | % Exhaustion | | | |
| | C.I. Acid Red 266 | | C.I. Acid Blue 113 | |
	A	B	A	B
5	84	61	45	37
10	94	75	69	52
20	96	88	87	73
30	98	90	94	84
60	98	94	99	97
∞ (equilibrium)	98	96	99	99

dye. The cationic agent, on the other hand, forms a salt with the dye, that is, it satisfies the dye anions and makes them less ready to react with the ionic sites (end amino groups) in the fiber.

These leveling agents have a quaternary ammonium group that forms the salt with the sulfonic acid group of acid dyes. The initial rate of dyeing is, thus, slowed down. As the temperature is raised, the salt dissociates into ions, the dye anion, and the cationic agent. The released dye anions are taken up by the fiber. At any given time, the concentration of free dye anions is very low and level dyeing is obtained as it would be from a very dilute solution.[33] The cationic leveling agent is used together with a suitable nonionic dispersing agent. While the depth of shade is not restricted by the use of these agents, the selection of dyes is critical for maximum levelness. Sometimes, the cationic leveling agent precipitates the acid dye as a complex and complicates the dyeing.

Nonionic Leveling Agents.   The dye aggregates are broken down into single molecules by these agents. Thus, they act as solubilizing and stripping agents for acid dyes. Because of this action, the rate of dyeing is retarded and the migration of dye is improved, which gives level dyeing. These agents have no affinity for the fiber. The nonionic agents can be used in combination with either anionic or cationic agents.

Mixed Leveling Agents.   A process of using a combination of anionic and cationic leveling agents claims to give barré-free dyeings.[29,36] The material is pretreated with anionic agent at pH 5.5 at 90°C. The temperature of the bath is rapidly raised to boiling and a solution of dye and cationic agent in admixture is added. The dye exhausts evenly within 45 min at boiling. This method gives improved coverage because of improved dye migration properties. The compatibility of dyes in admixture is also improved in certain cases. However, it is necessary to make experiments and to standardize the process since the dyeing results are not satisfactory with all the dyes.

9.5   HIGH-TEMPERATURE DYEING

The rate of dyeing increases with temperature. A simultaneous increase in the rate of migration of dye from heavily dyed portions to less heavily dyed portions also occurs with an increase in temperature.[37] Thus, the physical irregularities in the nylon yarn are almost completely covered if dyeing is carried out at temperatures above 100°C[38] (Table 9.5). The variations in dye uptake because of the variation in amino groups are not covered by the increase in dyeing temperature.[38] (See Table 9.5.) The ease of migration at higher temperatures is the cause of level dyeing (Table 9.6).[39] Even textured nylon yarns, which give depth variations on dyeing at 100°C, exhibit almost similar depth when dyed at 115°C. The mechanical properties of textured nylon 6 yarn are not significantly affected by the increase in the dyeing

TABLE 9.5   Dyeing of Nylon Yarns at 95°C and 130°C[38]

Dye on control nylon yarn = 100
Nylon A: Abnormal orientation
Nylon B: Abnormal end amino group concentration

| | Dye on Nylon Fiber | | | |
| | Nylon A | | Nylon B | |
C.I. Acid	95°C	130°C	95°C	130°C
Orange 33	126	113	76	87
Red 18	120	106	77	76
Red 85	111	107	72	72
Blue 113	112	105	85	85
Blue 138	126	99	86	76

TABLE 9.6   Effect of Temperature on Migration of Dyes on Textured Nylon[39]

| | % Migration | | | | |
Dyes[a]	100°C	105°C	110°C	115°C	120°C
C.I. Acid					
Yellow 29	74.2	82.8	100	88.0	72.8
Red 151	94.4	95.6	96.0	96.8	98.0
Blue 113	42.6	43.6	58.0	97.2	94.6
Blue 127	9.4	10.0	57.2	71.6	97.6
C.I. Direct					
Red 81	27.6	87.2	90.8	93.6	85.4

Note: 20 g/liter anionic agent, dyeing time: 30 min
[a] Some of the dyes probably hydrolize during migration test at high temperatures.

temperature (Table 9.7).[39] Similarly, fastness properties are also not affected by high-temperature dyeing.[37]

For high-temperature dyeing, the bath is set with

X g/liter	Acid dye
10–20 g/liter	Anionic leveling agent
	pH 7 with ammonia/acetic acid

TABLE 9.7 Effect of Dyeing Temperature on Mechanical
Properties of Textured Nylon 6 Yarn[39]

Dyeing Temperature	Tenacity (g/d)	Elongation (%)	Crimp Rigidity (%)
100°C	3.64	35.4	46.8
120°C	3.64	37.5	45.8

The goods are entered at 40°C and the temperature is raised to 115°C in 60 min. After dyeing for 45 min at 115°C, the bath is cooled (below 100°C) and drained. The goods are washed, soaped at 70°C for 15 min, washed, and finished as usual.

9.6 SWELLING AGENTS AND SOLVENTS

Barréness can be covered by the addition of swelling agents to the dyebath.[40] Yarn irregularities are well covered by the addition of 4% benzyl alcohol to the weight of the bath.[41] However, the cost of benzyl alcohol is prohibitive for its commercial utilization. Similarly, use of 5% n-butanol in HT dyeing gives excellent coverage of barréness.[42] Typical data on coverage of yarns with acid, direct, and reactive dyes are given in Table 9.8. The higher the rating, the better is the coverage.

Nylon 6 fabric shrinks but nylon 66 fabric is not affected by the presence of n-butanol during dyeing at 120°C. The dyeing cost is prohibitive unless the solvent is recovered. Attempts have been made to develop a solvent recovery system.

Due to the problems of barré dyeing, nylon is seldom dyed in long lengths in a continuous manner. The pad-steam process is probably the only

TABLE 9.8 Effect of n-Butanol on Coverage of
Barréness on Nylon Fiber (Assessment: Grey Scale
1–5)[42]

Dyeing Temperature:	100°C		120°C	
n-Butanol in Bath:	Nil	5%	Nil	5%
Dye	Grey Scale Rating			
C.I. Acid Orange 51	1–2	2–3	2–3	4
C.I. Acid Blue 105	1–2	2–3	2–3	4–5
C.I. Direct Blue 71	1	2	2–3	4–5
C.I. Reactive Black 12	1	2	2–3	3–4

method that can be used for continuous dyeing of nylons.[43,44] Nylon is dyed commercially in virtually all types of equipment. Dye becks, jigs, padding mangles, and liquor-circulating machines are used for stock, package, and beam dyeing.

Direct Dyes.   Direct dyes contain sulfonic acid groups and can behave as anionic dyes. However, their molecular weight and size are very high and their fastness properties on nylon are not good compared to those of acid dyes. Nevertheless, selected direct dyes are used for producing deep shades on nylon because of their low cost. In pale shades, direct dyes give barré dyeings because the big dye molecules hinder the migration process during dyeing. The methods of application of direct dyes to nylon are the same as those of acid dyes, except that 5–10% sodium sulfate may be added for satisfactory exhaustion. The wet fastness of direct dyes on nylon can be improved by the tannic acid–tartaremetic aftertreatment described for acid dyes.

Correction of Dyeing.   Uniform shades are not always easy to get with acid dyes. Pale shades are invariably uneven. The dye has a tendency to strike the fiber because of the attractive forces of ionic nature. Migration of dye during dyeing is poor, particularly, in pale shades. Thus, these dyes have a serious drawback in that they produce faulty dyeings. The dye on faulty dyeings on nylon can be partially stripped with ammonia (20–30 g/liter) and anionic wetting agent (1–2 g/liter) at boiling for 30–45 min. A treatment with sodium hydrosulfite (10–20 g/liter) at 70°C prior to the ammonia treatment is helpful in completing stripping.

The fastness of acid dyes on nylon to washing and light varies from dye to dye. Dyes with wet fastness 1–2 are also used if they give brilliant shades and the end use does not involve severe washing. At this stage, brilliant blue dyes with excellent wet fastness are missing. Wet fastness can be improved by a tannin process in which the dyed goods are treated with

10–20 g/liter	Tannic acid
10–30 g/liter	Formic acid (80%)

at 70°C for 15 min. Tartaremetic (10–20 g/liter) is then added and the treatment is continued at 70°C for 20 min. In place of the tannic acid–tartaremetic treatment, synthetic fixing agents such as Erionyl NW, Cibatex PA, Tanninol WR, and Nylofix P may be used with equivalent results. The brilliancy of the shade is partially lost by the treatment.

9.7   LOW-TEMPERATURE DYEING OF NYLON 66

Since the hydrosetting temperature of nylon is generally below boiling, nylon fabrics that have not been heat set previously will be heat set in the dyebath. If the fabric is wrinkled when it reaches the hydrosetting temperature, permanent crack marks may develop that cannot be eliminated by later heat setting. Nylon circular knitted goods are heat set in an autoclave before dyeing to avoid such marks. However, this may create moiré problems. Unset fabrics also have a permanent shrinkage problem if they are dyed without setting. Low-temperature dyeing is a potential solution to the moiré problem and can minimize the shrinkage problem as well as save energy.[45–47] By dyeing below the hydrosetting temperature, permanent cracks and crease marks can be prevented without the moiré and other problems associated with prior heat setting. Shrinkage problems can be minimized by setting the dyed fabric during framing. Low-temperature dyeing is done as far below the hydrosetting temperature as possible, consistent with the need to minimize barré dyeing. A temperature of 65°C has been reported for nylon 66.[48] It is still lower for nylon 6.

In dyeing nylon at 100°C, the migration of dye plays a very important role in deciding the uniformity of shade and barré-free dyeing. The higher the temperature, the better is the migration (Table 9.5). The migration is temperature-sensitive, even at low temperatures (Figure 9.9).[48] For low-temperature dyeing, it is necessary to obtain maximum possible levelness and barré coverage in the initial dye adsorption by the proper selection of dyes, auxiliaries, pH, and time–temperature profile. The dye must exhaust well (~90%) in low-temperature dyeing conditions within a reasonable time, giving uniform dyeing and an acceptable level of barré coverage. Typical dyes and their dyeing properties are shown in Table 9.9.[48,49] Most of the dyes have shown poor exhaustion at 65°C. Although premetallized dyes exhaust well, they do not penetrate into the fiber and accentuate the barré effect. The leveling dyes are monosulfonated acid dyes with good migration properties at boiling. Many such dyes give good dyeings free from barré at 65°C. However, their wet fastness is not very high. C.I. Acid Yellow 49, 159, 174, 198; C.I. Acid Orange 116, 128; C.I. Acid Red 151, 266, 299, 337; and C.I. Acid Blue 25, 40, and 78 give good results.[48] Dyes from other classes such as disperse and direct are also suitable. Barré coverage at 65°C is not as good as that obtained with the same dyes at boiling.

The exhaustion rate is controlled by manipulating the time–temperature profile so that the exhaustion rate of 1% per min can be obtained. The dyeing

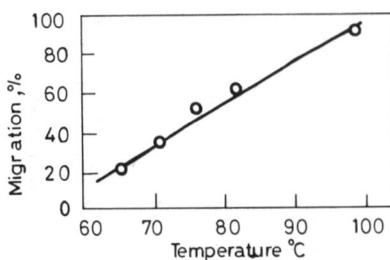

FIGURE 9.9. Migration (%) of an acid dye at different dyeing temperatures.[48]

TABLE 9.9   Exhaustion (%) and Barré Coverage Ratings
of Dyes on Nylon 66[48]

Dyeing: 65°C/1 hr
Ratings: 0–5 (worst to no visible Barré)

Class of Dye	Color Index (C.I.)	Exhaustion (%)	Barré Rating
Milling	Acid Green 25	65	1.5
Milling	Acid Blue 113	53	2.5
Premetallized	Acid Red 182	80	0.5
Acid dyeing	Acid Red 364	48	3.0
Direct	Direct Blue 8G	19	—
Leveling	Acid Red 266	98	4.0
Leveling	Acid Blue 277	91	2.5

is started at room temperature (27°C) and the temperature is raised only after holding the temperature at 27°C for 20–30 min. The slow exhaustion rate allows dye movement in the bath, so that level dyeings can be achieved in equipment such as beeks and paddle machines. A typical time–temperature profile is shown in Figure 9.10.[48] The total dyeing time is about 20 min longer than that for dyeing at boiling. However, the initial time of presetting is saved. The dye requirement is about 10–15% less because of ring-dyeing. The fastness properties of dyed goods are comparable to those dyed at 100°C.

9.8   DYEING WITH DISPERSE DYES

Disperse dyes are free from ionic water-solubilizing groups and are applied as a fine dispersion of dye particles. Variations in the amino groups in nylon

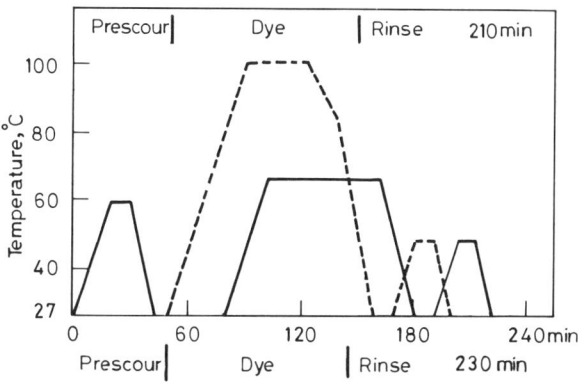

FIGURE 9.10.   Time–temperature profile of dyeing nylon at 100°C and 60°C.[48]

fiber have no effect on the shade of disperse dyes, since these dyes do not attract the amino groups. Since there are no ionic forces involved in the adsorption of disperse dyes on nylon, the dye uptake is very uniform and the migration of dye during dyeing is very good. Hence, the problem of barré dyeing is seldom serious with disperse dyes. But the wet fastness of disperse dyes on nylon is poor and not comparable to that of many acid dyes. The disperse dyes are therefore used only for light to medium depths. Disperse dyes are very easy to apply on nylon and cover all the orientational, physical, and chemical abnormalities in the fiber. However, the shades are dull. The method of application of disperse dyes on nylon consists of setting the dyebath with

X g/liter	Disperse dyes
1–2 g/liter	Dispersing agent
1–2 g/liter	Wetting agent
1 g/liter	Acetic acid (30%) to pH 5.5

and entering the goods at 60°C, raising the temperature to boiling in 30 min, and boiling for 45 min. The material is then washed and finished.

Dyeing with Reactive Disperse Dyes. Reactive disperse dyes such as Procinyl dyes (ICI) combine the advantages of disperse dyes and acid dyes. These dyes do not contain ionic solubilizing groups in their molecules and hence, they behave like disperse dyes producing level dyeings. They do, however, contain within their molecules a reactive group carrying a labile halogen atom which can react with amino groups. Because this chemical reaction forms a covalent bond between the dye and nylon, the reactive disperse dyes possess excellent wet fastness properties on nylon.

The dyeing is started at an acid pH so that the dye is taken up by nylon as a disperse dye. When the reaction between the dye molecules and amino groups in nylon takes place, the bath is then made alkaline. Thus, the bath is set at 40°C with

X g/liter	Reactive disperse dye
1 g/liter	Nonionic surfactant
2 g/liter	Acetic acid to pH 4.0

The goods are entered and the temperature is slowly raised to 85–100°C in 30 min and dyeing is continued for 30 min. Sodium carbonate (2.5–3 g/liter) is then added to the dyebath to get pH 10–10.5 and boiling is continued for an hour. The goods are then washed, soaped, washed, and dried.

The reactive disperse dyes react with the amino groups; since the number of amino groups is fixed and limited, the fiber gets saturated with the dye. The same dye on PET and cellulose triacetate acts only as a disperse dye and dissolves without any reaction. Thus, the equilibrium sorption isotherms for these fibers differ from those for nylon[26] (Figure 9.11). The linear isotherms on nylon 6 are obtained only by the Langmuir formulation, which involves adsorption (reaction) on specific limited number of sites (Figure 9.12). Nonreactive disperse dyes usually give linear isotherms on fibers including nylon.[50] Even though the reactive dye is adsorbed on specific sites, all the dye does not react, particularly, when the fiber is dyed from a solvent. The reaction is of a small magnitude as shown in Table 9.10.

The reaction of dyes with nylon under thermosol dyeing conditions depends on the presence of alkali on the fiber.[50] Reactive disperse dyes are still very expensive. The range is not complete and therefore they have not been widely used.

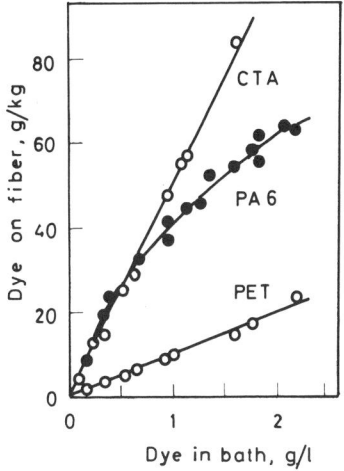

FIGURE 9.11. Equilibrium sorption of a reactive disperse dye on cellulose triacetate, nylon 6, and PET from perchloroethylene at 121°C.[26]

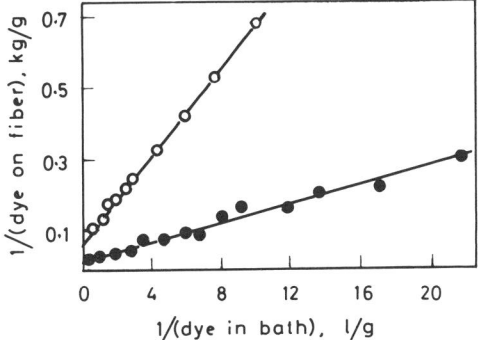

FIGURE 9.12. Langmuir adsorption isotherms of two reactive disperse dyes on nylon 6 from perchloroethylene.[26] ○: Dye 42 at 100°C; ●: Dye 41 at 121°C. (For formulas of dyes, see Ref. 26.)

**TABLE 9.10   C.I. Reactive Disperse Red 10 on
Nylon 6 from Perchloroethylene at 121°C[50]**

Dyeing Time (h)	Dye on Nylon Fiber			
	Total (g/kg)	Sorbed (g/kg)	Reacted Dye	
			(g/kg)	(millimole/kg)
3	28.0	23.6	4.35	6.1
6	39.9	34.5	5.35	7.5
12	67.4	59.7	7.50	10.5
18	61.4	52.8	8.55	12.0

9.9   DYEING OF NYLON FABRIC

Open-beam dyeing machines are used for dyeing nylon 66 and nylon 6 fabric in pale shades with disperse dyes.[51] Specially selected leveling acid dyes are used to get uniform barréfree dyeings. The selected dyes have similar exhaustion curves, are free from blocking tendencies, and are able to cover barréness when applied by the correct method. Such dyes are marketed by special brand names. A leveling agent (1%) is used in the pretreatment. For example, Preston et al.[51] have described the process with Tectilon (CIBA) dyes as follows.

The dyebath is set with

> 10 g/liter   Leveling agent
> 20 g/liter   Acetic acid (80%) to pH 4–4.5

Nylon 6 or nylon 66 is pretreated in this bath for 10 min at 30°C. The dye and a cationic surfactant (10 g/liter) are added, the temperature is raised to boiling at 2°C per min over 35 min, and the goods are boiled for 65 min (total time: 100 min). The bath is then cooled to 90°C and blown off to drain. A very high rate of exhaustion, even at low temperatures and complete exhaustion of dyes at boiling, demands very careful control of the temperature rise to avoid unlevelness. The latter cannot be corrected later because of the poor migration properties of all acid dyes except specially selected leveling dyes. The migration properties of dyes improve further if dyeing is carried out at 120°C.[51] The use of a leveling agent retards the dyeing and thus, improves uniformity of dyeing.

Milling acid dyes are applied in the same manner as acid dyes except that acetic acid is replaced by ammonium sulfate (50 g/liter) in the pretreatment. The pH of the dyebath falls from 6.5–7 to 5.5–6 as ammonia is liberated and expelled as dyeing proceeds. The anionic leveling agent is not used for metal-complex dyes and dyeing is carried out as above.[51]

In a continuous process with acid and metal-complex dyes, the method allows one to circumvent some levelness problems. The dye-liquors are padded or printed onto the fabric and are fixed by steaming.[52-54] The continuous dyeing methods have become popular for dyeing nylon carpets (See Chapter 16).

9.10 BARRÉNESS OF DYEINGS

Barréness is a defect of "regular" irregularity in the fabric in the form of horizontal coursewise lines of deeper or lighter depth of shade. The interspacing and frequency of these lines decides the degree of defect in the fabric. Barréness exists as the liveliest problem (in varying magnitude) in almost all dyed fabrics made from textured yarns of nylon and polyester. The barré effect occurs in knitted fabrics of nonsynthetic fibers such as wool and cotton but is not as critical as in synthetic yarns. A stitch length variation of 7% in cotton and wool yarns produces the same extent of barré effect as 3% variation in the synthetic fiber yarns. Because of their overall uniformity, the continuous filament yarns have accentuated the problem, which is further aggravated by wider use of circular knitting compared to flat knitting, warp knitting, or weaving. Barré is a fabric fault and not a yarn defect, although it can be produced by defective yarns. Barréness may occur during various stages of manufacture such as filament yarn production, texturing, knitting, dyeing, or garment production.

Gradation of Barréness. Visual inspection of the dyed fabric is done as a qualitative analysis by using convenient terms like "high" or "slight." It serves the purpose of classifying and identification in most cases, provided the appraiser remains the same. The method is subjective and liable to give inconsistent evaluations. The HATRA barréness scale has been proposed to quantify the degree of barréness and to insure reproducible measurements.[55,56] The principle of the HATRA scale is to match the fabric bars with standard frequency interval lines. The HATRA scale consists of five black frames with a grey card in each having horizontal lines at varying frequency intervals. Grade 5 is barré-free and grades 4 to 1 have progressively more barré.

Causes of Barréness.[48,49,57-59] There are configurational barré such as mechanical spacing in the fabric matrix. A variation in tension on individual feeders of knitting machines (apart from other reasons) produces tight or slack knitting at a few ends which produces differential stretch on the ends in the fabric thereby showing lighter or deeper shade when dyed. This visual effect is caused by the localized change in density of knitted fabric with differential stretch in the ends. With a number of packages running simultaneously, it is very likely that one or two feeders may be under abnormal, excessive, or less tension, which may produce barré effects. The variable tension developed during knitting can be due to a faulty package of yarn. The packages of yarn on a knitting machine are either in cone form or cheeses wound straight from texturing machines. Variations in package density from cone to cone or cheese to cheese because of the tension variations or differential storage history of yarns may sometimes modify yarns beyond recovery to their original qualities. These damaged yarns exhibit variations in the fabric during dyeing.

The texturing process, being a thermomechanical deformation of filaments is inherently prone to develop variations in the yarns with position to position variables in the texturing machines. Heater temperature, yarn tension, spindle speed, or twist, yarn speed, or contact time with the heater and so on are all critical factors that govern the dyeing properties of the yarn. If the yarn is twisted, the tension during twisting can influence the dye uptake. The higher the tension (due to lower overfeed), the lower is the dye uptake.[58] Thus, variations in twisting and doubling machines can introduce the barré effect.

The process of spinning the filaments and drawing (stretching) generates many irregularities in the yarns. The fine structure of filaments (orientation, crystallinity, crystalline packing, and so on) is decided during draw-down in melt-spinning and by a stretch ratio in drawing. Any variations mean a built-in yarn variation. Although texturing is a thermomechanical deformation of filaments, it still cannot wipe out such irregularities in the parent yarn. When the variations are beyond being covered up in texturing or dyeing, they may show up in the fabric as barréness. The uncovered structural irregularity of the yarn very readily shows up in the dyed fabric. Luster differences in the yarns also give barré effects.

Identification of Causes and Barré Effect.   After grading the barré fabric, courses showing abnormal deeper or lighter shade than the rest of the fabric are marked and such courses unroved in sufficient numbers (about 100). The depth of shade of the normal and defective courses is now compared either visually or by color-measuring instruments. Similar depth of shade on normal and abnormal courses indicates that the dyeing is normal and is not the cause of barréness. Knitting fault or stitch length variation in the fabric has probably introduced or enhanced the barréness.[59] If the stitch length gives identical values for normal and abnormal courses, there is no knitting fault. The possible cause could be differential yarn retraction behavior in the fabric, which can be measured by the single-yarn collapse method.[60] The difference in the yarn collapse (%) of the two would confirm if the observed barréness is due to a fluctuation in texturing variables.

If there is a difference in shade on normal and abnormal courses, it could be due to yarn production faults, texturing faults, or dyeing faults. Yarn textured at different temperatures exhibits different dye uptake as well as retraction differences. The measurement of the latter would confirm whether differences in heater temperature existed. Residual shrinkage values of the two yarns also give a direct idea about the heat treatment received by the yarn.[61] If it is concluded that there is no difference in heat treatment, then the following experiment is conducted.

The dye on the courses showing difference in depth of shade is stripped off, the stripped samples are then redyed under standard conditions, and the dyeings are compared. If the two samples exhibit similar depth of shade after dyeing, the fault lies in the dyeing process. On the other hand, if the difference in the depth of shade persists as earlier, there is no dyeing fault.

The only possibility now left is that of faulty parent yarns. The normal and abnormal yarns are examined for their birefringence and crystallinity. A difference in these indices confirms the origin of barréness in the fabric and indicates the area where the corrective steps are to be taken.[62] Typical analysis data are shown in Table 9.11.

The above-mentioned procedure is time-consuming and the cause of the defects is identified much too late to take corrective action. By the time the conclusion about the source of the faults is drawn, a substantial amount of production may have already gone into the "substandard" grade. The tendency of yarns or fabrics towards barréness can be quickly detected by the use of fugitive tints.[63,64] The tint solution is sprayed on the fabric to develop instantaneously fugitive images corresponding to those that will be generated after the dyeing process. These tints may be applied while the material is being knitted or woven so that corrective measures can be taken in time to avoid barré dyeing. Similarly, the dyed heated fabric is pressed on a PVC film at 140–160°C using a hot press or a hot domestic iron.[65] Because of heat and pressure, a matrix is produced on the PVC film. The impression is inspected for faults in the knitted structure of the fabric without any interference from the color and depth of the dye.

9.11   DYEING OF POLYAMIDE BLENDS

Polyamide–Cellulosic Blends.   In dyeing polyamide–cellulosic blends, it is possible to dye either fiber and reserve the other or to dye both to get solid shades. If cellulosic fiber is reserved, either disperse or acid dyes may be used, preferably the latter for the best reserve effect since disperse dyes usually stain cotton or viscose and a scouring treatment may be required. Typical acid dyes that in a neutral liquor will dye the polyamide without staining the cellulosic fiber are:

C.I. Acid Yellow 36, 72
C.I. Acid Red 88
C.I. Acid Blue 25, 67, 78, 138
C.I. Acid Black 48

TABLE 9.11 Analysis of Barré Sample[59]

No.	Test	Sample 1	Sample 2
1	Barré grading	2	1
2	Shade on normal (n) and abnormal (ab) courses	Similar (no dyeing fault)	Abnormal courses deeper
3	Stitch length	Same (no knitting fault)	
4	Yarn collapse (%)	n: 22.7 (Texturing fault) ab: 33.5	Similar (27%) (no difference in heat treatment)
5	Residual shrinkage	—	Similar (difference in heat treatment)
6	Redyeing	—	Same depth (no dyeing fault)
7	X-ray crystallinity (area)	—	n: 31 cm^2 ab: 61 cm^2 (Stretching fault) n: 12.1° ab: 15.2°

Polyamide fiber can be reserved by applying the following direct dyes below 90°C and under slightly alkaline conditions.

C.I. Direct Yellow 50
C.I. Direct Red 81
C.I. Direct Blue 1
C.I. Direct Black 71

Solid shades on polyamide–cellulosic blends can be obtained with a combination of direct and disperse or neutral dyeing acid dyes. In this case, direct dyes that do not stain polyamide and disperse or neutral dyeing acid dyes that do not stain cellulosic fiber are selected.[66]

The solid shades can also be produced using selected direct dyes only. For example, the following direct dyes are suitable for producing solid shades.

C.I. Direct Yellow 8, 9, 46
C.I. Direct Orange 1, 8, 12, 23
C.I. Direct Red 1, 21
C.I. Direct Brown 1, 25

These direct dyes are applied at boiling and pH 4–5.

Some of the vat dyes can give solid shades on polyamide–cellulosic blends. The dyeing temperature of a solid shade is always higher than that usually necessary for cellulosic fibers alone. A vat dye-liquor is circulated along with sodium hydroxide and a leveling agent (if required) at 20–30°C. Sodium hydrofulite is then added and after the bath is run for 15–20 min, it is gradually heated to the required temperature. The dyeing is carried out at maximum temperature for 30–45 minutes. The goods are then rinsed, oxidized, soaped, and washed. It is essential to subsequently dye polyamide fiber with acid dyes if vat dyes that do not dye both the fibers are used.[67]

Reactive dyes can also be used to produce solid shades on polyamide–cellulosic blends. Reactive dyes contain sulfonic acid groups and behave like acid dyes on polyamide fibers. Under acid conditions the polyamide fiber takes up reactive dyes. When the bath is made alkaline, these dyes react with the cellulosic component of the blend. The dyeing is carried out for 1 h at a temperature not exceeding 85°C in a liquor containing 100 g/liter common salt. Trisodium phosphate (15 g/liter) is then added to bring about the fixation of the reactive dyes. After dyeing for 1 h at boiling, the material is rinsed, soaped, washed, and dried.

Continuous dyeing methods for the application of disperse, acid, and metal-complex dyes to the polyamide component of the blend are based on pad-thermofix or pad-steam processes. The cellulosic component of the blend is generally dyed with vat, reactive, direct, and sulfur dyes.[54]

Dyeing of Polyamide–Wool Blends.   These blends are dyed with acid or metal-complex dyes. These dyes produce different tones, depths of shade, and fastness properties on the two fibers. This can be countered by using selected dyes and, if necessary, auxiliaries. In weaker shades, nylon usually absorbs most of the dye before the temperature is high enough for wool to be dyed.[68] In such a case, the addition of an anionic leveling agent is recommended to slow down the rate of dyeing on nylon. In heavy shades, wool dyes darker since the capacity of wool to take acid dyes is considerably greater than that of nylon[67] (Figure 9.13). Sometimes disperse dyes are used to fill up the shade on nylon but fastness properties are affected adversely. The leveling acid dyes are applied to polyamide–wool blends in the same way as that to 100% wool with the exception that formic acid must be used in place of sulfuric acid, because the latter would cause some degradation of polyamide. The dyebath is set with 10% Glauber's salt and 4% formic acid (85%). The leveling agent is added for pale shades. Dyeing is started at 60°C and the temperature is raised to boiling over 45 min. The dyeing at boiling is continued for 1 h. The goods are then washed, soaped, and washed. Mono and disulfonated acid dyes produce less disparity in pale shades compared to trisulfonated dyes.

In dyeings with superior fastness properties are required, it is essential to use 2:1 metal-complex dyes. These dyes are applied at pH 4–4.5. 1:1 metal-complex dyes are not used much for dyeing polyamide–wool blends because they require strongly acidic dyebaths. Similarly, chrome dyes are difficult to apply to polyamide fibers because of incomplete formation of metal-complex of the dye under normal conditions.

Production of reserve shades on polyamide–wool blends is difficult because most of the available dyes produce stain on other fibers. Some of the disperse dyes may give the reserve effect on wool, but the fastness properties are not good.

Continuous dyeing methods are available for dyeing polyamide fibers.[54] They are based on pad-thermofix or pad-stream methods but are not largely used.

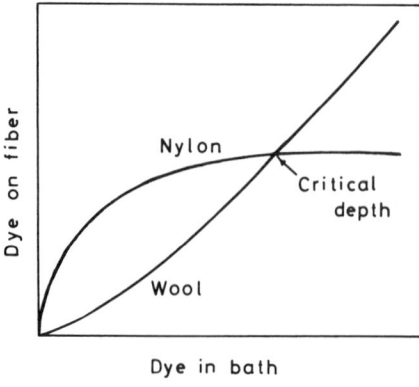

FIGURE 9.13.   Acid dye uptake by nylon and wool.[67]

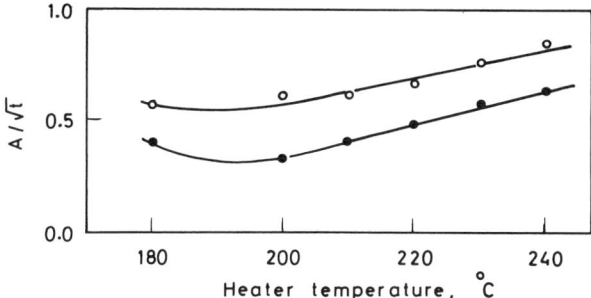

FIGURE 9.14.   Effect of texturing and heater temperature[69] on rate of dyeing of nylon $(A/\sqrt{t})$.
○: Yarn textured at different heater temperatures; ●: Flat yarn passed through only heater.

Polyamide–Polyester Blends.   These blends can be dyed with acid or metal-complex dyes to get the reserve effect on polyester. It is not possible to reserve nylon since disperse dyes will dye both fibers. For solid shades, disperse dyes must be used. Disperse dyes produce a different tone and a different depth of shade on nylon and polyester. The fastness properties on nylon are not good. During dyeing of polyamide–polyester blends with disperse dyes, the polyamide fiber is dyed heavily at the beginning of the dyeing process. The dye gets transferred to polyester on prolonged boiling. A carrier is required to get a level and solid shade.

Textured and Flat Nylon Blends.   Nylon fabric is constructed using warps and wefts of flat and textured yarns. The rate of dyeing of the two yarns may differ significantly even if both receive the same thermal treatment (Figure 9.14).[69] This may create a problem in getting a solid shade.

REFERENCES

1. Holfeld, W. T., and Shepard, M. S., *Can. Text. J.,* **94** (April 1977), 77.
2. Butterworth, J., Gordy, D. L., and Hyder, W. L., *ADR,* **55** (1966), 989.
3. Newby, W. E., Hug, G. T., and Thomas, R. J., *ADR,* **57** (1968), 113.
4. Peters, R. H., *JSDC,* **61** (1945), 95.
5. Rush, J. L., and Miller, J. C., *ADR,* **58**(4) (1969), 37.
6. Thomas, R. J., and Holfeld, W. T., *TCC,* **4** (1972), 216.
7. Dorset, B. C. M., *Text. Mfr.,* **99** (Aug. 1972), 30.
8. Sookne, A. M., and Harris, M., *J. Res. Natl. Bur. Stds.,* **26** (1941), 289.
9. Remington, W. R., and Gladding, E. K., *JACS,* **72** (1950), 2553.
10. Daruwala, E. H., and Bhatt, N. J., *TRJ,* **34** (1964), 435.
11. Brody, H., *TRJ,* **35** (1965), 844, 895.
12. McGregor, R., and Harris, M., *JAPS,* **14** (1970), 513.
13. Newport, J. F. L., McGregor, R., and Peters, R. H., *JSDC,* **86** (1970), 408.
14. Hopper, M. E., McGregor, R., and Peters, R. H., *JSDC,* **86** (1970), 117.

15. Suganuma, K., *TRJ,* **49** (1979), 536; **51** (1981), 626.

16. Carlene, P. W., Fern, A. S., and Vickerstaff, T., *JSDC,* **63** (1947), 388.

17. Atherton, E., Downey, D. A., and Peter, R. H., *TRJ,* **25** (1955), 977.

18. Iijima, T., and Sekido, M., *Sen-i Gakkaishi,* **15** (1959), 911.

19. Sekido, M., Iijima, T., and Takahasi, N., *Sen-i Gakkaishi,* **21** (1965), 524.

20. Bell, J. P., *TRJ,* **38** (1968), 984.

21. Takazawa, H., Kuroki, N., and Katayama, A., *Sen-i Gakkaishi,* **24** (1968), 185.

22. Vickerstaff, T., *Physical Chemistry of Dyeing,* Oliver and Bold, London (1950), p. 381.

23. Richardson, G. M., *Mod. Text.,* **53** (April 1972), 21.

24. Doremus, R. H., *Poly. Letters,* **4** (1966), 755.

25. Munden, A. R., and Palmer, H. J., *JTI,* **41** (1950), 609.

26. Datye, K. V., *Colorage,* Sp. Supplement—Reactive Dyes (Jan. 22, 1976).

27. Peters, H. W., and Turner, J. C., *JSDC,* **74** (1958), 252.

28. Greider, K., *JSDC,* **90** (1974), 435.

29. Blackburn, S., and Dowson, T. L., *JSDC,* **87** (1971), 473.

30. Turner, G. R., Newby, W. E., and Speck, S. B., *ADR,* **56** (1967), 998.

31. Hughes, J. A., Sumner, H. H., and Taylor, B., *JSDC,* **87** (1971), 463.

32. Korchagin, M. V., and Pavlova, V. V., *Text. Prom.,* **7** (July 1971), 57.

33. McGrew, F. C., and Sharkey, W. H., *TRJ,* **21** (1951), 875.

34. Hindle, H. W., *ADR,* **45** (1956), 972.

35. Richardson, G. M., *Mod. Text.,* **53** (March 1972), 34.

36. Dowson, T. L., *Text. Mfr.,* **99** (Sept. 1972), 30.

37. Brooks, J. A., and Reith, J. E., *ADR,* **44** (1955), 698.

38. Hadfield, H. R., and Seaman, H., *JSDC,* **74,** (1958), 392.

39. Vaidya, A. A., and Sunder, R., *Colourage,* **25**(7) (1978), 19.

40. Hobday, C., and Siegrist, G., *MTB,* **41** (1960), 1119.

41. Bittles, J. A., Brooks, J. A., Iannarone, J. J., and Landerl, H. P., *ADR,* **47** (1958), 183.

42. Derbyshire, A. N., Harvey, E. D., and Parr, D., *JSDC,* **91** (1975), 106.

43. Robin, J. B., *Teintex,* **25** (1960), 7.

44. Somm, F., Oschatz, C. H., and Lehmann, H., *MTB,* **58** (1977), 228.

45. Beirtz, H., *ADR,* **68**(6) (1979), 22.

46. Stakelbeck, H. P., and Engeller, E., *MTB,* **62** (1981), 579.

47. Petty, J. B., *ADR,* **70**(6) (1981), 34.

48. Evans, B. A., and Holfeld, W. T., Book of Papers, AATCC Natl. Tech. Conf., Atlanta, USA, (1977), 158.

49. Holfeld, W. T., and Hallada, D. P., *Can. Text. J.,* **92** (1975), 37; **92** (1975), 67; 84.

50. Datye, K. V., Pitkar, S. C., and Purao, U. M., *Teinture Apprets,* **128** (Feb. 1972), 7.

51. Preston, D. T., Skelly, J. K., and Wilson, P., *JSDC,* **82** (1966), 176.

52. AATCC Rhode Island Sec., *ADR,* **40** (1951), 75.

53. Tszi, G. S., and Korchagin, M. V., *Tekhnol Tekstil Prom.,* **23** (1963), 105.

54. Herboult, G. A., *ADR,* **54** (1965), 552.

55. HATRA Production Note No. 17 (1967).

56. Jackel, S. M., *Dyer,* **137** (1967), 335.

57. Denton, M. J., and Crum, R. J., Shirley Inst. Pamphlet No. 103, (Oct. 1970).

58. Gupta, V. B., and Amritharaj, J., *TRJ,* **46** (1976), 785.

59. Vaidya, A. A., and Nigam, J. K., *Text. Inst. Ind.,* **15**(3) (1977), 96.

60. Munden, D. L., Knapton, J. J., and Frith, C. D., *JTI,* **52** (1961), 488.

61. *Textured Yarn Technology,* Monsanto Publication **1** (1968), 268.

62. Station, W. O., *JAPS,* **7** (1963), 803.

63. Hendrix, J. E., Farmer, L. B., and Kunh, H. H., *Text. Industries,* **9** (1974), 138.

64. SPRC (unpublished work).

65. Schonrock, G. M., and Weigold, S., *TCC,* **1** (1969), 579.

66. Neubert, J., *MTB,* **32** (1951), 708.

67. Hug, G. T., *ADR,* **47** (1958), 452.

68. Luttringhaus, H., *ADR,* **44** (1955), 194.

69. Warwicker, J. O., Shirley Inst. Publication **S25** (1976), 49.

10 DYEING OF ACRYLIC FIBER

The first acrylic fiber produced from 100% acrylonitrile was extremely diffi-cult to dye.[1] It had no affinity for dyes and a high T_g (104°C). Initially, a small proportion of basic monomer was incorporated into the fiber to impart acid dyeability.[2,3] The cuprous ion technique was also developed for applying acid dyes.[4–7] However, a major breakthrough was achieved with the produc-tion of fibers containing anionic dye sites and the subsequent development of special basic (cationic) dyes. These dyes proved to be so successful that about 80–85% of the total acrylic fiber in the world today is dyed with cationic dyes.[8] Thus, acrylic fiber is available as cationic dyeable, deep dyeable, and acid dyeable. The fiber may be bright, semi-dull or full-dull. Dope-dyed and gel-dyed acrylic fibers are also marketed. The shrinkage of the fiber can be 2–40% depending on whether or not the fiber has been preshrunk. Bicomponent fibers are used to get a three-dimensional crimp effect with good bulk. Modacrylic fibers containing less than 85% acryloni-trile are also available.

It is important to know the stress–strain behavior and thermoplasticity of acrylic fiber under dyeing conditions. Stress–strain curves below and above the glass-transition temperature are shown in Figure 10.1.[9] Diagrams of acrylic fibers are characterized below the glass-transition temperature (50–95°C with decreasing water content) by a steep, glasslike starting zone A, followed on the other side of the elasticity limit by a zone B of greater resilience triggered by stress softening. Zone A does not exist anymore above the glass-transition temperature (when dyeing takes place). The fiber is more extensible against a resistance that gradually increases until it reaches the yield stress where structural resistance to plastic slip processes are overcome. Thus, under dyeing conditions, acrylic fiber may get easily and permanently deformed with only a small stress.

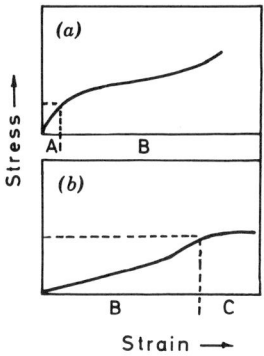

FIGURE 10.1. Stress–strain diagram of acrylic fibers.[9] (a) Below the glass-transition temperature. (b) Above the glass-transition temperature.

Unlike polyester, acrylonitrile fiber structure is like an amorphous chain packing; the fiber does not have any clearly marked crystal structure that can be defined by X-ray analysis. Thus, the crystalline–amorphous two-phase model cannot be employed for characterizing the structure of acrylic fibers (see Chapter 2). Nevertheless, all the sites in the fiber are not available for dye sorption as is shown by the following:

Material	Fiber Saturation Value
Acrylic polymer	2–2.2
Wet-as-spun fiber	1.8
Spun, stretched, and dried fiber	1.1
Annealed fiber	1.5–1.7

All sites are available for dyeing except for the unaccessible sites. The differences in the dyeability of various acrylic fibers are thus only due to the differences in the number of available sites in the fiber. On the other hand, the rate of dye sorption of acrylic fibers is affected even by minimal changes in the structure. With increasing temperature of steam-annealing, the rate of dyeing increases considerably.[9]

Mechanism of Dyeing. The dyeing of acrylic fiber with cationic dyes is best described as an ion exchange—the maximum dye uptake is given by the number of acid groups in the fiber,[10-14] even though a small amount of dye may be held by a dissolution process.[10] It is generally believed that cationic dye diffuses through the acrylic fibers by a site-to-site mechanism,[15,16] even though none of the existing theories of ionic dyeing provides a completely satisfactory account of the ion sorption equilibria in cationic dye–acrylic fiber systems.[17]

Cationic dyes give brilliant shades with excellent fastness properties on acrylic fiber.[18-20] These dyes are chemically bonded to the acidic (anionic) sites present in the fiber. The number of such sites vary from fiber to fiber depending on the manufacturing process. The dye that is not chemically bonded to the anionic sites exhibits poor wet fastness properties. Oversaturation of cationic dyes is therefore avoided by not exceeding the saturation limit of a fiber for a given dye combination.

10.1 DYEING WITH CATIONIC DYES

10.1.1 Fiber Saturation Value

The fiber saturation value A is given for a number of acrylic fibers in Table 10.1.[21] These values are determined by using C.I. Basic Green 4 (Malachite

TABLE 10.1 Fiber Saturation Value of Acrylic Fibers[21]

Acrylic Fiber	Fiber Saturation Value (A)
Acribal	2.1
Acrilan 16	1.5
Acrilan 70	1.4
Anilana	2.2
Beslon	2.7
Cashmilon FH	2.3
Chemfalon 31	1.7
Courtelle	2.2
Creslan 61	1.8
Creslan 61 BF	1.7
Crylenka	1.9
Crylor 20	2.0
Dolan	2.8
Dralon	2.1
Euroacril	2.1
Exlan DK	2.2
Exlan L	1.1
Jaykrylic[a]	1.7
Leacril 16	1.8
Malon	2.2
Melana	1.8
Orlon 42	2.1
Orlon 21	4.2
Orlon 23	1.7
Redon F	2.1
Toraylon	2.8
Velicren	2.4
Vonnel V 17	1.4

[a] Unpublished SPRC work, 1980.

Green) under specific dyeing conditions.[22] The method for determining A is given in detail in a proposed standard by the SDC.[22] Recently, C.I. Basic Green 4 dye was replaced by C.I. Basic Blue 9 (Methylene Blue) because the green dye is not stable under dyeing conditions. The commercial acrylic fibers contain sulfonic and carboxyl groups; the sulfonic groups are present as a sodium salt. During dyeing with a cationic dye, sodium ions (and protons) are replaced by dye cations.[23,24] A typical sorption curve for a model dye and a simultaneous desorption curve for sodium ions are shown in Figure 10.2.[24] Thus, dyeing is a stoichiometric exchange of ions, and equivalent amounts of dyes are adsorbed by a given fiber to exchange the sodium or hydrogen ions in the fiber.[10–14] For each cationic dye, the saturation concentration C will depend on its molecular weight (or equivalent weight if the dye is polyvalent). The sorption of dyes on a given fiber are proportional to their molecular weight ratios or

$$F = \frac{A}{C}$$

where F is the dye saturation factor and A is the fiber saturation value with a known dye. Many times, the constitution of the dyes is not known and therefore, the dye saturation factors are determined experimentally. The dye saturation factor F for typical cationic dyes is given in Table 10.2.[21,25] Since F is the ratio of equivalent weights of the dyes, it does not depend on the fiber and is the same for all acrylic fibers. Knowing the dye saturation factor F and the fiber saturation value A, it is possible to calculate the saturation concentration C of a given dye on a given fiber and to determine whether or not a particular dyeing recipe will cause oversaturation of the fiber. The following example illustrates the calculations.

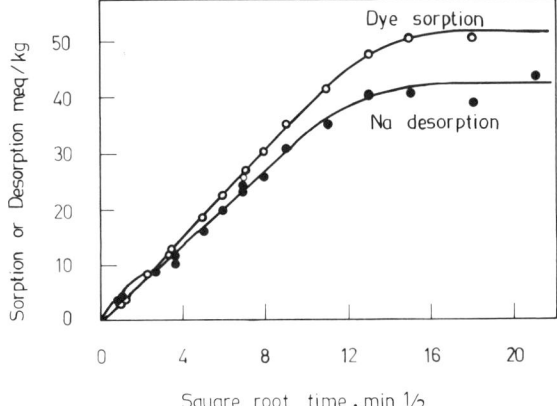

FIGURE 10.2. Dyeing acrylic fiber with a cationic dye at 98°C.[24] Fiber: Orlon 75.

TABLE 10.2   Dye-Saturation Factor F of Typical Cationic Dyes[21,25]

Name of the Dyestuff	F
Astrazon Yellow GRL	0.27
Astrazon Red GTL	0.47
Astrazon Blue RL	0.17
Basacryl Orange FL	0.39
Basacryl Blue GL	0.21
Maxilon Red 3 GL	0.31
Maxilon Black NL	0.40
Remacryl Red BL	0.33
Remacryl Blue B	0.35
Sandocryl Brill. Yellow B6 GL	0.50
Sandocryl Red BF	0.38
Sandocryl Black BBL	0.57
Sevron Blue B	0.21
Synacryl Red B	0.39
Synacril Blue R	0.22

An acrylic fiber (Jaykrylic; $A = 1.7$) is dyed with the following recipe:

1.5% Yellow dye $(F_1 = 0.30)$

0.5% Red dye $(F_2 = 0.45)$

0.6% Blue dye $(F_3 = 0.40)$

The saturation factors for yellow, red, and blue dyes are F_1, F_2, and F_3, respectively. If the summation of % shade $\times F$ of dyes is greater than the A value of the fiber, the dyes will cause oversaturation of the fiber.

Thus, $P_1F_1 + P_2F_2 + P_3F_3$ must be greater than A for oversaturation, where P_1, P_2, and P_3 are percentage shades of the three dyes, respectively. Thus,

$$(1.5 \times 0.30) + (0.50 \times 0.45) + (0.6 \times 0.40) = 0.915$$

where the A of the fiber is 1.7. The above recipe will not cause oversaturation of the fiber.

The mechanism of sorption of cationic dyes is equally applicable to any other cationic compound. A colorless cationic compound called a *cationic retarder* is usually added to the dyebath to avoid the "strike" of the dye on fiber and to get even dyeing. These compounds will also exchange with sodium ions or protons in the fiber and get adsorbed on the acrylic fiber. A typical sorption curve of a retarder is shown in Figure 10.3.[24] It is necessary to use an appropriate concentration of the retarder in order to avoid oversa-

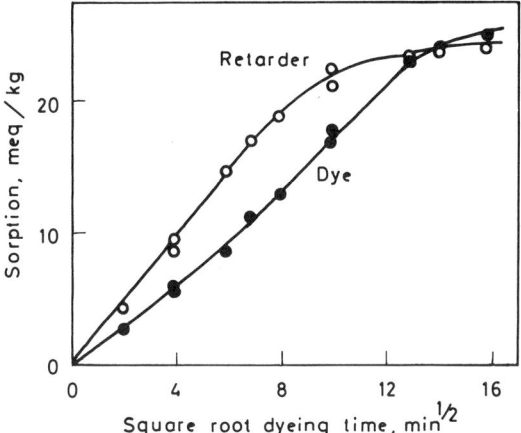

FIGURE 10.3. Competing uptake from a dye-retarder mixture on Orlon 75 at 95°C.[34]

turation of the fiber with the dye plus retarder, that is, the retarder is treated as a colorless dye for the above calculations. In deep shades where the dyes themselves are going to almost saturate the dye sites in the fiber, the retarder is not used. The saturation factor F for a retarder must be known—either obtained from the manufacturer or determined. The saturation factors of typical retarders are given in Table 10.3, using an acrylic fiber of saturation value $A = 1.8$.[25] By proper selection of the concentration of cationic dyes and the retarder it is possible to get an exhaustion of about 90–97% of the cationic dyes on acrylic fiber. This dye–fiber system has the advantages over other dye–fiber systems of economy and low water pollution.

10.1.2 Level Dyeing of Acrylic Fiber

Acrylic fiber has a glass-transition temperature of 80–85°C below which the fiber does not absorb any dye and the diffusion of dye into the fiber is very

TABLE 10.3 Saturation Factor F of Cationic Retarders[25]

Retarder	F
Astragol PAN (Bayer)	0.67
Crinolo (Mostti)	0.65
Dyasist (Arkansas)	0.23
Eagalil AD (La Tessilchemica)	0.56
Sale Basacryl G (BASF)	0.63
Tinegal TCI (Ciba-Geigy)	0.60
Tinegal TSG (Ciba-Geigy)	1.0

Note: Fiber : Leacryl 16 ($A = 1.8$).

slow. However, when the temperature exceeds the glass-transition temperature, the rate of dye uptake increases very rapidly, even with a small increase in temperature (Figure 10.4).

Because of the sudden increase in the rate of dye uptake, there is a tendency for the dye to strike unevenly on the fiber, giving unlevel results.[26-33] The problem is further aggravated by the formation of strong ionic bonds between the cationic dyes and anionic sites in the fiber. The dye tends to remain at the site where it is adsorbed and the migration of dye during dyeing is very poor. Once the dye is adsorbed unevenly, it is very difficult to level it out.

The pH is controlled at 4.5 in dyeing acrylic fiber with cationic dyes with acetic acid or preferably sodium acetate–acetic acid buffer mixture. The rate of dyeing is usually controlled by the temperature and the amount of retarder employed, not by the pH. However, with acrylic fiber that contains many weakly acidic dye sites, the pH can be used to control dye uptake.[11,12] The cationic dyes may change color when applied above the optimum pH level. The pH is controlled to within 0.5 pH units to avoid unacceptable color variations when dyeing hydrolysis-prone dyes.[14]

An antifoaming agent is used to control the foaming caused by the desorption of lubricant from the acrylic fiber during the initial heating stage in order to avoid the foam's interference with the liquor circulation in package dyeing. The load of packages after dyeing with inadequate amounts of a antifoaming agent may show significant distortion of package shape due to the unsatisfactory flow conditions in the initial temperature rise and subsequent hold-up period. This type of fault (nonuniform package densities and, in turn, flow rates within the column of packages) manifests itself as a variation

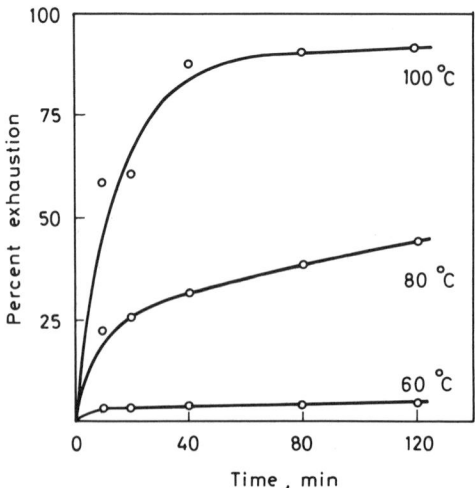

FIGURE 10.4.   Exhaustion of Methylene Blue 2B on acrylic fiber (unpublished SPRC work).

in dye concentration in different regions of the substrate, giving unlevel dyeings.[34]

Cationic dyes can be used with one another in mixtures, depending on the required depth to be dyed. The dyes used in mixtures should possess similar affinities and rates of adsorption.[30] This condition is particularly important for the production of delicate pale shades. The dye manufacturers provide a combination index K for their dyes. The method suggested by Datye and Mishra for dyeing combination shades may be used to identify the behavior of cationic dyes in mixture.[35] Cationic dyes with the same or nearly the same combination indices have a similar rate of adsorption in compound shades. In the case of a mixture of dyes with different combination indices, the dye with higher K value is always taken up slowly and therefore requires prolonged dyeing time to develop the tone. This creates difficulties in matching and may give unlevelness and hue variations, even under the best conditions of agitation–circulation of the liquor.

Various approaches have been explored to solve the problem of uneven dyeing of acrylic fiber. They are the use of (*a*) a cationic retarder, (*b*) an anionic retarder, (*c*) a polymeric retarder, (*d*) an optimum dyeing cycle, (*e*) a HT dyeing method, and (*f*) special cationic dyes.

Cationic Retarder.   These agents have an affinity for the acrylic fiber and compete with the cationic dyes for adsorption on available sites in the fiber. The rate of dyeing is thus slowed down with an increased concentration of the retarder which results in level dyeing (Figure 10.5).[36] If the retarder is not added, about 40% of the dye is exhausted within the few minutes required to raise the temperature from about 80°C to boiling. The remaining dye is exhausted within 15–20 min. Although the cationic retarder slows down the rate of dyeing, it does not alter the saturation value of a dye. The retarder adsorbed by the fiber is replaced totally by the dye during the boiling stage of dyeing because the fiber–cationic retarder bonds are much weaker than fiber–dye bonds.[37–41] The amount of cationic retarder required depends on

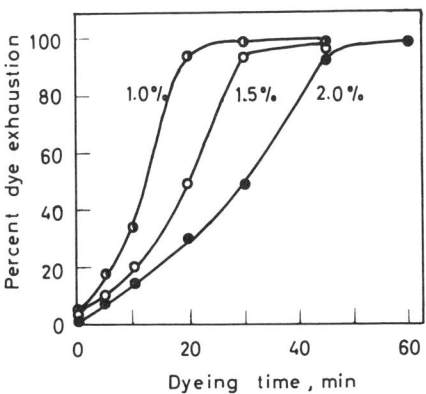

FIGURE 10.5. Rate of dyeing of a cationic dye on acrylic fiber.[36] Retarder DS (DuPont) 1%, 1.5%, and 2%. Dye: Severon Brill. Red 4G, 0.2%.

the depth of shade. Higher amounts of retarder are required for pale shades than for deep shades. The dyebath is set at 60°C with acetic acid (0.5 g/liter) and cationic retarder in the required amount is added. The goods are run for 10 min before the cationic dye is added. The temperature is slowly raised to boiling in 30 min and dyeing is carried out at boiling for 60–90 min. The dyebath is then slowly cooled, dropped, and the material is washed, soaped, and dried.

Anionic Retarder.[41–43] The cationic dye reacts with the anionic retarder to form a complex. The insoluble complex is maintained in a finely disperse form by a nonionic dispersing agent. This complex is adsorbed on the fiber rather loosely thereby permitting migration and leveling on the surface of the material. As the temperature is raised, the evenly distributed complex breaks down, releasing the dye to diffuse into the fiber while anionic retarder and nonionic dispersing agent remain in the bath. By this method, the rate of dyeing of different dyes is approximately the same since the rate of adsorption is no longer wholly dependent on the inherent characteristics of the dye. Instead, it is more dependent on the rate of surface adsorption of the dye–retarder complex and the rate of dissociation of the complex at boiling. The net result is that the dyes are adsorbed together. The dyebath is set at 60°C with anionic retarder (0.5–2%), nonionic dispersing agent (2–3%), sodium acetate, and acetic acid. The material is run for a short time to ensure uniform conditions throughout the dyebath before the dye is added as a solution. The temperature is raised quickly to 80°C and then slowly to boiling, where it is maintained for 60–90 min. The goods are finally soaped, washed, and dried.

Polymeric Retarder.[44] These retarders are cationic polyquaternary ammonium compounds with high molecular weight (1000–20000) compared to conventional cationic retarders (300–500). The polymeric retarders cannot enter the fiber and remain at the surface of the fiber because of their big size. Since these retarders have multiple charges along the chain, they have a high affinity for the fiber surface. The rate of dyeing is thus very well controlled by these retarders, even at a very low concentration of 0.1–0.2%. A comparative evaluation of cationic, anionic, and polymeric retarders is done in Table 10.4.

Optimum Dyeing Cycle. In this technique, attempts are made to regulate the dyeing process by controlling the temperature rise with time. The technique claims to give uniform dyeings even in the absence of a retarder. A typical method,[45] which depends on a stepwise rise in the temperature of the dyebath and the use of Glauber's salt to assist migration, may run as shown in Figure 10.6.[21]

The dyeing is started at 40°C, the dye is added after 15 min, and the temperature is raised to boiling (in steps) in 90 min. After dyeing at boiling

TABLE 10.4   Comparison of Different Retarder Systems[44]

Criteria	Cationic	Anionic	Polymeric
Economy	+[a]		++
Selectivity with dyes of different rates of exhaustion		+	
Strength and evenness of retarding action		+	+
Dyebath exhaustion	+		+
Compatibility with anionic dyeing system		+	+
Tendency to block fiber sites	Nonblocking types available	+	+
Tendency to block dyes	+		+
Effect of excess retarder	Nonblocking types available	+	+
Staining of other fibers	+		+
Possible precipitation during dyeing	+		+

[a] + = Preferred System.

for 90 min, the bath is cooled before dropping. This method has the advantages of economy in retarder use and ease in redyeing to heavy shades, since the fiber is not saturated with a retarder. However, the process is suitable only for selected dyes and should be initially tried on a laboratory scale.[46]

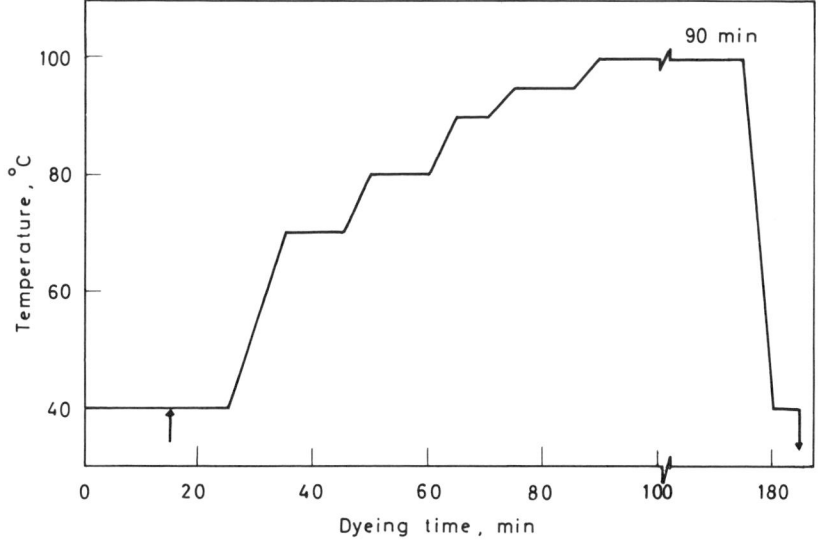

FIGURE 10.6.   Temperature–time cycle for acrylic fiber dyeing.[21]

In the Defitherm process (BASF),[47,48] the dyes are added to the bath at the actual uptake temperature which can be predetermined for a given combination of dyes. However, in this method, special auxiliary Defithermol TR has to be added to the weight of dyes used. The method also gives even dyeing in the presence of cotton or wool in the blends.[49-51]

In another method,[46] dyeing is carried out at a predetermined constant temperature in the range of 85–95°C, thus, omitting the slow rise of temperature through the critical range. An improvement is to commence dyeing at 90°C and to raise the temperature to 95°C in 10 min. Dyeing is continued at 95°C for 30 min, and is completed by raising the temperature to boiling, and dyeing for 15 min before sampling. For deep shades, dyeing is started at 95°C, the temperature is raised to boil in 10 min, and is maintained for 45 min. A weakly cationic retarder may be added to further control the dye uptake.[52] Alternatively, a uniform level dyeing on acrylic fiber is obtained using NaCl (5%) or Glauber's salt (10%) as a migrating agent in the dyebath.[13]

High-Temperature Dyeing Method. The poor migration properties of cationic dyes are the cause of uneven dyeing of acrylic fiber. Ingham and Lemin[53] found that the migration of these dyes increases remarkably if the temperature is raised above 100°C (Table 10.5). However, there is an objectionable degradation of acrylic fiber, particularly above 110°C. The cationic dyes are usually fairly stable at 110°C. Addition of a retarder is not required, thus the cost of dyeing is low. The reproducibility of shade from lot-to-lot is excellent at 110°C. After dyeing, the bath is carefully cooled. Well-penetrated deep shades can be easily produced at 110°C.

Special Cationic Dyes. The main difference between conventional cationic dyes and special cationic dyes is the excellent migration properties of the latter. A typical dye is Maxilon M (Ciba–Geigy). Migration properties of cationic dyes are given in Table 10.6.[54-56] The rate of rise in temperature is not important with special cationic dyes as any unlevelness is corrected during the boiling phase. This reduces the total time of dyeing considerably. Furthermore, the cost of retarder is an additional saving since no retarder is required for dyeing medium and deep shades. The dyebath is set with:

5–10%	Glauber's salt
0.5–1%	Sodium acetate
0.5–1%	Acetic acid to get pH 4–4.5

The goods are run at 80°C for 5 min. A solution of Maxilon M dye in water is then added along with a retarder. The dyeing is carried out at 80°C for 5 min. The temperature is raised to boiling in 15 min and the dyeing is carried out for 60–90 minutes. The bath is then cooled and drained, and goods are washed, soaped, and washed.

The wet fastness of the special leveling-type cationic dyes is similar to

TABLE 10.5   Migration (%) of Cationic Dye on Acrylic Fiber[53]

Migration Dyeing Time (min):		15	30	60	90	120
Synacryl Fast Dye	°C			Migration (%)		
Yellow R	100	6	8	12	—	16
	110	14	36	40	—	56
	120	40	42	64	—	68
Yellow Brown G	100	8	8	12	—	24
	110	18	28	32	—	44
	120	24	26	36	—	52
Scarlet G	100	—	—	2	2	4
	110	—	6	8	10	16
	120	—	6	8	14	22
Red 2G	100	4	6	8	—	14
	110	8	14	16	—	32
	120	16	22	28	—	40
Blue 5G	100	—	4	10	14	18
	110	—	12	22	28	40
	120	—	38	52	60	66

Note: Acrylic Fiber: Orlon.
Migration (%): $2 \times \%$ dye on originally undyed fiber.
(100% Migration: Undyed and originally dyed samples match.)

that of conventional cationic dyes. The light fastness of the yellow and red dyes is very good, but the blue dye is not as good.

Carrier Dyeing.   Some commercial acrylic fibers have difficulties in producing deep and penetrative dyeings. By incorporating certain carriers such as benzyl alcohol in the dyebath, it is possible to get deeper shades.[57,58] These carriers decrease the T_g of the acrylic fiber.[59] Carrier dyeing has not become popular because of its higher cost. Furthermore, there is a limit to getting deeper shades because the maximum shade obtainable on acrylic fiber depends on its fiber saturation value; the carrier can increase the rate of dyeing and may allow penetration to reach all the dye sites in the fiber. Once these sites are filled, there can be no more sorption of dye. This is not the case in polyester, since the mechanism of dye sorption is a dissolution of a nonionic dye in the solid polymer and not Langmuir adsorption on a specific limited number of sites that are exhausted during dyeing (see Chapter 3.)

Continuous Dyeing Method.[60–62]   Continuous dyeing methods have not been widely used for dyeing acrylic fabrics because continuous fixation of cationic dyes on acrylic fabric is not economical. For example, although

TABLE 10.6   Migration Properties of Cationic Dyes on Acrylic Fiber (100°C)[54]

Dyes	% Migration in 60 min	Time to Reach Levelness (min)
Maxilon Yellow M-4 GL	100	60
Maxilon Red M-RL	90	75
Maxilon Blue M-2G	86	80
C.I. Basic Yellow 21	46	290
C.I. Basic Yellow 45	42	350
C.I. Basic Orange 53	42	350
C.I. Basic Brown 14	34	530
C.I. Basic Red 22	62	160
C.I. Basic Red 46	52	230
C.I. Basic Red 54	32	600
C.I. Basic Red 59	36	470
C.I. Basic Violet 35	36	470
C.I. Basic Blue 3	46	290
C.I. Basic Blue 22	44	320
C.I. Basic Blue 41	42	350
C.I. Basic Blue 80	40	380

Note: % Migration = 2 × % dye on initially undyed fiber.
(100% Migration = Undyed and originally dyed samples match.)

cationic dyes fix significantly more rapidly from a short liquor, they require more fixing time to be on par with acid dyes on nylon.

In the thermosol process, the acrylic fabric is padded with a liquor consisting of

X g/liter	Cationic dye
0.1–1 g/liter	Anionic agent
0.1–1 g/liter	Nonionic dispersing agent
1–5 g/liter	Antimigrating agent

The padded goods are dried at 100–110°C and thermofixed at 190–200°C for 30–60 sec. A final washing and soaping treatment completes the process. The function of the anionic agent in the padbath is to form a complex with the dye and to prevent uneven dyeing because of preferential affinities of the dyes in the mixtures. Nonionic dispersing agent is added to keep the complex in dispersed form. All the cationic dyes are not stable at thermofixation temperatures and only a stable few are selected.

Dyeing in Package Form.[34]   Significant improvements in productivity can be achieved by adopting a faster rate of temperature rise when dyeing acrylic yarn in package form, provided other factors are satisfactorily controlled. A relationship between package density, flow rate, temperature rise, and acceptable levelness has been proposed. For a given package preparation technique and a specific dyeing machine, the package density of the yarn and flow rate will usually remain constant from batch to batch. It is also essential to make a selection of dyes with a proper combination index (capable of covering variation in the yarn), and to calculate the correct quantity of a suitable retarder in order to ensure a controlled rate of dyeing and a satisfactory degree of exhaustion.

Stripping of Cationic Dyes. The partial stripping of cationic dyes from acrylic fiber can be carried out by treating the dyed material with a boiling solution of a detergent (0.5–1 g/liter) and acetic acid at pH 4 for 1–2 h. Use of a reducing agent is not suggested since it either dulls or flattens the shade so that little stripping with much dulling results.[63]

For better stripping, a treatment with sodium chlorite (0.5 g/liter) and nitric acid (1 g/liter) at pH 4 for 30–60 min at boiling is recommended.[64] This treatment should not be used for acrylic–wool blends because it seriously damages the wool. The following method for stripping cationic dyes from acrylic fiber has been patented.[65]

The faulty dyeings are treated for 1 h at boiling in a bath containing

5 ml/liter	Monoethanolamine
5 g/liter	Sodium chloride

They are then rinsed and bleached for 30 min in a bath containing

5 ml/liter	Sodium hypochlorite (150 g/liter available chlorine)
5 g/liter	Sodium nitrate (corrosion inhibitor)
	Acetic acid to get pH 4–4.5

Finally, the goods are treated with sodium bisulfite (3 g/liter) at 60°C for 15 min and washed.

Fastness Properties of Cationic Dyes on Acrylic Fiber. The wash fastness of cationic dyes on acrylic fiber is excellent because the dye–fiber bond is extremely stable. The classic basic dyes have somewhat better light fastness on acrylic fibers than on natural fibers. It has been suggested that the polymer is so hydrophobic that moisture and oxygen, which are known to participate in fading, cannot gain access.[66] Greatly improved light fastness has been obtained with recently developed cationic dyes. The development of new dyes has been so successful that dyed acrylic fibers are currently considered to have the best combination of light fastness, brightness, and wash fastness of any dyed textile fiber and set the standard for the rest of the industry.

Wegmann[67] has shown that there is a close relation between the light fastness and the chemical constitution of the cationic dyes. The light fastness of cationic dyes on acrylic fiber increases with the decreasing basicity of the dye. The more the dye–fiber bond is of a purely electrostatic nature, the lower the light fastness. With increasing homopolarity, the light fastness increases. Thus, the new cationic dyes have the positive charge isolated from the rest of the dye molecule.

10.2   DYEING WITH DISPERSE DYES

Acrylic fiber can be dyed with disperse dyes at 95–110°C but the rate of dyeing is low and exhaustion is poor, particularly in heavy shades compared to cellulose acetate or nylon fiber (Table 10.7).[68] Thus, only pale to medium shades are dyed at boiling, even if excess dye is taken in the bath (Figure 10.7).[69] The exhaustion is somewhat better at higher temperatures; dyeing at 110°C should, however, be regarded as the limit because many acrylic fibers have excessive shrinkage above this temperature.

The outstanding feature of disperse dyes, which makes them especially attractive for package dyeing, is their excellent leveling properties.[70] The lower affinity for the fiber is responsible for better leveling. These dyes are generally employed for dyeing pale shades, for which large amounts of retarder and careful temperature control are necessary if cationic dyes are employed. The fastness to light and washing of disperse dyes on acrylic fiber is good. These dyes are free from gas fading on acrylic fiber but tend to sublime and to mark off on the adjacent material during heating and hot pressing. Bleeding during heat treatment is caused by dyes with low molecular weight and high volatility that are only used for dyeing acrylic fiber. In heavy shades, the rubbing fastness is not all that could be desired. The brightness of shade of disperse dyes on acrylic fiber is quite inferior to that obtained with cationic dyes. The shades are not even as bright as those on PET with the same dyes.

TABLE 10.7   Saturation Dye Uptake of Disperse Dyes at 95°C[68]

	Dye Uptake (%)		
C.I. Disperse Dye	Acrylic Fiber	Nylon	Cellulose Acetate
Yellow 1	3.0	5.0	16.0
Yellow 3	1.4	4.8	7.4
Orange 3	1.1	1.8	7.4
Red 15	1.8	4.5	11.0

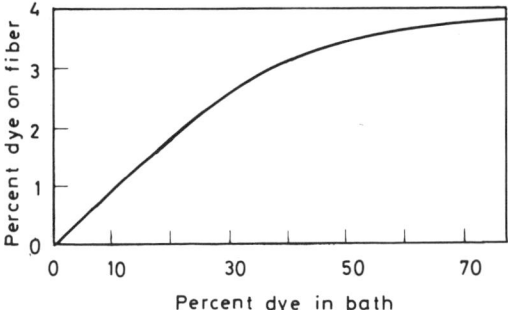

FIGURE 10.7.   Exhaustion of C.I. Disperse Red 20 on Acrilan 16.[69]

Method of Dyeing.   The dyebath is set at 50°C with sodium dihydrogen phosphate (1 g/liter) or acetic acid (0.5 g/liter (80%)) and nonionic dispersing agent (1 g/liter). The material is run for 10 min. The dye dispersion is added and the material is run for 10 min more. The temperature is raised to boiling in 20 min and dyeing is continued at boiling for 60 min. The bath is cooled gradually and drained off. During dyeing, alkaline pH is avoided because the fiber becomes discolored.

Continuous Dyeing Method for Disperse Dyes.   Acrylic fabrics can be dyed with disperse dyes by the thermofixation process.[71] Selected disperse dyes give excellent color yields. The addition of urea and dicyandiamide in the padbath is recommended.[72]

10.3   ACRYLIC–CELLULOSIC BLENDS

Blends of acrylic and cellulosic fibers are used in outerwear fabric and carpets. The dyes for acrylic fiber are from the cationic or disperse class while those for cellulosic fiber are from the direct, vat, reactive, or sulfur class.[51,73–80]

Dyeing with Cationic and Direct Dyes.   The direct (substantive) dyes are anionic in nature and tend to precipitate with cationic dyes. It is therefore safer to apply these dyes by the one-bath, two-stage or the two-bath method. First, the acrylic component has to be dyed with cationic dyes. Some cationic dyes and acrylic fibers are susceptible to alkali and hence, the pH of the dyebath is not allowed to exceed 5 pH units during dyeing with direct dyes. In the one-bath, two-stage method, the dyebath is set with

$X\%$	Cationic dyes
3%	Acetic acid (30%) to get pH 4–5
0–5 g/liter	Glauber's salt anhydrous
0.05–0.1 g/liter	Potassium bichromate (if necessary)
0.5–1%	Leveling agent

The dyeing is started at 70°C and the bath is raised to 80°C in 10 min. The temperature is slowly raised from 80°C to boiling in 40 min. The dyeing is continued at boiling for 60 min. The bath is then cooled to 50–60°C and a solution of the direct dye is added. The temperature is raised to boiling in 40 min and is maintained at boiling for 60 min. The bath is then cooled to 60°C and drained. The goods are finally washed. In order to improve the wash fastness of direct dyes, an aftertreatment with a cationic agent may be given.

Dyeing with Cationic and Vat Dyes. Vat dyes are used in place of direct dyes where high fastness properties are required. A one-bath, one-stage process is not possible because cationic dyes are not stable in the presence of caustic soda and hydrosulfite.

One-Bath, Two-Stage Dyeing Method. The dyebath is set with

$X\%$	Cationic dye
3%	Acetic acid (30%) to pH 4–5
1%	Dispersing agent
$Y\%$	Vat dye pigment

The dyeing of cationic dyes is completed as in the earlier case. The bath is then cooled slowly by not more than 1°C/min to 70°C. The required amounts of caustic soda and hydrosulfite are added. This is followed by the addition of Glauber's salt (anhydrous) (15–20 g/liter) if required for vat dyes and dyeing at 70°C for 20 min. Draining the bath, oxidizing, soaping, and washing complete the process.

Two-Bath Dyeing Method. The safest way to dye with cationic dyes and vat dyes is to use separate baths. It is necessary to apply the cationic dyes first. If the sequence is reversed, the cellulosic fiber is liable to be stained by the cationic dyes because many vat dyes act as a mordant for cationic dyes and induce their absorption by cellulosic fiber.

Dyeing with Cationic Dyes and Reactive Dyes. This dye combination is useful for getting very bright shades with good fastness properties. Since reactive dyes are anionic in nature, there is a danger of precipitation with cationic dyes and hence, a one-stage process is not recommended.

In the one-bath, two-stage process, the dyebath is set with

$X\%$	Cationic dyes
3%	Acetic acid (30%) to pH 4–5
1%	Dispersing agent

After dyeing cationic dyes as usual, the bath is cooled to 50–60°C. A solution of reactive dye is added, followed by the addition of common salt. The temperature is then raised to the required level. Alkali is added and dyeing is carried out for 60 min. The bath is then drained, and the goods are washed, soaped, and washed. The two-bath dyeing process is the safest process for this combination of dyes. Cationic dyes are applied first, followed by the application of reactive dyes.

Dyeing with Cationic and Sulfur Dyes. The one-bath dyeing process cannot be used with the cationic and sulfur dye combination, since sulfur dyes require sodium sulfide for reduction, and this has an adverse effect on the cationic dyes. Furthermore, sulfur dyes act as a mordant for cationic dyes. It is therefore necessary to follow a two-bath dyeing process in which the cationic dyes are applied first, followed by the sulfur dyes. Selected sulfur dyes are used for dyeing both the component fibers in the blend.[81]

Dyeing with Disperse and Direct or Vat Dyes. When high brightness is not required, acrylic fiber can be dyed with disperse dyes. The cellulosic portion of the blend is then dyed with direct or vat dyes.

10.4 ACRYLIC–WOOL BLENDS

Blends of acrylic fiber with wool have assumed increasing importance, particularly in knitting outlets. The proportion of the constituents varies widely, but the most important mixtures are those containing 50% or more acrylic fiber. The presence of acrylic fiber confers on the resultant blends greater dimensional stability, strength, and abrasion resistance. In addition, the majority of blends have excellent bulk and heat-retention properties. Blends of acrylic fiber with wool are dyed with a combination of cationic dyes and acid or metal-complex dyes.[82–88] The acid or metal-complex dyes do not stain the acrylic component of the blend and hence, where contrasting light and dark colors are required (e.g., black with yellow), it is preferable to dye the wool to the darker color. On the other hand, since wool contains carboxylic acid groups that are ionized under the mildly acid conditions used in the dyeing of the blend, sites are available for the adsorption of cationic dyes. There will always be sorption of cationic dyes by wool. In dyeing acrylic–wool blends, the cationic dyes first go on to the wool. During the boiling stage, most of the cationic dye migrates to the acrylic portion. However, some dye remains on the wool even after prolonged boiling. This staining of the wool gives low quality dyeings with poor light and wet fastness properties. The reproducibility of shade is also impaired because the amount of cationic dye that migrates from the wool to the acrylic fiber depends on the dyeing temperature and time. Thus, the depth of shade obtained on the acrylic fiber in the blend is subject to fluctuation from one batch to another.

The extent of the staining of the wool depends on the nature of the cationic dye and the dyeing process. Most of the manufacturers of dyes give information about the staining properties of cationic dyes, which is referred to when selecting dyes for acrylic–wool blends.

Dyeing with Cationic and 1 : 1 Metal-Complex Dyes. The metal-complex dyes have at least one anionic group and one positively charged metal atom that simultaneously gives the molecule a cationic character. These dyes can be considered to be amphoteric in nature and are therefore particularly suitable for dyeing acrylic–wool blends. These dyes do not cause any precipitation with the cationic dyes in the dyebath.

Dyeing Method. The bath is set with

$X\%$	Cationic dye
3%	Sulfuric acid (96%) or
5%	Formic acid (85%) to get pH 2–3
0.03–0.1%	Potassium bichromate (if necessary)

The goods are entered at 70°C and the bath is heated to 80°C in 10–15 min before adding

$X\%$	Metal-complex dye
1–2%	Dispersing agent

The dyebath is heated within 40 minutes from 80°C to boiling and dyeing at boiling is continued for 90–120 min. The dyebath is slowly cooled to 60°C at the rate of 1°C/min before the liquor is drained. Potassium bichromate is incorporated in the dyebath to prevent the reducing action of metal in 1 : 1 metal-complex dyes on cationic dyes. The goods are washed, soaped, washed, and dried.

Alternatively, the cationic dyeing is completed without the addition of the metal-complex dye. After dyeing at boiling for 30 min, the bath is cooled to 80°C before adding the solution of a metal-complex dye and a dispersing agent. The bath is heated within 20 min to boiling and dyeing is completed in 60 min at boiling before dropping the bath to 60°C as described above.

Dyeing with Cationic and 2 : 1 Metal-Complex Dyes. This dye combination is used where dyeings of high fastness properties are required. 2 : 1 Metal-complex dyes, however, tend to form precipitates with cationic dyes. For this reason, one-step dyeing of acrylic–wool blends is possible only for the production of pale shades.

In the one-step process, the dyebath is set with

$X\%$	Cationic dyes
3%	Acetic acid (30%) to get pH 4–5
0.03–0.1%	Potassium bichromate (if necessary)

the goods are entered at 70°C, and the temperature is raised to 80°C in 10–15 min before adding

$X\%$	2 : 1 Metal-complex dye
1–2%	Dispersing agent
2–3%	Ammonium sulfate or
2–3%	Sodium acetate

The temperature is raised to boiling in 40 min and maintained for 90–120 min before it is cooled slowly to 60°C. Draining, washing, soaping, and washing complete the process.

Alternatively, a two-step process may be followed in which cationic dyeing is completed, the bath is cooled to 60–70°C before adding dispersing agent, ammonium sulfate, and sodium acetate, and after 5–10 min, the 2 : 1 metal-complex dye is added. The temperature is raised to 100°C within 20 min and is maintained for an hour before it is cooled as described earlier.

The two-bath dyeing method is the safest method of dyeing acrylic–wool blends but it is a time-consuming process. The acrylic component is dyed first as in the one-bath, two-step dyeing methods described earlier. The bath is then drained and the wool component is dyed with 2 : 1 or 1 : 1 metal-complex dyes in a fresh separate bath.

Dyeing with Cationic and Acid Dyes. Most of the acid dyes precipitate in the presence of cationic dyes. It is therefore necessary to use the cationic dye first followed by the acid dye, either by the one-bath, two-step method or by the two-bath method.

A typical method involves dyeing with the cationic dye as described above. After dyeing at boiling for an hour, the bath is cooled to 50°C before adding the following:

$X\%$	Acid dye
5–10%	Glauber's salt (anhydrous)
1%	Dispersing agent
1%	Acetic acid to get pH 5.0

The temperature is raised to 100°C and is maintained for 1–2 h before it is cooled to 60°C at a rate of 1°/min. The bath is drained, and goods are rinsed, washed, soaped, washed, and dried.

10.5   OTHER ACRYLIC BLENDS

10.5.1   Acrylic–Polyester Blends

It is difficult to get a reserve effect on either of the fibers because cationic dyes stain polyester and disperse dyes dye both the fibers. Solid shades are obtained using a combination of cationic dyes and disperse dyes. These dyes can be applied by either the one-bath, one-step method, the one-bath, two-step method, or the two-bath method. The one-bath, one-step process is the most economical one but there is a danger of precipitation of cationic dyes and anionic carriers or dispersing agents present in the disperse dyes. Dyeing is carried out at pH 4.5 in the presence of a carrier at boiling. In a one-bath, two-step method either polyester or acrylic fiber is dyed first. If polyester is dyed first (with disperse dyes) in the presence of a carrier, the danger of precipitation of cationic dyes is there; although disperse dye is taken up by the polyester, the dispersing agent is still present in the dyebath during the dyeing of the cationic dyes. However, the matching of shade is easy because the polyester fiber gets its final shade at the end of first step. When the acrylic fiber is dyed first, the possibility of precipitation is not there but shade matching is a problem. This is because the acrylic fiber will be dyed further with disperse dyes and the shade will change.

The two-bath method is the safest method of dyeing acrylic–polyester blends but involves additional cost. The polyester is dyed with disperse dyes in the presence of a carrier at boiling. The goods are rinsed and cationic dye is then applied to the acrylic fiber in a separate bath.

The high-temperature (125–130°C) dyeing method for polyester cannot be used for dyeing acrylic–polyester blends because acrylic fiber turns yellow and its handle becomes harsher and flatter. However, a temperature of 105–106°C may be used and dyeing is carried out in the presence of a small amount of carrier.

10.5.2   Acrylic–Polyamide Blends

The reserve effect can be obtained by using cationic dyes that will reserve polyamide fiber. Similarly, acid and metal-complex dyes will reserve acrylic fiber. When the dyeing temperature is restricted to 70°C, it is also possible to reserve acrylic fiber with disperse dyes.

It is necessary to use a combination of cationic dyes and acid or metal-complex dyes for solid shades. These dyes can be applied by any of the three methods described earlier. In the one-bath, one-step process there is a danger of precipitation of dyes. It is essential to use a suitable dispersing agent. The method is generally used for dyeing pale to medium shades. In all the cases, dyeing is carried out at a temperature near 100°C.

10.5.3   Acrylic Fiber–Silk Blend Yarns[89]

A two-bath treatment is necessary for the dyeing of acrylic–silk knitting yarns. Cationic dyes are used for the acrylic component and acid or direct dyes are used for silk in all but the very light depths. The bath is set with the cationic dye at pH 4.0 and dyed at 98°C for 2 h in the usual manner. The silk is stained with cationic dyes and is cleared in an intermediate acid rinse before being dyed in a second bath with acid dyes at 85°C or with direct dyes at 65°C.

REFERENCES

1. Laucius, J. F., Clarke, R. A., and Brooks, J. A., *ADR,* **44** (1955), 362.
2. Field, T. A., and Fremon, C. H., *TRJ,* **21** (1951), 531.
3. Mahoney, H. F., *ADR,* **54** (1965), 1050.
4. Blaker, R. H., and Laucius, J. F., *ADR,* **41** (1952), 39.
5. Szlosberg, E., *ADR,* **41** (1952), 510.
6. Schmitt, C. H., Sealfrunk, C. W., and Walker, H. R., *ADR,* **44** (1955), 904.
7. Khachoyan, J., and Niederhauser, J. P., *JSDC,* **74** (1958), 133.
8. Kellet, H., *Dyer,* **140** (1968), 801.
9. Falkai, B. U., International Conference on Man Made Fibers for Developing Countries, SASMIRA, Bombay, India, (1982).
10. Glenz, O., and Beckmann, W., *MTB,* **38** (1957), 296, 783, 1152.
11. Balmforth, D., Bowers, C. A., and Guion, T. H., *JSDC,* **80** (1964), 577.
12. Cegarra, J., *JSDC,* **87** (1971), 149.
13. Voltz, J. T., *TCC,* **9** (1977), 113.
14. Dowson, T. L., and Roberts, B. P., *JSDC,* **95** (1979), 47; also Dowson, T. L., *JSDC,* **97** (1981), 115.
15. Rosenbaum, S., *TRJ,* **34** (1964), 159, 291.
16. Goodwin, F. L., and Rosenbaum, S., *TRJ,* **35** (1965), 439.
17. Guin, T. H., and McGregor, R., *TRJ,* **44** (1974), 439.
18. Carbonell, J., *Dyer,* **135** (1966), 249.
19. Hobson, R. H., *TCC,* **4** (1972), 232.
20. Siepmann, E. J., *TCC,* **7** (1975), 28.
21. Mayer, U., and Wurz, A., *Dyeing and Finishing of Acrylic Fibers Alone and in Blends with Other Fibers,* BASF pp. 268 and 277.
22. SDC, *JSDC,* **89** (1973), 292.
23. Vaidya, A. A., *Text. Dyer Printer,* **8**(8) (1975), 47.
24. Biedermann, W., and Ischi, A., *JSDC,* **95** (1979), 4.
25. Leacril, Monte fiber, Italy, *Dyeing of Acrylic Fiber,* pp. 10 and 11.
26. Vetter, H., *Text. Prax.,* **14** (1959), 409, 513.
27. Beckmann, W., *JSDC,* **77** (1961), 616.
28. Rosenbaum, S., *TRJ,* **33** (1963), 899; **34** (1964), 52.
29. Jowett, A. M., and Cobb, A. S., *Rev. Prog. Color,* **3** (1972), 81.

30. Zimmermann, C. L., *TCC*, **7** (1975), 208.

31. Asquith, R. S., Blair, H. S., and Spencer, N., *JSDC*, **94** (1978), 49.

32. Heane, D. J., Hill, T. C., Park, J., and Shore, J., *JSDC*, **95** (1979), 125.

33. Doyle, T., *ADR*, **68**(3) (1979), 30.

34. Park, J., and Shore, J., *JSDC*, **97** (1981), 223.

35. Datye, K. V., and Mishra, S. N., *Textilveredlung*, **18** (1983), 211.

36. Clarke, R. A., and Bidgood, L., *ADR*, **44** (1955), 631.

37. Stump, W., *ADR*, **52** (1963), 289.

38. Carbonell, J., *ADR*, **55** (1966), 956.

39. Teossier, L. R., and Boynes, R. L., *ADR*, **56** (1967), 1016.

40. Blackburn, D., *Dyer*, **153** (1975), 418.

41. Fath, H. J., Moritz, R., Kasten, V., and Mieth, W., *Textilveredlung*, **13** (1978), 309.

42. Landerl, H. P., and Bear, D. R., *ADR*, **54** (1965), 222.

43. Leddy, J. A., *ADR*, **49** (1960), 272; **55** (1966), 1003.

44. Dullaghan, M. E., and Ultee, A. J., *TRJ*, **43** (1973), 10.

45. Iannarone, J. J., and Thackrah, J. S., *ADR*, **55** (1966), 133.

46. Cusak, P., *Dyer*, **141** (1969), 25.

47. Mayer, U., *ADR*, **58**(6) (1969), 25.

48. Mayer, U., and Sulfow, M., *ADR*, **58**(7) (1969), 48; **58**(8) (1969), 25.

49. Weigold, S., *ADR*, **58**(8) (1969), 29.

50. Fleischer, H., *ADR*, **58**(9) (1969), 15.

51. Berndt, F., *ADR*, **58**(9) (1969), 22.

52. Stevens, C. B., *Text. Month*, (Aug. 1973), 47.

53. Ingham, D. J., and Lemin, D. R., *Dyer*, **145** (1971), 532.

54. Koller, J. A., and Motter, M., *ADR*, **66**(3) (1977), 34.

55. Biedermann, W., *Rev. Prog. Color*, **10** (1979), 1.

56. Beal, W., *Dyer*, **157** (1977), 261.

57. Gur-arieh, Z., and Ingamells, W. C., *JSDC*, **90** (1974), 8.

58. Gur-arieh, Z., Ingamells, W. C., and Peters, R. H., *JAPS*, **20** (1976), 41.

59. Ingamells, W. C., and Peters, R. H., *Polym. Eng. Sci.*, **20** (1980), 276.

60. Edwards, H. D., *JSDC*, **79** (1963), 15.

61. Kellet, H., *JSDC*, **84** (1968), 257.

62. Capponi, M., *Textilveredlung*, **6** (1971), 1.

63. Ward, A. M. V., *ADR*, **62**(9) (1973), 30.

64. Vogel, T., Debruyne, J., and Zimmermann, C., *ADR*, **47** (1958), 581.

65. Geigy, B. P., 956, 389 (1964).

66. Jones, F., *Rev. Text. Prog.*, **14** (1962), 303.

67. Wegmann, J., *MTB*, **39** (1958), 408.

68. Walls, I. M., *JSDC*, **72** (1956), 262.

69. Croxson, T. E., *ADR*, **49** (1960), 314.

70. Iannarone, J. J., and Thackrah, J. S., *ADR*, **52** (1963), 357.

71. Narkar, R. K., and Narkar, A. K., *Text. Dyer Printer*, 2 (Dec. 1969), 8.

72. Kern, R., *ADR*, **54** (1965), 580.

73. Hindle, H. W., *ADR*, **44** (1955), 191.

74. Iannarone, J. J., and Wygand, W. J., *ADR*, **47** (1958), 585.

75. Burgess, R. L., *ADR,* **48**(24) (1959), 19.
76. Edwards, H. D., and Jackson, N., *JSDC,* **75** (1959), 383.
77. Oguin, L., *ADR,* **51** (1962), 555.
78. Keaton, J., and Preston, D. T., *JSDC,* **80** (1964), 312.
79. Hallada, D. P., *ADR,* **55** (1966), 172.
80. Berndt, F., *ADR,* **58**(9) (1969), 22.
81. Tobin, H. M., *ADR,* **70**(9) (1981), 32.
82. Hindle, H. W., *JSDC,* **74** (1958), 151.
83. Lister, G. H., *JSDC,* **74** (1958), 158.
84. Rawicz, J. A., *ADR,* **50** (1961), 415.
85. Cheetham, R. C., *JSDC,* **83** (1967), 320.
86. Fleischer, H., *ADR,* **58**(9) (1969), 15.
87. Lemin, D. R., *JSDC,* **91** (1975), 169.
88. Cusak, P., *Rev. Prog. Color,* **6** (1975), 13.
89. Kreckwitz, E., *Tinctoria,* **68** (Sept. 1971), 322.

11 | DYEING OF POLYPROPYLENE AND OTHER FIBERS

Polypropylene fiber has proved a most difficult fiber to dye to commercially acceptable standards of brightness of shade, color fastness, and levelness. These difficulties still persist and have limited the growth of this fiber. Otherwise, polypropylene is comparable to the other main synthetic fibers: polyester, nylon, and acrylic. It has some advantages such as low specific gravity which gives good covering power for a given weight of fiber, good thermal insulation, and low cost. However, it has poor draping qualities, negligible moisture content, and low melting point (see Chapter 1). The major problem with polypropylene is in dyeing and printing.

11.1 DYEING PROBLEMS WITH SYNTHETIC FIBERS

The problem of dyeing new fibers was first felt with cellulose acetate when it was introduced in 1920.[1] The fiber has only a few hydroxyl groups, a tendency to saponify at high temperatures of dyeing particularly under alkaline conditions, low water sorption, and low swelling. The dyes suitable for natural fibers are of no avail since the direct cotton and vat dyes have no affinity for cellulose acetate, the azoics and vat dyes are used under alkaline conditions, and although the basic dyes have affinity, they have poor light and washing fastness on cellulose acetate. Initially, the problem was solved by partial saponification of cellulose acetate to cellulose and the cellulose thus formed was then dyed. However, this was not a solution since the fiber lost its smooth handle and luster and gave uneven results. Another approach was to introduce compounds with diazotizable amino groups into cellulose acetate by converting them into temporarily soluble compounds. Thus, amino compound is treated with formaldehyde and sodium bisulfite to get the soluble ionamines. The original amine is slowly generated during dyeing and is readily taken up by cellulose acetate. This amine is diazotized and then coupled with naphthol to get colors with good fastness to washing. However, the dyeing process is too long and reproducible shades are not obtained. Compounds with an affinity for cellulose acetate are almost insoluble in water but are readily soluble in organic solvents. If insoluble amino or hydroxy azo compounds are ground in a solution of sodium sulforecenolic acid to get an aqueous dispersion, they can be adsorbed by cellulose acetate. Such compounds are called *disperse dyes*. Thus, the problem of dyeing cellulose acetate was solved by developing a new class of dyes—disperse dyes.

The first fully synthetic fiber, nylon, could use dyes already on the market. Acrylic and polyester fibers, on the other hand, were both extremely difficult to dye when they were introduced in the early 1950s. Various methods were explored, for example, the cuprous ion method for acrylic fibers, a carrier method for PET, and so on. Within a short time, the coloring of these fibers also became possible. For acrylic fiber, new comonomers were developed so that the copolymer fiber has an affinity for acid or cationic dyes with excellent fastness properties. Cationic dyeable polyester, high-temperature dyeing, thermosol dyeing, and solvent dyeing are additional solutions for the problem of dyeing PET materials. New machinery was developed to carry out dyeing above 100°C under pressure. Various temperature programs, use of retarders, and reduction-clear treatments are other developments that improved the evenness and fastness of dyeings on these fibers. Thus, the problem of coloring the new synthetic fibers has been solved. The new disperse and cationic dyes introduced to suit the newer dyeing techniques can dye, respectively, PET and acrylic fibers to the most stringent specifications of the commercially accepted dyeing methods described earlier.

In the case of polypropylene fiber, the problem of coloration still persists more than 20 years after their commercial introduction. Successful coloration will probably depend on the modification of the polymer to develop sites for dye sorption, use of polymer alloys instead of polypropylene alone and so on. It is still difficult to predict the nature of the final solution to this problem. The absence of dye sites in the fiber, solubility parameters matching those of the dry-cleaning solvents, and high crystallinity are some of the properties that make coloration in full depth with good fastness properties difficult to achieve.

11.2 DYEING OF POLYPROPYLENE

Because of the difficulties in dyeing polypropylene fiber, the most widely used coloring process is the mass-coloration process (see Chapter 4). More than 50% of polypropylene is colored by this process. Such colored fibers are used in floor coverings and upholstery materials.[2] However, the limitations of the mass-coloration process are too serious to use it for other textiles where fashion requirements are continuously changing.

11.2.1 Unmodified Polypropylene

Polypropylene PP, is very difficult to dye or print because the fiber contains no chemical groups that can form a covalent, ionic, or hydrogen bond with the dye molecules. Such a fiber can be colored with nonionic dyes that are soluble in this hydrocarbon fiber. These dyes are applied at or above 100°C from aqueous baths or by the dry-heat thermosol process.

Bird and Patel studied the dyeing of polypropylene with commercial disperse dyes.[1] Disperse dyes capable of dyeing polypropylene fiber have high diffusion coefficients and low saturation values. Azo and anthraquinonoid disperse dyes containing hydrophilic hydroxyalkyl groups have no substantivity for polypropylene and the substantivity increases with the hydrophobic character of the dye. Thus, incorporation in the dye molecule of a long alkyl chain gives disperse dyes with good fastness properties on polypropylene fiber. In spite of all the claims in various dye patents, efforts to produce dyes to color polypropylene have been discouraging. The depth of shade is never more than a medium shade, the dyeings have poor fastness (particularly to light), and dry cleaning removes nearly all the dye from the fiber. There is also a tendency for the dye to migrate to the surface of the fibers during storage.

Typical rate-of-dyeing curves for disperse dyes on polypropylene are shown in Figure 11.1.[1] The disperse dyes show poor exhaustion. The rate of dyeing increases steadily with the rise in temperature and the activation energy of dyeing is 28.9 kcal/mole. The adsorption isotherms of disperse dyes are linear in shape up to fiber saturation. The partition coefficient decreases with the rise in temperature with heat of dyeing of 11.8 kcal/mole for Dimethyl Yellow dye. There is no fundamental difference between polypropylene and other hydrophobic substrates except that the former is highly

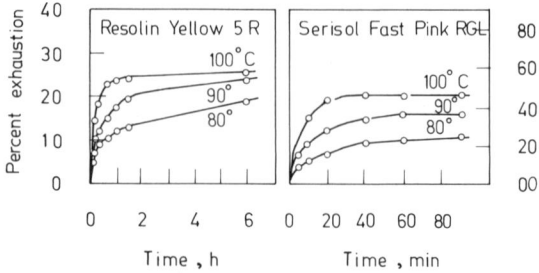

FIGURE 11.1. Typical rate-of-dyeing curves for disperse dyes on unmodified polypropylene.[1]

TABLE 11.1 Dye Saturation Values on Polypropylene and Cellulose Acetate at 80°C

	Saturation Value (g/100 g)	
Disperse Dye	Polypropylene	Cellulose Acetate
Dimethyl Yellow	0.67	2.0
1-Anilinoanthraquinone	0.32	0.60

crystalline and therefore, has a low saturation value (Table 11.1).[1] The hydrophobic dye dissolves in the polypropylene which acts as an aliphatic hydrocarbon solvent.[1] The absence of hydrogen-bonding groups in polypropylene polymer is also the cause of poor sorption of disperse dyes, since it is these groups that probably account for the major part of the substantivity of disperse dyes for cellulose acetate and polyamide fibers.

Substantivity for polypropylene depends on the absence of hydrophilic groups in the dye molecule, since such groups cause the dye molecules to be attracted to the aqueous phase, that is, to the dyebath, resulting in poor exhaustion. On the other hand, the dye must exert strong nonpolar van der Waals forces. Very few commercial disperse dyes meet these requirements.

Aliphatic hydrocarbons are very poor solvents for disperse dyes containing hydrophilic groups. Only dyes soluble in such solvents because of similar solubility parameters are capable of dyeing polypropylene. In Table 11.2, the solubilities of the three dyes in n-hexane and polypropylene are compared at different temperatures; the order is the same in both solvents.[1]

The rates of diffusion of disperse dyes in unmodified polypropylene are higher than those in secondary cellulose acetate (Table 11.3).[1] This is because the accessible regions in polypropylene have a more open structure

TABLE 11.2 Solubility of Dyes in n-Hexane and Polypropylene[1]

	Solubility (g/liter)	
Disperse Dye	n-Hexane (20°C)	Polypropylene (80°C)
p-Aminobenzene → o-Cresol	0.08	0.75
1-Anilinoanthraquinone	0.30	1.70
Dimethyl Yellow	6.20	3.50

TABLE 11.3   Diffusion Coefficients of Disperse Dyes in Polypropylene and Secondary Cellulose Acetate[1]

Dye	Temperature (°C)	Diffusion Coefficient ($cm^2\ s^{-1} \times 10^4$)	
		Polypropylene	Acetate
Dimethyl Yellow	60	4.81	2.3
Dimethyl Yellow	80	33.7	2.9
1-Anilinoanthraquinone	80	6.74	1.12

than those in the secondary acetate. Owing to the absence of interchain attraction arising from, for example, the presence of hydrogen-bonding groups, the fiber is effectively in a permanently swollen state, which accounts for the fact that disperse dyes are readily removed from the fiber by dry-cleaning solvents. This type of "swollen" state accounts for the ability of disperse dyes to migrate to the surface of the fibers during storage at room temperature.

The fine structure of polypropylene fiber influences the dyeing behavior of disperse dyes. Polypropylene forms a monoclinic crystal unit cell containing four chains, each chain in a 3-helix conformation. The crystalline structure of polypropylene is observed as spherulites, that is, three-dimensional, radiating fibril structures. There is a significant increase in the spherulite radius with extrusion temperature. The diffusion coefficient exhibits an approximate linear relationship with spherulite radius (Figure 11.2).[3] Spherulites are not perfectly crystalline and the interstices and defects between the microfibrils could be accessible to dye molecules. The diffusion coefficient of a disperse dye decreases with the amorphous and crystalline orientation of the PP–fiber during drawing (Table 11.4).[3]

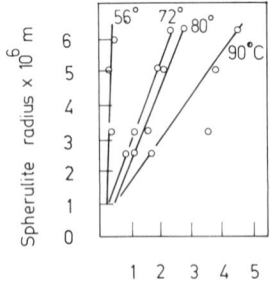

FIGURE 11.2.   Effect of spherulite radius on diffusion coefficient at different diffusion temperatures.[3]

TABLE 11.4   Effect of Extrusion Temperature on
Spherulite Radius and Activation Energy of
Diffusion of C.I. Disperse Yellow 3[3]

Extrusion Temperature (°C)	Spherulite Radius $\times 10^6$ m	Activation Energy of Diffusion (kcal/mole)	
		Undrawn	Drawn
210–240	2.56	10.2	9.3
230–260	3.25	12.3	16.3
250–280	5.06	14.5	23.5
260–290	6.31	15.3	25.4

The polypropylene polymer molecules end with a terminal vinylidine group that could interact with a conjugated dye molecule by the following mechanism:[3]

The number average molecular weight ($M_N = 10^5$) would correspond to the saturation value of 0.2–0.8-mg dye/g PP–fiber if molecular weight of dye is 200–400. The saturation values of many solvent dyes are found to be between 0.2 and 0.6 mg/g. Thus, with certain dye structures, an ionic attachment to the polymer chain ends is possible. Furthermore, the dye molecules penetrate into the spherulite structures.

Disperse dyes are applied to PP–fiber from an aqueous dispersion under neutral or slightly acidic conditions. Only a very select few could be dyed to any appreciable depth.[4] The range of shades that is available is limited and deep shades cannot be dyed.

Vat dyes are also nonionic dyes with a higher molecular weight than the conventional disperse dyes. They can be applied to polypropylene by the vat acid process to get pastel shades. However, range and depths of shade are limited, shade duplication is difficult, and fastness properties are not satisfactory.[4] Solvent dyes that are also nonionic in nature can be applied to get a wide range of shades including deep shades. The dispersion of oil-soluble disperse dyes (more hydrophobic dyes) is difficult to prepare in stable forms for dyeing and color build-up. Their fastness to dry cleaning is very poor. Light fastness and sublimation fastness are also poor in most of the cases. The development in the fiber of an azoic pigment using acidified naphtol-base dispersions with subsequent diazotization in situ also fails to give dyeings with good light fastness properties.[4]

Sulfur dyes and their prereduced solutions yield fairly deep surface dyeing of polypropylene without much penetration in the fiber. The rubbing fastness is therefore very poor. The shade range and light fastness are also not acceptable for most uses.[5]

The dyeing of standard polypropylene with all classes of dyes led to the general conclusion that no existing class of dyes can provide an adequate range of shades with acceptable fastness properties. It became obvious that new dyes must be designed for use on polypropylene, or that polypropylene must be modified for dyeability, or that both measures must be taken. Very little success has been achieved in developing new dyes for dyeing unmodified polypropylene. Most of the research work is now concentrated in developing modified polypropylene fiber.

11.2.2 Modified Polypropylene

Various approaches have been tried by both fiber manufacturers and finishers to impart dyeability to polypropylene.[6–11] Some of the approaches are mentioned here.

Copolymerization with Other Monomers. Many types of comonomers have been proposed to increase the accessibility of PP to disperse dyes and the specific dyeability of acid or cationic dyes.[11] In general, copolymerization with other monomers causes poisoning of the stereospecific polymerization catalyst and gives a polymer with inferior mechanical properties. Some of these problems have been overcome and better dyeable fibers with good mechanical properties are commercially produced.

Incorporation of Metals. Metal salts and organometallic complexes of nickel or aluminum were initially incorporated into polypropylene to reduce the degradation of polypropylene by light. However, the presence of metals was found to impart dyeability to the fiber for metal chelating dyes. Special metallizable disperse dyes are now available for the dyeing and printing of metal-containing polypropylene.

The metal-containing polypropylene gives dyeings that have satisfactory fastness properties. However, the special metallizable disperse dyes strike the fiber rapidly and irreversibly, which leads to uneven dyeings. They are more useful in printing.[13,14]

Blending with Other Polymers. Polypropylene is blended with dyeable polymers before melt-spinning. The polymers for blending are based on polyamides like nylon 6 and nylon 66. Such a polypropylene can be dyed with acid dyes.[15] Blending polypropylene with polyester imparts improved disperse dyeability.

Grafting. Graft copolymerization represents a very convenient way of attaching a dyeable polymer chain to the polypropylene.[8] The monomer systems grafted onto polypropylene are vinyl sulfonic acids, vinyl pyridines, vinyl morpholines, styrene, acrylic and methacrylic acids, acrylonitrile, vinyl acetate and so on. The grafted polypropylene can be dyed with acid, cationic, or disperse dyes, depending on the monomer used for grafting.

Chemical Modification. Dyeability with cationic or disperse dyes is imparted by partial modification of polypropylene by sulfonation, chlorosulfonation, nitration, phosphorylation, and halogenation. However, these modifications cause deterioration in the physical properties of the fiber. It is also difficult to control the reaction with reagents that are toxic or corrosive in nature.[16,17]

Dyeing of Modified Polypropylene. The dyeing behavior of modified polypropylene fibers depends on the method and extent of modification. No general method of dyeing is applicable to all types of modified fibers.

Dyeing with Acid Dyes. The procedures used for dyeing modified polypropylene fibers with acid dyes are similar to those used in the dyeing of nylon, except that the pH is maintained at a lower level.[15] In a typical procedure, the bath is set at 40°C with an anionic retarding agent and the prescoured fabric is run for 5 min. Formic acid is added to get a pH of 2–2.5, and the temperature is raised to boiling over 45 min and is maintained for 45 min. The bath is dropped and the goods are afterscoured with soda ash and detergent. The fastness properties of the dyeings are satisfactory (Table 11.5).[15] Ratings for fastness to washing, dry cleaning, and gas yellowing for dyes in Table 11.5 are 5.

TABLE 11.5 Fastness Properties of Acid Dyes on Modified Polypropylene[15]

C.I. Acid	Light Fastness (Xenon)	Wet and Dry Crock Fastness
Yellow 151	6	5
Yellow 174	6	5
Red 191	5–6	5
Red 337	5–6	4–5
Blue 40	5	3–4
Blue 80	5	4

Mechanism of Acid Dyeability.   The dyeability of acid-dyeable polypropylene is decided by two factors: (*a*) introduction of cationic dye-sites into the fiber and (*b*) diffusion of an anionic dye from the aqueous bath into the fiber in order to reach the dye-site and to form an ionic bond. The introduction of cationic functionality in many forms will produce a fiber with some acid dyeability, but not necessarily a commercially dyeable fiber. The presence of a functional dye-site as an isolated particle in a matrix of polypropylene will not impart acid dyeability. A mechanism for the transport of dye into the fiber must be furnished so that the highly polar acid dye can enter the nonpolar hydrocarbon fiber to reach the dye-site. Furthermore, the hydrophilic dye prefers water to the hydrocarbon fiber and thus, the fiber will not dye. The diffusion of dye through the polypropylene material is negligible. Electromicrographic studies of the fiber showed that the dye-sites as microfilaments are dispersed within the polypropylene. The surface microfilaments readily swell with water and allow the diffusion of dye through them. The crossover points between these microfilaments serve as the path for a complete penetration of the fiber. Since the dye-sites are highly polar with low crystallinity, diffusion occurs at a reasonable rate with the dye penetrating the fiber completely.[15]

Dyeing with Metal-Complex Dyes.   Metal-complex dyes give a wide range of shades with good fastness properties. Combination shades can be dyed without any difficulty. An efficient prescour with mild alkali and nonionic detergent is given to the material. The goods are then rinsed well with water. The dyebath is set to pH 4 with a 1–2% nonionic wetting agent and formic acid. The goods are run at 60°C for 15 min. The dye is then added, the temperature is slowly raised to boiling in 45 min, and is maintained for 90 min. The goods are then soaped with 2 g/liter soda ash and 2 g/liter nonionic wetting agent at 60°C for 20 min. This is followed by washing and drying. The temperature of the dyebath is kept at as near to boiling as possible, since the exhaustion of the dye depends on the dyebath temperature, as can be seen from Figure 11.3.[18] For deep shades and black, it is essential to add 1–2% of a carrier of the butylbenzoate type.

Dyeing with Disperse Dyes.   The prescoured material enters the dyebath at 40°C. The dyebath contains disperse dye, nonionic wetting agent, and buffer to get pH 6–7 during dyeing. The temperature is slowly raised to boiling in 30 min and is maintained for 60–90 min. The goods are then soaped, washed, and dried. The temperature of the dyebath is maintained at as near to boiling as is possible since the dyeing temperature has a considerable influence on the exhaustion of the disperse dye (Figure 11.3).[18] The light fastness of many disperse dyes is good but the wash fastness is only adequate; in the case of deep shades, it is often unsatisfactory.

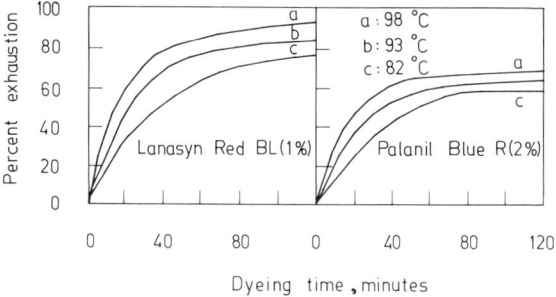

FIGURE 11.3. Exhaustion of metal-complex and disperse dyes on modified polypropylene at different temperatures.[18]

Dyeing of Fibers Having Metal Dye Sites. A mechanism for dyeing metal-modified fibers is dye absorption and diffusion followed by chelation or complexing *in situ* with the metal in the fiber.[4,7,12] Thus, a dye–metalorgano complex is formed. An example of such a coordinate compound is amino thiazole compound coupled to an orthohydroxy intermediate giving an imino-azo link.[12] Orthohydroxy azo compounds or azo dyes with carboxylic acid groups in the ortho position do not give good dyes.

The brightness of many shades is similar to that of disperse dyes on polyester fiber. Light fastness increases with the depth of shade. On the other hand, fastness to washing and dry cleaning are as good in dark shades as in light shades. Resistance to gas fading is generally good. Sublimation fastness is excellent at all temperatures. Crock fastness is generally good, provided the dyeing is adequately scoured.

Modified polypropylene with metal–dye-sites has been designed to be receptive to dyes containing groups capable of chelating with metal. The dyes can be classified by the method of application as disperse type. They are applied from an acidic dyebath containing a dispersing agent as well as a sequestering agent. The use of 0.1–0.2% (owf) ethylenediamine tetracetic acid salt or other suitable sequestrants is essential where metal ions are present (from hard water or steam lines). Ions of metals, such as copper, iron, or manganese, if not chelated by a suitable sequestrant, will complex with the dye causing weak or off-shade dyeings. Pastel and light shades are particularly sensitive to this condition.

Polypropylene-containing metal–dye-sites can be dyed on conventional dyehouse equipment such as that used for stock, skein, package, and piece dyeing. The dyeing temperature is increased slowly to obtain controlled exhaustion and maximum leveling. Exhaustion and penetration of dyes into the fiber are enhanced by dyeing at boiling. For optimum fastness properties, an effective afterscour is mandatory. In general, scouring with a detergent, alkali, and sequestering agent at a temperature ranging from 80–90°C is effective in cleaning the dyed goods. A typical dyeing procedure is as follows.

Prescouring. Alkaline prescouring of the goods is essential to remove such foreign matter as dirt, oil, size, and fiber-processing finishes that may interfere with subsequent dyeing. The following general conditions are recommended.

$X\%$	Alkali (soda ash or tetra sodium pyrophosphate) to pH 8–9
1–2%	Nonionic or anionic detergent
0.3–1%	Sequestering agent
60–70°C	Temperature
20–30 min	Time

After treatment, the goods are cooled, rinsed until clear, and dried below 130°C if necessary. When dyeing fibers that contain metal–dye-sites, sequestering agents are used in the prescour to prevent any trace of metal (from steam or water) from being carried over into the dyebath. Carpets made of polypropylene with jute backing are scoured at a minimum pH of 9.0 to minimize the jute stain on the polypropylene.

Dyeing. The dyebath is set at 40°C with 0.25–1.0% of a nonionic detergent and acetic acid to adjust pH of the bath to 3.5–4.5. Previously dispersed and strained dye solution is added and the bath is brought to boiling in 30–45 min. An additional adjustment of pH between 2.5–3.0 may be advantageous for heavy shades or a long liquor ratio or with specific fibers. The goods are run at boiling for 1–2 h or until maximum exhaustion. The bath is cooled to 60°C and shading dyes are added. If the shade is satsifactory, the dyed goods are overflow-rinsed to cool the bath slowly. The goods are then afterscoured with 0.5–2.0% of a nonionic detergent and 1.0–3.0% soda ash for 15 min at 70°C.

Stripping. For metal-containing fibers, the stripping recommendations of the fiber manufacturer are followed very closely. Improper stripping procedures may degrade the fiber, decrease its redyeability, or adversely affect its stability. For the cationic dye-site fiber dyed with premetallized, acid, or disperse dyes, conventional stripping methods using sodium or zinc formaldehyde sulfoxylate or titanous sulfate are effective. Thus, the stripped fiber may be satisfactorily redyed. With some acid dyes, a preboiling with 5.0% (owf) ammonia improves the degree of stripping. Sodium chlorite is not recommended.

Polypropylene Blends. Blends of polypropylene with acrylic fiber are easily dyed by initially dyeing the acrylic component, dropping the dyebath, and then dyeing the polypropylene fiber. The polypropylene is stained while the acrylic fiber is dyed and the shade of the acrylic fiber is altered during the dyeing of polypropylene fiber. On the other hand, with polypropylene–cotton blends, the polypropylene is dyed first with only a slight stain on cotton. The temperature of the bath is lowered to 60–70°C and the pH is raised to 7–8 with sodium pyrophosphate. The cotton is dyed with direct dyes.[15] Continuous dyeing of these blends is also done using selected dyes for polypropylene followed by vat dyes for cellulose. Polypropylene is dyed by pad-thermofix procedures in the same manner as polyester blends (see Chapter 7), with a thermofixation temperature of 138°C for high polypropylene ratio and 154°C for fabrics with only 10–20% polypropylene. The thermofixed fabrics are scoured and dried, and vat dyes are applied to the cellulose by the customary continuous vat dyeing procedures.[4]

Blends of wool and polypropylene result in fabrics with improved crease retention, stability, and abrasion resistance. The metallizable dyes stain wool; the dyes that give minimum stain are selected for their cross-dyed effects. Excessive stain may mar these effects or may create a different color on the wool which makes union dyeing difficult. The stains usually have an adverse effect on fastness properties. A two-bath dyeing procedure, though expensive, offers assurance of optimum results. After the normal prescouring procedure (as is used for wool), the polypropylene is dyed, the dyebath is then dropped, and the wool cleaned by scouring with a liquor containing nonionic detergent (0.5%) at pH 8.5 (with ammonia) at 75°C for 15 min. After rinsing, the wool is overdyed as desired.

Polypropylene fabric may be dried on a stenter frame at a temperature up to 125°C. Heat setting may be carried out at 132–138°C for 1–3 min depending on the weight of the fabric and the type of machine. (The temperature is proportionally higher for blends but is still below the softening point of polypropylene (160°C).[4]

Present Status of Coloration of Polypropylene. At present, no satisfactory method except mass coloration is available for the coloration of polypropylene. Although various modified polypropylene fibers are under investigation, they do not have satisfactory mechanical properties with good dyeability.

11.3 DYEING OF POLYVINYL ALCOHOL FIBER

The dyeing of PVA is essentially similar to that of cellulosic fibers since these fibers contain free hydroxyl groups, a part of which is blocked with formaldehyde during the production of these fibers to make the latter water-insoluble. The dyeability of these fibers depends on the extent of formalization.[19] They can be dyed with typical dyes for cotton such as direct, vat, sulfur, reactive, and azoic dyes. These fibers can also adsorb disperse and metal-complex dyes.[20,21]

Dyeing with Direct Dyes.   The unblocked free hydroxyl groups in the PVA are the source of its dyeability with direct cotton dyes. The accessibility of the PVA is decided by the extent of orientation and crystallinity of the chain molecules; it is much less than cotton. A number of dyes fail to give satisfactory dyeing on PVA. The constitution of direct cotton dyes, the number of sulfonic acid groups and amino groups, the complexity, and the molecular weight affect the dyeing of PVA. Direct dyes with simple structure and low molecular weight are particularly suitable for dyeing PVA. The method of dyeing PVA with direct dyes is the same as that for cotton except that the maximum dyeing temperature is less than 90°C. The feel of the fabric turns harsh above 90°C. The fastness properties of dyed goods are similar to those of dyed cotton and can be improved by the usual aftertreatments given to dyed cotton.

Vat Dyes.   These dyes have a markedly lower degree of substantivity on PVA than on cotton. Only pale to medium depths of shade can be dyed with these dyes. Dyeing methods are similar to those for cotton but the temperatures can be higher (up to 80–90°C). Oxidation of leuco-vat dyes on PVA is slow and is carried out with hydrogen peroxide. The vat-acid process is also used for dyeing PVA. The dyed goods exhibit excellent all around fastness properties.

Sulfur Dyes.   The dyeing method is the same as that for dyeing cotton but the temperature is higher (80–90°C). Since the saturation value of dyes on PVA is low, it is difficult to produce full black. The wash and light fastness is good but the chlorine and crock fastness of the dyeings in deep shades are not satisfactory.

Azoic Dyes.   PVA goods are worked in an alkaline bath containing both diazo and coupling components at 70–80°C. The color is developed in a cold bath (10–12°C) containing sodium nitrite and acetic acid (or formic acid or inorganic acid). All depths of shade are produced by this method. The brightness and fastness properties of the dyed goods are satisfactory except the rubbing fastness which is poor.

Other Dyes.   Disperse dyes are adsorbed by PVA but the wash fastness of the dyeings is unsatisfactory which greatly limits their use. Similarly, cationic dyes have a limited affinity to PVA and poor light and wash fastness. Metal-complex dyes (2 : 1 type) have good dyeing properties and give pastel to deep black shades of good all around fastness properties. The methods of application are similar to those used for wool, although 20–30% more dye is required for a given shade on PVA. Acid dyes can give light staining to dark shades of poor wash fastness. Acid-mordant dyes are used to produce pale to medium shades which are afterchromed to get good wash fastness.

11.4 DYEING OF POLYVINYL CHLORIDE FIBER

PVC can be dyed with disperse, disperse metal-complex, and cationic dyes. The dyeing temperature is restricted to 60–70°C since the fiber has a low softening point and it shrinks up to 50% in boiling water. Carrier has to be used to increase the rate of dyeing and the exhaustion of disperse dyes. Chlorinated benzene, ethyl acetate, butyl salicylate, and dibutyl phthalate are recommended as carriers.[22] The PVC can be stabilized by a heat treatment in the relax state and such fibers can then be dyed with disperse dyes at boiling without a carrier.[23,24] Selected disperse 2 : 1 metal-complex dyes can also be used but their light fastness is not satisfactory. In this respect, cationic dyes are better and the shades are brighter.

11.5 DYEING OF POLYURATHANE FIBER

Acid, metal-complex, and chrome dyes can be used for dyeing polyurathane fibers. The dyebath is set at pH 4–5 with acetic acid and nonionic wetting agent (1%) apart from the dye. The goods are entered at 40°C and the temperature is slowly raised to boiling in 1 h. The bath is drained after dyeing, and the goods are washed and soaped. An aftertreatment with tannic acid–tartarematic is given to improve the fastness properties of heavy shades. In the case of chrome dyes, the chroming is done in the exhaust bath with 1–3% sodium dichromate and formic acid at pH 4–5 at boiling for 30–45 min. Sodium thiosulfate is then added to reduce the dichromate ions and the treatment at boiling is continued for 30 min. If the dyebath is not exhausted, the treatment is given in a fresh bath. After the chroming treatment, the goods are scoured with 0.5–1% nonionic detergent and 1–1.5% tetrasodium pyrophosphate at 70°C for 15 min. Pale shades are produced by dyeing at boiling using a disperse dye and an anionic dispersing agent at pH 6–7.

11.6 DYEING OF CELLULOSE ACETATE

The original disperse dyes developed for cellulose acetate suffered from three major drawbacks: (a) they were volatile, (b) they were affected by nitrous oxides, and (c) the dispersion was unsuitable for package dyeing. All these limitations have been overcome and the new disperse dyes are the only dyes that are used for dyeing cellulose acetates.[25–27]

Cellulose acetate is dyed with disperse dyes from a solution of a nonionic or anionic dispersing agent at pH 6.5–7. The goods are entered at 40–50°C, and the bath is heated up to 80°C in 30 min and is maintained at 80°C for 60 min. The goods are washed, soaped, washed, and dried.

The disperse pinks, violets, and blues with anthraquinonoid structure were found to have poor fastness to oxides of nitrogen liberated from gas heaters and furnaces. A great deal of work has been done to overcome this problem, which has largely been solved by the development of new dyes. Protective compounds have also been developed that can be incorporated in the spinning dope or applied before or after dyeing. The wash fastness of disperse dyes increases with increasing molecular complexity and, depending on the end-use requirement, satisfactory dyes can be selected from the range of dyes. The light fastness of the disperse dyes on cellulose acetate is good. The sublimation fastness also depends on the molecular weight and structure of dyes. For navy blue and black shades, disperse dyes that can be further diazotized and coupled with 3-hydroxy naphthoic acid are used. After dyeing, the diazotizeation is carried out at 20°C in 30 min by treating the goods with nitrous acid ($NaNO_2$, 2 g/liter; HCl, 30%, 6 ml/liter). After rinsing, the goods are treated with a solution of a coupling agent at 60°C for 30 min. The goods are then washed and soaped (see Chapter 7).

Because of low affinity and the danger of saponification under strongly alkaline conditions, vat dyes are applied to cellulose acetate by only the vat-acid method.[26] The vat acids—leuco compounds of vat dyes—have a considerable affinity for cellulose acetate and may be applied to the fiber virtually as disperse dyes. The dye is vatted with caustic soda and hydrosulfite in a normal manner. A dispersing agent is then added. The prepared vat-dye solution at 50–60°C is poured into water containing acetic acid (amounting to 2.2 times the weight of caustic soda used for vatting). Dyeing is carried out for 60–90 min at 70–80°C.

Solubilized vat dyes produce pale shades of good fastness properties on cellulose acetate.[27] The dye is applied in the presence of 2–3 g/liter anionic dispersing agent, 25–50 g/liter common salt, and 1–2% reducing agent (Formusul) to prevent premature oxidation. The temperature is raised to 80°C and 1 ml/liter sulfuric acid is added. The dyeing is carried out for 20 min more. This is followed by the development at 40°C for 15 min in a solution containing 5 ml/liter sulfuric acid, 1 g/liter sodium nitrate, and 1 g/liter nonionic dispersing agent. Final soaping completes the process.

Acid dyes can dye cellulose acetate in the presence of a swelling agent. These dyes give bright shades of good fastness properties. However, the method is expensive as the dyebath should contain about 30–80% swelling agent. Alcohols, recorcinols, cyclohexanol, and other compounds have been suggested as swelling agents.[25]

11.7   DYEING OF CELLULOSE TRIACETATE

Cellulose triacetate fiber (CTA) is hydrophobic in nature and swells much less in water than cellulose acetate. It is nearer to polyester in its dyeing properties than to cellulose acetate. It can be dyed at boiling in 1–2 h. The rate at which disperse dyes diffuse into the fiber can be substantially in-

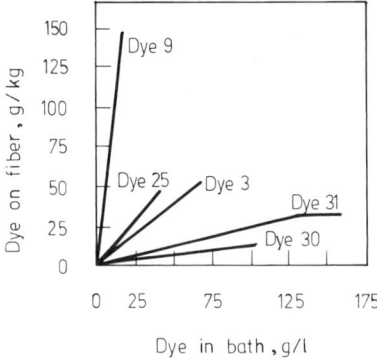

FIGURE 11.4. Sorption isotherms of nonionic dyes from perchloroethylene on cellulose triacetate.[28] (For the formulas of the dyes, see Ref. 28.)

creased by the use of a carrier; the dyeing time is shortened, deeper shades can be obtained, and better fastness properties are achieved.

Dyeing with a carrier is carried out in the same manner as that described for polyester (see Chapter 7). Similarly, high-temperature (HT) dyeing is carried out as described earlier but the temperature is lowered to 110–115°C for 60–90 min.[21]

Faulty dyeings are corrected by treatment with a carrier (dichlorobenzene type) and a nonionic leveling agent at 100°C. The dye can be stripped by treatment with 2 g/liter sodium dithionate at pH 8–9 at 60–70°C or with zinc formaldehyde sulfoxalate at pH 3–4 at 80°C. This may be followed by bleaching with sodium chlorite at 80°C.

The dyeing of cellulose triacetate from perchloroethylene at 121°C has been studied by Datye et al.[28] Sorption isotherms for nonionic dyes are found to be linear in shape (Figure 11.4). The partition coefficients of dyes on cellulose triacetate fiber are higher, the same, or lower than those on PET. The vapor-phase dyeing results for cellulose triacetate are discussed in Chapter 3 (see Table 3.1), where it is seen that the dyeing properties of cellulose triacetate resemble those of PET. Thus, cellulose triacetate can be thermofixed in the same manner as PET (see Chapter 7).

REFERENCES

1. Bird, C. L., and Patel, A. M., *JSDC,* **84** (1968), 560.
2. Ward, D., *Text. Month,* (Dec. 1974), 63.
3. Gardner, K. L., and McNally, G. M., *JSDC,* **93** (1977), 4.
4. Metropolitan Section, AATCC, *ADR,* **54** (1965), 107.
5. Anderson, N. L., Dawson, R., Sievenpiper, F. L., and Stright, P., *ADR,* **52**(2) (1963), 48.
6. Farber, M., *ADR,* **55** (1966), 536.
7. Maury, L. G., *TCC,* **4** (1972), 143.
8. Shore, J., *Rev. Prog. Color.,* **6** (1975), 7.

9. Burdett, B. C., in *Theory of Coloration of Textiles,* Bird, C. L., and Boston, W. S., Eds., Dyers Company Publication Trust, U.K. (1975), p. 156.

10. Lee, J. G., *Dyer.,* **155** (1976), 209.

11. Ahmed, M., *Polypropylene Fibers–Science and Technology,* Elsevier Scientific Publishing Co., Amsterdam (1982), p. 100.

12. Turbak, A. F., *TRJ,* **37** (1967), 350.

13. Haynes, R. R., Mathews, J. H., and Heath, G. A., *TCC,* **1** (1969), 74; **2** (1970), 279.

14. Lee, J. G., Paper Presented at Plastics and Rubber Institute, International Conference in York, U.K., Sept. 30, 1975.

15. Levine, M., and Weimer, R. P., *TCC,* **2** (1970), 269.

16. Schork, C., *Ciba Review,* **3** (1964), 29.

17. Fumoto, I., *Sen-i Gakkaishi,* **22** (1966), 184.

18. Curtis, R. G., Dellis, D. D., and Bryant, G. M., *ADR,* **53** (1964), 380.

19. Tenabe, K., and Morimoto, O., *J. Soc. Text. Cellulose Ind. Japan,* **11** (1955), 86.

20. Nomura, S., and Tanabe, K., *JSDC,* **74** (1958), 359.

21. Hindle, W. H., *ADR,* **49** (1960), 463.

22. Shore, J. C., in *The Dyeing of Synthetic Polymer and Acetate Fibres,* Nunn, D. M., Ed., Dyers Company Publication Trust, U.K. (1979), p. 404.

23. Gord, M., *Teintex,* **26** (1961), 107.

24. Lokhande, H. T., and Kalontarov, G., *Indian J. Tech.,* **12** (1974), 154.

25. Fortess, F., *ADR,* **44** (1955), 524.

26. Olpin, H. C., and Wood, J., *JSDC,* **73** (1957), 247.

27. Campbell, B., *JSDC,* **82** (1966), 303.

28. Datye, K. V., Pitkar, S. C., and Purao, U. M., *Textilveredlung,* **6** (1971), 593.

12 | DIFFERENTIALLY DYEABLE SYNTHETIC FIBERS

Fibers with essentially the same chemical origins but different affinities for a given class of dyes are called *differentially dyeable fibers*. This difference in their affinity for dyes is usually produced by incorporating a small amount of an additional chemical during the polymerization which imparts the desired affinity. The differentially dyeable fibers (DDF) are used in producing multi-colored fabrics and carpets.[1-3]

DDF fibers have many advantages over normal fibers. Attractive, fancy color effects on fabrics made from various DDF can be produced by a simple piece-dyeing process. The same fabric can be used to produce a number of color combinations. This is impossible with fabrics made from normal fibers. This also eliminates the use of dyed yarns and, in turn, the problems connected with their production. As described elsewhere (see Chapters 6 and 7), yarn dyeing is a very costly process that gives a large volume of "seconds," entangled material, and waste. Bleeding of colors and dulling of shades on dyed yarns during the subsequent scouring and bleaching of fabrics, the limitations on hues, the high cost of winding–rewinding, a large inventory of different dyed yarns, and many other such problems and limitations are absent in DDF fabrics. Because of this and since various color effects are produced from the same raw fabric (at a late stage of processing), the orders for delivery can be attended to easily and quickly. There is a large savings in the inventory of yarns and fabrics. Furthermore, the cost of producing color effects is low, since dyeing in a single bath can produce such effects. The limitation of DDF is that the property of special dye affinity is usually imparted during the manufacture of the fiber. Once the DDFs are blended, the processor cannot modify the dyeing properties to produce color effects that were not envisaged during blending.

12.1 CATIONIC DYEABLE POLYESTER (CD–PET)

The first cationic dyeable polyester, Dacron T-64 (DuPont), was introduced in 1962. Within a few years, the production of the cationic dyeable polyester (CD–PET) increased to about 15% of the total world production of polyester fibers.[4] Cationic dyeable polyester is now a well-established commercial product and is now marketed under various trade names,[5] for example, Dacron T-64, Fortrel 402, 404 (Celanese), and Hystron 440 (Hystron Fibers Inc.).

There are many patented processes that describe a variety of compounds that can be incorporated during polymerization so that the polyester will be cationic dyeable. Many of these compounds are alkali metal salts, for example, of 5-sulfodimethyl isophthalate, 5-sulfoisophthalic acid,[6] *m*- or *p*-phenylsulfonic acid, 2-naphthol-8 sulfonic acid, hydroquinone sulfonic acid,[7] sulfonated pyrrole,[8] 2-bromoethane phophonic acid,[9] 4-sulfophenyl 3,5 dicarbomethoxybenzene sulfonate,[10] aliphatic dicarboxylic acid,[11,12] and alkylbenzene sulfonic acid.[13,14]

Preparation. Cationic dyeable polyester is prepared by adding a compound containing an anionic group in the molecule during the preparation of the normal polyester (see Chapter 1). The addition can be done at different stages of polymerization, for example, dimethyl-sodium-5-sulfoisophthalate is added to the oligomeric (transesterified) PET rather than to the monomer mixture.[15] The DMT used for normal PET cannot be easily sulfonated and 2-sulfoterephthalic acid is not available for the addition to PET. Therefore, the nearest match of this product, 5-sulfoisophthalic acid, is widely used. A typical cationic dyeable polyester containing the above-mentioned compound is shown below

$$PET-OCH_2CH_2OOC-\text{\bigcirc}-COOCH_2CH_2O-PET$$
$$SO_3Na$$

where PET is the polyethylene terephthalate polymer.

Fiber Properties. Addition of a compound containing an anionic group disturbs the regularity of the PET polymer chain molecule which makes the structure of cationic dyeable polyester fiber less compact than that of normal PET. This is reflected in the lower mechanical properties of the fiber (Figure 12.1).[5] When 5-sulfoisophthalic acid is used, the drop in the tensile strength of PET is small and the cationic dyeable polyester thus produced has a tensile strength that is sufficient for its textile use. The lower tenacity of cationic dyeable polyester helps to reduce the pilling problem (Figure 12.2).[5] A slightly more open fine structure of cationic dyeable polyester results in higher dye uptake and deeper shades with disperse dyes than normal PET

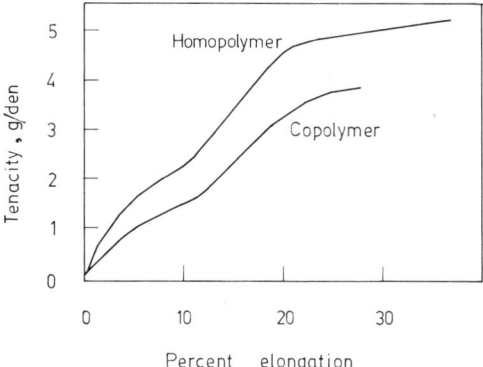

FIGURE 12.1. Tenacity and elongation (%) of PET (homopolymer) and cationic dyeable PET (copolymer).[5]

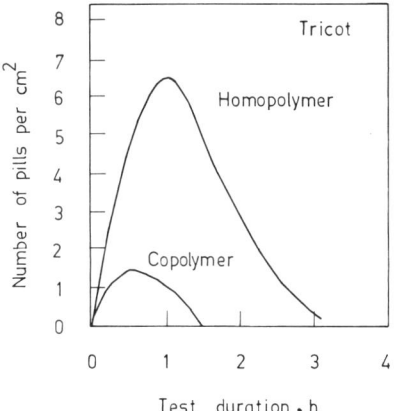

FIGURE 12.2. Pilling behavior of PET and cationic dyeable PET.[5]

under identical dyeing conditions. The uptake of typical disperse dyes on the two types of PET fibers is given in Table 12.1.[16]

In order to get desired contrast or tone-in-tone effect in blends of two types of PET fibers, it is necessary to dye a lighter shade with disperse dyes and to overdye the cationic dyeable polyester to a very deep dark shade with cationic dyes. Thus, the cationic dyeable polyester is dyed with both disperse and cationic dyes. The exhaustion of cationic dyes at boiling on CD–PET in the absence of carriers is quite low compared to acrylic fiber or cationic dyeable nylon (Figure 12.3).[5] It is essential to use the carrier dyeing or the HT dyeing process to get deep and brilliant shades (see Chapter 7).

Dyeing of CD-PET. Two important limitations of CD–PET must be kept in mind while processing these fibers and their blends. First, cationic dyeable polyester fiber is more sensitive to heat than normal PET, so that the CD–PET material must be treated below 180°C if the handle is not to be detrimen-

TABLE 12.1 Uptake (%) of Disperse Dyes by PET Fibers (100°C, 2 g Carrier/liter, 2% Shade)[16]

C.I. 'Disperse	Normal PET	CD–PET
Yellow 80	100	118
Red 11	100	120
Blue 28	100	121
Blue 56	100	117

tally changed. Second, CD–PET is subject to hydrolysis, particularly under alkaline conditions.

Cationic dyeable polyester, like any other polyester, has a strong tendency to retain any traces of oily material.[17] Therefore, it is necessary that the spinning oils be removed carefully from the fiber. The pH of the dyebath in dyeing CD–PET greatly influences the color yield, shade, exhaustion, fastness, and physical properties.[3] Generally, as the dyebath pH increases, the cationic dyes become more and more unstable at higher temperatures and decompose rapidly. This results in low color yield, alteration of shade, lower light fastness, and staining of adjacent fibers. On the other hand, when the pH is low, CD–PET is hydrolyzed and loses its strength. This adverse reaction can be checked by the addition of Glauber's salt. A pH of around 4 with a buffer of acetic acid–sodium acetate gives optimum results. Glauber's salt increases the migration of cationic dyes. It, however, slows down dye absorption and exhaustion is poor. Shade reproduction under these conditions become difficult since the dye in bath is subject to decomposition at high temperatures.

The addition of 6 g/liter Glauber's salt is sufficient to prevent fiber hydro-

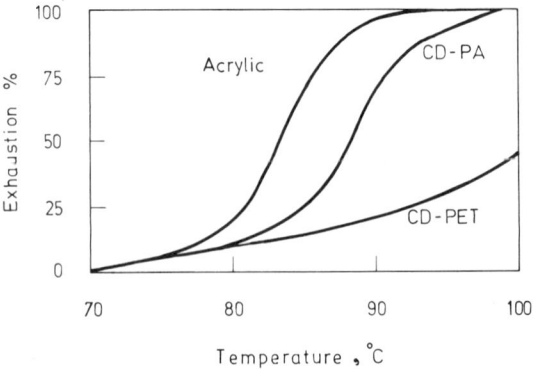

FIGURE 12.3. Dyeability of acrylic, cationic dyeable nylon (CD–PA), and cationic dyeable polyester (CD–PET).[5]

lysis during dyeing at 120°C.[1] However, when a blend with normal PET is dyed, 2–3 g/liter is quite adequate to avoid staining the normal fiber. The critical step in the mechanism of hydrolysis is the exchange of sodium ions from the sulfonate groups in the fiber for hydrogen ions.[18] The free sulfonic acid thus produced catalyzes the protonation of ester bonds.[19]

The mechanism of dyeing CD–PET with cationic dyes has been studied by Rossbach et al.[20] Fibers containing monosodium 5-sulfoisophthalic acid as a comonomer are dyed with purified C.I. Basic Blue 3 at pH 1.8 and 4.5. The adsorption isotherms at 80–100°C are of the Langmuir-type. The saturation values agree well with the sulfonic acid content at pH 1.8 and with the sum of the carboxyl and sulfonic acid groups content at pH 4.5. The kinetics and equilibrium dyeing with cationic dyes have been described.[21,22] A method for the assessment of the dyeing conditions of cationic dyeable polyester with cationic dyes using an isoreactive dyeing technique has been suggested.[23]

An Isoreactive Dyeing. In isoreactive dyeing, the dye is adsorbed by the fiber at a constant rate. Isoreactivity curves for different initial concentrations of C.I. Basic Yellow 13, C.I. Basic Red 22, and C.I. Basic Blue 41 for a total dyeing time of 40 min and using cationic dyeable polyester (Dacron T-64 of DuPont) were determined by Cegarra et al. (Figure 12.4).[23] About 20%

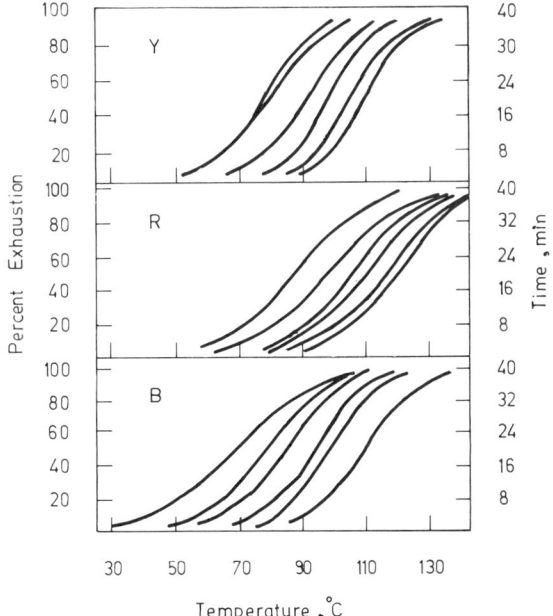

FIGURE 12.4. Isoreactivity curves for different initial concentrations of dyes.[23] Curves from left to right. Dye on wt. of fiber: 0.1%, 0.2%, 0.5%, 1%, 2% and 4%. Y: C.I. Basic Yellow 13. R: C.I. Basic Red 22. B: C.I. Basic Blue 41.

of the dye is assumed to be sorbed by the fiber without producing any visible change in the color at the end of dyeing. Consequently, dyeing is started at a constant temperature that is maintained until 20% exhaustion is reached after which it is increased in accordance with the isoreactivity curve. The time required is usually very small and of the order of 2–4 min. The initial and final temperatures become higher as the initial concentration of dye increases. For the final stage of the dyeing process, the dyebath is maintained at a constant temperature in order to fix the adsorbed dye inside the fiber.

12.2 CATIONIC DYEABLE POLYESTER BLENDS

Cationic dyeable polyester is used by itself, particularly in the field of double knits. It is blended with cotton, wool, and normal polyester to get multicolor effects.[17] It offers the possibility of dyeing dark browns and blacks economically. The blend fabrics are dyed by the simple piece-dyeing method to produce the multicolor effects.

12.2.1 PET, CD–PET, and Cellulosic Fiber Blend

Fabrics prepared using blend yarns consisting of normal PET, cationic dyeable PET, and cellulosic fibers can be dyed by the piece-dyeing method using a combination of disperse, cationic, and reactive or direct dyes to produce three-color effects.[17] Such cross-dyed shades on these blends are free from frosting. The latter is very common with such shades on normal PET–cellulosic fiber blends. The shade changes during wear at abraded places where the cotton portion of the two-component blend fabric is removed because of its poor abrasion resistance. The shade on the PET then predominates which gives the frosting effect.[24,25] The frosting becomes more evident when the normal PET is dyed a lighter shade than the cellulosic fiber. This problem is minimized by the use of three-component blend yarns. The fabric is then dyed in the following manner.[24]

The normal PET is dyed to the lightest shade, the cationic dyeable polyester to the deepest shade, and the cellulosic portion to an intermediate shade. For example, when contrasting shades of deep brown and light blue are to be produced, the blend is dyed to a light blue shade with a disperse dye (normal PET), deep brown shade with a cationic dye (cationic dyeable PET), and a composite of deep brown and light blue shade with reactive or direct dye (cellulosic fiber). Even if the cellulosic portion is lost during wear, the garment made out of such a blend fabric will not have a patchy appearance. Such blends are of particular importance in the production of durable-press garments, where the abrasion resistance of the cellulosic portion is very low.

Selection of Dyes. The cationic dyes are selected for their fastness properties and dyeing behavior (such as exhaustion and staining on other fibers).

Renard[5] has reported the fastness properties of some cationic dyes. Wygand[25] has given a list of cationic dyes suitable for dyeing blend fabrics. The disperse dyes are selected from a range with high sublimation fastness. This is particularly necessary when a durable-press finish is given to the fabric. The dyes should only minimally stain the cellulosic portion since such stains have a tendency to dull the shade and to lower the fastness properties of the dyeing.

Carrier Dyeing Method.   The three-fiber blend fabric is dyed by the two-bath dyeing method in which disperse and cationic dyes are applied in the presence of a carrier. The cellulosic portion is dyed later in a fresh bath. The selected carrier may precipitate the cationic dyes, therefore, it is preferable to use a carrier such as biphenyl.[26] The dyebath is set with

0.5–1%	Nonionic wetting agent
0.5–1%	Acetic acid to pH 5.5
$X\%$	Carrier

The fabric is treated for 10 min before adding filtered disperse dye dispersion in water. After 10 min at 60°C, cationic dye solution in hot water containing acetic acid is added. After it is run for 10 min more, the temperature is raised to boiling in 30 min and is maintained at boiling for 2 h. Cationic dyeable polyester and cationic dyes hydrolyze under hot alkaline conditions, pH above 7 is avoided,[27] and dyeing is carried out at pH 5–6. The bath is then drained and cellulosic fiber is dyed in a fresh bath using reactive or direct dyes.

In the single-bath dyeing method, the dyebath is set as described above and 2–5% sodium sulfate (anhydrous) is added before the goods are run for 10 min at 40°C. Direct dye solution is then added followed by the disperse dye dispersion. After the goods are run for 10 min, the temperature is raised to 70°C in 1 h and is maintained for 10 min before adding the solution of cationic dye in water containing acetic acid. After a further 10 min, the temperature is raised to boiling or up to 110°C if the HT-dyeing equipment is available and maintained for 2 h. The bath is cooled to 80°C before draining and the goods are washed.

HT-Dyeing Method.   The temperature of dyeing blends containing CD–PET cannot be higher than 120°C since CD–PET hydrolyzes rapidly at higher temperatures. Even below 120°C, Glauber's salt (3–5 g/liter) is added to the HT dyebath to minimize the degradation of CD–PET.[17,28,29] In a typical two-bath dyeing process in the HT-beam dyeing machine, the dyebath is set as in carrier dyeing, the material is treated for 10 min at 60°C before adding the dye dispersion and the solution of cationic dyes described earlier. The temperature is then raised to 115°C in 60 min and is maintained for 1 h. The bath is cooled below 100°C and drained. The cellulosic portion is then dyed in a fresh bath with direct or reactive dyes. A one-bath HT method is available but is rarely used.

Continuous Dyeing Method. Since three-fiber blend fabrics are produced in large volumes, continuous processing of these fabrics becomes a commercial proposition.[25] In the continuous dyeing process, the cationic dye is applied by the pad-steam process and disperse and vat dyes by the thermosol process.

The goods are padded with a liquor consisting of

X g/liter	Cationic dye
1–2 g/liter	Wetting agent
1–2 g/liter	Acetic acid to pH 5.5

The goods are then steamed for 1 min at 100–105°C, and then washed, soaped, washed, and dried.

The fabric is again padded with a liquor consisting of

X g/liter	Disperse dye
Y g/liter	Vat dye
1–5 g/liter	Antimigrating agent
1–2 g/liter	Wetting agent
	Acetic acid to pH 5.5

It is then dried at 100–110°C and thermofixed at 180°C for 1 min. The goods are then padded with a liquor consisting of

40–150 ml/liter	Caustic soda (30%)
15–30 g/liter	Hydrosulfite
20–50 g/liter	Sodium chloride

steamed at 100°C for 1 min, oxidized with hydrogen peroxide or sodium perborate, washed, soaped, washed, and dried.

In another process, cationic dye is applied by the pad-steam method described earlier. The goods are then dyed with a combination of disperse and reactive dyes by the thermofixation method. The fabric is padded with a liquor consisting of

X g/liter	Disperse dye
Y g/liter	Reactive dye
1–5 g/liter	Antimigrating agent
1–2 g/liter	Wetting agent
	Acetic acid to get pH 5.5

dried at 100–110°C, and thermofixed at 180°C for 1 min. This is followed by repadding with a solution containing

5–20 g/liter	Sodium carbonate
50–150 g/liter	Sodium chloride

The material is then steamed at 100°C for 1–2, washed, soaped at boil, washed, and dried.

12.2.2 PET, CD–PET, and Wool Blends

The dyeing of blends of normal PET, cationic dyeable polyester, and wool is complicated by the fact that the wool component is easily degraded above 110°C. The maximum temperature of dyeing is therefore 106°C. Exhaustion of disperse and cationic dyes under these conditions is not complete unless a carrier is used. Furthermore, wool is stained by disperse and cationic dyes in the presence of a carrier.[30] The addition of formaldehyde (1–3%) is recommended to minimize the degradation of the wool. Disperse dyes are selected from a range that gives minimal staining on wool under the above-mentioned dyeing conditions. Similarly, cationic dyes are selected so that the fastness to light of dyed blend fabrics is not low. Many such dyes are now marketed.

Dyeing Process. The dyebath is set with

0.5–1%	Wetting Agent
5–10%	Sodium sulfate
0.5–2%	Acetic acid
3%	Formaldehyde (40%)

The goods are run at 50°C for 10–15 min. An aqueous dispersion of a disperse dye and a solution of wool dyes are added and the goods are run for 10 min. The temperature is raised to 80°C within 10 min and is maintained for 15 min for the exhaustion of dyes on wool. Carrier is then added, after 10 min, a solution of cationic dyes is added. The bath is heated to boiling after 10 min (or up to 106°C if a pressure vessel is available) and dyeing is continued for 2 h. The bath is dropped at 80°C and the goods are washed, soaped, washed, and dried. The advantages and limitations of various dyeing methods have been described in Chapter 7.

12.3 CARRIER-FREE DYEABLE POLYESTER

The steep increase in crude oil prices has substantially increased the cost of dyeing by the carrier method since most of the carriers are obtained from crude oil. The imposition of water pollution regulations has created the need for a dyeing process without a carrier. For normal polyester, dyeing is a long process that consumes considerable energy. Carriers add to the water pollution since many good carriers are not biodegradable (see Chapter 7). In the absence of carriers, it is essential to use the costly HT dyeing method. Many fibers in blends are damaged under HT conditions. PET fiber that can be dyed at boiling without a carrier may be a solution to these difficulties.

12.3.1 Commercial Fibers

Many such fibers are now commercially produced.[31–36] The basic structure of the fiber remains the same as that of normal PET. However, the internal structure is less compact. The fiber has low pilling and soiling tendencies. Typical commercial fibers are Diolene 742 and Diolene 42 (Enka Glanzstoff AG),[32] Kodel V (Eastman Chemical Products Co),[33] F11 type 405 (Celanese Fiber Marketing Co.),[34] Trevira 210, 310, 630, and so on (Hoechst).[35,36] These fibers do not differ significantly in density or melting point from normal PET. However, their glass-transition temperature T_g (see Chapter 2) is about 10°C lower. The diffusion of the disperse dyes in these fibers is faster and hence, they give a deep shade at boiling, even in the absence of carriers.

Three types of carrier-free dyeable polyester fibers have been developed. One is based on polybutylene terephthalate and is particularly suitable for use in carpets.

$$HO—(CH_2)_4—O—[CO-\langle\bigcirc\rangle-COO—(CH_2)_4—O—]_n$$

$$—CO-\langle\bigcirc\rangle-COOH$$

Polybutylene terephthalate

The second is a polyglycol-modified fiber that is recommended for use in clothing.

$$[—CO-\langle\bigcirc\rangle-COOCH_2CH_2—O]_m \quad [CH_2—CH_2O]_nH$$

Block A   Block B

Polyethylene terephthalate–polyethylene oxide–block copolymer

The fiber is suitable for blending with wool or cotton. Since blends of carrier-free dyeable polyester fiber and wool can be dyed at boiling, the degradation of wool, which occurs when dyeing is carried out at 115–120°C, is eliminated. The blends of carrier-free dyeable polyester and cotton have a better feel than blends of normal polyester and cotton.

The third fiber is dicarboxylic-acid-modified polyethylene terephthalate fiber.

$$R—O—[CO—\underset{\text{(benzene ring)}}{\bigcirc}—COOCH_2CH_2—O]—$$

$$[CO—(CH_2)_x—COOCH_2CH_2O]_nH$$

Dicarboxylic-acid-modified terephthalate fiber

12.3.2 Dyeing

Carrier-free dyeable polyester fiber gives the same depth of shade by dyeing at boiling without a carrier as that obtained by dyeing normal polyester at 125°C or at boiling in the presence of a carrier (Table 12.2).[36]

The disperse dyes exhaust faster on carrier-free dyeable polyester than on normal polyester. The dyeing time is thus about 20% lower for the carrier-free dyeable polyester. Disperse dyes with very good sublimation fastness show a much stronger color build-up on this type than on the normal type of PET. The penetration of the dyes in the carrier-free dyeable PET is also better. The fixation of disperse dyes in the continuous thermosol dyeing method can be carried out at a temperature 20°C lower than that used for normal polyester. The time of contact is also reduced by about 40–50%. In the dyeing of blends of carrier-free dyeable polyester with cotton or wool, the cross-staining of cotton or wool by disperse dyes is substantially low. Thus, a rigorous aftertreatment to remove the surface dye from cotton and

TABLE 12.2 Comparative Depth of Shade on Normal PET and Carrier-Free Dyeable PET[36]

| Fiber | Comparative Depth | | |
	100°C with Carrier	100°C without Carrier	125°C without Carrier
1	97	—	100
2	—	100	108
3	—	105	110

Note: Fiber 1 = Trevira 220 (Normal PET)
Fiber 2 = Trevira 210 (Carrier-free dyeable PET)
Fiber 3 = Trevira 310 (Carrier-free dyeable PET)
Carrier: Orthophenylphenol (2 g/liter)

wool is not required. Because of all these advantages, the use of carrier-free dyeable polyester fiber will increase in the near future.

12.4 DIFFERENTIALLY DYEABLE POLYAMIDE

Differentially dyeable polyamides were introduced in 1960 in the carpet industry. They are also well-established in warp knitting.[37-42] Three types of differentially dyeable polyamide yarns are produced, namely, (a) normal dyeable, (b) deep acid dyeable, and (c) cationic dyeable. The affinity of these fibers for different classes of dyes is modified by manipulating the terminal groups of the polymer chain molecules.[43-46] Normal dyeable polyamide may have about 40 meq of free amino groups and about 65 meq of free carboxyl groups/kg fiber (see Chapter 1). Deep dyeable polyamide has a high number of amino groups (about 100 meq/kg) and a very low number of carboxyl groups (about 10 meq/kg). The reverse is the case with cationic dyeable nylon (100 meq COOH and 10 meq NH_2/kg). Reactive dyes and acid dyes are fixed on the basic (amino) groups while cationic dyes are absorbed on the acidic (carboxyl) groups. When nylon has a preponderance of amino groups, it becomes deep acid-dyeable nylon and exhibits a high affinity for acid and reactive dyes. When carboxyl groups are plentiful, cationic dyeable nylon with a high affinity for cationic dyes results. When carboxyl groups are absent or in a low concentration, the fiber exhibits poor uptake of cationic dyes. The behavior of different types of polyamide during dyeing can be summarized as shown in Table 12.3.[43] Since disperse dyes are nonionic in nature and are not influenced by the concentration of ionic amino or carboxyl groups, all the polyamide fibers are equally dyed. This property is used to cover faults in the blend polyamide fabrics. Such fabrics are dyed with disperse dyes to a solid shade.

TABLE 12.3 Dyeing Behavior of Different Polyamides[43]

| | | Polyamide | |
Dye	Normal Dyeable	Deep Acid Dyeable	Cationic Dyeable
Acid dyes			
Monosulfonated	2	4	0
Polysulfonated	1	4	0
Cationic dyes	0–1	0	4
Disperse dyes (nonionic)	4	4	4

Ratings: 0   Reserved
         0–1 Staining
         1–4 Dyeing with increasing substantivity

Preparation.   A chemical is incorporated during the polymerization of caprolactam or nylon 66 salt to impart the desired dyeability to the fiber. A large number of chemicals have been claimed in patents, some of which are listed in Table 12.4 and Table 12.5.[44,46] These additives act essentially as chain terminators and are present at the end of the chain molecules. The polymer thus obtained is processed in the conventional manner to produce filament yarns (see Chapter 1). The fibers are tinted for identification.

Fabrics.   The blend fabrics of polyamide are constructed by circular knitting, warp knitting, or weaving in the usual manner, taking into consideration the dyeing characteristics of individual component yarns in the fabric design. Once the fabric is woven, nothing can be done to modify the latent design knitted or woven by the combination of differentially dyeable yarns. Any unwanted mixing or wrong choice of yarns can ruin the fabric. The fabric is scoured with a nonionic detergent (1–2 g/liter) and trisodium phosphate (1 g/liter) or soda ash (0.5 g/liter) at 60°C for 30 min. If necessary, sodium hydrosulfite (0.5–1 g/liter) is added for the complete removal of the tints. The fabric is then thoroughly washed and dried before it is heat set at 180°C for 30 sec or at a lower temperature of 140–150°C if it is made of crimped yarns.[47,48] The fabric is rarely bleached. If at all required, a mild

TABLE 12.4   Additives for Deep Acid Dyeable Polyamide[44,46]

Hexamethylene diamine
Ethylenediamine, decamethylenediamine, triethylene-
tetramine
Diethylenetriamine, triethyltetramine, tetraethylpentamine
Diaminoalkylpiperazines (with phosphorus acid)
Aliphatic amines with benzenedisulphonic acids
Polyamino compounds based on *S*-triazine

TABLE 12.5   Additives for Cationic Dyeable Polyamide[44,46]

Mono- and dicarbomethoxy benzene sulfonic acid
2-Aminomethyl benzene sulfonic acid
4,4'-Diamino-2-2'-dibenzyl disulfonate with adipic acid
Sodium vinyl or styrene sulfonate
Acrylic acid, methacrylic acid
Sodium *p*-hydroxyphenyl sulfonate or 1 : 2 dihydroxy phenyl sulfonate
Sodium salt of *N,N'*-bis(*p*-sulfoethyl)-*m*-benzenedisulfonamide
Sodium-1-naphthyldisulfimide
2,4-Dichloro-6-sulfoanilino-*S*-triazine
2,4-Diphenoxy-6-sulfoanilino-*S*-triazine
2,4,6-Tris(aminoethane sulfonate)-*S*-triazine

sodium chlorite (0.1–1 g/liter)–acetic acid (for pH 4) bleaching treatment is given at 70°C. Peracetic acid alters the dyeability of yarns and is never used for bleaching.

Dyeing. The dyeing of fabrics made from differentially dyeable nylons can be carried out in winch, jigger, beam dyeing machines, or jet dyeing machines.[49-55] In textured materials, the best results are obtained on jet dyeing machines since there is no tension on the fabric during dyeing. The feel of the textured material is preserved best on jet dyeing machines. A combination of cationic dyes and acid dyes or occasionally, reactive dyes is employed to produce color effects. The usual three processes of differential dyeing are used for these fibers. In the two-bath process, the acid dyes are applied, the bath is drained, and the cationic dyes are applied. This is followed by the usual soaping and washing treatments. The process is lengthy and time-consuming. The one-bath, one-step process in which acid dyes and cationic dyes are applied in one bath under controlled conditions is more economical and time-saving. The bath is first set with a nonionic dispersing agent (to avoid mutual precipitation of acid and cationic dyes) and a buffer to get pH 6.5–7. The dissolved acid dyes and cationic dyes are added and dyeing is started at 40°C. The temperature is slowly raised to boiling in 30 min. The dyeing is carried out at boiling for 45–60 min. This is followed by the usual soaping, with 1–2 g/liter nonionic detergent at 60°C for 20 min, washing, and drying. A one-bath, two-step process is safer for getting satisfactory results. In this process, the dyebath is set with acid dyes and a buffer to get pH 6.5–7. The goods are run at 40°C for 15 min. Nonionic dispersing agent (1 g/liter) is then added, followed by a well-dissolved solution of the cationic dye. Dyeing is carried out at 40°C for 15 min. The temperature is raised to boiling in 30 min. Dyeing at boiling is continued for 45–60 min followed by soaping, washing, and drying.

The pH of the dyebath is important in the dyeing of differentially dyeable polyamide yarns. Variations in the pH affect the exhaustion greatly (Figure 12.5).[56] With regular nylon, the exhaustion of acid dyes is almost complete at

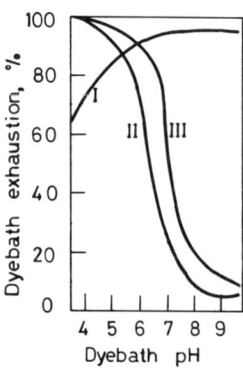

FIGURE 12.5. The effect of pH on the exhaustion of dyes on regular and cationic dyeable nylon yarns.[56] I: C.I. Basic Red 23 on cationic dyeable nylon. II: C.I. Acid Red 30 on regular nylon. III: C.I. Acid Yellow 19 on regular nylon.

low pH values (3–4) and decreases to about 10% as the pH is increased to 9. Consequently, for economical dyeing with acid dyes, a low pH is required. However, at a low pH, the exhaustion of a cationic dye on the cationic dyeable polyamide is low. It is therefore necessary to use a compromise pH in the range of 5–7. The effect of the pH of the dyebath on the dyeing of differentially dyeable nylon also depends on the number of sulfonate groups present in the dye molecules. As the degree of sulfonation increases, the concentration of dye on the deep dyeable and regular polyamide decreases at a given pH (Figure 12.6).[56] For instance, the concentration of trisulfonated dye on the regular polyamide at pH 7.0 is almost zero, which is not the case with monosulfonated dye. When these two dyes are used together, the dyeing curves are approximately superimposed, which gives a combination color of the mono and trisulfonated dyes on the deep dyeing polyamide yarn and a pale color of the monosulfonated dye on the regular polyamide yarn. Two color effects can thus be obtained.

Selection of Dyes. The selection of acid dyes for the blend of deep dyeable polyamide and cationic dyeable polyamide or normal dyeable polyamide can be made from those dyes that reserve the cationic dyeable nylon or that give high contrast on deep and normal dyeable polyamides. The acid dye is applied from a neutral bath (pH 7). For dyes that are recommended for low or medium contrast, the dyeing of acid dyes is carried out in mildly acidic bath (pH 6.5).

The combination of reactive dyes and cationic dyes gives brilliant shades and good fastness properties. Reactive dyes based on monochlorotriazine are of particular interest. Like acid dyes, the reactive dyes produce deep shades on deep (acid) dyeable polyamide and light shades on normal dyeable polyamide and reserve cationic dyeable polyamide.

The dyeing process consists of setting the dyebath with a nonionic dispersing agent and a buffer to get pH 6.5. Separately predissolved reactive and cationic dyes are added. The dyeing is carried out at 40°C for 15 min. The temperature is gradually raised to boiling in 30 min and dyeing is continued at boiling for 45–60 min. The bath is then drained and goods are scoured

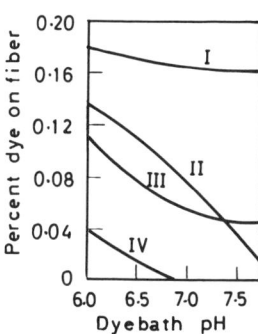

FIGURE 12.6. The effect of pH on dye uptake for mono and trisulfonated dyes on deep dyeing and regular nylon.[56] I: Monosulfonated on deep dyeing nylon. II: Trisulfonated on deep dyeing nylon. III: Monosulfonated on regular nylon. IV: Trisulfonated on regular nylon.

with 1–2 g/liter nonionic detergent at 60°C for 20 min. This is followed by washing and drying.

The various color effects produced by using differentially dyeable polyamide yarns and anionic and cationic dyes are summarized in Table 12.6.[56] For more than three colors, it is essential to use either mass-colored nylon (see Chapter 4) or some other fiber material such as PET or cellulose triacetate. Both these fibers remain white in the presence of ionic dyes. The white portions in the design can be further brightened by an optical brightening agent.[57]

Differentially Dyeable Effects by Printing Technique.   The principle of this technique is to alter the dye affinity of normal dyeable polyamide by introducing a compound that reacts with the end amino groups thus decreasing their number by the simultaneous introduction of sulfonic acid groups.[58–61] The compound may be of the following type (A):

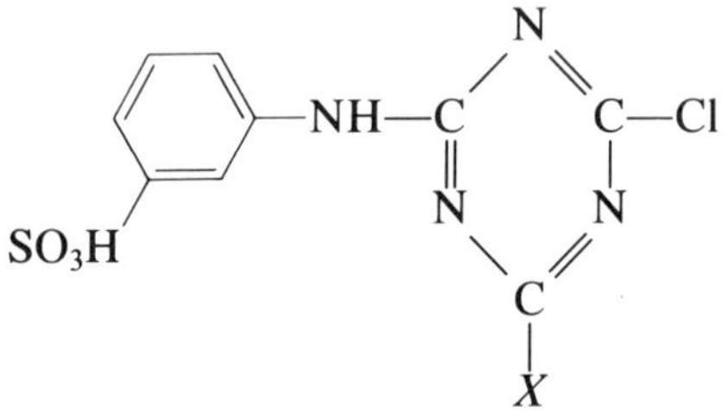

Compound A

The amino group at the chain terminal reacts with the chlorine in *S*-triazine residue to form a covalent bond to give (B).

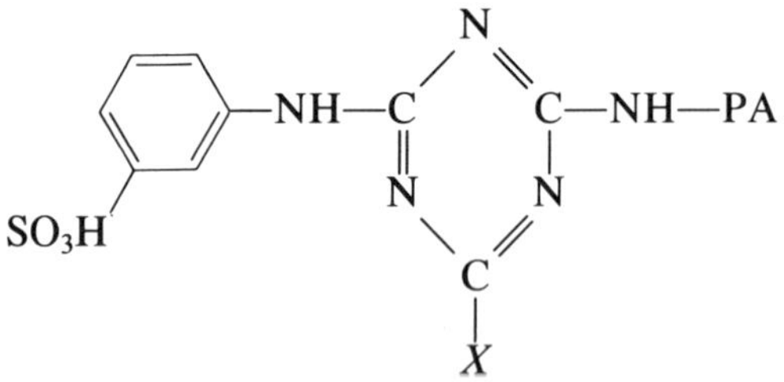

Compound B

Sandospace R (Sandoz) is such a compound. It reacts with nylon 6 under weakly acid to neutral pH conditions. The sulfonic acid groups impart affinity to cationic dyes. Thus, printed portions absorb cationic dyes and remain reserved with anionic (acid-reactive) dyes. This process does not affect the affinity of disperse dyes and the shade on printed portions. The rest would be the same if disperse dyes are used.

TABLE 12.6   Color Effects with Combinations of Differentially Dyeable Polyamide Yarns[56]

1. Normal Dyeable and Deep (acid) Dyeable

Effects: White and color, tone-in-tone, two related tones
Advantages: Low-cost yarns, 100% polyamide fabric
Limitations: Strict pH control necessary, complimentary colors not possible, deep dyeable polyamide only in pale to medium shade for white color effect.

2. Normal Dyeable and Cationic Dyeable

Effects: Any two colors, color and white
Advantages: Easy dyeing method, less strict pH control, complimentary shades possible, cheaper than combination No. 3
Limitations: Very deep shades and solid shades not possible.

3. Deep (acid) Dyeable and Cationic Dyeable

Effects: Any two colors, color and white
Advantages: Easiest combination to dye, minimum pH control, deep shades and complimentary colors possible.
Limitations: Expensive combination, darkest shade must be on deep (acid) dyeable yarn.

4. Normal Dyeable, Deep (acid) Dyeable, and Cationic Dyeable

Effects: White and two colors, three color effects
Advantages: White and two complimentary colors possible, keeping normal dyeable yarn white; white and two related colors possible, keeping cationic dyeable yarn white; two related colors and one complimentary color.
Limitations: Strict pH control, pale to medium shades if normal dyeable yarn white, limitation in depth on normal and cationic dyeable yarns.

Sandospace R is applied from a paste containing

30 g/liter	Sandospace R
60 g/liter	Urea
10 g/liter	Phosphate buffer (pH 6.5–7.5)
100–300 g/liter	Alginate thickening (4%)
0.1–1 g/liter	Lyogen V (Sandoz)

The prints are steamed at 100°C for 2–5 min. The goods are then rinsed and dried. The fabric is now ready for dyeing, either with acid dyes alone or with the combination of acid dyes and cationic dyes, acid dyes and disperse dyes, or mixtures thereof. It is claimed that this process is more versatile, cheaper, and simpler than the conventional methods of using differential dyeable polyamide yarns.

12.5   DIFFERENTIALLY DYEABLE ACRYLIC FIBERS

Three types of acrylic fibers are used by piece dyeing for the production of multicolored effects.[1,62–67] The regular dyeable fibers have normal substantivity for cationic dyes and disperse dyes and virtually no substantivity for acid dyes. The acid dyeable fibers have good substantivity for acid dyes, normal substantivity for disperse dyes, and virtually no substantivity for cationic dyes. The modacrylic fibers have varying substantivity for cationic and disperse dyes and usually no substantivity for acid dyes. Various combinations of these three types of fibers and appropriate dyes are used to produce color effects.

Selection of Dyes.   In general, regular dyeable and modacrylic fibers exhibit similar fastness properties with cationic dyes under neutral dyeing conditions. But the regular dyeable fiber is dyed darker than modacrylic fiber and at a very low pH, the entire cationic dye is taken up by the regular dyeable fiber, leaving the modacrylic fiber almost completely reserved. Thus, it is possible to produce color and white effects using regular dyeable and modacrylic fibers.

The acid dyeable acrylic fibers are dyed at a dyebath pH of around 2 for proper exhaustion of acid dyes. Consequently, acid dyes that are stable at such a low pH are selected. The selected acid dyes should give minimum staining of cationic dyeable acrylic fiber. The light and wet fastness properties of disperse dyes on acid dyeable and modacrylic fibers are generally poor and these dyes are usually used for pale and medium shades only.

Dyeing by One-Bath Method.   The dyebath is set at 50°C with

0.1–0.5%	Wetting agent
2–5%	Sodium sulfate anhydrous
3%	Sulfuric acid to pH 2–2.5
$X\%$	Acid dye

The goods are treated for 15 min and the temperature is raised to 70°C before adding the cationic dye and a retarder. After the bath is run for 10 min, the temperature is further raised to boiling in 60 min and is maintained at boiling for 1–2 h. The bath is cooled slowly before dropping. The goods are washed, soaped, washed, and finished with a softener.[63] Both acid and cationic dyes have poor migration properties and create leveling problems if proper care is not taken to follow the dyeing program from the beginning of dyeing.

REFERENCES

1. Liddiard, A., *Rev. Prog. Color.*, **1** (1970), 74.
2. Haigh, D., *Rev. Prog. Col.*, **2** (1971), 27.

3. Beutler, H., and Bohnert, E., *TCC*, **1** (1969), 574.

4. Sumitomo Chemical Co., *Japan Text. News*, **257** (1976), 99.

5. Renard, C., *MTB*, **54** (1973), 1328.

6. DuPont, USP, 2,895,986 (1959).

7. Toray Industries, BP, 1,214,087 (1968).

8. Celanese Corporation, USP, 3,661,504 (1970).

9. Goodyear Tire and Rubber Co., USP, 3,661,856 (1971).

10. Eastman Kodak Co., USP, 3,528,947 (1968).

11. Fischer, P., Drechsier, H., et al., *Faserforschung Textiltechnik*, **29**(6) (1978), 418.

12. Berger, N., Flath, H. J., et al., Proceedings 10th Congress IFATCC, **1** (1975), 111–38.

13. Allied Chemical Corp., USP, 3,705,878 (1972).

14. Gulrajani, M. L., *Colourage*, **22**(14) (1975), 23.

15. Cleary, J. W., USP, 3,922,250 (1975).

16. SPRC, Kota, (unpublished work).

17. Abrahams, D. H., *ADR*, **59**(10) (1970), 36.

18. Ingamells, W., Lilov, S. H., et al., *JAPS*, **26** (1981), 4087.

19. Burgert, L., Prikryl, J., et al., *Faserforschung Textiltechnik*, **29**(6) (1978), 426.

20. Rossbach, V., Muller, H., and Nissen, D., *Textilveredlung*, **9** (1974), 339.

21. Renard, C., *Teintex*, **36** (1971), 845.

22. Beckmann, W., and Hamacher, H., *Chemiefasern*, **23** (1973), 436; *Bayer Farben Revue*, **22** (1972), 22.

23. Cegarra, J., Puente, P., Valldeperas, J., Posse, R., and Carbonell, J., *JSDC*, **98** (1982), 151.

24. Vaidya, A. A., and Chipalkatti, V. B., *Text. Dyer Printer*, **7**(6) (1974), 33.

25. Wygand, W., *TCC*, **3** (1971), 96.

26. Tiefenbacher, E., *Textilveredlung*, **5** (1970), 100.

27. Strelecky, M., *Textilveredlung*, **5** (1970), 9.

28. DuPont Tech. Bull. **D 191** (April 1967).

29. Hoffman, D. E., *ADR*, **58**(19) (1969), 50.

30. Wygand, W., *TCC*, **1,** (1969), 447.

31. Burnthall, E. V., and Lomartire, J., *TCC*, **2** (1970), 218.

32. Hurten, J., and Marxmeir, H., *Text. Prax. Int.*, **33** (1978), 712.

33. Eastman Chemical Products, *ADR*, **63**(6) (1975), 28.

34. Moussalli, F. S., and Brown, C. L., *TCC*, **3** (1971), 202.

35. Braun, P., Muller, S., Osterloh, F., and Zimmermann, H., *Colourage Annual* (1978), 92.

36. Muller, S., *MTB*, **62** (1981), 795.

37. Schork, D., *Textilveredlung*, **2** (1967), 650.

38. Egli, H., *Textilveredlung*, **2** (1967), 856.

39. O'Mahoney, G. M., *ADR*, **57** (1968), 853.

40. Kuhn, R., *Dyer.*, **144** (1970), 446; **144** (1970), 456.

41. Dowson, T. L., *Rev. Prog. Color.*, **2** (1971), 7.

42. Dorset, B. C. M., *Text. Mfr.*, **97** (1971), 111.

43. Vaidya, A. A., Ravishankar, S., and Agnihotri, V. G., *Colourage*, **22**(22) (1974), 31.

44. Gulrajani, M. L., *Colourage*, **21**(8) (1974), 25.

45. McGregor, R., *TCC*, **9** (1977), 98.

46. Vaidya, A. A., *Colourage Annual,* (1978), 67.

47. Rush, J. L., and Miller, J. S., *ADR,* **58**(4) (1969), 37.

48. Warwicker, J., *JSDC,* **86** (1970), 303.

49. Lux, E., *Text. Prax.,* **22** (1967), 650.

50. Egli, H., *ADR,* **57** (1968), 1099.

51. Tiefenbaher, C., *Textilveredlung,* **5** (1970), 100.

52. Newton, E. J., *Dyer,* **146** (1971), 447.

53. Perrig, G., *MTB,* **52** (1971), 101.

54. Narker, R. K., and Narker, A. K., *Text. Dyer Printer,* **4**(9) (1971), 44.

55. Muller, H., and Rossbach, V., *TRJ,* **47** (1977), 44.

56. Chantler, M. D., Partlett, G. A., and Whiteside, J. A., *JSDC,* **85** (1969), 621.

57. Vaidya, A. A., *Colour and Chemical Weekly, India, Annual* (1979), 26.

58. Kuhlamann, U., BP., 1,189,726 (1966).

59. Egli, H., and Ulshoefer, H., *TCC,* **3** (1971), 31.

60. Nair, S. S., and Sudan, R. K., *Text. Dyer Printer,* **7**(8) (1974), 42.

61. Achwal, W. B., and Nagar, M. R., *Indian J. Text. Res.,* **2**(3) (1977), 82.

62. Frey, F., and Siegrist, G., *Textilveredlung,* **1** (1966), 70.

63. Beal, R. P., *ADR,* **56** (1967), 418.

64. Schork, D., *Textilveredlung,* **2** (1967), 659.

65. Steward, S., *Can. Text. J.,* **85** (1968), 37.

66. Beautler, H., *Chemiefasern,* **19** (1969), 108.

67. Pedrick, R., *Mod., Text. Mag.,* **50** (Oct. 1969), 60.

13 | INTRODUCTION TO PRINTING

Printing, that is, localized coloring, was first employed by ancient Indians and Egyptians as a method of ornamentation and to satisfy a need for artistic expression. This basic desire for design and color has kept the printing industry growing for centuries. Among all the processes that are used to decorate textile fabrics, printing is the most important. The figures for world production of printed fabric are given in Table 13.1.[1] Out of 9.5×10^9 m^2 of broad woven fabrics finished in the United States in 1974, about 2.1×10^9 m^2 (i.e., more than 22%) fabrics were printed.[2]

Centuries of experience in printing animal and natural fibers helped to develop the initial printing processes and techniques for synthetic fibers. Novel printing techniques and processes have been developed in the last two decades, especially for synthetic fibers. Methods for printing any synthetic fiber material or its blend to get beautiful, fancy color effects are available. In this chapter, the techniques of printing are described and their application to printing polyamide and other fibers is discussed.

13.1 STYLES OF PRINTING

Textile prints are classified by their method of production. Thus, a style indicates a type of process involving particular types of mechanical operations and chemical reactions. Some of these operations may form a part of another process but when they are performed in a definite sequence they give rise to a style. The important styles of printing of synthetic fibers are the same as those used for natural and animal fibers and are described below.

TABLE 13.1 World Production of Printed Fabrics[1]

Year	Production (m^2)
1970	12.5×10^9
1975	19×10^9
1980	22×10^9

Direct Style. The color is applied directly to the fabric by printing with a paste containing dyes, thickener, and other ingredients required for the fixation of dyes. After printing, the color is fixed by steaming or a dry-heat treatment. This method is most commonly used for printing synthetic fibers and their blends. The limitations of this style of printing are:

1. It is extremely difficult if not impossible to produce perfect large blotches. With disperse dyes, speckiness is sometimes a problem.
2. It is difficult to incorporate small, illuminated colored motifs in a dark blotch without a heavy overlap onto the blotch.
3. With knitted goods that require good penetration of the print paste, it is difficult to print fine designs with sharp outlines in an accurate repeat. Even the white unprinted areas may not be distinct.

Discharge Style. In discharge printing, the fabric is first dyed by any suitable method. It is then printed with a printing paste containing a discharging agent that destroys the color. The prints are developed by steaming or dry-heat treatment. For the production of colored or illuminated discharges, dyes that are stable to the discharging agent are incorporated in the discharge printing paste.

Resist Style. In resist printing, the fabric is padded with dyes and dried carefully. At this stage, the dye is only deposited superficially and is not fixed. The goods are then printed with a resisting agent, and the dye and the resist are developed by steaming. The dye is not developed at the printed portions, thus giving white effects.

Alternatively, instead of padding, the ground shade can be produced by printing with a blotch roller or a cylindrical screen. In this case, there are two possibilities. The first is the *preprint process* in which the resist–discharge paste is printed first, the fabric is dried, and the ground shade is printed. This is followed by steaming and washing. The second method, known as the *overprint process,* consists of printing the ground shade, drying carefully, and then printing with the resist–discharge paste. This is followed by steaming and washing.

The resist–printing method (also known as *resist–discharge printing method*) is simpler and has a more universal application than the discharge printing process. Discharge printing involves the complicated procedure of dyeing the ground shade and the discharge finishing. It is generally difficult to get reproducible results in all the operations. Furthermore, complete removal of dyes after fixation is difficult and only a limited number of dyes are available for the discharge printing. Because of these reasons, the resist–printing style is most suitable for the production of fancy prints, particularly on polyester fabrics.

Other Styles of Printing.   Special techniques such as tie and dye, batik, and flock printing are also suitable for printing synthetic fibers. However, these techniques are seldom used for synthetic fiber materials. Similarly, printing of yarn is uncommon with synthetic fibers. On the other hand, new techniques such as transfer printing, carbonized style, and Brosso printing have been especially developed for synthetic fibers and their blends.

13.2   PRINTING EQUIPMENT

A variety of equipment from handmade wooden blocks to computerized rotary-screen printing machines is used in the printing industry. Some of the most commonly used equipment is described below.[3]

Block Printing.   This is the oldest and simplest method of printing. A design or part of a composite design is cut in relief on a wooden block. The block is placed on a color pad that contains the necessary printing color paste and the raised pattern on the block becomes uniformly charged with the printing color. The block is then placed on the fabric and is smartly tapped on the back, either with the fist or a wooden mallet. The colored impression is thus transferred on to the fabric. Only one color is applied at a time by a single block. Block printing gives very low production but is still in use by the small-scale producers of printed fabrics.

Screen Printing.   This is a relatively simple method of printing. It can be carried out without the use of complicated and costly equipment and gives much higher production than block printing. In its simplest form it is carried out on tables of 50–100 m in length and 0.75 m in height. The width of the table is 1.2–1.6 m. The table is made of seasoned wood, reinforced concrete, or a combination of iron frames and iron stands with wooden tops. It is usually covered with woolen felt. If the goods to be printed are fastened with pins, then the felt is covered with a cotton-backed grey cloth; however, if the goods are to be fastened with an adhesive, a waterproof resin cover is used. The registration of the design is effected by a guide rail with adjustable metal stops to fit the width of each repeat, along the side of the table. Adjustable screws or bolts fastened to the base of the screen frame make contact with the guide rail. A bracket or angle iron is fitted on the end of the screen which makes contact with the stops fastened on the guide rail. The screen is prepared by a photographic or a photochemical method using a silk, nylon, or polyester gauge mounted and stretched on a wooden or metal frame. In the photochemical method, the gauge is painted with a sensitizing solution, such as gelatin-dichromate or PVA-dichromate, and dried in dark. The pattern to be printed is painted on a transparent (tracing) paper with a dense ink: this is known as a *positive*. The positive is placed on top of the sensitized screen and the whole assembly is exposed to light. Gelatin-dichromate on the portion exposed to light becomes insoluble, while from the unexposed portion of the screen, it can be washed off by a jet of water. The screen is allowed to dry slowly and the layer of gelatine is reinforced by immersing it for 30 min in a solution containing one part ammonium dichromate and one part of 40% formaldehyde solution per 100 parts water. The screen is then washed and dried.

For screens sensitized with PVA-dichromate solution, a reinforcing treatment of 5 parts acetaldehyde, 5 parts isobutyraldehyde, and 2 parts sulfuric acid per 100 parts of solution is given. The screen is then rinsed and dried.

In the printing operation, the paste is poured in at the end of the screen frame and is transferred to the fabric underneath by drawing it to and fro across the screen with the squeegee, with two to four strokes at a time. In order to avoid marking off, one screen is printed in alternate position (1,3,5, etc.) and after the whole table is printed, the remaining positions are printed. The same procedure is followed for the subsequent colors. The fabric is dried after printing and then processed further in a suitable manner.

Automatic-Screen Printing. Most of the automatic-screen prin₁ ₁g machines consist of cast-iron machined plates with elastic backing and a continuous conveyer belt running over the elastic backing and returning to the feeding end from below the table top. On its return journey, the conveyor belt gets washed in a washing tuckle consisting of a spurt pipe, a brush, and a squeezer. The adhesives and printing colors present on the belt are thus removed. In-between the table top and belt, there can be hot pipes which serve to dry the washed belt as well as the printed fabric. At the feeding end, there is an arrangement for applying adhesive to the underside of the fabric. The screens are carried by iron frames that are automatically lowered or raised. The belt travels exactly one repeat of the design each time to an accuracy of 0.1 mm. It is possible to print a number of colors in one passage of the fabric through the machine.

The screen is lowered when squeegees start working automatically and scrap the paste once or twice or more as required across the screen. The screen goes up, the fabric moves ahead up to one repeat, the screen comes down, and this cycle of operation continues each time. When the fabric reaches the other end of the table, it moves ahead for complete drying and the conveyor belt returns to the feeding end. A typical machine is shown in Figure 13.1. There are also other modified machines being continuously developed.

Screen printing is useful in printing blotches and large repeats which is not possible by roller printing. A crush effect that is very commonly observed in roller printing is not present in screen printing because there is hardly any pressure on the fabric during screen printing. Screen printing also gives more intense and brighter prints than roller printing and shorter runs can be economically produced by screen printing. However, the production capacity of a screen printing unit is much lower than a roller printing unit and it is difficult for screen printing to imitate the half-tone and shading effect of roller printing.[3]

Roller Printing. Roller printing gives a high rate of production, technical excellence, and can be used for any style of printing with a very large number of colors. In its simplest form (Figure 13.2), roller printing consists of (1) a hollow copper roller engraved with a pattern and carried by (2) a steel shaft called a mendril. The roller (1) revolves in contact with (3) a cast-iron pressure bowl above and (4) a brush furnisher below that is partly immersed in a color paste contained in (5) a box. A sharp-edged metal blade (6) called the cleaning doctor rests on the printing roller and serves to scrape the

FIGURE 13.1. Buser flat-bed screen printing machine, Hydromag H-V (Courtesy of Fritz Buser AG.)

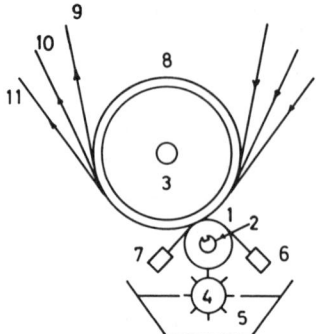

FIGURE 13.2. Schematic diagram of roller-printing machine. (For details see text.)

superficial color from the smooth part of the roller. At the opposite side, there is (7) a second doctor blade called the lint doctor that removes the loose filaments and naps that detach themselves from the fabric and stick on to the moist surface of the roller. In multicolor printing machines, the lint doctor also performs the function of cleaning the surface of the roller by removing the color that it picks up from those parts of the fabric that have already been printed by other rollers. To impart elasticity to the cylinder (3), it is covered with a fabric called (8) lapping cloth. It has also an endless (9) rubber or woolen blanket that circulates around the bowl and imparts additional spring. A (10) back grey is interposed between the blanket and (11) the fabric to be printed to prevent the blanket from being soiled by the color that penetrates through the fabric. The whole machine is worked by one large cogwheel that gears into a small pinion at the end of the mendril and is driven by an electric motor or a main driving shaft.

The engraved roller (1) first receives the charge of the color from the furnisher (4). Excess color is scraped off by the doctor (6) and what remains in the gravure lines of the engraved pattern is transferred to the fabric (11) which along with the blanket (9) and the back grey (10) passes through the machine between the central bowl (3) and the roller (1). The roller (1) then touches the lint doctor (7) where the naps and other extraneous matter are cleared off. The whole process is continued until the entire batch of fabric is printed.

There are as many engraved copper rollers working around the pressure bowl as there are colors in the design. Each engraved roller has a separate furnisher, color box, and color and lint doctors. If one of the rollers prints its part above or below its proper place, it is adjusted by momentarily increasing or decreasing its speed. Some of the modern roller printing machines have washable blankets, push-button setting arrangements and so on. A roller printing machine has certain limitations such as the size of the pattern, the number of colors to be printed, the fullness or bluminess of the print, the soiling of the bright colors by other colors from the previously printed roller, the crushing of the fabric and so on.[4]

Rotary-Screen Printing.   This method of printing is increasingly used for high-quality printed fabrics.[1-8] Thus, the number of roller and screen printing machines in the United States has slightly decreased from 1969 to 1975, whereas rotary-screen printing machines have markedly increased (Table 13.2).[2] Rotary-screen printing can be considered an improvement on flat-bed screen printing, where most of the defects are removed by sophisticated workings and by the conversion of flat screens into cylindrical screens.[9] In a typical rotary printing machine (Figure 13.3), perforated seamless circular metal (nickel) screens are used. The printing is usually effected on conveyer blanket, significantly shorter than those required for flat-bed printing, and speeds of 25–100 m/min are attained. Adhesion of fabric to the blanket is not as critical as in flat-bed printing because the tendency to lift the fabric and

TABLE 13.2 Various Types of Textile Printing Machines in the United States[2]

	Number of Machines		
Year	Roller	Flat-Bed Screen	Rotary Screen
1969	394	109	42
1970	378	108	68
1973	374	99	164
1975	345	97	179

FIGURE 13.3. Rotary screen printing machine. (Courtesy: Harish Engineering Works, Bombay, India.)

splashing of printing paste are greatly reduced. The squeegees inside the rotary screen may be the electromagnetically held rod type or the flexible-blade type (Figure 13.4). The print paste is pumped into the screens as required using automatic level controllers. A high-efficiency dryer is provided to dry the fabric immediately after printing. The changing of pattern can be achieved in about 3 min/screen because of the use of low-weight screens and simple drive couplings. The preparation of screens, their handling, and storage are easy. Rotary-screen printing machines can be used for printing all sorts of woven and knitted fabrics. Fashionable designs with fine lines, accurate fits, running stripes, and large blotches in a number of colors can be easily printed.

Comparison of Printing Machines. The relative advantages and limitations of roller, flat-bed screen, and rotary-screen printing machines are summarized in Table 13.3.

Selection of Design. The selection of a design for printing is decided by the machine to be used and the style of printing. In roller printing machines, there are severe limitations on the selection of designs. The machine can take 6–8 color rollers, but, in practice, it is difficult to produce good color effects with more than four colors because of problems of fitting of design and carryover of colors. For heavy shades and large coverage designs, dummy positions of rollers have to be introduced to minimize the soiling of colors by the carryover. A repeat of the design is limited to the size of the roller (maximum roller size = 45–46 cm). In flat-bed screen printing ma-

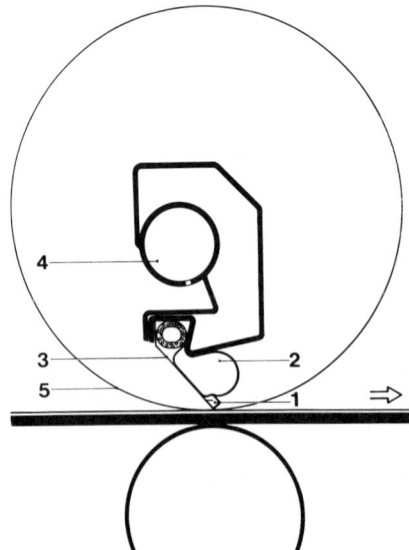

FIGURE 13.4. Stork air-flow squeegee. 1: Long-life squeegee-tip; 2: Air-activated bellow ensures uniform squeegee pressure over the full width of the fabric; 3: Rust-proof squeegee; 4: Rigid squeegee body; 5: Rotary screen (Courtesy of Stork Brabant BV.)

TABLE 13.3 Comparison of Printing Machines

	Roller	Flat-Bed Screen	Rotary Screen
Production speed (m/min)	100–150	10–15	25–100
Minimum production range per design (m)	10,000	1000–10,000	1000–10,000
Space requirement	Small	Large	Small
Consumption of color and chemicals	Low	Medium	Low
Color depth and brightness	Fair	Very good	Good
Maintenance of white ground	Fair	Good	Good
Crush effect	High	Low	Low
Smudging	Maximum	Minimum	Medium
Dot effect	Good	Good	Very good

chines, the screen size is about 1 m and a repeat of 90 cm is possible. The design-repeat can be increased threefold by taking more than one screen for the same color. The number of colors can be more than four. Rotary screens give 64 cm (24.2 in.) repeat, even though there is an arrangement for printing larger repeat as well as for cross-border saree. (It is, however, not popular because of the high cost, complicated working, and low production.)[9] For short runs and an unlimited number of colors, table printing is most suitable. It can employ designs of a complicated and intricate nature with respect to pattern and tone. Semiautomatic and automatic-screen printing machines do not have this freedom in design and number of colors.[9]

Blotch printing is not successful on roller printing machines unless it is carried out by the discharge style. Rotary machines are well-suited for this purpose. Flat-bed machines have the problem of color spreading when the prints are wet because of the pressure of subsequent screens. With proper precautions, blotches are printed on flat-bed machines. Screen printing gives bright print effects compared to roller printing. The latter appears crushed because of heavy pressure. If fine-mesh (150–200) screens are used, sharp-line effects can be produced, even with screen printing. However, the disperse dyes used on such machines must be microdisperse or liquid brands so that they do not choke the screen. Even with all these precautions, fine-mesh screens can give choking problems. For unbroken designs with encapsulated dyes, the mesh size must be less than 120. The best sharpness among all the screen-printing machines is easily obtained by table printing. The roller printing machine cannot print horizontal lines (along the weft) because of the doctor-blade cutting effect. Vertical lines are produced on roller and rotary machines. Half-tone effects are best produced by roller machines, but can also be produced on rotary and flat-bed machines. A similar effect is partially achieved on screen printing machines by reducing the number of dots/cm. Joint marks are absent in rotary and roller machines, but can pose serious problems in table printing and flat-bed machines.

13.3 PRINTING PASTE

The printing paste may contain, apart from dyes, ingredients such as wetting agents, dispersing agents, solvents, hygroscopic agents, antifoaming agents, acids and alkalis, oxidizing and reducing agents, carriers, mild oxidants, thickeners, and binders.[10]

13.3.1 Ingredients

Dyes. The dyes are selected from a chemical class depending on the type of fiber to be printed. In general, dyes in paste form are preferred for printing because they give fewer agglomeration problems than dyes in granular form.

Wetting Agents. In the preparation of printing paste, dye is usually pasted with a small amount of a wetting agent followed by the addition of water. The function of the wetting agent is to get a smooth paste without the formation of lumps of dye powder.[10] If lumps are formed they get deposited on the fabric and produce dark specks. The wetting agent functions by lowering the surface tension of the water, thus facilitating better dissolution or dispersion of dyes in the printing paste. Turkey Red Oil was used as a wetting agent for centuries. A number of anionic and nonionic surfactants are presently available for use as wetting agents in a printing paste.

Dispersing Agents and Solvents. The printing paste is a highly concentrated dispersion or solution of a dye. It is essential to use a dispersing agent and/or a solvent to prevent dye agglomeration, crystallization, or precipitation in the printing paste. Dispersing agents compatible with the dye and other ingredients in the paste are used in the minimum necessary quantities. Solvents like diethylene glycol, cellosolve (thiodiglycol), and sodium benzyl sulfanilate (Solution Salt SV) help to keep the dye in solution. Hydrotropic compounds such as urea facilitate the dissolution process.

Hygroscopic Agents. The function of hygroscopic agents in the printing paste is to take up a sufficient amount of water during steaming. This gives mobility to dye molecules in the printed dried paste and enables them to diffuse into the fiber. Glycerine, diethylene glycol, and urea are used as hygroscopic agents.

Antifoaming Agents. Many printing pastes have a tendency to froth during printing because of the presence of wetting agents and continuous agitation by the printing roller and brush furnisher. This causes the paste to overflow onto the floor or into another color box. The print becomes specky and lighter in shade. Frothing can be reduced by adding antifoaming agents to the printing paste. Benzene, pyridine, turpentine, and some silicone compounds are employed as antifoaming agents. (Benzene, being carcinogenic, should be avoided.)

Acids and Alkalies. Depending on the class of dyes used and the type of fiber to be printed, acid or alkali is added to the printing paste. The acids commonly used in printing paste are organic acids such as citric acid, lactic acid, acetic acid, formic acid, and tartaric acid. Strong acids are used as are their ammonium salts, such as ammonium chloride, sulfate, nitrate, sulfocyanide, and dihydrogen phosphate. These salts liberate corresponding acids after ammonia evolves out during steaming or dry-heat treatment. The alkalis used in printing include sodium or potassium hydroxides, carbonate, bicarbonate, or silicate, ammonium hydroxide, and triethanolamine.

Oxidizing and Reducing Agents. Oxidizing and reducing agents are required for printing certain classes of dyes and in discharge and resist styles of printing. The most commonly used oxidizing agents are chlorates, chromates, dichromates, nitrates, nitrites of sodium and potassium, and potassium ferrocyanide.

The commonest reducing agents are those based on sodium hydrosulfite. Sodium hydrosulfite decomposes and gets deactivated at temperatures prevailing in pressure steaming. Its derivative—sodium sulfoxylate formaldehyde—which is widely used in the printing of cotton, has a limited utility in the discharge printing of synthetic fibers because of its sensitivity to acids. It is also hygroscopic and causes haloes. Hence, zinc sulfoxylate formaldehyde (water-soluble type), for example, Arostit ZET (Sandoz), is generally used for discharge printing on synthetic fibers. It is nonhygroscopic and effective at high temperatures. Because of its limited solubility in water, there is no danger of halo formation.

Thiourea dioxide is another reducing agent suitable for synthetic fibers, but it is much milder in its reducing action. It cannot be employed with success on dark grounds. A mixture of thiourea dioxide and Arostit ZET (1 : 1) is an ideal reducing agent, particularly where wool is involved.

Stannous chloride or tin salt is also used for producing discharge prints on synthetic fabrics. Stannous chloride is highly acidic and can adversely affect the brightness of some colors used in color discharges. It is therefore necessary to add sodium acetate as a buffer. Tin salt has a tendency to cause a yellowish discoloration in the printed areas. This yellowing can be minimized by adding a suitable nonvolatile organic acid like citric or tartaric acid alone or preferably mixed with sodium or potassium thiocyanate. Potassium sulfite, sodium bisulfite, glucose, and ferrous sulfate are also used as reducing agents.

Carriers. In the printing of polyester fiber, carriers are often employed to facilitate the uptake of dye by the fiber (see Chapter 7). Carriers like benzyl alcohol are employed in the printing of acrylic fiber with cationic dyes.

Mild Oxidants. In the alkaline printing pastes, a mild oxidizing agent, sodium-*m*-nitrobenzene sulfonate (sold commercially as Resist Salt or Ludigol), is incorporated to prevent the reduction of dyes during steaming. Mild oxidant is also used in discharge printing with a reducing agent. If the dyed cloth to be printed is padded with a solution of Resist Salt, it prevents the adverse effect of the reducing agent on the unprinted dyed ground during steaming. Sodium chlorate can be used to protect the brightness of prints against any reduction of disperse dyes under prolonged steaming conditions at elevated temperatures.

13.3.2   Thickeners

The main functions of a thickener in a printing paste are (a) to act as a vehicle for carrying the dye on to the fabric and (b) to prevent the spreading of the color on the fabric by capillary action beyond the limits of the defined portion in the pattern. The essential qualities of a thickener are:

1. It should have the desired physical and chemical properties such as viscosity, flow property, and ability to wet and to adhere to the internal surface of the etchings of the engraved roller.
2. The storage stability of the thickener paste must be good.
3. It should be compatible and inert to dyes and other ingredients of the printing paste.
4. It should absorb water during steaming without causing flushing.
5. It should have good thermal stability. The thickener film should not break during high-temperature steaming or thermosoling.
6. The removal of the thickener from the fabric after fixation of the prints should be easy.
7. It should be biodegradable.
8. It should be available at reasonable price.

A large number of thickeners are available for printing. The choice of thickener will depend on the class of dyes to be printed and the style of printing. The following materials are commonly used as thickeners.

Starches and Cellulose Derivatives.   *Starches.*   The starch is cooked in water to get a thick paste. This is suitable for printing pastes that are not acidic in nature. Maize starch withstands alkali better than wheat starch. The latter has a good binding quality but it imparts a harsh feel to the printed fabric. This drawback is overcome by mixing gum tragacanth with the starch.

Modified Starches.   The most commonly used starch derivative in printing is dextrin or dark British gum. It is a modified product obtained by heating starch with mineral acid or roasting it at 160°C until it becomes completely soluble in water. In the case of light British gum or yellow dextrin, the roasting process is stopped before complete solubilization takes place. The main features of these gums are their good leveling properties and solubilities, which make them easily washable after the fixation of prints. Starch ethers and esters are also used as thickeners. Many starch derivatives are readily soluble in cold water, have good storage stability, and are stable under alkaline or acidic conditions in the printing paste.

Methyl Cellulose. It is prepared by reacting soda cellulose with methyl halide. The most important property of methyl cellulose is that it is soluble in cold water but insoluble in boiling water and in the presence of alkali. The removal of methyl cellulose after the fixation of prints is thus very easy. About 5–6% methyl cellulose usually gives a workable paste. It has a good storage stability.

Carboxymethyl Cellulose. It is prepared by reacting alkali cellulose with monochloro acetic acid. CMC is stable to alkali and can be used for printing vat dyes. It gives sharp prints and can be readily washed away after the fixation of prints.

Hydroxyethyl Cellulose. It is obtained by treating soda cellulose with monochlorohydrin. It is soluble in water and gives prints with sharp outlines.

Gums. *Gum Tragacanth.* It is obtained from leguminous plants like Astragalus Gummifer as a dried-up exudate. It is soluble in water and gives a thick smooth paste. About 4–6% gum is used in the printing paste. Gum tragacanth (or Gum Dragon) is stable under mild alkaline conditions. It gives a soft handle to the fabric. Gum tragacanth works well with the starch.

Gum Arabic. It is obtained as an exudate from the Acasia plant. A 30–50% solution of the gum is used for preparing printing paste. It gives prints with sharp outlines and most level blotches. Gum arabic (or gum senagal) is stable under both strongly alkaline and strongly acidic conditions. It can be easily removed from fabric after the fixation of prints.

Gum Karaya. This gum is obtained from the Karaya tree. It is acidic in nature and relatively insoluble in water. About a 40–50% solution of gum Karaya is used in preparing the printing paste. The printing paste has good storage stability.

Nafka Crystal Gum. This is a highly purified form of natural gum with a high thickening power. It is soluble in cold water. The Nafka crystal gum paste is very stable under acidic and alkaline conditions. The gum gives clear prints.

Locust Bean Gum. It is obtained from the seeds of the locust bean of the carob tree. Only a 2–3% solution of the gum gives a print paste of required viscosity. Derivatives of locust bean gum are extensively used in the printing of synthetic fibers. These gums (marketed as Indalca gum) are nonionic and exhibit an outstanding compatibility with dyes over a wide pH range. However, the supply of locust bean gum is very limited because it originates from a slow-growing tree. The price of these gums is therefore quite high.

Guar Gum.   Locust bean gum and guar gum have closely related properties, both being galactomannans. Guar gum was developed in 1953 as a substitute for locust bean gum.[11-15] The supply of guar gum is abundant because the gum originates from an annual Guar plant. Its capacity of giving high viscosity pastes is unique and generally, a 1–3% solution of guar gum gives satisfactory viscosity. The paste is stable for many days. However, guar gum is generally modified to improve its flow properties and dispersibility. The type of reactions that are commonly used to modify guar gum include etherification, esterification, oxidation, reduction, and formation of cyclic derivatives.

Derivatives of guar gum are extensively used in the printing of synthetic fibers. The Meyprogums belong to this class. These gums are nonionic in nature and are stable over both acid and alkaline conditions. The gums are unaffected by the presence of electrolytes in the paste.

Sodium Alginate.   The sodium salt of alginic acid is extracted from seaweed.[16] It is soluble in water and hence, preparation of its paste is easy. About a 4–6% solution of sodium alginate gives a paste with satisfactory viscosity. Because of its good adhesion properties, the dry film of sodium alginate does not get detached from the fiber or cracked in the mechanical handling of the dried fabric. Sodium alginate has good wetting power and thus gives better penetration of the dye in the fabric. This makes sodium alginate particularly suitable for the printing of hydrophobic synthetic fibers. Removal of sodium alginate from the fabric after printing is easy and it gives fabric with soft hand. Sodium alginate does not react with reactive dyes and hence, is an ideal thickener for printing these dyes. Being anionic in nature, sodium alginate is unsuitable for printing with cationic dyes or in the presence of cationic surfactants. It may get precipitated with acids. Sodium alginate is not compatible with soluble calcium salts or with salts of heavy metals such as iron, zinc, and chromium. If these salts are present, it is necessary to add a sequestering agent in the print paste. The major problem with sodium alginate is its limited availability and high price.

Emulsion Thickeners.   Emulsion thickeners came into existence with the rapid development of pigment colors. However, they are also employed for printing other classes of dyes.[17,18] In an emulsion system, there is a dispersed phase and a continuous dispersing medium. When very stable emulsions are required, emulsification is brought about in the presence of a third component called the *emulsifier*. For textile printing, oil-in-water types of emulsions are used. Kerosene–water emulsions have also been widely accepted. Different types of emulsifiers are available such as anionic, cationic, and nonionic; the nonionic type of emulsifier is preferred and alkyl-phenol-ethylene oxide condensates have proved most useful.

It is possible to get printing pastes with the desired viscosity with these thickeners at very low solid contents. Better penetration of dyes and higher color yields are therefore obtained. These thickeners dry very rapidly after printing because of the low solid content and the volatile nature of the oil phase. This characteristic is an important advantage in screen printing. Materials printed with emulsion thickening are very soft after drying and are much less liable to crack or mark off. Removal of thickener after printing is very easy and thus the handle of the fabric is soft. The drawbacks of the emulsion thickening are (*a*) considerable fire hazard, (*b*) serious problem of air pollution, and (*c*) increasing price of hydrocarbon oils.

Synthetic Thickeners. The impetus for the development of synthetic thickeners has come partly to augment the continuously dwindling supply of natural gums and thickeners and partly because of the substandard rheological properties of natural thickening agents and their modified products. The need to replace mineral oil in the pigment printing emulsions and sodium alginate in the printing of disperse and reactive dyes are also cause for this development.[19-21] With the development of sophisticated machines like the rotary printing machines and the need for printing photographic designs using fine lines and dot screens, the demand for consistent purity and rheological properties of the thickening agents has become very severe.[17] Apart from the technical properties, ecological factors such as biodegradability and toxicity have to be considered. In this respect, however, natural thickeners and gums have an advantage.

Synthetic thickeners are long chain polymer derivatives obtained from substituted vinyl compounds that are easily polymerized by a free-radical mechanism or by the addition of an ionic substance. They are compatible with disperse, reactive, acid, and cationic dyes. The paste of synthetic thickeners can be prepared in 5–10 min and there is no need to add any preservative because the paste has good storage stability. Synthetic thickeners give a paste of suitable viscosity at low concentration and give better definition of the prints, even at low solid content. Synthetic thickeners have good thermal stability and can be used for all types of fixation processes such as high-temperature steaming and thermosoling. They usually give about 15–20% higher color yield compared to that of natural thickeners. Removal of these thickeners after the fixation of prints is very easy.

Synthetic thickeners suffer from two major drawbacks: (*a*) they are not biodegradable, and (*b*) their viscosity severely decreases in the presence of electrolytes. Since most of the conventional dyes contain electrolytes, it is necessary to develop special electrolyte-free dyes.

Pectin. Pectin is a polyuronide that occurs in all plant tissues. In some cases it accounts for as much as one-third of the dry matter as in the peels of citrous fruits. Pectin has a low viscosity and does not react with reactive dyes. The feel of the fabric is not affected by pectin. Since it is obtained from agricultural waste, a regular supply is ensured and it is quite cheap. Pectin gels at pH above 9 and hence, cannot be used under highly alkaline conditions.[18]

13.3.3 Binders

A binder used in pigment printing is a substance that can form a film on the printed portion. The fastness properties of a binder film largely determine the fastness of the pigment prints and the quality of a print thus depends on the quality of the binder. The binder film formed on the fiber surface must be colorless and clear, of even thickness, smooth, and neither too hard nor too soft. It should be elastic and have good adhesion to the substrate without being tacky. Furthermore, it should have good resistance to mechanical and chemical stresses, but should be readily removable from the printing rollers, screens, back-greys, and blanket during the printing operation or shortly afterwards.[19]

Albumin, casein, and glue were used earlier as binders-cum-thickeners. They were replaced by cellulose esters such as cellulose acetate and cellulose nitrate. The fastness properties of all these binders are not very good. A large number of synthetic binders have been developed, the most important among them being vinyl resins, acrylic resins, melamine-formaldehyde and urea-formaldehyde precondensates, and chlorinated rubber. Helizarin Binder UD (BASF) is formed from acrylic esters with styrene and gives prints with excellent fastness properties. Helizarin Binder NTA (BASF) is a butadiene copolymer. The newest development in the field of pigment printing is the production of substances that combine the functions of binders, thickening agent, white spirit, emulsifier, leveling agent, and the catalyst. The typical trade products are Lutexal HD (BASF) and Acramin CA 3241 (FBY).

Special binders are developed for printing polyester–cotton blend fabrics. They are based on styrene-butadiene and acrylonitrile-butadiene and have good adhesion power even for polyester.

13.4 FIXATION OF PRINTS

After printing, the goods are dried and the prints are fixed. Four methods are available for fixation of the prints on the fabric.

Steaming without Pressure. This is the oldest method of fixation of prints. It consists of treating the fabric with saturated steam at a temperature of 100–

101°C under atmospheric pressure. The method is extensively used for the fixation of acid dyes on nylon and cationic dyes on acrylic fibers. In the case of polyester, it is essential to incorporate a carrier in the printing paste. Even after prolonged steaming treatment, the color yield is not satisfactory. Hence, this method is not used for the fixation of disperse dyes on polyesters.

Steaming under Pressure.   Saturated steam at a pressure of 0.5–2 kg/cm² is used for the fixation of dyes in this method. Very good fixation of prints is obtained when polyester fabric printed with disperse dyes is steamed for 20–30 min. The feel of the fabric is also not affected since there is no tension on the fabric during steaming. The pressure steaming is usually done on a batchwise cottage steamer. Fabric interleaved with a grey fabric to prevent marking off is arranged in the form of loops and suspended from wooden rollers carried on the framework of a portable wagon that is run on rails into the cottage. The wagon is charged into the cottage steamer, which is then closed, and the fabric is steamed under pressure. Continuous pressure steamers are also available for high production. For the fixation of disperse dyes on nylon, steam at 0.5 kg/cm² pressure is used for 30 min. Disperse and cationic dyes on acrylic fiber can be fixed at a steam pressure of 0.2–0.5 kg/cm² and time of steaming is 20—30 min. The method is also suitable for the fixation of disperse dyes on cellulose triacetate.

Dry-Heat Fixation.   This method of continuous fixation of prints is used mainly for polyester. The prints of disperse dyes are fixed by the dry-heat treatment at 180–200°C for 30–120 sec in a pin stenter or in other thermosoling equipment (see Chapter 7). The method gives high productivity and does not produce flushing of the prints. However, the brightness of the prints is not as good as that obtained by the steaming methods. The feel of the fabric is harsh. There is also a severe restriction on the use of disperse dyes. Dyes with low sublimation fastness stain the white ground.

Superheated Steam Fixation.   The dry steam is passed through a heat exchanger to raise its temperature from 150°C to 220°C to get dry high-temperature or superheated steam. The pressure of the steam after superheating may drop down to atmospheric pressure, and for its use for fixation, a pressure chamber is not required. Thus, the construction of a superheated steamer is simple because the problem of constructing seals for the fabric inlet and outlet which prevail in pressure steamers do not arise. Continuous superheated steamers give high productivity.[22–31] Compared with dry-heat fixation with hot air, superheated steam fixation requires lower temperatures (170–180°C) for the fixation of disperse dyes. Thus, there is less restriction on the selection of disperse dyes and dyes of medium sublimation fastness can also be used. The method is of particular interest for textured materials because low temperatures are used and there is no tension on the fabric during steaming. The addition of urea or other auxiliary such as Luprinton ATP or Luprinton HDF in the printing paste is recommended to get good color yield.

High-temperature dry steaming can also be used for fixation of disperse dyes on triacetate.[27] Acid dyes on nylon can also be fixed using high-temperature dry steaming but the color yield is slightly lower than that obtained by using saturated steam at 102°C.[27] Cationic dyes on acrylic fiber do not give good color yield when fixed by the superheated steam.[27,30]

13.5 DIRECT STYLE OF PRINTING

13.5.1 Polyamide Textiles

A considerable proportion of polyamide textiles is printed to produce fancy colored fabrics.[32–42] Barréness, which is a serious problem in the dyeing of polyamide fabrics (see Chapter 9), does not exist in printing. Dyes from different classes can be applied in conjunction with one another without any blocking effect. Acid and metal-complex dyes are usually employed for direct printing of polyamide textiles. At times, disperse dyes are also used. Acid dyes give brilliant prints and a few exhibit all around fastness properties on nylon 6 and nylon 66. It is, however, necessary to use metal-complex dyes to get very high fastness properties. These dyes usually produce dull prints.

Because of the smooth cylindrical fiber structure and the low swelling properties of polyamide fibers, sorption of aqueous solutions and pastes by the fabric is very poor. It is therefore necessary to use printing rollers with shallow engravings so that all the paste in the engraved design is taken up by the fiber. Thickening agents of high solid content are best suited for printing polyamide textiles. Crystal gum is a very good thickener. Locust bean gum (Gum Indalca), guar gum, or sodium alginate are also used. Thiodiethylene glycol acts as a solvent, a hygroscopic substance, and a swelling agent for polyamide fibers.

Printing with Acid and Metal Complex Dyes.   The printing paste consists of:

5–50 parts	Acid or metal-complex dye
50 parts	Urea
30–50 parts	Thiodiethylene glycol
45 parts	Water
200–265 parts	Boiling water
600 parts	Crystal gum thickening (1 : 3)
5 parts	Ammonium sulfate or tartrate (1 : 2)
0.1 part	Antifoaming agent
1000 parts Total	

Goods are printed, dried hard, and steamed at 103–105°C for 20–40 min. After steaming, the goods are rinsed cold and hot, and then soaped at 40–60°C with a mixture of nonionic detergent and sodium carbonate. For maximum wet fastness, an aftertreatment with a suitable agent (e.g., Erional NW or Metexil FA-SW) is recommended.

One of the main difficulties when printing polyamide with acid and metal-complex dyes is the white ground being stained during washing off. It is possible to prevent this staining of the ground, for example, by treating the goods with Cibatex PA before printing. Alternatively, Cibatex PA may be added to the washing-off liquor. The heat-set polyamide goods are pre-treated with Cibatex PA (0.5–2% on owf) at pH 3.5–4.0 (acetic or formic acid) at 80°C for 30 min, rinsed cold, and dried before printing with acid and metal-complex dyes.

Printing with Disperse Dyes.   Disperse dyes are used principally for producing prints of medium and pale shades on polyamide fabrics.[43] They have good leveling properties and hence, are of special interest for blotch prints. Disperse dyes are not suitable for producing heavy shades because they have insufficient wet fastness and do not always give the desired brightness of prints. The print paste consists of

X parts	Disperse dyes
Y parts	Water
200–300 parts	Urea
400–600 parts	Thickening
1 part	Wetting agent
1000 parts Total	

Thickening based on crystal gum, gum indalca and so on, can be used. The fixation of the prints on polyamide fiber can be carried out by any of the following three methods.

1.   *Pressure Steaming.* The printed and dried goods are steamed using steam at a pressure of 0.5 kg/cm^2 for 30 min. Higher steam pressures do not increase the depth of color while steaming at atmospheric pressure results in lower color yields.

2.   *Superheated Steaming.* This method gives the best color yields and a pleasant feel to the fabric. The unprinted white ground is not tinted. The method is of particular interest in fixing disperse dyes on textured polyamide materials, since the texture is not adversely affected during the steaming process. Fixation with superheated steam is carried out at a temperature of 160–190°C for a period ranging from 1–6 min.

3.   *Dry-Heat Fixation.* The fixation is carried out by heating the fabric to 170–200°C for 1–2 min in a pin stenter. The feel of the fabric becomes harsh on thermofixation.

Washing Off. The goods are first washed with cold water. This is followed by washing with a detergent solution at 60°C. A final washing with water completes the process. The white ground is usually stained during the washing-off process.

Printing with Reactive Disperse Dyes. Reactive disperse dyes, for example, Procinyl Dyes (ICI), are developed to get prints on polyamides with good wet fastness properties. Procinyl dyes show good compatibility in admixture with one another and have the advantage of retaining the printed mark under moist steaming conditions.

The print paste consists of

10–60 parts	Reactive disperse dye
250–300 parts	Water
10 parts	Acetic acid

The suspension is stirred into

600 parts	Thickening
10 parts	Sodium chlorate
20 parts	Perminal KB
1000 parts	Total

The thickening can be crystal gum, Indalca gum, or Maypro Gum AC. The printing paste is stable for about four weeks at 15°C. After printing, the goods are dried, steamed at 100°C for 30 min, rinsed in cold water, and soaped for 5–10 min at 50°C in 0.2% solution of a nonionic detergent. Finally, the goods are rinsed in water and dried.

13.5.2 Acrylic Textiles

Cationic dyes give bright prints with satisfactory fastness properties on acrylic textiles.[44-54] Various recipes for these dyes contain a solvent or an accelerant to increase the rate of diffusion of dye into the fiber during fixation. A retarder which is essential for dyeing acrylic fiber (see Chapter 10) is not required in the printing paste.

Benzyl Alcohol Recipe. The printing paste made according to this recipe exhibits very good wetting power, good stability, and good reproducibility of the prints. The prints are very bright and well penetrated.

The high wetting power shown by the paste enables blotch prints of perfect levelness and suppresses frosting on high-bulk acrylic yarns. The paste contains, apart from the dye and water, a crystal gum or Indalca gum thickening (500 parts), benzyl alcohol (20–50 parts), tartaric acid (1:1, 5 parts), and Luprintan PED (BASF) (10–30 parts). Anionic thickening reacts with the cationic dye and is not used. Benzyl alcohol attacks certain types of screen lacquors, pump packing-rings, tubes, and doctor blades of rubber and due account must be taken of this fact.

Recipe (ICI)

20–100 parts	Cationic dye
20 parts	Solution assistant (Matexil PN-3BN)
20 parts	Acetic acid (30%)
5 parts	Citric acid
10–25 parts	Luprintan PED (BASF)
X parts	Water
500 parts	Thickening
1000 parts	Total

After printing, the fabric is dried at temperatures of not more than 100°C to avoid yellowing of the fiber. The prints are fixed in saturated steam at 100°C or under pressure of 0.2–0.5 kg/cm^2 at 102–105°C for 20–30 min. Overheating leads to unlevel prints at the selvages. Superheated steam at 150°C for 1–2 min is sufficient for satisfactory fixation of prints. The printed and steamed fabric is then rinsed in cold water, soaped at 60°C for 5 min with a nonionic detergent, washed, and dried.

Printing with Disperse Dyes. The disperse dyes give satisfactory pale to medium shades on acrylic fibers. The maximum depth that can be reached depends a great deal on the type of acrylic fiber. Disperse dyes are easy to print on acrylic fabrics; however, the prints are not as bright as those produced by cationic dyes. The wet fastness of many disperse dyes is also not very high and hence, proper selection of dyes has to be made.

Printing Recipe

X parts	Disperse dye
Y parts	Water
500 parts	Thickening (crystal gum or Indalca gum)
1–2 parts	Wetting agent
10 parts	Resist salt
Z Parts	Carrier (optional)
1000 parts	Total

Addition of thiourea (20 parts/1000 parts paste) improves the color yield. The printed fabric is dried carefully and the prints are fixed by steaming for 30 min at a pressure of 0.5–0.7 kg/cm^2. Alternatively, superheated steam at 125–150°C for 1 min can also be used. In such a case, an addition of 100–200

parts urea, 20 parts thiourea, 30 parts Glyzine PFD (BASF), and 50 parts
Lyogen TG/1000 parts of printing paste is made. Fixation by dry-heat treatment is not recommended since the acrylic fiber turns yellow and hard at
temperatures of 190–210°C.

After fixation, the goods are washed and given a reduction-clear treatment with

1–2 parts	Lyogen DFT
2 parts	Sodium hydrosulfite concentration 85%
4 parts	Sodium carbonate
1000 parts	Total (with water)

at 50–60°C for 10 min. The fabric is then washed and dried.

Printing with Vat Dyes. Selected vat dyes can be printed on acrylic textiles
by the conventional alkali-sulfoxylate formaldehyde process. Vat-dye paste
(50–200 parts) is stirred in the thin neutral thickening (100–200 parts) which
is further stirred into thickening (500–600 parts) containing glycerine (up to
50 parts), sodium carbonate (50–60 parts), and sodium sulfoxylate formaldehyde (100 parts) to get a printing paste (1000 parts). Thickening is based on a
crystal gum. Goods are printed, dried, and steamed for 10 min at atmospheric pressure. This is followed by washing off and oxidation with hydrogen peroxide (5 parts) and ammonia (specific gravity 0.880, 3 parts) in water
(1000 parts) at 90°C for 10–15 min. The goods are then washed, soaped with
Lissapol ND (ICI), washed, and dried.

13.5.3 Acrylic Blend Textiles

Printing of Acrylic–Wool Blend Fabrics. The combination of cationic dyes
and acid or metal-complex dyes, which is most commonly used for dyeing
these blends, is not suitable for printing. This is because dyes from these two
classes interact and precipitate in the printing paste. Selected vat dyes,
which can give solid shades on acrylic and wool fiber, can be used for
printing these blends by using a special reducing agent called Manofast. This
reducing agent is acidic in nature and does not require sodium carbonate
which damages wool.

Printing Recipe

50–150 parts	Vat dye paste
100–200 parts	Water
590 parts	Crystal gum thickening
90 parts	Glydote BN
50 parts	Glyzine PFW (BASF)
60 parts	Manofast
1000 parts	Total

The printed goods are dried and steamed for 15–20 min at atmospheric pressure. Steaming is followed by rinsing in cold water and oxidation with hydrogen peroxide and ammonia at 60°C. The goods are finally soaped with Lissapol ND at 60°C followed by washing and drying.

A combination of disperse dyes with acid or metal-complex dyes can also be used for printing on acrylic–wool blend fabrics. However, the brightness of the prints is not very good.

Printing of Acrylic–Cellulose Blend Fabrics.   A combination of cationic dyes with reactive dyes cannot be used because of the danger of precipitation of these dyes in the printing paste. Further, fixation of reactive dyes require alkaline conditions, whereas the fixation of cationic dyes require acidic conditions. Vat dyes or a combination of disperse and reactive dyes give prints on these blend textiles.

Vat Dyes.   Selected vat dyes give solid shades on acrylic–cellulosic blend fabrics. The printing method is similar to one described earlier for printing vat dyes on 100% acrylic fabric.

Disperse and Reactive Dyes.   Selected disperse dyes that are stable under mild alkaline conditions are used.

Printing Recipe

X parts	Reactive dye
Y parts	Disperse dye
Z parts	Water
500 parts	Thickening (sodium alginate)
1–2 parts	Wetting agent
50–100 parts	Urea
5–20 parts	Sodium bicarbonate
1000 parts	Total

The fabric is dried carefully after printing and steamed for 30 min at a pressure of 0.5–0.7 kg/cm^2. The goods are finally washed, soaped, washed, and dried.

13.5.4 Polypropylene Textiles

Unmodified polypropylene fiber does not give satisfactory printed effects. The metal (nickel)-modified polypropylene fibers can be printed with chelating dyes to get prints with high light fastness.[55] The chelating dyes are structurally similar to disperse dyes except that they have functional groups (e.g., *o*-hydroxy imino-azo) in the proper configuration to form cyclic chelate structures with the nickel in the organonickel dye-sites.[56] During steaming or aging of the prints, a chemical reaction between the dye and the organonickel dye-sites occurs, producing the dye–metal–chelation product. The fastness of these prints is the result of the irreversible chemical bonding during chelation.

The printing paste consists of a thickener such as Polygum 260 (Polymer Ind. Inc., U.S.A.), a carrier (to get better crockfastness), acid for pH control at 4.0–5.5, a sequestrant, and water.[55] The goods after printing are dried, aged at 102–106°C, scoured to remove thickener and other ingredients, washed, and dried. Judicious selection of proper dye-sites and print paste is necessary to get maximum brightness, color value, and fastness properties.

13.5.5 Cellulose Acetate Textiles

Disperse dyes are most widely used for printing both cellulose acetate and triacetate fibers. They give sharp, bright prints with good fastness properties. Heat setting of triacetate is carried out after printing. A typical recipe for printing acetate fiber is given below:

Stock Color

20–70 parts	Disperse dye
320–270 parts	Water
500–600 parts	Thickening
50 parts	Albatex BD (10%) (Ciba–Geigy)
10 parts	Silvatol 1 (Ciba–Geigy)
1000 parts	Total

Albatex BD (sodium metanitrobenzene sulfonic acid) is used to prevent reduction during steaming. Silvatol 1 is a leveling agent that also prevents foaming. Addition of propylene carbonate helps to solubilize and increase the dispersion of dyes, thus, intensifying the prints. The addition of urea improves fixation by HT steaming and is sometimes included in the paste.

Reduction Thickening

370 parts	Water
600 parts	Thickening
25 parts	Albatex BD (10%)
5 parts	Silvatol 1
1000 parts	Total

The most suitable thickening is a mixture of gum arabic (1:1) and crystal gum (1:2) in the ratio 1:1. It is also possible to mix this thickening with locust bean derivatives or British gum. Emulsion thickening can also be used.

After printing, the cellulose acetate goods are steamed with saturated steam (100–102°C) for 45–60 min. In the case of cellulose triacetate, steaming at a pressure of 1–1.5 kg/cm^2 for 30 min gives good results. Dry-heat fixation at 180–200°C for 30–60 sec can also be done with disperse dyes with good sublimation fastness. Use of superheated steam at 170–180°C for 2–4 min is suggested to give good results.[57] When dry heat or superheated steam is used, it is necessary to select a thickener that does not turn yellow and is easy to wash. Sodium alginate, crystal gum, and modified locust bean gum give good results.

Printing with Acid Dyes. The procedure is to print the acid dye paste containing urea (150 parts/1000 parts) or thiourea (30 parts/1000 parts) as a swelling agent. Thiourea lowers the strength of the fabric to a certain extent. The prints obtained have excellent fastness to washing and light. Fastness to burnt gas fumes varies from dye to dye and many anthraquinonoid dyes exhibit poor fastness.

13.6 DISCHARGE STYLE OF PRINTING

13.6.1 Polyamide Textiles

The polyamide fabric is dyed with dischargeable acid dyes and printed with a paste containing zinc sulfoxylate formaldehyde (e.g., Arostit ZET, Sandoz) for the production of a white discharge. For color discharges, the ground shade can be dyed with either acid or disperse dyes. In the case of color discharge on an acid-dyed ground, it is possible to use a discharge paste based either on Arostit ZET or tin salt, along with discharge-resistant acid dyes. For the production of color discharges on a disperse-dyed ground, a printing paste based on tin salts is used. The recipes (Sandox) for different classes of dyes are below.

For a white discharge:

X parts	Thickening (crystal gum or locust bean gum)
Y parts	Water
100 parts	Glycerine
100–250 parts	Arostit ZET granulated
20 parts	Citric acid
3–5 parts	Leucophor PC liquid (Optical Brightener)
1000 parts	Total

For a color discharge using sulfoxylate formaldehyde:

X parts	Discharge-resistant acid dye
50 parts	Urea
50 parts	Solvent (Lyocol TG liquid)
Y parts	Boiling water
Z parts	Thickening
100 parts	Glycerine
100–250 parts	Arostit ZET granulated
30 parts	Citric acid
5–10 parts	Leveling and defrosting agent (Lyogen V Liquid)
1000 parts	Total

For a color discharge using tin salt:

X parts	Discharge-resistant acid dyes
50 parts	Urea
50 parts	Solvent for dyes (Viscontin FB liquid, Sandoz)
Y parts	Boiling water
Z parts	Thickening
5–10 parts	Leveling and defrosting agent
100–200 parts	Tin salt 1 : 1
1000 parts	Total

For a color discharge with disperse dye:

X parts	Thickening
Y parts	Water
20 parts	Wetting agent (Sandozin NIT liquid)
100–200 parts	Tin salt (1 : 1)
30–70 parts	Sodium acetate
Z parts	Discharge-resistant disperse dye
1000 parts	Total

In all these cases, the goods are dried after printing and steamed for 15–20 min at a pressure of 0.3–0.5 kg/cm^2. The goods are then washed, soaped at 60°C, aftertreated with a suitable agent to improve the wash fastness in the case of acid dyes, rinsed, and dried.

13.6.2 Acrylic Textiles

White discharges can be produced on acrylic fabrics using dischargeable cationic dyes for the ground shade and printing with discharge paste based on zinc sulfoxylate formaldehyde. The color discharges on acrylic fabric can be produced by dyeing with dischargeable cationic dyes and printing with discharge paste containing discharge-resistant cationic dyes along with tin salt. The color discharges can be also produced using disperse-dyed ground and printing with tin salt along with discharge resistant disperse dyes. The recipes (Sandoz) for different dyes are given below.

Resist for basic dyes on acrylic fibers is obtained by treating the material in an aqueous bath or paste containing a water-soluble polyamide before coloring or printing to get the resist effect under cationic dyes.[58]

For a white discharge on cationic-dyed ground:

X parts	Thickening (crystal gum or locust bean gum)
Y parts	Water
40–80 parts	Arostit ZET granulated
100 parts	Solvent for dyes (Lyocol TG liquid)
1000 parts	Total

For a color discharge using cationic dyes:

X parts	Discharge-resistant cationic dye
20–30 parts	Isopropyl alcohol
Y parts	Boiling Water
Z parts	Thickening
5–10 parts	Tartaric acid
0–50 parts	Sodium acetate crystals
100–200 parts	Tin salt (1 : 1)
1000 parts	Total

Benzyl alcohol (30–50 parts) or a leveling and defrosting agent (5–10 parts, e.g., Lyogen V liquid, Sandoz) and a stabilizer for cationic dyes (5–10 parts, e.g., Imerol NCP liquid, Sandoz) are added to the paste.

For a color discharge using disperse dyes:

X parts	Thickening
50 parts	Polyethylene glycol, molecular weight 200
100–200 parts	Tin salt (1 : 1)
25–50 parts	Sodium acetate crystals
20 parts	Wetting agents
4 parts	Discharge-resistant disperse dyes
2 parts	Water
1000 parts	Total

The fabric after printing is steamed at a pressure of 0.5 kg/cm^2 for 20 min. In order to get better fixation, Luprinton PFD (BASF) (10–20 g/kg) may be incorporated in the printing paste. The goods are then washed, soaped, washed, and dried.

13.6.3 Cellulose Acetate Textiles

A white discharge on cellulose acetate and triacetate can be produced by printing with zinc sulfoxylate formaldehyde, whereas for a color discharge, tin salt is used. The fabric is first dyed with dischargeable disperse dyes and then printed with the following printing paste.

For a white discharge:

X parts	Thickening
50 parts	Glycerine
Y parts	Water
30–50 parts	Sandotherm ACS Liquid
150–250 parts	Arostit ZET granulated
30 parts	Citric acid (1 : 1)
1000 parts	Total

For a color discharge:

W parts	Thickening
X parts	Water
30 parts	Glycerine
30–50 parts	Sandotherm ACS Liquid
20 parts	Wetting agent
Y parts	Discharge-resistant disperse dye
Z parts	Water
100–200 parts	Tin salt (1 : 1)
0.35–0.75 parts	Sodium acetate crystals
1000 parts	Total

After drying the prints, the goods are treated with saturated steam at 102°C for 20 min or with pressure steam at 0.2 kg/cm² for 15–20 min. This is followed by rinsing, soaping, washing, and drying. In the case of cellulose triacetate materials, steaming is carried at a pressure of 0.5 kg/cm² for 15–20 min followed by washing, soaping, reduction-clearing if necessary, washing, and drying.

REFERENCES

1. Spruyt, J., *ADR,* **66**(9) (1977), 48.
2. Ward, D. D., *ADR,* **65**(5) (1976), 20.
3. Mock, G. N., and Jacumin, E. R., *TCC,* **14** (1982), 46.
4. Sahakari, V. D., in Book of Papers of Second Annual Symposium on Textile Printing, Chavan, R. B., Ed., IIT, Delhi, India (1979), p. 158.
5. Howarth, A., *Rev. Prog. Color,* **1** (1970), 53.
6. Schwaebel, R., and Nordmeyer, K., *TCC,* **3** (1971), 133.
7. Aaron, R., *TCC,* **3** (1971), 13.
8. Miles, L. W. C., *Rev. Prog. Color.,* **4** (1973), 44.
9. Koshti, R. S., in Book of Papers of Second Annual Symposium on Textile Printing, Chavan, R. B., Ed., IIT, Delhi, India (1979), p. 76.
10. Vaidya, A. A. and Trivedi, S. S., *Textile Auxiliaries and Finishing Chemicals,* ATIRA, India (1975), p. 47.
11. Cooney, J. A., *ADR,* **64**(6) (1975), 20.
12. Mudki, J. P., and Warty, S. S., *Man Made Text. in India,* **19**(11) (1976), 578.
13. Mehta, H. U., Patel, R. S., Mehta, K. S., and Trivedi, S. S., *Colourage,* **25**(24) (1978), 35.
14. Rosenbaum, J., and Shelso, J., *TCC,* **11** (1979), 220.
15. Barnhardt, G., *TCC,* **11** (1979), 224.
16. Hilton, K. A., *Ciba Review,* **1** (1969), 19.
17. Schwindt, W., Faulhaber, G., and Moore, A. J., *Rev. Prog. Color.,* **2** (1971), 33.
18. Gulrajani, M. L., in Book of Papers of Second Annual Symposium on Textile Printing, Chavan, R. B., Ed., IIT, Delhi, India (1979), p. 45.
19. Holst, L. T., *TCC,* **11** (1979), 53.
20. Hughes, D. W., *JSDC,* **95** (1979), 381.
21. Habereder, H., *MTB,* **61** (1980), 165.
22. Lockett, A. P., *JSDC,* **83** (1967), 213.
23. Marsen, J., *ADR,* **58**(8) (1969), 35.
24. Alsberg, F. R., *JSDC,* **89** (1973), 117.
25. Cockroft, H., *Colourage,* **20**(21) (1973), 36.
26. Hofsteller, R. S., *TCC,* **6** (1974), 156.
27. Badertscher, W., *TCC,* **6** (1974), 156.
28. Host, L. T., *ADR,* **65**(11) (1976), 47.
29. Eible, J., Afkahmi, R., Mohsen, N., and Faruk, M., *MTB,* **57** (1976), 663.
30. Schoepblin, H. P., *TCC,* **10** (1978), 225.
31. Chandavarkar, S. P., in Book of Papers of Second Annual Symposium on Textile Printing, Chavan, R. B., Ed., IIT, Delhi, India (1979).

32. Ruf, R., *MTB,* **42** (1961), 317.
33. Ruf, R., *ADR,* **51** (1962), 885.
34. Manderla, H. J., *MTB,* **43** (1962), 1310.
35. Sansone, R., *Dyer,* **129** (1963), 439.
36. Sohaub, A., and Berthound, R., *MTB,* **145** (1964), 286.
37. Dowson, T. L., *TCC,* **2** (1970), 273.
38. Beal, W., and Corbishley, G. S., *JSDC,* **87** (1971), 329.
39. Mehta, Y. R., and Shirali, P. M., *Colourage,* **21**(7) (1974), 21.
40. Sumitomo Chem. Co. Ltd. *Japan Text. News,* **280** (1978), 78.
41. Gulrajani, M. L., and Chauby, S. S., *Colourage Annual* (1978) 105, 111, and 116.
42. Bass, P. H., *TCC,* **11** (1979), 96.
43. Schoefplin, H. P., *TCC,* **10** (1978), 225.
44. Wirth, H., *MTB,* **41** (1960), 738.
45. Freker, G., *Z., Ges. Text. Ind.,* **65** (1963), 276.
46. Manderla, H. J., *MTB,* **45** (1964), 101.
47. Courtaulds Ltd., *Text. Mfg.,* **90** (1964), 35.
48. Meyer, G., *MTB,* **50** (1969), 698.
49. Howarth, A., *Rev. Prog. Color.,* **1** (1970), 53.
50. Hofsteller, R., *Teintex,* **35** (1970), 691; *MTB,* **51** (1970), 955.
51. Ruf, R., *Teintex,* **37** (1972), 203.
52. Meyer, G., and Dietz, H., *Text. Prax.,* **27** (1972), 294.
53. Miles, L. W. C., *Rev. Prog. Color.,* **4** (1973), 44.
54. Davidson, B. M., *ADR,* **66**(6) (1977), 29.
55. Hayness, R. R., Mathews, J. H., and Heath, G. A., *TCC,* **1** (1969), 74; **2** (1970), 279.
56. Turbak, A. F., *TRJ,* **37** (1967), 350.
57. Alsberg, F. R., *JSDC,* **87** (1973), 117.
58. FBY, BP, 1,244,454 (1969).

14 | PRINTING OF POLYESTER AND ITS BLENDS

Polyester textiles are printed by either the direct style or the discharge-resist style of printing. In the direct style of printing, the following steps are involved: (*a*) application of printing paste, (*b*) fixation of dye on fiber, and (*c*) aftertreatments.

The textile material is usually undyed to get white ground or dyed in pale shades on which overprints are obtained. In the discharge style of printing, the textile material is dyed with selected dyes by conventional dyeing methods. The dye on the fiber is then destroyed by printing in order to get a white discharge and by simultaneous printing of other colors in order to get a color discharge. Many times, the discharge is not perfectly white because of the colored decomposition products of the dyes. The resist style of printing may be used to produce the white effect. In this style of printing, dye transfer to the fiber is hindered so that the fabric remains uncolored at the resisted portions and colored at the other portions. Many other printing effects are produced by a variety of techniques as is described in this chapter.

14.1 DIRECT STYLE OF PRINTING POLYESTER TEXTILES

Polyester textiles are printed with disperse dyes.[1,2] The printing paste consists of

10–100 parts	Disperse dye
150 parts	Water
5 parts	Sodium chlorate
700 parts	Thickening
X parts	Acid solution for pH
1000 parts	Total

The above paste is adjusted to pH 5–6 by the addition of a nonvolatile organic acid such as citric acid or tartaric acid. If acetic acid is used to get the pH, then ammonium sulfate (5 parts) has to be added to maintain the pH while steaming. Sodium chlorate protects the brightness of the print against any possible reduction of the dyes under prolonged steaming conditions at elevated temperatures. Resist salt is not effective in this paste.

The disperse dyes for printing should have good fastness to sublimation and light and good dispersion properties. The liquid disperse dyes are preferred because they contain less dispersing agent and are very finely dispersed. The liquid dye is thoroughly shaken before it is weighed out. Alternatively, the disperse dye in powder form is strewn on water that is constantly stirred to get a dispersion in liquid form. This dispersion is then vigorously stirred into the thickening.

Suitable thickeners for printing polyester are (*a*) esterified gum with low-viscosity, locust bean gum or guar gum, (*b*) low-viscosity sodium alginate, and (*c*) a mixture of emulsion thickening and sodium alginate thickening. Low-viscosity carboxymethyl cellulose (CMC) as such or in admixture with any of the above thickeners is sometimes used. CMC does not give an ideal yield of color, although its use in the mixture gives prints with superior sharpness. The disperse dyes are printed under mild acidic conditions (pH 5.5–6.0) and the thickener should be stable and soluble under these conditions.

Printing of polyester with conventional disperse dyes and thickeners gives color yields of only 60–80% despite the widespread use of high-temperature steamers. The washing-off procedures are lengthy and consume large volumes of water and chemicals. Disperse dyes that are free from diluents will fix almost 100 percent on a scoured PET material, given enough time and energy. However, the fixation of commercial disperse dyes is incomplete because of the presence of a large amount of dispersing agent in the dye (see Chapter 7). The dye fixation is further affected by the presence of thickeners with a high solid content. The addition of carriers does not increase the color yields appreciably, particularly in deep shades. Recently, a number of dyestuff manufacturers have introduced disperse dyes that contain very little diluent.[3–6] Some of these trade products are Terasil X dyes (Ci–Gy) and Foron P dyes (Sandoz). These dyes are applied by printing, using a synthetic thickener based on polymerized acrylic or maleic acid. These thickeners are effective at low concentrations and are easy to remove after printing. The synthetic thickeners are sensitive to electrolytes and hence, the use of electrolytes in the printing paste is avoided. Disperse dyes plus a synthetic thickener offer the advantage of very high fixation of dye, which gives bright shades. No sublimation of dyes in HT steaming and good overall fastness properties are added advantages.

The acid solution is added to the thickening while the mixture is being stirred to avoid localized precipitation of the thickener by the acid. The wetting agent helps in wetting the hydrophobic surface of the PET with the printing paste. A mild oxidizing agent in the printing paste avoids degradation of disperse dyes by reducing action (due to cellulosic fiber or thickener) during steaming.

It is not necessary to print the PET fabric under conditions that will help penetration of the paste in the fabric. In fact, it is preferable to deposit the dye paste on the surface of the fabric so that only the dye diffuses into the fabric during the fixation step. Deeper penetration of the dye paste during printing will deposit more dye in the interstices of the fabric and may cause back-grey marks on the cloth in roller printing, thus increasing the consumption of paste. Sometimes, the paste in the interstices dries up and while lifting the fabric piece from the printing table in flat-bed printing units, the color film sticks to the resin at the bottom (used for fixing the fabric to the table while printing) and leaves the fiber surface. When such a detached printed film is developed, the color appears faded out from that printed portions. To avoid deep penetration of prints, the mesh of the table screen or the rotary and flat-bed printing machines is extremely fine. Similarly, the rollers of roller printing machines are engraved shallow.

The printed goods are dried carefully on a stenter. Since disperse dyes are most susceptible to flushing during pressure steaming, even if the slightest moisture remains in the film of the print paste, it is advisable to slightly overdry the fabric.

Fixation of the Dye. The disperse-dye prints are fixed on the PET by wet (saturated) steaming or dry-heat fixation. Steaming at 100°C at atmospheric pressure is a time-consuming process that requires 2–3 h to get satisfactory color yields. Furthermore, it is essential to use a nonvolatile carrier (which should also not get steam-distilled) in the printing paste. Therefore, steaming under pressure is preferred. The dye can be fixed within 20–30 min when the steam pressure is 1.4–2.1 kg/cm². The pressure-steaming method gives satisfactory color yield and smooth fabric feel. The quality of steam must be good to get satisfactory results. Carried (liquid) water droplets in the steam should be minimum. A cyclone type of steam inlet may remove some droplets. Pressure steaming is carried out in an autoclave in which the fabric is mounted on a frame in a number of layers. A grey cloth is wrapped along with the printed fabric to avoid any contact between the layers. (Padding the grey cloth with a mild oxidizing agent protects it from rapid degradation in steaming.)

The fixation by dry heat is carried out at 165–180°C with superheated steam[7-12] or at 180–200°C by hot air.[13] The time of contact with the dry steam is 1–6 min, while that with hot air is 0.5–2 min. Fixation accelerants up to 10–20 g/kg paste have been claimed to improve the yield.[14]

The superheated steam has an advantage over hot air in its high heat capacity. It gives good dye fixation at lower fixation temperatures and thus, dyes with lower sublimation fastness can be used. The steaming is carried out under tensionless conditions and therefore the feel of the fabric is very soft. Furthermore, the prints are brighter in the steaming process than in the hot-air thermofixation process. The superheated steaming is of particular value for printing knitted goods made from textured yarns. The hot-air fixation process using a pin stenter or other thermosoling equipment imparts a harsh feel to the fabric. The disperse dyes selected for hot-air fixation should have good sublimation fastness. The dry-heat fixation method has the advantages of high productivity and no flushing of prints.

The prints steamed at atmospheric or high pressure improve in depth and brilliancy if they are further treated for a short time by dry air or superheated steam after removing the thickening and ingredients in the paste by cold rinsing. The dye diffuses deeper into the fiber by this post-heating and thus, gives a deeper shade with improved brightness.[15]

Aftertreatments.   In order to get the desired brightness and fastness properties of prints as well as satisfactory handle, it is essential to remove all the unfixed dye and the ingredients of the printing paste from the surface of the fiber by a washing treatment. The printed goods are rinsed with cold and hot water in an open soaper to remove dispersing agent, residual dye, thickening and so on. A reduction-clear treatment is given to the fabric printed in deep shades at 50°C for 10–20 min with 2 parts of caustic soda flakes, 2 parts of sodium hydrosulfite, and 2 parts of nonionic wetting agent/1000 parts of water. An initial cold wash is essential to remove unfixed dye under conditions when it is not readsorbed on the white ground thus, avoiding staining of the ground in later hot-water washing and soaping. Special disperse dyes containing aliphatic carboxyl groups are recommended for printing (described later).

14.2 DIRECT STYLE OF PRINTING POLYESTER–CELLULOSIC BLEND TEXTILES

With the large-scale acceptance of polyester–cotton (PET/CO) blend fabrics for apparel end-uses, a number of printing methods for these fabrics have been developed.[16–19] A number of difficulties have been encountered in printing these blends because of the entirely different characteristics of the two component fibers. Dyes that can fix on both fibers are not fully developed and only a selected mixture of dyes from the disperse, vat, direct, or reactive classes is used. The conditions in the printing paste and on the fiber during fixation are not always congenial with the mixture of the two types of dyes. These problems have been overcome to a certain extent and satisfactory methods are now available to produce fancy colored prints on these blend fabrics. Nevertheless, pigment printing is still the most common method for printing these blends.

14.2.1   Pigment Printing

Resin-bonded pigments are used to print PET/CO blend fabrics because of the following advantages:[20–22] (a) the same pigment is suitable for any blend fabric, (b) PET and CO do not need separate pigments, (c) the low cost of printing, (d) simple process, and (e) the sighting of pattern is immediate. Some of the drawbacks and limitations of pigment printing are (a) poor rubbing fastness, particularly in heavy shades; (b) objectionable stiff handle of the fabric if the amount of binder is not minimal; (c) the dry-cleaning fastness of prints of some pigments is poor.

The selection of the binder and pigment is the key to the success of pigment printing. The most commonly used binders for cotton fabrics are not suitable for PET/CO blend fabrics because of their poor adhesion to the polyester due to the circular cross-section and passivity of PET. Therefore, specially developed styrene-butadiene and acrylonitrile-butadiene binders are employed. Only an optimum amount of binder is used because too much stiffens the fabric and too little gives prints with poor fastness. The curing conditions are controlled so that proper fixation of the prints by the binder occurs. Radiation curable binders such as polyether, polyurathanes, and polycaprolactone have been developed. Curing is done with these binders by an electron beam. This consumes significantly less energy than the conventional process.[23] We still do not have a binder that will penetrate deep into the polyester fiber; most of the pigment–binder composition remains on the fiber surface. Thus, it is not possible to get satisfactory prints in heavy shades with good rubbing fastness.

A typical printing paste consists of the following composition:

X parts	Pigments
Y parts	Binder
500–700 parts	Emulsion thickening
1–5 parts	Acid-liberating catalyst
1000 parts	Total

The printed fabric is dried and cured at 160–170°C for 4–5 min. A final washing off completes the process.

In order to overcome the poor rubbing fastness in deep shades, a mixture of pigment and disperse dyes is used. The disperse dye diffuses into the polyester fiber during curing and effectively dyes the PET component of the blend. Insolubilization of the resin binder around the pigment on both component fibers is simultaneously achieved. The printing paste then contains disperse dyes as well as the ingredients commonly used for pigment printing. The prints are dried, cured at 190–200°C for 45–60 sec, and washed. This process is costlier than the normal pigment printing process and the prints are not very bright. Furthermore, the impairing effect of the binder on the feel of the fabric is not improved.

14.2.2   Printing with a Mixture of Dyes

The following classes of dyes in mixture are used for printing PET/CO blend fabrics: (*a*) disperse and reactive, (*b*) disperse and vat, and (*c*) special brands of dyes.

Disperse Dyes and Reactive Dyes.   This dye combination gives bright prints with excellent wet fastness without affecting the handle of the fabric.[20,24,25] The cost of prints is, however, high and the light fastness of the prints is not always very good. The sequence and number of operations to develop the prints of disperse and reactive dyes in mixture may be summarized as shown in Scheme 14.1. From an economy and ease of operation viewpoint, a single high-temperature fixation process involving simultaneous diffusion of the disperse dye into the PET and reaction of the reactive dye with CO is most desirable. For this purpose, the fabric is printed with a paste of the following composition:

X parts	Disperse dye
Y parts	Reactive dye
600–700 parts	Sodium alginate or emulsion thickening
2–10 parts	Sodium bicarbonate
50–100 parts	Urea
5–10 parts	Resist Salt
1–2 parts	Wetting agent
1000 parts	Total

The prints are dried and thermofixed at 190–210°C for 60–90 sec, soaped, and finished. The disperse dye should not be alkali-sensitive in the paste and should be stable to alkali at 190–210°C as well as possessing the usual properties required for the thermosol process (see Chapter 7).

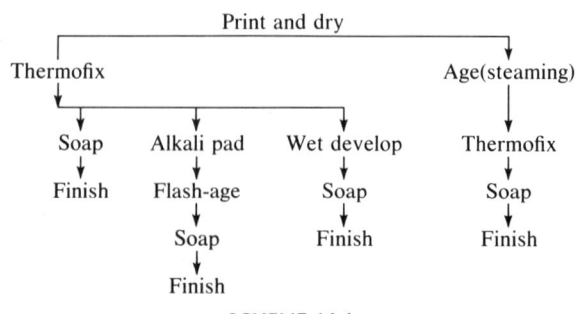

SCHEME 14.1

In this process, the color yield is not very good because of the incomplete fixation of reactive dyes under the dry-heat conditions of fixation. Furthermore, very few disperse dyes can withstand alkali at elevated temperatures and give poor color yield. By adding a steam-aging operation, it is possible to improve the yield of the reactive dyes. It is therefore preferable to have a two-stage fixation process in order to get excellent color yield, brightness of shade, and absence of staining of the white ground. The composition of the printing paste is as follows:

X parts	Disperse dye
Y parts	Reactive dye
600–700 parts	Thickening
5–10 parts	Resist Salt
1–2 parts	Wetting agent
1–2 parts	Acetic acid
1000 parts	Total

The PET/CO blend goods are printed, dried, and thermofixed at 190–210°C for 60–90 sec before they are chemical-padded with a liquor containing

0–6%	Sodium hydroxide
10–15%	Sodium carbonate
15–20%	Sodium chloride

The padded (printed) goods are flash-aged, soaped, and finished. For this method, the use of acid-stable reactive dyes of the vinyl sulfone type is of special interest.[26] The flash-aging operation can be avoided for dyes with very high reactivity towards cotton. Thus, the sequence of operations in such a case is print, dry, thermofix, chemical pad, wash off, and finish.

The mixtures of conventional reactive and disperse dyes create the problem of tinting of unprinted (white) ground, particularly during the washing-off–soaping operations. In order to overcome this problem, special dyes, for example, Dispersol PC ICI, Resocotton dyes, Bayer, have been developed.[27,28] These dyes form an ionic (sodium) salt in the alkaline wash liquor and are easily removed. The ionic salt of the dye has no affinity for either the PET or the CO component of the blend and leaves the white ground of the prints untinted. The best combinations of disperse and reactive dyes (Dispersol PC/Procion PC, ICI) are identified for printing on PET/CO blend fabric. The disperse dyes are supplied preferably in liquid form. After fixing the prints, the goods are rinsed in cold water to remove the thickener and the other auxiliary chemicals on the fiber. The goods are then treated with caustic soda flakes in water (2 parts/1000 parts) at 80°C for at least 2 min before washing in a solution of a nonionic detergent at 80°C for 1–2 min. A final rinse with cold water completes the process. A conventional eight-tank, open-width, soaper-range is found to give satisfactory results. Bayer's Reso-cotton dyes are mixtures of disperse dyes (Resolin) and reactive dyes (Leva-fix P or PA) that can also be used to give untinted white grounds during alkaline scouring as described above.[29]

Liquid brand mixtures of disperse and acid-fixing reactive dyes are also marketed as simple-to-use, one-pack liquid dyes.[28] The reactive dye used in the mixture can be fixed under acidic conditions and hence, both the disperse and the special acid-fixing reactive dyes give very good color yields. The stability of the printing paste over a long period is very good since the acid-fixing reactive dyes such as Procion T (ICI) dyes do not hydrolyze during storage under acidic conditions. Since the special disperse dye is converted into an ionic salt as described above and the acid-fixing reactive dye does not fix on CO under alkaline conditions, the washing-off operation with alkali leaves white (unprinted) ground untinted. Another advantage of the special liquid brand formulations is that they can be applied without prior causticization or mercerization treatment of the PET/CO blend fabric. The typical ICI recipe for such liquid-brand mixtures is given below.

Stock paste with ammonium alginate thickening:

15 parts	Matexil FN-T (fixing agent)
555 parts	Water
30 parts	Manutex PC-MV (ammonium alginate thickening)
600 parts	Total

Stock paste–half-emulsion method:

102 parts	Water
13 parts	Matexil PN-PR
25 parts	Matexil FN-T
510 parts	White spirit
25 parts	Slurry of Manutex PC-MV
25 parts	White spirit
1000 parts	Total

Print Paste

100–400 parts	Procilene PC liquid dye (ICI)
600 parts	Water
0–390 parts	Stock paste
1000 parts	Total

The fabric is printed and dried. The prints are developed either by dry-steaming at 175–185°C for 10 min or by thermofixation at 210–220°C for 30–

45 sec. The goods are then washed in an open soaper as follows: One rinse with cold water, two washes with 4 parts caustic soda flakes/1000 parts of water at 80–85°C, two rinses at 85°C, one wash with water at boiling, and a final cold-water wash.

Another variation of the process of printing with a mixture of disperse and reactive dyes is Hoechst's sodium formate printing method.[30,31] This method is applicable to blends of PET and CO or wool. In this method, a mixture of dyes (Ramaran® Printing Dyes) is printed on the blend fabric using sodium formate as a fixing agent for reactive dyes. Since sodium formate gives a neutral pH, the printing paste stability is unlimited. There are fewer restrictions on the choice of disperse dyes since the fixation of disperse dyes is very good because there is no alkali present in the printing paste. Washing is done in neutral liquors so that the dyes are not fixed on white grounds to tint them during washing. The process may be described as follows: (The dye is sprinkled in the thickening.)

Stock thickening:

345 parts	High-viscosity alginate thickening (4%) (containing 5 g/kg Calgon T)
115 parts	Starch ether thickening (1%)
354 parts	Water
25 parts	Sodium formate
40 parts	Emulsifier DMR (10%)
20 parts	Ludigol (1 : 2)
1 part	Silicone antifoaming-emulsion SLE
400 parts	White Spirit
1000 parts	Total

Reduction thickening:

400 parts	Stock thickening
300 parts	Alginate thickening
300 parts	Water
1000 parts	Total

Printing pastes

800 parts	Stock thickening
X parts (powder)	Ramaron Printing Dyes
—	Water (80–90°C)
Y parts	Reduction thickening
1000 parts	Total

Printed and dried fabric is thermofixed with superheated steam at 175°C for 7 min. The goods are washed by spraying cold water on the goods followed by washing in an open soaper at 40°, 60°, and 80°C with Leomin OR (2 g/liter) and Calgon T(0.5–1 g/liter) in the first two baths.

Disperse Dyes and Vat Dyes.   This combination of dyes satisfies the requirements for high-quality printing on PET/CO blends.[32,33] The overall fastness properties of prints are good. The color yield and penetration of dyes are very good with a wide choice for the selection of disperse dyes. The printing paste of these dyes is almost indefinitely stable and the prints on the fabric may also be developed as and when convenient. The staining of the ground is very low. The large printed areas do not affect the handle of the fabric or the uniform application of the resin finish. The light fastness of most of these combinations is rarely affected by the subsequent resin finishing. The main drawback of this combination of dyes is the complicated process of application and the very high cost of printing compared to pigments.

Printing paste recipe:

X parts	Disperse dye
Y parts	Vat dye
50–100 parts	Water
600–700 parts	Thickening
1–2 parts	Nonionic wetting agent
1000 parts	Total

Disperse dyes that have satisfactory sublimation fastness and that give minimal staining of cotton are selected. The excessive concentration of disperse dyes in the paste gives prints with poor rubbing fastness. Since certain vat dyes can also be adsorbed by PET, such dyes may be selected to partially replace disperse dyes.

Starch and starch ethers, gum tragacanth, locust beam gum, or oil-in-water emulsion can be used as a thickening agent. After printing and drying, the disperse dye is fixed on the PET portion of the blend by the usual thermosoling treatment (190–210°C/30–60 sec) on a stenter. Some of the vat dyes also diffuse into the polyester fiber. The thermofixed printed fabric is chemical-padded with a liquor of

75–100 parts	Caustic soda (50%)
50–60 parts	Sodium carbonate
50–60 parts	Sodium hydrosulfite
50–60 parts	Sodium chloride
	Water to make liquor
1000 parts	Total

and flash-aged for 30–45 sec at 120–125°C. The printed goods are then given a cold overflow rinse and are oxidized (preferably) with hydrogen peroxide and acetic acid. Hydrogen peroxide does not stain the white ground (like sodium bichromate which makes stains that are very difficult to wash out). Peroxide also helps to clear the cross-staining on cotton. After oxidation, the goods are hot washed, soaped, and dried.

Special Brands of Dyes. Polyester–cellulosic blends are generally printed either with a single dye system like pigment colors or by a mixture of dyes of two different classes, such as disperse and reactive dyes or disperse and vat dyes. The disadvantages of pigments were described earlier. When a mixture of dyes is used, a large part of the dyes goes on to the wrong fiber. Unlike dyeing, printing has a small possibility of any leveling out process that could cause a migration of dye from one fiber to another. The degree of fixation that can be achieved in printing PET/CO blends with a mixture of dyes is usually less than 50%. This leads to the following problems: (*a*) dye is wasted, (*b*) the white grounds are tinted during washing, and (*c*) the effluents are loaded with dyes. It is thus necessary to develop a single dye system for PET/CO blend fabric.

Special Vat Dyes. A vat dye alone can be used in place of the disperse and vat dye combination by the above-mentioned methods if the dye can diffuse into PET fiber under thermofixation conditions. Casella has marketed such dyes under the trade name, *Polyestren Dyes*.[34] Such dyes are only suitable for producing pale to medium shades. The shade on polyester is rather dull and the dyes are expensive.

Dybln Dyes.[35] These special-brand disperse dyes (DuPont) give a solid shade on both polyester and cotton fiber from a specially formulated printing paste. These dyes are applied by the simple print-dry-thermofix-wash off sequence. Because no binder is required, the handle of the printed fabric is not impaired and the fastness to rubbing, washing, and dry-cleaning is superior to that of pigment prints. Furthermore, the staining of the white ground is absent, the color yield is very good, and the process of printing is very simple.

The printing paste is similar to that used for the usual disperse dyes on polyester except that a high boiling solvent which swells cotton is added. The disperse dye dissolves in this solvent in swollen cotton during the thermofixation. During the washing-off treatment, the solvent (which is polyethyleneglycol-based) is removed by rinsing it in cold water to collapse the swollen structure of the cotton and thus, to trap the disperse dye in the fiber. For polyester, the Dybin dye behaves as a conventional disperse dye and diffuses during the thermofixation.

Usually, the wet fastness and brightness of prints on cotton is not as high as that of reactive and vat dyes. It is possible to use reactive disperse dyes under conditions that facilitate the reaction of the dye with cellulose.[15] This improves the fastness properties.

Cellestren Dyes.   BASF developed Cellestren dyes that can be fixed on both the cotton and the polyester of the blend fabric in the presence of a special auxiliary, Glyezin CD.[36,37] The auxiliary may either be incorporated in the printing paste or padded to the fabric from its aqueous solution prior to printing. In the latter case, about 20–22% Glyezin CD is deposited on the weight of the cotton component of the blend and the fabric is dried prior to printing. Since Glyezin CD is not added to the print paste in this case, the paste has better properties. Instead of the two-stage process, Glyezin CD may be incorporated in the printing paste to avoid the initial pad-dry step. The printing process may be described as follows:

Stock thickening:

400 parts	Alginate thickening (10%)
10 parts	Resist Salt
5 parts	Citric Acid or monosodium phosphate
5–10 parts	Calgon T
200 parts	Starch ether thickening (10%)
5–10 parts	Luprintan HDF
	Water to make paste
1000 parts	Total

Printing paste:

X parts	Cellestren Dyes
700 parts	Stock thickening
50–100 parts	Glyezin CD

The goods are printed, dried, thermofixed (210–215°C/1 min in hot air or 190°C/5 min in superheated steam), and washed off with cold water, and soaped with soda ash and a detergent. These dyes have all the advantages of the Dybin process, including low effluent in the afterwash.

14.2.3   Carbonized Prints

Burnt-Out Style.   Fine-count 100% PET printed sarees are manufactured from PET/CO blend fabrics by removing the CO component by a treatment of 70% sulfuric acid after printing with suitable disperse dyes.[38–41] The process of removing the cellulosic component with sulfuric acid is called *carbonization*. The prints are produced on the blend fabric using selected disperse dyes by the conventional print, dry, fix by wet or dry-fixation process, and wash.

The carbonization process can be carried out either on a jigger or by a pad-batch method with 70% sulfuric acid. Addition of urea or sulfamic acid to the sulfuric acid bath is suggested to prevent dulling of the prints during carbonization. In the jigger process, the goods are run in sulfuric acid at 20–30°C for 30–45 min. The goods are then washed, neutralized with sodium carbonate, and optionally bleached with sodium hypochlorite and hydrogen peroxide. A final washing treatment completes the process.

In the pad-batch process, the printed fabric is padded with 70% sulfuric acid containing 0.5% urea at 20–30°C. The goods are then batched on a rotating batching trolley for about 2 h, washed, neutralized, bleached to remove the brownish tint imparted during the carbonization, washed, and dried. To get good results, it is necessary to control the concentration of sulfuric acid.

Brosso Style of Printing. In this style of printing, the cellulosic component of the PET/CO blend fabric is burnt out from certain portions by printing with a paste containing acid-liberating agents.[42] Generally, aluminum sulfate or sodium bisulfate are used as acid-liberating agents.

Recipe with aluminum sulfate:

20 parts	Aluminum sulfate
5 parts	Tartaric acid
X parts	Water
2 parts	Noigen EL-40
Y parts	Maypro KN (10%) thickening
1000 parts	Total

Recipe with sodium bisulfate:

20 parts	Sodium bisulfate
15 parts	Water
2 parts	Noigen EL-40
63 parts	Maypro KN (10%) thickening
1000 parts	Total

The goods are printed, dried at 110°C, and cured at 150°C for 4–5 min. The goods are then washed, bleached, washed, and dried.

14.3 DIRECT STYLE OF PRINTING POLYESTER–WOOL BLENDS

Polyester–wool blends are usually printed with a mixture of disperse and metal-complex dyes. Ready-made mixtures are also marketed, for example, Lanestren dyes (BASF). The printing recipe involves the use of a carrier.

The fixation of prints is carried out in saturated steam under pressure (2.5 kg/cm²). The use of reactive dyes in place of the metal-complex dyes gives brighter prints.

Stock thickening:

400 parts	Thickener (12–14%)
50 parts	Lyoprint G (Ciba–Geigy)
50 parts	Urea
30 parts	Acetic acid (40%)
40 parts	Carrier
430 parts	Water
1000 parts	Total

The dye is strewn into the stock thickening. The goods are printed, dried, and steamed for 20 min at 2–2.5 kg/cm². Alternatively, steaming may be carried out at atmospheric pressure (100°C/15 min) followed by thermofixation at 180°C for 3 min. The latter process gives paler prints.

The wet fastness properties of prints are decided by the efficiency of the washing-off treatment. A cold-water rinsing is essential in removing the unfixed dye without staining the white ground. The temperature is raised to 90°C in the final washing-off bath followed by cold rinsing. The addition of ammonia (2 g/liter) and a wetting agent (2 g/liter) in all the washing-off baths improves the results.

14.4 DISCHARGE STYLE OF PRINTING POLYESTER

Polyester fabric is dyed with selected dyes suitable for discharge printing[43-47] by the carrier or HT-beam dyeing method (see Chapter 7). Zinc sulfoxylate formaldehyde is used as a discharging agent for white discharges and tin salt is used for color discharges. The paste (1000 parts) for a white discharge consists of a thickening of crystal gum or locust bean gum, a carrier (50–100 parts), sodium thiocyanate (1 : 1, 40 parts), solvent (e.g., Lyocol TG (Sandoz), 100 parts), zinc sulfoxylate formaldehyde (200–300 parts), and citric acid (1 : 1, 30 parts), or tartaric acid (1 : 1, 20 parts). Discharge-resistant disperse dye is added to the printing paste for color discharges and tin salt (1 : 1, 100–200 parts) is used as a discharging agent in place of zinc sulfoxylate formaldehyde.

The processing to be carried out after printing is the same as that for the production of white and color discharges. This includes drying the prints, fixation by steaming with steam at a pressure of 1.4 kg/cm² for 20–30 min, washing with a suitable detergent at 60°C, cold washing, and reduction-clearing at 60°C, followed by a final rinsing with cold water.

Benzoyl Peroxide Method. A method for discharge printing on polyester fabrics was developed by Asahi Chemical Co. (Japan). In this method, a combination of benzoyl peroxide and sodium chlorate in a ratio of 3:1 is used as the discharging agent. These agents are particularly effective on anthraquiononoid purple, blue, and violet disperse dyes. A paste containing these two chemicals, a chlorate stable optical brightener, a thickener, and other ingredients described earlier is printed on polyester fabric dyed with anthraquinonoid disperse dyes. The goods are steamed under pressure at 130°C for 20 min. This is followed by washing, soaping, washing, and drying.[48]

14.5 RESIST STYLE OF PRINTING POLYESTER

For the production of the white effect by resist printing, active carbon and zinc sulfoxylate formaldehyde are used; tin salt is used as the resisting agent for color effects. Resist printing by active carbon is the oldest method and requires no specifically selected dyes. Being a physical resist, the process gives defects like overlapping of colors at the edge of the frame and staining by contact with a screen in the screen printing. It is not suitable for roller printing because of the large particle size of the active carbon. Because of all these reasons, resist printing using active carbon is rarely practised on a commercial scale.

White and color resist-discharge prints are produced by overprinting the padded unfixed grounds. The fabric is padded with a disperse-dye liquor having a nonionic dispersing agent (0.5–1%), migration inhibitor (0.5–1.5%), and acetic acid (0.1–0.2%), and carefully dried below the temperature of 120°C to avoid premature fixation of the dye which is harmful in getting white effects. It is then printed with white or colored resist paste.

Resist-Discharge Paste:

White Resist	Color Resist	
X parts	X parts	Thickening
Y parts	Y parts	Water
Nil	Z parts	Discharge-resistant disperse dye (optional)
50 parts	50 parts	Solvent (for dye)
100–200 parts	Nil	Zinc sulfoxylate formaldehyde or
Nil	100–200 parts	Tin salt
Nil	100 parts	Urea
30 parts	Nil	Hexametaphosphate (Calgon T)
20 parts	20 parts	Tartaric acid
20–40 parts	20–40 parts	Dispersing agent
40 parts	40 parts	Sodium thiocyanate (1:1)
10 parts	10 parts	Defoaming agent
1000 parts	1000 parts	Total

After the goods are printed with a resist paste, they are thermofixed at 170°C for 5 min with superheated steam. These goods are then washed; soaped at 60°C; reduction-cleared with a liquor containing 1–2 g/liter caustic soda, 1–2 g/liter sodium hydrosulfite, and 1 g/liter detergent at 60°C; washed; and dried.

Alternatively, resist-discharge prints are produced on the printed unfixed ground. For printing of the ground, the conventional printing recipes for dischargeable disperse dyes using emulsion or semiemulsion thickening and nonionic dispersing-leveling agent (20 parts) are used. For the overprint method, a nonionic wetting agent (10 parts) is added to the printing paste. The resist-discharge paste described above is printed before (preprint method) or after (overprint method) the ground shade is printed (see Chapter 13). The material is dried after printing the ground shade at a temperature below 120°C. The fixation and washing-off processes are the same as those for overprinted padded grounds described above.

Modified Stannous Chloride as Resisting Agent. Sakaoka et al. have developed a modified stannous chloride known as Unistan CRX (MeO · SnCl$_2$ × H$_2$O, where MeO is the metal oxide) for resist-printing on polyester fabrics.[48] Unistan CRX is available as a white paste. It has a pH of 3.65 compared to a pH of 2.18 of a 1% aqueous solution of stannous chloride. Unistan CRX starts the reduction at 80°C and reaches the peak at 100°C compared to the reduction action of stannous chloride at 40°C and its peak at 60°C. It has a strong reducing power without any adverse effect on the undischargeable dyes in the paste or ground. The prints have no halation. The storage stability of the printing paste is good since the reducing action of the product is revealed only by the steaming of the printed fabric. Thus, the reducing action is not lost on storage. The fastness properties of prints are not adversely affected. The undercloth and the steamers are least damaged.

The resist-discharge paste of modified stannous chloride (100–300 parts/1000 parts depending on the depth of shade on the ground) is overprinted on a fabric dried after padding or printing with a dischargeable dye. If the paste contains undischargeable disperse dye, color–resist-discharge prints can be obtained. The printed fabric is dried, steamed under pressure at 130°C for 30 min, and washed off as usual.

Resist Effect by Chelate Formation. An interesting method of resist printing on polyester has been described[49] in which azo disperse dyes capable of forming a water-soluble chelate are used. The method consists of printing the fabric with bivalent-metal salts such as cupric formate, cupric acetate, copper sulfate, and cobalt chloride. After drying, the fabric is overprinted with selected disperse dyes with chelate-forming groups such as OH, COOH, NH$_2$, and C=O. After the fabric is dried at 100°C, it is thermofixed at 190–200°C for 30–60 sec. During the thermofixation, the dye on the printed patterns forms a water-soluble dye chelate that has no affinity for polyester whereas the ground dye gets fixed. The printed pattern becomes white during the subsequent washing-off treatment when chelated dye dissolves in the alkaline reduction-clear solution. The final acid rinsing removes the metal-chelated dye remaining on the printed fabric.

The method gives a sharp image-edge by the wet process, does not cause either tin burning or fiber deterioration as with tin salt resist-printing, and gives little problem of bleeding ground colors. The method is particularly useful for producing bright turquoise blue shades which are difficult to produce by the discharge printing method. The method is limited only to a few chelate-forming disperse dyes. The stability of the paste containing chelating metal salts is limited and clogging of the screen-mesh by the paste can be a problem.

Resist Printing with Alkali. The conversion of a nonionic dye into an ionic form can be done by protonating an amino group to form ammonium or converting aliphatic carboxyl group into its sodium salt.[31,50,51]

$$D-N\begin{matrix} R' \\ \diagdown \\ R \end{matrix} + HCl \rightarrow D-N^+\begin{matrix} R' \\ | \diagdown \\ H \quad R \end{matrix} + Cl^-$$

$$D-COOH + NaOH \rightarrow D-COO^- Na^+ + H_2O$$

where D–NRR' and D–COOH are nonionic dyes. Even though both these approaches are possible, dyes that can be easily protonated are not yet commercially available. On the other hand, introduction of a range of dyes (Dispersol PC of ICI) containing a COOH group provides a unique basis for the economical production of a variety of dischargeable ground shades on polyester fabric.[50] The process eliminates the use of expensive stannous chloride which is also a source of environmental pollution. The process involves the use of a cheap, readily available sodium carbonate or caustic soda. The printing can be carried out either by the "wet-on-wet" method or the "pad-print" methods. In the wet-on-wet method, the fabric is first printed with the resist paste with or without alkali-resistant dye to get color or white resist. The fabric is overprinted with the Dispersol PC dye paste containing citric acid (2 parts), dried rapidly at 100–120°C, thermofixed at 160–170°C for 6–8 min with superheated steam, washed, reduction-cleared, washed, and dried.

The resist paste consists of the following:

600–650 parts	Thickening (CMC and starch ether)
75–100 parts	Glycerine (as a humectant)
75–100 parts	Polyethylene glycol (molecular weight 300)
50 parts	Sodium carbonate
50 parts	Alkali resistant-disperse dye (optional, for colored discharge)
150 parts	Water
1000 parts	Total

The pad-print method is preferred because it gives better wetting of the ground. The goods are padded with a liquor containing

X parts	Dischargeable dye (Dispersol PC dyes)
800–1000 parts	Water
2–5 parts	Citric acid
100 parts	Thickener (CMC)
1–3 parts	Wetting agent
1000 parts	Total

The fabric is dried at 100°C and printed with the resist-printing paste given above. This is followed by quick drying at 120–140°C, thermofixation in superheated steam at 160–170°C for 6–8 min, washing, reduction-clearing, washing, and drying.

14.6   DISCHARGE PRINTING ON POLYESTER–COTTON BLEND TEXTILES

It has been almost impossible to produce discharge or resist prints on PET/CO blend textiles with the conventional disperse and reactive or vat dyes. However, with the newly developed special disperse and reactive dyes as well as the new printing recipes, these styles of printing are now feasible for blend fabrics. By selecting disperse-reactive dyes that are not fixed on both the PET and cellulosic fiber under alkaline conditions, discharge-resist prints can be produced using an alkali paste. For color discharges, alkali-stable disperse dyes and conventional reactive dyes that are fixed under alkaline conditions are used.[44] The goods are padded with the dye-liquor containing sodium alginate (500 parts, 0.8%), ammonium dihydrogen phosphate (4 parts), a wetting agent (e.g., Metexil FN-PC, ICI, 30 parts), and the mixture of the two dyes (total liquor: 1000 parts), dried at 100–120°C, and printed with the following discharge printing paste.

40–80 parts	Sodium carbonate
40–80 parts	Glycerine
40–80 parts	Polyethylene glycol (as a dye solvent)
100 parts	Reducing agent
600 parts	Thickening (CMC, starch, or a mixture thereof)
	Water to make liquor
1000 parts	Total

For color discharges, alkali-resistant disperse dyes and the conventional reactive dyes (e.g., Procion P dyes, ICI) are added to the paste and the reducing agent is not incorporated in the paste. The prints are fixed by steaming at atmospheric or higher pressures followed by thermofixation. The goods are finally washed with a caustic soda solution followed by a nonionic dispersing agent solution.

In another method, sodium formate is used as a fixing agent for reactive dyes and sodium bisulfate-alkali is used as a resisting agent.[31] The fabric is padded with a mixture of disperse and reactive dyes (Samoron, Ramazol dyes) with a mild oxidizing agent (sodium-*m*-nitrobenzene sulfonic acid 1 : 2, 20 parts), sodium formate (20 parts), and Solidokoll N (1 : 2, 60 parts, a padding auxiliary). Samoron dye can be destroyed by alkali. The goods are dried at 100–120°C and printed with the resist-printing paste with or without a stable reactive resist agent (150 parts) and thickening (600 parts/1000 parts paste). The resist agent has the following composition:

310 parts	Sodium bisulfate (38° Be)
65 parts	Glyoxal (40%)
125 parts	Water
500 parts	Sodium bicarbonate
1000 parts	Total

The prints are dried and the goods are thermofixed in the HT steamer at 175°C for 7 min. The final washing is done with water at 40°, 60°, and 80°C, containing 2 g/liter Leomin OR and 1 g/liter Calgon T.

REFERENCES

1. Eisenlohr, R. H., *TCC,* **2** (1970), 266.
2. Kamat, S. V., *Colourage,* **23**(9B) (1976), 34.
3. Baderstscher, W., *TCC,* **6** (1974), 156.

4. Schoepblin, H. P., *TCC,* **10** (1978), 225.

5. Baderstscher, W., Kunz, W., and Warwick, J., Proceedings of 19th Biannual Symposium of SDC, Sheffield, England (Sept. 13, 1979), p. 1.

6. Hofstetter, R., and Robert, G., *Textilveredlung,* **14**(2) (1979), 51; *Dyer,* **162** (1979), 225.

7. Lockett, A. P., *JSDC,* **83** (1967), 213.

8. Marsden, J., *ADR,* **58**(8) (1969), 35.

9. Alsberg, F. R., *JSDC,* **89** (1973), 117.

10. Hofsteller, R. S., *TCC,* **6** (1974), 156.

11. Hoist, L. T., *ADR,* **65**(11) (1976), 47.

12. Eibl, J., Afkashmi, R., Mohsen, N., and Faruk, M., *MTB,* **57** (1976), 338.

13. Mehta, Y. R., *Colourage,* **23**(12B) (1976), X XVI.

14. Cockroft, H., *Colourage,* **20**(21) (1973), 36.

15. Datye, K. V. (unpublished work).

16. Vaidya, A. A., *Text. Dyer Printer,* **10**(1) (1976), 34.

17. Glover, B., *Rev. Prog. Color.,* **8** (1977), 36.

18. Jenkinson, K., *JSDC,* **95** (1979), 384.

19. Fees, E., *MTB,* **60** (1979), 586.

20. Fortress, F., Szilagyi, G., and Thorton, D. B., *ADR,* **52** (1963), 403.

21. Saville, A. K., Larson, O. S., and Meunier, P. L., *ADR,* **57** (1968), 938.

22. Schwindt, W., *Text. Prax. Inter.,* **1** (1979), 47.

23. Park, K., Frame, R. L., and Bryant, G. M., *TCC,* **11** (1979), 107.

24. Zimmermann, H., *Textilveredlung,* **12** (1977), 43.

25. Koth, K., *MTB,* **59** (1978), 47.

26. Hyeckle, M. O., *ADR,* **57** (1968), 479.

27. Milne, S. W., *Colourage Annual* (1976–1977), 57.

28. Glover, B., *ADR,* **67**(12) (1978), 47.

29. Schwaebel, R., *Colourage Annual* (1976–1977), 63.

30. Spier, K., and Roth, K., Internationales Textil-Bulletin Wettansgabe Farberie Druckerie/ Ausrustung (1977), 333.

31. Weyer, H. J., Proceedings of 19th Biannual Symposium of SDS, Sheffield, England (Sept. 14, 1979), p. 1.

32. Schwidth, W., and Sommer, C., *MTB,* **48** (1967), 1338.

33. Hart, B., *TCC,* **2** (1970), 253.

34. Musshoff, H. J., *JSDC,* **77** (1961), 89.

35. DuPont, *ADR,* **60**(3) (1971), 25.

36. Blum, A., *Chemiefasern/Textilindustrie,* **28/80** (1978), 648.

37. Miksovsky, F., *JSDC,* **96** (1980), 347.

38. Schwab, H., *Dyer,* **146** (1971), 637.

39. Patel, P. T., *Colourage,* **23**(9B) (1976), 39.

40. Prabhu, C. N., and Shah, J. K., *Colourage,* **23**(11) (1976), 29.

41. Parikh, M. R., and Moonim, S. M., *Colourage Annual* (1976–77), 69.

42. Narkar, R. K., *Text. Ind. and Trade J., India,* (Nov.–Dec. 1975), 29.

43. Chandavarkar, S. P., *Text. Ind. and Trade J., India,* (Nov.–Dec. 1975), 25.

44. Kunnel, W., *MTB,* **57** (1976), 346.

45. Mtsui Toatsh Chemical Inc., *Japan Text. News,* **274** (1977), 101; **275** (1977), 79.

46. Sakoka, H., and Haffori, K., *Japan Text. News,* **271,** (1977), 63.
47. Narkar, A. K., *Text. Dyer Printer* **11**(9) (1978), 45.
48. Mitsubishi Chemical Ltd., *Japan Text. News,* **296** (1979), 85.
49. Bass, P. H., *TCC,* **11**(5) (1979), 96.
50. Glover, B., and Hansford, J. A., *JSDC,* **96** (1980), 355.
51. Kangle, P. J., Ramanathan V., and Argay, R., Ger. P. 2,027,952 (1970).

15 | TRANSFER PRINTING

15.1 HEAT-TRANSFER PRINTING

One of the most important developments in the printing of synthetic fiber textiles was the introduction of heat-transfer printing, also known as Sublistatic, Colorstatic, vapor-phase, dry-heat, or sublimation-transfer printing.[1] In this printing technique, a design printed on paper with a suitable volatile dye is transferred to a receptive fabric under controlled conditions of temperature, pressure, and time.[2-4]

The concept of transfer printing appears in two British patents for cellulose acetate in the 1930s.[5,6] The transfer of dye from cotton to polyester under dry-heat conditions has been reported since the early 1950s.[7] The first patents on transfer printing were filed in the 1960s by de Plasse.[8] However, the real breakthrough for transfer printing came only after 1965 when the Sublistatic Corporation developed the required technology for the commercial success of the process.[9] The process caught on like wild fire; by 1971 worldwide usage of the process was estimated at over 20 million m^2.[10] In less than 10 years, it was estimated at 1 billion m^2. The process has thus become an important part of the printing business.[11]

Heat-transfer printing is a two-step process.[12] First, the desired pattern is printed on paper using special inks containing volatile nonionic dyes. The printed paper is then placed on the fabric, and heat and pressure are applied to the back of the paper. The dye on the paper sublimes, diffuses into the gap between the paper and the fabric to condense onto the fiber surface, and subsequently diffuses into the interior of the fiber.[9,12] After the dye is transferred, the paper is separated from the fabric. The printed fabric needs no aftertreatment and is ready for use. Advantage can be taken of the high speed and accuracy of multicolored printing on a substrate (paper) of high rigidity and smoothness compared to typical fabrics. Polyester and cellulose triacetate fabrics are ideally suited for this process because of the good color range, the fastness to washing of the prints obtained, and the stability of these fibers at the temperature employed. Color depth and washing fastness of the prints on polyamide and acrylic fabrics are restricted by the low substantivity of volatile dyes and the time and temperature limitations for the minimum effect on handle and color of acrylics.[13] Fibers with a low softening point are totally unsuitable for the process. Blends with cellulosic and animal fibers will have only pale effects. Thus, in principle, the heat-transfer process is basically an adaptation of the thermofixation process, where through the use of heat and little pressure, designs are transferred from rolls of paper to fabric.[14-16] The preparation of the printed paper is a vital step in this process.

15.1.1 Transfer Printing Inks

The papers are printed with an ink made up of dye formulations, a vehicle (media), and a thickener. Nonionic (disperse) dyes with suitable heat-transfer properties are used in specially prepared formulations.[1] The dyes used in the mixtures should have similar sublimation rates, diffusion rates, and fastness properties. Typical such dyes are listed in Table 15.1.[12,14,17,18] The dyes should not bleed in durable press finishes on blends and should not exhibit catalytic fading in mixtures. They should not be very sensitive to minor differences in temperature during heat transfer, which could cause variations in depth and tone of shade. The dyes should show only a gradual loss of depth when reduction in color and temperature are made so that the tone of the shade in any depth remains the same. The dyes should not interact with each other and the substrate.

Considering the stringent requirements for the dye properties, the number of suitable dyes for the heat-transfer process cannot be very large.[11-21] In fact, in the early days of heat-transfer printing, only four or five dyes were available, which were mixed to produce thousands of shades. However, jet-black shade was not available and only deep grey could be produced. The range of dyes has slowly increased. The formulation of dyes for heat-transfer inks involves various techniques. Ethyl cellulose, acrylic resins, polyvinyl acetate and so on have been used for the purpose. The dye must form a "solid" solution into the formulation so that it remains in a finely divided or even a monomolecular state. The inks are usually prepared by dissolving such formulations in a suitable solvent. After use, the inks are stored for reuse. The vehicle or media for the ink is a liquid such as water, spirit, glycol, toluene, ethyl methyl ketone and so on. The dye formulation must dissolve or disperse in the media. A thickener may be added to give body

TABLE 15.1 Typical Dyes for Transfer Printing[12,14,17,18]

		Chemical Class
Yellow		
C.I. Disperse Yellow	3	Monoazo
	42	Nitrodiphenylamine
	54	Quinophthalone
	60	Monoazo
Red		
C.I. Disperse Red	4	Anthraquinone
	11	Anthraquinone
	13	Monoazo
	59	Anthraquinone
	60	Anthraquinone
Violet		
C.I. Disperse Violet	1	Anthraquinone
	4	Anthraquinone
	23	Anthraquinone
Blue		
C.I. Disperse Blue	3	Anthraquinone
	14	Anthraquinone
	19	Anthraquinone
	24	Anthraquinone

and viscosity to the ink. Ethyl cellulose, polyvinyl acetate, other cellulose ethers, acrylic resins, and shellac are common thickeners. The dye formulation also brings in some thickener.

15.1.2 Preparation of Printed Paper

Good-quality bleached craft paper is clay coated and true-blade machine-glazed so that the dye does not diffuse into the paper, thus contaminating the machines and giving lower transfer efficiencies. The surface of the paper should be nonfibrous and nonabsorbent so that the ink remains on the surface. However, the ink should stick to the paper so that smearing or blurring does not occur. After printing, the paper is dried at a minimum temperature with low air circulation.

Since the cost of paper is high, attempts have been made to use low-cost paper. A coat of gelatine or sodium alginate on the paper makes the surface impermeable to the dye. Reusing the paper after stripping the residual ink is not possible because the paper is cut to the required size during heat transfer. Printing on a continuous rubber belt and using it for heat transfer has been a failure. After the end pieces of printed paper have been cut into various sizes they have been claimed for random printing effects by the heat-transfer technique.

Methods of Printing Paper.   The paper for transfer printing can be printed by several methods used in the graphics industry,[1,22] for example, gravure, flexography, lithography, deep print, and offset methods. Recently, rotary-screen printing machines have been used that can increase the flexibility in the volume of dye applied and the length of economic run. With these machines, textile printers can prepare their own paper in the printing shed.[23] However, at present, more than 90% of the paper is printed with rotogravure printing machines.[24] These machines employ etched metal rollers similar to those used for textile printing. Half-tone effects are obtained by varying the depth and width of the engraved cells. Engraving depths are usually about 0.05 mm. Very deep engravings may not release the ink and heavier layers of ink may flake off as the dried paper is flexed. Since a thin film is applied to the paper, the ink must be fairly concentrated to build-up depth of shade on the fabric. This is best done with solvent-based inks and dye formulations free of dispersing agents or any other diluent. During the printing process, the shallow indentations are filled by rotating the roll in an ink trough and removing the excess ink by a doctor blade. Each color ink is applied by a separate roll and the paper is dried between each color application.

The advantages of rotogravure printing are as follows:[12] (*a*) The textile industry is already familiar with rotogravure printing (i.e., roller printing), (*b*) It measures an exact amount of ink, (*c*) Half-tones and fine lines are reproduced exactly, (*d*) Best registration of design from very fine half-tone to full-tone and complete solid shades, (*e*) Quick drying inks are used, (*f*) No contamination of printing plates or rollers, (*g*) No need of stopping the machine to correct off-registration, (*h*) An unlimited number of color units can be coupled together, and (*i*) Minimum use of ink, low wastage, and low printing cost. The limitations and disadvantages of the process are: (*a*) the high cost of engraving, (*b*) it needs very smooth paper, (*c*) the ink is solvent-based and (*d*) doctor blade defects such as streaks and scumming.

Rotary-Screen Printing.   This printing technique is the newest in the textile industry. Conventional machines can be adapted to printing papers by fitting new entrance and exist systems to these machines. The quality of prints is perhaps better than that obtained by direct (wet) printing of fabrics. Thus, the entire transfer-printing operation can be carried out on the mill premises. This enables better coordination and quick delivery of the printed goods to the customers.

In rotary-screen printing, the stencil is a hollow perforated metal cylinder through which the ink is forced by an internal squeegee. Aqueous print pastes are used to prevent the screen plugging that would occur if fast-drying, solvent-based pastes were used. The compositions are similar to the formulas used for printing fabrics. However, a thermoplastic resin is added to give adhesion between the paper and fabric while the dye transfer is made.

The limitations of rotary-screen units are (a) uniform ink application is difficult, (b) unsuitable for fine lines in designs, (c) contamination cannot be easily prevented, (d) screens are fragile, (e) high viscosity inks are needed, maximum ink usage, and waste, and (f) printing cost is high.

Two other printing methods of some interest are flexographic printing and lithographic printing.

Flexographic Printing. This method has the advantages of lower initial cost, quick changeover of designs, and suitability for printing short lengths of paper.[25,26] However, it suffers from design limitations, lower printing speed, and difficulty in building-up heavy shades.

In the flexographic method, the pattern is raised in relief on a flexible plate, usually of rubber. To prepare the flexo patterns, a negative mold is made from an engraving etched in zinc, manganese, or copper. The rubber is then poured into this mold and vulcanized to the desired hardness. The plate is mounted on the print-roller and the ink is fed in by a metering roller. Since many ink solvents can play havoc with rubber rollers and plates, it is recommended that only alcohols and glycol-ethers are used as solvents.

Lithographic Printing. Lithographic printing is a process based on the principle that oil and water mutually repel each other. Offset printing refers to the transferring of a print image to an intermediate rubber roller or plate which then prints it onto the paper. This permits a more uniform printing of fine lines and details. The method has the advantages of lower investment, lower cost of color plate preparation, and the possibility of soft, subtle shadings. However, the process is not a continuous one and gives lower production. It is also difficult to get heavy shades.

Tack. After printing, the paper is coated with resins or lacquers. The coat is called a *tack*. This tack is essential for the paper to stick to the fabric during transfer for true contact and without any relative movement of the paper and the fabric during the dye transfer. The contact has to be as perfect as possible because the rate of dye transfer decreases with the gap between the paper and the fabric.[16] The tack material should have thermal stability and should not possess any affinity to the dye, the paper, or the fabric. When the paper is lifted from the printed fabric, the tack helps to restore the feel of the crushed fabric by pulling out the surface of the fabric. Another advantage of the good contact is that all the dye sublimed from the paper is taken up by the fabric and does not diffuse away. The resistance to abrasion of the printed paper during storage is improved by the tack.

The cost of printing paper thus depends on a number of factors including the complexity of design, the number of colors, and the exclusive rights for the design.

15.1.3 Machines for Transfer Process

Transfer printing of fabrics and garments can be carried out in a continuous or batchwise manner using flat-bed processes or drums and calenders.[12,22,27–32] Thermopresses of different designs are available in which the transfer of dye to fabric, garments, or garment-blanks is effected with very uniform temperature distribution. They are employed for printing fully fashioned garments, knitwear, cut-and-sew panels, dress lengths, and motif printing. The fabric is placed on the pressbed, the paper placed on the fabric, and the hot-press head is then lowered onto the assembly (Figure 15.1a and 15.1b). Printing time and pressures are varied according to the fabric from 15–30 sec at 180–220°C. The production rates are of the order of about 1500 garments per shift.

Continuous Transfer Machines. In these machines, the fabric and the printed paper pass around a large cylinder (drum); the contact is maintained by an endless blanket under tension (Figures 15.2a, 15.2b, and 15.2c). Control of the blanket tension is vital to minimizing the flattening and glazing of the fabric which is to some extent inevitable because of the operating temperatures and the thermoplastic nature of the fibers. The cylinder diameter ranges from 0.5–2 m and the width ranges up to 3 m depending on the machine. The cylinder is heated by circulating hot oil or by electrical heating. Assuming a required contact time of 20 sec, the larger machines will give a production speed of 15 m/min.

Vacuum calenders were developed in order to increase the speed of these machines, to allow a substantial reduction in temperature of the machine, and to effect better dye penetration. An important machine belonging to this group is the Kannegiesser Vacumat Machine[12,29] (Figure 15.3). In this machine, the paper and the fabric are held against a perforated drum by the development of an internal vacuum. Heating is from the external side and is done by infrared elements. The control of the heating is done by measuring the temperature of the emerging paper. No pressure blanket is used so that flattening of the fabric is eliminated. The penetration of the dye into the fabric is excellent. The speed is 150–200 m/h. The Stork machine has a conventional cylinder type of construction but the whole unit is enclosed in a vacuum chamber with entry and exit seals.

15.1.4 Fabrics for Transfer Printing

Polyester. The transfer-printing process has achieved its largest success with polyester fabrics. Over 90% of heat-transfer printing is carried out on 100% polyester fabrics. However, only about 8% of all printed polyester is produced by this route, so that there is still considerable growth potential for this method of printing.[33]

The success of transfer printing on polyester is due to the fact that dis-

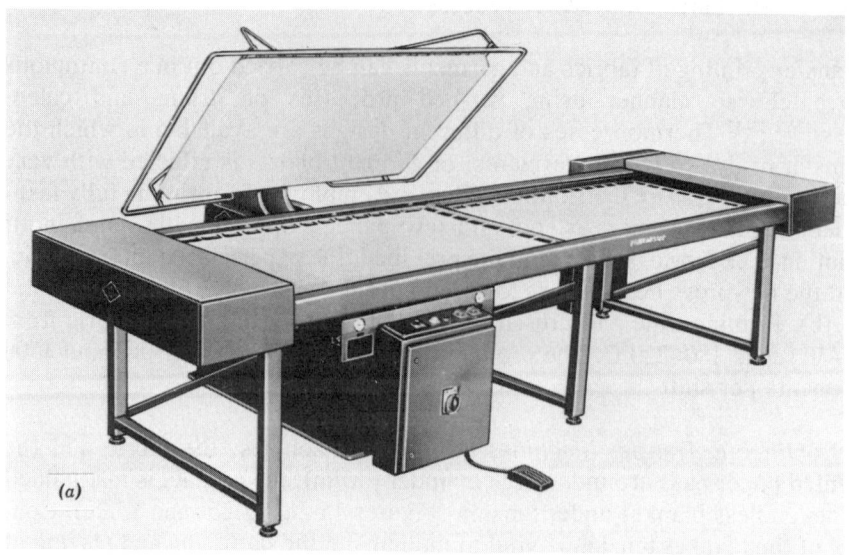

FIGURE 15.1.   Flat-bed transfer printing machines (*a*) and (*b*).

perse dyes, which are ideally suited for the coloration of polyester, are also suitable for transfer printing. Holland and Litherland[1] have studied the suitability of a large number of dyes belonging to different classes for transfer printing. They observed that 139 dyes are suitable for the transfer process at around 200°C. Of these 139 dyes, 91 are disperse dyes, 34 are cationic dyes, 2 are acid dyes, 2 are mordant dyes, and 9 are solvent and vat dyes. Out of these dyes, only 37 dyes are suitable for commercial application because the others do not have satisfactory fastness properties or build-up of shade. Most of these suitable dyes belong to disperse class.

Another reason for successful transfer printing on polyester fabrics is that, unlike nylon 6 or acrylic fiber, polyester fiber does not turn yellow or degrade when it is heated to 200°C for 1–2 min.[34] Since the transfer of disperse dyes is best carried out by heating the printed paper and fabric together to 190–210°C for 20–40 sec, polyester fabric gives the best color yields without any yellowing.

A wide variety of fabric constructions can be transfer-printed including woven, nonwoven, and knitted goods. The biggest success of transfer printing has been in the field of polyester double-knit jersey fabrics.

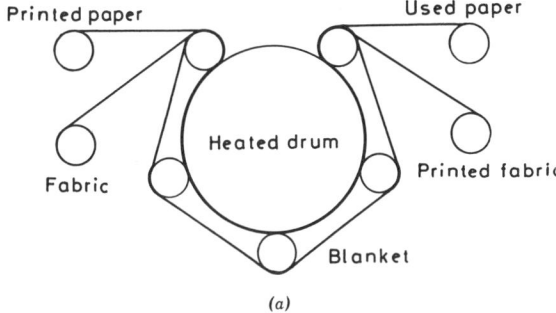

(a)

(b)

FIGURE 15.2. Heat-transfer calenders. (*a*) Schematic drawing; (*b*) Stork large-width transfer calender (courtesy of Stork Brabant BV); (*c*) Kannegiesser continuous thermoprint transfer machine (courtesy of Kannegiesser.)

FIGURE 15.2 (continued)

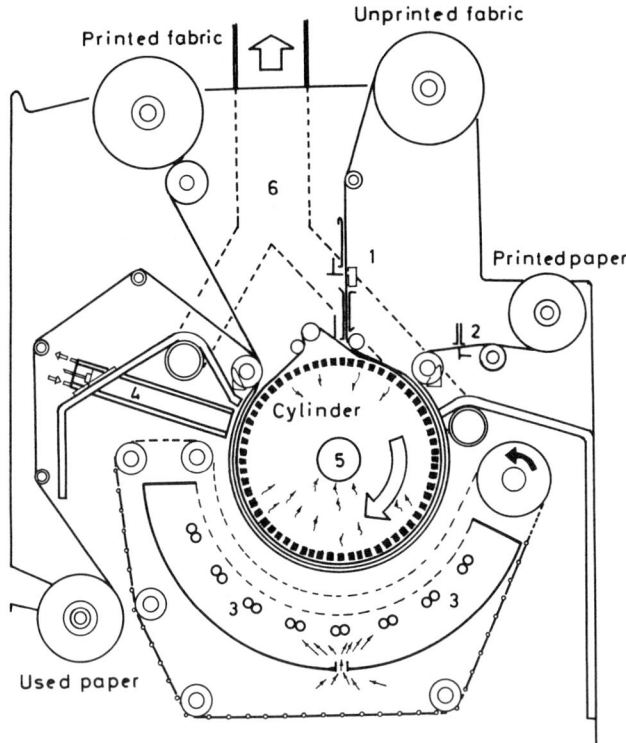

FIGURE 15.3. Kannegiesser vacumat machine. 1: Fabric cutter; 2: Paper cutter; 3: Infrared heaters; 4: Infrared thermometer; 5: Vacuum cyclinder; 6: Heat outlet.

Polyester–Cellulosic Blends. Blends in which the polyester component is more than 70% can be printed by the transfer-printing technique in a manner identical to that for a fabric of 100% polyester. The prints although skittery to the expert are generally acceptable. Since cotton and other cellulosic fibers in the blends remain undyed, blends with less than 70% PET are not suitable for transfer printing by the conventional method.

Transfer printing of 50 : 50 polyester–cellulosic blends is difficult because suitable dyes are not available for printing the cellulosic component of the blend. This is because almost all dyes with an affinity for cellulosic fibers also have an affinity for the transfer paper which is cellulosic in nature. Furthermore, the dyes used for the coloration of cellulosic fibers are not sufficiently volatile to get transferred from the paper to the fabric.

The major research work on transfer printing of polyester–cellulosic blends is directed towards developing methods to modify cellulosic fibers in the blend by a suitable pretreatment in such a way as to develop affinity for disperse dyes. Such blend fabrics can then be conveniently transfer printed with disperse dyes.

Blanchard et al.[35] have developed a series of pretreatments for imparting disperse dyeability to cellulosic fibers. The most suitable among them is the partial acetylation of cellulose. The acetylation is carried out at 25°C by treating the fabric first with 1.5% perchloric acid in glacial acetic acid followed by a treatment with 40% acetic anhydride again in glacial acetic acid. The fabric is then washed and dried.

Nishida et al.[36] have recommended a liquid-phase as well as a vapor-phase pretreatment for acetylation of cellulosic fibers. The vapor-phase pretreatment, which claims to give good results, consists of suspending the fabric in the vapors of a boiling mixture of trichloroethylene and acetic anhydride for 1.5–120 min. This is followed by washing with a dilute ammonia solution and then water. The fabric is subsequently dried and transfer printed. Prints with satisfactory fastness properties are obtained by this method. However, the strength of the fabric is considerably lessened by the pretreatment. Pretreatment with polyethylene glycol (mol. wt. ranging from 135 to 2000) along with dimethylol dihydroxy ethylene urea (DMDHEU) and an acid-liberating catalyst by the pad-dry-cure process is also reported.

Sublistatic SA and Heberlien have jointly developed the Hecowa Print Process. In this process, woven and knitted fabrics of 100% cotton or PET/ CO blend are treated with cross-linking agents prior to transfer printing. The method claims to give good printed effects.

Weiland and Robin[37] have suggested a pretreatment with N-methylol carbamate, polyethylene glycol, and a cross-linking latex. When this mixture is applied by the pad-dry-cure process it imparts disperse dyeability to cellulosic fibers. Another method consists of padding the fabric with a cyclic reactant-resin along with a catalyst and the other usual auxiliaries, drying at 100°C, and curing at 160°C for 4–5 min.[38] The goods are then transfer printed with disperse dyes. A similar method of pretreating polyester–cellulosic blends with cross-linking agents followed by transfer printing is suggested by Vellins.[13,39,40]

An interfacial polymerization technique was used to form polymers on the cellulose fiber surface which increased their affinity for disperse dyes. In a typical method, a polyester–cellulosic blend fabric is padded with an aqueous solution containing Bisphenol A (5.7%) and sodium hydroxide (2%). The moisture on the fabric is reduced to 20–25% after which the fabric is padded with sebacyl chloride (6%) in toluene. The fabric is dried, washed with water, dried, and transfer printed to get fairly durable prints.

In general, although a large number of pretreatments are reported to impart disperse dyeability to cellulosic fibers, none of them is satisfactory. The transfer printing of polyester–cellulosic blend fabrics is therefore not carried out on a large commercial scale.

Polyamide.   Nylon fabrics can be printed with disperse dyes by the transfer-printing process. The wet fastness of disperse dyes on nylon is relatively poor, particularly with the low-molecular-weight dyes used for transfer printing.[13] The prints on the nylon 66 exhibit superior wash fastness and, consequently, heat-transfer printing is used for printing nylon 66, which also has better thermal stability than nylon 6 (the latter gets discolored within 15–20 sec at 200°C). Attempts to improve the wet fastness of prints by using reactive dyes have failed primarily because of the lack of sublimation-transfer properties of these dyes. Even if a volatile reactive disperse dye is transferred to nylon, the reaction is not complete within the short time of dye transfer and unreacted dye remains on the fabric giving poor wet fastness[41] (Figure 15.4). Even with various pretreatments of nylon 66, the reaction cannot be completed in a short time. Such volatile reactive disperse dyes are not yet marketed and the process is not fully developed for commercial exploitation.

Acid and metal-complex dyes which are commonly used for dyeing nylon are unsuitable for heat-transfer printing because these dyes have high melting points and low vapor pressures and hence, do not get vaporized and transferred at about 200°C. However, the recently developed Dew Print machine enables wet-transfer printing of acid and metal-complex dyes on nylon[42,43] (see Sec. 15.6).

Acrylic Fabrics. Acrylic fabrics can be printed with either disperse dyes or cationic dyes. In the case of disperse dyes, the dye transfer is carried out at 195°C for 10–20 sec. Higher temperatures are not recommended because they cause yellowing of the fabric. Vacuum calenders are preferred for carrying out the transfer of disperse dyes at lower temperatures. Disperse dyes, in general, give dull shades and their wash fastness is not very good.

Some of the cationic dyes are also used for transfer printing. The transfer is carried out at 185°C for 20 sec. These dyes produce extremely bright prints. The light fastness of these dyes, however, is not very good (Table 15.2).[1] Cationic dyes which are widely used for dyeing acrylic fiber cannot be used in transfer printing because these dyes do not sublime under normal transfer conditions. A method to modify cationic dyes to make them suitable for heat-transfer printing has been suggested. In this method, the cationic

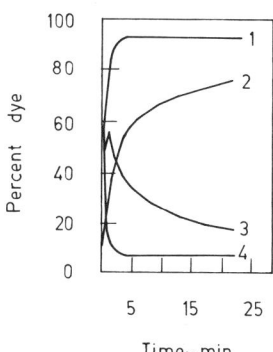

FIGURE 15.4. The transfer of a reactive disperse dye on PA 66 at 200°C.[41] 1: Total dye on fiber; 2: Reacted dye on fiber; 3: Unreacted dye on fiber; 4: Residual dye on paper.

TABLE 15.2   Fastness Properties of Cationic Dyes on Acrylic Fibers[1]

Dye	Wash Fastness	Light Fastness
C.I. Basic Yellow 11	5	4–5
C.I. Basic Yellow 13	5	4–5
C.I. Basic Yellow 29	5	4–5
C.I. Basic Red 1	4	1
C.I. Basic Orange 21	4–5	2
C.I. Basic Violet 16	4–5	1
C.I. Basic Green 1	4–5	2–3
C.I. Basic Blue 5	4–5	2
C.I. Basic Green 4	5	3

dyes are first treated with triethanolamine under controlled conditions and are then used for printing.[44] Modified cationic dyes that are suitable for printing on acrylic and cationic dyeable polyester are reported.[45]

15.2   ADVANTAGES AND LIMITATIONS OF HEAT-TRANSFER PRINTING

The advantages of the process[1,9,12,46,47] are:

1. *Completely dry system:* No steaming; no washing-off treatments; no drying, therefore, no migration problems and no effluent problems.
2. *Low cost:* Labor, capital, training of personnel, inventory of printed fabrics and dyes, and low (less than 2%) substandard and misprinted material. (The cost of storing printed paper is 20–50% that of printed fabric.)
3. *Better control:* Stock, repeat order (samples in 30 sec), no color matching, minimum run for test market, quick delivery, almost no downtime, and continuous or batch process.
4. *Better product:* Limitless design variations, outstanding definition with perfect pattern fitting, freedom for designers, brilliant prints, printing on both sides, garments or blanks can be printed, and better coverage of poor quality and irregular material.

The limitations of the process are:[12]

1. *Fiber:* Hydrophobic fibers only, paler skittery prints on blends, and pretreatment for cellulosic fibers essential.
2. *Dyes:* The color range and fastness of prints are limited, special formulations are required, and all conventional dyes are useless for the process.

3. *Substrates:* Expensive paper needing special preparation and finish, slow deliveries (advantages in avoiding pollution during transfer are counterbalanced by the pollution during the manufacture of the paper), no recycling of paper possible, and limitations on the width of paper (and fabric).

4. *Print-runs:* The longer the run, the lower the cost; slow rate of production (not much savings in time over conventional printing of textiles); color and design cannot be modified on printed paper at a later stage; exclusive designs are expensive; and skills and machinery in the printing of paper are required.

5. *Printed material:* Handle becomes glassy, penetration of color into the fabric is poor, textured material loses some crimp and bulk, and combination with other styles of printing is not possible.

6. *Responsibility:* Divided between paper printer and processor, shifts from printing house to finishing house, and no role for the skilled textile printer.

15.3 NOVEL EFFECTS

Although heat-transfer printing has attained popularity in a short period of time, it is very likely that many of the above limitations will be overcome in the near future. Many new ideas have exploited heat-transfer printing to produce different color effects. Some of these are mentioned below.[9,48–50]

1. Any absorbent paper (filter paper, tissue paper, unfinished paper) may be dipped into the disperse-dye liquor [containing a mild oxidizing agent (e.g., sodium metanitrobenzene sulfonic acid, 2 g/liter)], dried, cut into motif shapes, and placed between two layers of a fabric. On heating, the dye is transferred to the fabric, which gives sharp reproduction of the motif. By this technique, prints with a variety of motif and colors can be easily produced from many conventional disperse dyes. If the designed cut paper is stacked in a holder and allowed to drop on the running fabric at the desired position, it is possible to produce a floral or geometric design. If the cut paper pieces are mixed and dropped on fabric through a hopper, any random design can be produced. Such processes have no limitations on width, length, color, design and so on. The paper may be produced by padding, dipping, painting, screening, or printing in order to get the desired texture. The process can be adapted to suit the local needs. The two sides can be printed in two designs.[48]

2. A thin (muslin) cloth is padded with a disperse dye, placed on the PET fabric, and heated, if necessary under a vacuum or by transpirating hot air,[9,50] so that the disperse dye is transferred to the PET fabric. By placing a mechanical resist between the two fabrics, it is possible to produce designs. Mechanical resists can be cut paper, foil, metal mesh, another fabric printed

with a resistant film of thickening and so on. Alternatively, by heating the fabric locally, it is possible to transfer the dye as desired. Alternatively, by the selective suction (or compression) of hot air through the assembly, patterns can be produced. By using the porous, continuous substrate rolls, any number of colors can be transferred by this technique.[50] If the transpiration process is used, disperse dyes with good fastness properties can be transferred within a few seconds at 200°C. The substrate-fabric carrying the dye is washed free of residual dye and dried before the dye is applied again. Alternatively, the cotton fabric may be printed on the conventional machines and used as described above. Conventional disperse dyes in liquid form or granules can be used.

3.   For checks and lines, cotton threads may be used as substrate for the dye. Continuous threads, belts, or ribbons may be used for the purpose. Similarly, such continuous lengths may act as mechanical resists between conventional paper, random printing motif transfer from perforated or porous support and so on.

4.   The cut-outs of the conventional heat-transfer paper may be used to produce polka dots and random prints, saree borders, and designs.

5.   Paper for heat-transfer printing can be prepared with ''running'' colors or by mixing colors by the following technique: A porous paper (e.g., filter paper) is padded in a solvent. The dye preparation is sprayed on it when the preparation dissolves and flows down the length of paper. The paper is then dried and used for the transfer printing. In place of paper, cotton fabric also gives good results.

6.   A paper padded with a solvent is used for producing unibroken designs (similar to the color effects using encapsulated dyes). Thus, the paper (in a horizontal position) is sprayed with a mixture of dye formulations (or disperse dyes) and dried before use.

It is seen from the above examples that the heat-transfer process can be extended to produce a variety of color effects. Understanding the mechanism of dye transfer in heat-transfer printing will enable one to exploit the process fully.

15.4   MECHANISM OF HEAT-TRANSFER PROCESS

A number of articles and reports have appeared qualitatively describing dyes and fibers, conditions used on different machines, production rates, economics of the process and so on. Studies directed towards the elucidation of the mechanism of dry-heat printing by these processes, however, are very few.[9,12,14,41,49–51] It is now generally accepted that adsorption of dyes occurs mainly through the vapor phase by volatilization of the nonionic dye, even though it is still believed that dye particles directly dissolve in the fiber. Thus, the choice is restricted to dyes that have a comparatively high volatility (vapor pressure) at the application temperature. The other theories of dye transfer, such as dissolution of dye in auxiliaries or the physical transfer of dye by pressure, can now be rejected.

The stages in the migration of dyes on the paper to the fiber surface are: (a) sublimation of dye (Figure 15.5),[15] (b) diffusion of dye vapor to the fiber surface, and (c) condensation of dye vapor on the available fiber surface. These stages are controlled essentially by the properties of the dye. The stages in which the dye is received by the fiber material are: (a) adsorption on the fiber surface, (b) diffusion into the fiber phase towards the core, (c) diffusion along the fiber surface into the body of the thick fabric, and (d) saturation and equilibrium of the dye-sorption process.[50] These stages are also influenced by the properties of the fibers and the yarn and fabric constructions. For example, if the arrival of dye vapor to the surface is faster than its diffusion into the fiber phase, the rate of dye transfer is lower than when a slow diffusion process is not involved.[12,14] Because of this, the time and temperature of dye transfer depends on the fiber material (Table 15.3). The same dye on paper under identical transfer conditions may not give the same depth of shade on different fibers or on the same fiber material in different physical forms or weaves (Figure 15.6).[15]

The vapor pressure of a dye at a given temperature is an inherent property of the dye molecule and cannot be changed without changing the constitution of the dye. Increasing the volatility of a dye may improve the sublimation but it adversely affects the fastness properties of the prints.[52] The sublimation rate, however, can be easily modified by bringing down the size of the dye particles or by the transpiration of an inert hot gas over the surface of dye particles.[50] The sublimed dye vapor is saturated near the paper up to a distance that is equivalent to the mean free-path length of the dye molecules in the vapor phase at a given pressure and temperature. Since the mean free-

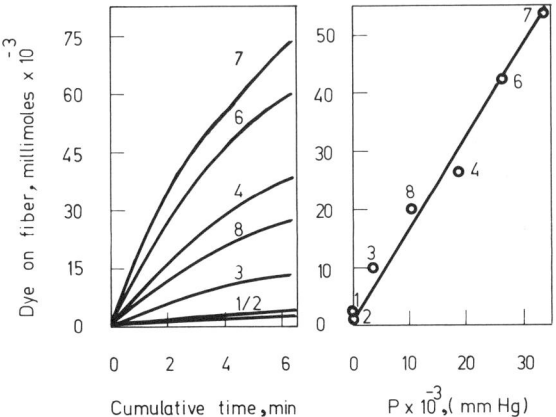

FIGURE 15.5. Transfer of dyes by sublimation to polyester and the relationship of vapor pressure (P) of dyes and sublime dye on fiber.[15] (For the formulas of the dyes, see Ref. 15.)

TABLE 15.3   Conditions of Heat Transfer on Textile Materials[12]

Fiber	Temperature Range (°C)	Time (sec)
Secondary acetate	190–210	15–30
Triacetate	190–210	20–40
PA6	190–200	20–40
PA66	190–210	20–40
Polyester	200–230	20–40
Courtelle	190	15–30
Acrilan, Orlon	200–210	15–30
Wool–PET	200–220	20–40
Cotton–PET	200–230	20–40

path length is very small, and as the dye vapor diffuses away from the paper, the concentration of the dye in the vapor phase decreases. The vapor thus becomes unsaturated, its concentration becomes dependent on the concentration of dye on paper, and the dye transfer to the fiber decreases with the distance between the paper and the surface of the fiber.[51,53] In the conventional heat-transfer printing, there is very little penetration of the dye vapor into the body of the fabric and, therefore, transfer printing essentially colors only the surface. The mean free-path-length of the dye molecules in vapor increases considerably by reducing the pressure. This property of the vapors has been exploited to improve the penetration of the dye into the body of fabric by application of a vacuum during dye transfer.[50] A vacuum increases the rate of sublimation and the penetration of dye vapor into the fabric.[9,50] However, it adversely affects the surface adsorption-condensation of the

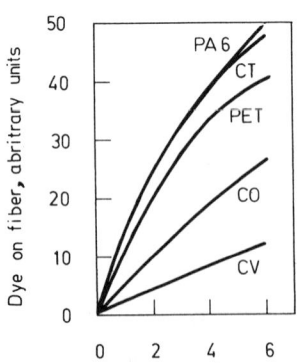

Cumulative time, min   **FIGURE 15.6.**   Transfer of dye on various fibers at 200°C.[15]

dye on the fiber. In fact, it facilitates the desorption process if and when the vapor concentration falls, giving unsaturated vapor. Thick materials of thermoplastic fiber-forming polymer get permanently crushed by the excessive load generated on the fiber by the vacuum at high temperatures. The vacuum helps to generate true contact between the paper and the fiber surface, thus helping the dye transfer. If the dye is fixed on the fiber, for example, by a chemical reaction, vacuum technique will give improved yields. In the future, reactive nonionic dyes with good sublimation rates will have to be developed for transfer printing. Alternatively, transpiration processes will be exploited to increase the sublimation rates. Research is aimed at getting bright prints with all-around good fastness properties on synthetic fibers and their blends.

15.5 FILM RELEASE SYSTEM OF PRINTING

The biggest limitation of heat-transfer printing is that it cannot be used equally well on all textile fiber materials. To overcome this difficulty, Rattee[54–56] has developed a process known as the Fabprint Melt Transfer Process. In this system, the paper is printed with a highly flexible polymeric film containing dyes and dye-fixation agents. The fabric and paper are passed through the nip of a heated calender at 60–150°C (generally at 100°C) with a nip pressure of 25 kg/cm^2. The coloring matter and the thermoplastic binder are transferred to the fabric under pressure and heat. After the release of the film, the paper is removed when the design is completely transferred to the fabric. The prints are then fixed by steaming or dry-heat treatment. The polymeric film breaks up during these treatments. Generally, washing of the prints is not necessary. However, for certain dyes such as reactive dyes which do not give 100% fixation, washing-off is required to remove the unfixed dye.

The Fabprint process gives prints with excellent color definition. Mark-off problems during fixation are negligible because of the nonoccurrence of melting. In this process, the transfer does not depend on any property of the dye such as volatility or solubility. It is therefore possible to use dyes of all classes. Disperse dyes of high sublimation fastness can be conveniently printed on polyester by this method, which is not possible by the heat-transfer printing method. Similarly, acid dyes can be printed on nylon and wool, and reactive dyes on cellulosic fibers.

15.6 WET-MIGRATION PROCESS

Migration printing in which water-soluble dyes migrate to the fabric with fixing chemicals in the presence of water was claimed by Datye et al.[57,58] Two principles are involved in these processes:[9] (a) the dye gets deposited where water as a liquid phase carrying the dye is converted into steam (i.e., the gaseous phase) and (*b*) the steam does not carry anything nonvolatile— solid or liquid—with it. A fabric is padded with alkali, salt, urea, and other auxiliary chemicals and is placed on a hot plate. A printed paper with a reactive dye and a hydrotropic compound such as urea which lowers the temperature on dissolution is placed on top of the wet fabric which is, in turn, covered with a perforated plate. Water in the fabric gets converted into steam which diffuses up and condenses on the paper. The dye on the paper dissolves in the condensed water and the solution returns to the fabric. This cycle continues until all the water evaporates. This process gives sharp prints on cellulose, nylon, acrylic, and wool. Similar processes based on this principle are called *wet-transfer processes*.[33,59]

Dew Print Wet-Transfer Machine.[33,59]   This machine is based on the Fastran Process of Dawson International and is suitable for printing different classes of dyes on various textile fibers. The wet-transfer process consists of three stages: (*a*) printing the required design on paper using a water-soluble acid, cationic, and reactive dyes dispersed in a suitable paper printing medium; (*b*) impregnating the fabric (nylon, wool, acrylic, cotton, and blends) with an aqueous solution containing a dye-fixation catalyst and a thickener, the latter acting as a dye migrating controller and film stabilizer; and (*c*) bringing the printed paper and the impregnated fabric together by applying pressure for a period of a few seconds to several minutes. The paper and the fabric are kept under moist conditions at a temperature of at least 100°C. Under these conditions, the dyes on the printed paper are almost completely transferred to the fabric and, provided sufficient time is allowed, are fixed in the same operation.

A schematic diagram of the Dew Print machine for carrying out stages (*b*) and (*c*) is shown in Figure 15.7.[33] The machine consists of a padding mangle in which the fabric is impregnated with a suitable dye-fixing agent, a thickener and so on. The wet fabric along with the printed paper are then passed through a transfer printing unit, which consists of a typical blanket calender

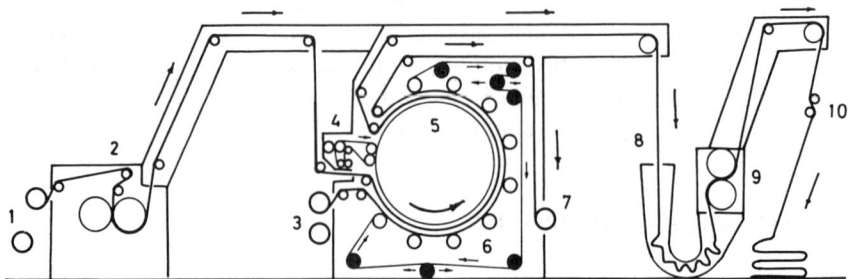

FIGURE 15.7.   Dew print wet-transfer machine.[33] 1: Fabric unwind; 2: Impregnation mangle; 3: Transfer paper unwind; 4: Bowl cleaning device; 5: Main transfer cylinder; 6: Pressure action rollers; 7: Transfer paper unwind; 8: Tensionless washer; 9: Wash-off nip; 10: Plaiter.

with a steam-heated cylinder and a water-impervious blanket or belt of high tensile strength. A number of pressure rollers are provided around the circumference of the heated cylinder so as to apply pressure to the fabric and paper sandwich in addition to the pressure afforded by the tension of the blanket. For a machine with a cylinder of 1.5-m diameter, the running speed can range from about 5 m/min for a heavy acrylic fabric to about 20 m/min for a cotton fabric. After wet-transfer printing, the fabric is washed to remove thickener, catalyst, and a small amount of residual (unfixed) dye.

Using the Dew Print wet-transfer machine, it is possible to carry out wet-transfer printing on acrylic fabrics with cationic dyes. Thus, very bright prints with good fastness properties can be obtained on acrylic fabrics.[33,36] Special printing papers based on metal-complex dyes have been produced.[42] With the introduction of this machine, the wet-transfer printing of nylon has become feasible. Ciba–Geigy Corp have also patented a wet-transfer process for printing acid dyes on nylon.[43]

REFERENCES

1. Holland, G., and Litherland, A., *JSDC*, **87** (1971), 488.

2. Jones, F., and Leung, T. S. M., *JSDC*, **90** (1974), 286.

3. Consterdine, K., *Rev. Prog. Color.*, **7** (1976), 34.

4. Demsey, E. P., and Vellins, C. E., *Heat Transfer Printing*, Interprint, U.K. (1977).

5. British Celanese Ltd., BP 293,022 (1929).

6. British Celanese Ltd., BP 349,683 (1931).

7. Anon, *DuPont Dyes and Chemicals Tech. Bull.*, **7**(4) (1951).

8. de Plasse, M. N., French P. 1,223,330 (1960).

9. Datye, K. V., *Textile Printing*, Book of Papers, Chavan, R. B., Ed., 2nd Annual Symposium, IIT, Delhi, India (Sept. 14–15, 1979), p. 15.

10. Jones, F., *Textile Printing*, Miles, L. W. C., Ed., Dyers Co. Publication Trust, Bradford, UK (1981), 65.

11. Gorandi, I. J., and Larson, O. S., *ADR*, **64**(2) (1975), 53.

12. Datye, K. V., *Colourage*, **22**(13) (1975), 29.

13. Vellins, C. E., *ADR*, **64**(2) (1975), 41; **64**(7) (1975), 29; **67**(2) (1978), 52; **68**(2) (1979), 38.

14. Datye, K. V., and Pitkar, S. C., Seminar on *Contribution to Chemistry of Synthetic Dyes and Mechanism of Dyeing*, UDCT, Bombay (1967), p. 47.

15. Datye, K. V., *Textilveredlung*, **4** (1969), 562.

16. Datye, K. V., Pitkar, S. C., and Purao, U. M., *Textilveredlung*, **8** (1973), 262.

17. Vellins, C. E., *ADR*, **63**(2) (1974), 18.

18. Aihara, J., *ADR*, **63**(7) (1974), 20.

19. Eibl, J., Gencer, O., Kerber, A., and Trauzettel, H. S., *MTB*, **54** (1973), 161.

20. Aihara, J., Nishida, K., and Miyataka, M., *ADR*, **64**(2) (1975), 46.

21. Kawamura, Y., *Japan Text. News*, **270** (1977), 89.

22. Burtonshaw, D., *MTB*, **54** (1973), 168.

23. Leijdekkers, P. E. J., *ADR*, **65**(10) (1976), 73.

24. Filatures Prouvost Masurel and Cie, B.P. 1,189,026 (1970).

25. Ball, M., *ADR,* **64**(7) (1975), 32.

26. Lund, G. V., *ADR,* **64**(7) (1975), 25.

27. Dowson, T. L., *JSDC,* **89** (1973), 474.

28. Fenoglio, R., and Gorondy, E. G., *TCC,* **7** (1975), 25.

29. Murphy, J. M., and Dowson, T. L., *TCC,* **7** (1975), 202.

30. Miles, L. W. C., *JSDC,* **93** (1977), 161.

31. Moore, N. L., *Rev. Prog. Color.,* **9** (1978), 73; *JSDC,* **90** (1974), 318.

32. Mock, G. N., and Jacumin, E. R., *TCC,* **14** (1982), 46.

33. Wild, K., *JSDC,* **93** (1977), 185.

34. Griffiths, J., and Jones, F., *JSDC,* **93** (1973), 176.

35. Blanchard, E. J., Bruno, J. S., and Gautreaux, G. A., *ADR,* **65**(7) (1976), 26.

36. Nishida, K., Katoh, T., Minekawa, K., Hotsuta, T., Iwamoto, H., and Toda, H., *ADR,* **64**(1) (1975), 36; **64**(4) (1975), 39.

37. Weiland, H. G., and Robin, A., *ADR,* **66**(7) (1977), 34.

38. Heywood, D. W., and Lambert, D. S., *JSDC,* **93** (1977), 195.

39. Vellins, C. E., *Text. Inst. Ind.,* **16** (1978), 367.

40. Vellins, C. E., *Text. Dyer Printer,* **13**(3) (1979), 33.

41. Datye, K. V., *Colourage,* **23**(2) (1976), 16.

42. Transprint Co., *Text. Month* (Jan. 1976), 67.

43. Ciba Geigy Corp., USP, 4,155,707 (1979).

44. Metropolitan Section, *TCC,* **6** (1974), 242.

45. Wallenberger, F. T., and Lauderback, S. K., *TCC,* **10** (1978), 127.

46. Schaub, J. H. W., *ADR,* **67**(4) (1978), 28.

47. Steinberg, D. J., *ADR,* **67**(2) (1978), 59.

48. Datye, K. V., et al., German Offen, 2,049,912 (October 20, 1969).

49. Datye, K. V., *Colourage,* **24**(4) (1977), 27.

50. Datye, K. V., *JSDC,* **96** (1980), 434.

51. Datye, K. V., *JSDC,* **94** (1978), 415.

52. Datye, K. V., Kangle, P. J., and Millicevic, B., *Textilveredlung,* **2** (1967), 5: **2** (1967), 263.

53. Datye, K. V., Pitkar, S. C., and Purao, U. M., *Colourage Annual* (1974), 47.

54. Rattee, I. D., and Lewis, D., BP (Application), 41,474 (1975).

55. Rattee, I. D., *Dyer,* **156** (1976), 393.

56. Rattee, I. D., *JSDC,* **93** (1977), 190.

57. Datye, K. V., et al., BP 1,227,271 (March 22, 1968).

58. Venkataraman, K., Ed., *The Chemistry of Synthetic Dyes,* Vol. VIII, Academic, New York (1978), pp. 191–220.

59. Anon, *Dyer,* **154** (1975), 312.

16 | CHEMICAL PROCESSING OF CARPETS

The art of carpet weaving is very old. Oriental carpets were made on vertical handlooms. The loom was strung with woolen warps and the pile was made by looping short pieces of dyed yarn around the warp strings in knots. When a row of knots was tied, the weft was passed across, combed, and hammered down to compact the finished portion.

The last three decades have seen a revolution in the pattern of use of fibers for carpets and in the manufacturing and coloration methods for floor coverings. The greatest development in carpet manufacturing was the tufting process. The principle of the tufting process is to pierce a carrier or backing fabric with a needle containing yarn, a looper then catches the yarn, the loop is released immediately or retained for later cutting, and then the process is gone through again and again endlessly (Figure 16.1). The main structural difference is that tufted carpeting lacks true interlacing of the piles within the backing. The rubberized latex used in the bond, however, gives both adequate rigidity and tuft bind. Higher speed and larger output account for the great expansion of the tufting process. The pile heights can be varied or cut or uncut yarns can be combined to produce curved and embossed textures. Multicolor effects are also possible. Carpets are produced by some other techniques such as knotting, needle felt and so on.[1]

The carpet industry is producing an ever-increasing proportion of world textiles. The two prime factors in the rapid development of the carpet industry are the rapid rise in tufted carpeting (Table 16.1)[2] and the advent of synthetic fibers.[3] Synthetic fibers can be produced in uniform staple lengths and diameters with mass-colored, differentially dyeable, antistatic, soil-releasing, and flame-retardant properties.[4-7] Wool is therefore giving way to

417

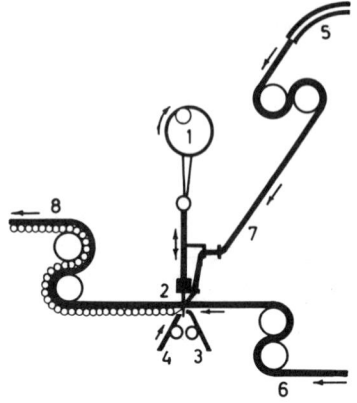

FIGURE 16.1. Schematic drawing of a tufting machine. 1: needle-drive cam; 2: needle; 3: looper; 4: knife; 5: pile yarn guide; 6: backing; 7: pile yarn; 8: carpet.

TABLE 16.1   Estimated Size of the Printed Carpet Market in 1974 and 1980[2]

Country	Production of Tufted Carpets ($m^2 \times 10^6$)		Printed in Piece Form (%)	
	1974	1980	1974	1980
United States	750	950	14	15
West Germany	125	140	7	12
United Kingdom	100	140	33	40
Belgium	5	100	2	15

synthetic fibers and, in 1979, the consumption of various fibers in the U.S. carpet industry was:[8]

79%	: Nylon
6%	: Polypropylene
4–5%	: Acrylic
4–5%	: Polyester
Negligible	: Wool

Nylon exhibits good floor performance properties and versatility of coloration and is the main fiber in carpets and floor coverings.

16.1   PRETREATMENT OF CARPET MATERIALS

Loose fibers, yarns, or carpet materials are scoured with nonionic surfactants to remove spin-finishes, coning oils, and impurities from the fiber materials before dyeing. Since carpets are invariably sold in colored form, bleaching is not required. Yarns are heat set in hank form or as a package using saturated steam.[9] A large number of machines for continuous setting of carpet yarns are available.[10,11]

In the Perfecta Set system, the yarns are fed from a tag-ending creel with 12 ends. A coiler lays a predetermined weight of yarn on a perforated stainless-steel conveyor in blanket form. The ends are laid out in layers so that after heat setting and during rewinding there is no risk of disturbing the layers. When the conveyor is full and the blanket prepared, the conveyor enters the autoclave for thermal treatment for a programmed period (normally about 4 min with intermittent evacuations). After heat setting, the blanket moves out of the autoclave into the uncoiling and cooling zones where the individual yarn ends are separated at a high speed. Assisted by nip rollers, the ends are then passed on to the take-up winder with 12 spindles. Processing a 1300 dtex/twofold filament nylon yarn at 400 m/min gives a production rate of 80 kg/h.

The TVP system (Superba SA, France) consists of two side-by-side processing units each capable of handling six ends of yarn at a rate of 80–100 kg/h. Yarns are relaxed and bulked at 98°C in a presteaming zone and are then heat set under pressure in saturated steam up to 150°C. The throughput-speed ranges up to 600 m/min.

Six ends of yarn can be processed simultaneously in a GVA continuous system. The setting temperature can be up to 220°C with superheated steam. The Relset (Gilbos NV, Belgium) method uses a dry-heat system in which yarn is heated inside metal tubing.

16.2 DYED YARNS FOR CARPETS

A carpet without color is almost a contradiction. The richness of coloring in a pile fabric such as a carpet in contrast to flat fabrics is due to the absence of specular reflection from the tufts and the consequent lack of dilution of the hue with white light. Color makes the design possible in a woven carpet because it gives the classic appearance that distinguishes a woven pattern. With a large number of colors, intricate designs and fine graduation can be achieved. In some Axminster designs, as many as 50 colors are introduced, although the current tendency is to hold it to 10 or 20. Still, the carpet manufacturer may have several thousand standard shades in his range. The carpet dyer has to obtain an accurate match to a particular shade, which has to remain a match under different lighting conditions.

Carpets are used as floor coverings and therefore, are prone to soiling. They are thick materials and cannot be easily washed. Once wet, they lose their appearance and texture unless they are brushed, combed, and refinished. Therefore, bright and deep colors are preferred. The carpet colored in uniform solid shades easily brings out any extraneous matter deposited on the surface. The dust and soil become prominently visible on such a carpet. It is therefore a practice to have a unibroken design on the carpet so that dirt is not easily noticed. Thus, an apparently monocolored carpet is in fact dyed in a unibroken pattern.

Coloring can be done at various stages in the making of the carpet. The important methods of dyeing and printing are (*a*) mass coloration, (*b*) fiber dyeing, (*c*) space dyeing, (*d*) yarn dyeing, (*e*) piece dyeing, and (*f*) piece printing.

Mass Coloration.   Carpet yarns of regenerated and synthetic fibers can be produced in various shades by mass-coloration techniques. The uniformity and reproducibility of shades, the absence of the barré effect, and the possibility of using stuffer-box crimping are some of the advantages. Carpets made out of mass-colored yarns are not wetted after manufacture and hence, have a better appearance and handle.

The limitations of mass-colored fibers are (*a*) only a small number of shades are possible; (*b*) permanence of color and no correction of unevenness, shade variation, or shades that have gone "out of style" is possible; and (*c*) if pigments are used as colorants, the shades are dull and less brilliant. (For details, see Chapter 4.) Almost all types of synthetic and regenerated fibers including polyester, nylon, acrylic, polypropylene, and rayon are mass colored to produce premium carpet products with outstanding color performance.

Fiber Dyeing.   Fibers are dyed by the loose-fiber dyeing method and are blended with different fibers to get stripe-free colored blend yarns.[12,13] The fiber dyeing method offers a wider scope for getting multicolored carpets and, therefore, the standard loose-stock batch dyeing machines are still widely used in the carpet industry. Stock dyeing of staple fibers involves uniform packing of the material into a pressure vessel and circulating the dye-liquor through the packed material as the temperature is increased to maximum. Dyeing is continued until the dyebath is exhausted. Dyeing is carried out at atmospheric or higher pressures. Alternatively, the fiber can be dyed in the densely packed bales supplied by the fiber producers using the Obermaier Bale Dyeing System[14] (see Chapter 6).

Continuous processes of loose-fiber dyeing involve impregnation of the fiber with dye-liquor on a brattice-fed padding mangle followed by steaming on a perforated drum.[15,16] Alternatively, the filaments can be continuously dyed in tow form before they are cut into staples (see Chapter 4).

16.2.1   Yarn Dyeing

Yarn dyeing offers a simple method for producing yarns in different colors for multicolored carpets. Small and medium-scale carpet manufacturers usually prefer the yarn-dyeing technique to the more complicated, capital-intensive carpet-dyeing technique and printing processes for multicolored effects. The yarn is dyed either in hanks or in a package form.

Hank Dyeing. The method has two main advantages:[12,17,18] (a) it produces yarn with good bulk and appearance and (b) it handles small lots efficiently. Thus, it is an ideal method for producing a large number of shades in small quantities. The drawbacks of the hank dyeing method are (a) a high liquor-to-goods ratio which entails a high cost in dyes, chemicals, and energy; (b) difficult control, which gives shade variations in a lot and from lot to lot; (c) entanglement of yarn which gives a high percentage of waste; and (d) hank reeling and rewinding onto cones after dyeing, which entails a high labor cost.

Hanks are hung on a perforated arm through which the dye-liquor is forced. The hanks are slowly rotated or rolled over the arms to improve the uniformity of the shade. Alternatively, they are fully immersed in the dye vessel or the dye is applied by the spray-jet technique. After dyeing, the yarn is centrifuged and dried before it is rewound on to cones.

Package Dyeing. The package-dyeing process offers the following advantages: (a) high-speed package-to-package winding, which results in increased production at a considerable savings in winding–rewinding costs; (b) low liquor-to-material ratio, thus savings in chemicals, water, and dyes as well as steam and energy; low effluent volume; (c) compact unit with low floor space requirements; and (d) versatility in handling any type of fiber. However, the handle and appearance of package-dyed yarns are quite inferior to hank-dyed yarns unless improved winding and dyeing machines are used. The package-dyeing units are capital-intensive, which entails a high initial investment.[19-21]

The conventional package-dyeing machines with capacities of up to 1000 kg can be used. Batches up to 4000 kg can be dyed in one operation by coupling a few dyeing units by a figure-eight arrangement of interchanging pipes or by the use of independent mixer units.[19] The rapid package dryer in which the carrier containing wet packages is loaded and hot air is circulated through the package, is used with package-dyeing machines. This dyeing method is not practiced on a wide scale for carpet yarns because of unevenness caused by channeling of the dye-liquor, shade variations from the inside to the surface of the package, small size of package, and high cost.

16.2.2 Space Dyeing

The process of applying different dyes at intervals along a yarn's length is called *space dyeing*.[22-26] The oldest method of producing such multicolored yarns is to immerse a part of the hank in a dyebath of a given color, to remove the hank, and to immerse another part of the hank in a dyebath of another color. Alternatively, colors may be applied on rotating spray bars.

Thus, the hank is dyed in three or more colors along its circumference. This method is time-consuming and the reproducibility and wet fastness are not satisfactory. Uniformity and consistency are almost always wanting in all space-dyeing processes.

The large-scale production of space-dyed yarns began in the early 1960s when they were employed in the manufacture of tufted carpets. The most important methods of production of space-dyed yarns are: (*a*) knit–deknit process, (*b*) random printing of the yarn web, (*c*) printing a yarn warp, and (*d*) yarn-package impregnation methods.

Knit–Deknit Process.   This space-dyeing method consists of knitting the yarn into a plain circular tube of fabric (8–65-cm circumference), printing the fabric with three or more colors, steaming, washing off, and pulling back the yarn so that it can be wound on to cones. Many variations of this basic principle are in use. The process is fairly simple. However, it is a lengthy process and may cause distortions in the crimp of the yarn during the deknitting operation. It is difficult to process staple spun-yarn by this technique.

Random Printing of the Yarn Web.   This is a development of the original hank-printing process that is aimed at improved productivity. About 400 ends of yarn are taken from a creel to form a web that passes between 3–4 pairs of rollers. The lower roller of each pair is run in a dye-liquor. A mechanical control device actuates the top roller to bring the web in contact with the lower roller so that the color is transferred to the yarn intermittently. The dye is then fixed as usual.

Printing of Warp Yarn.   In this process, the color is applied by printing stands. Each stand consists of two cylinders—one is mounted directly above the other; the lower one rotates in the dye-liquor. The lower cylinder consists of holder bars equally spaced around its periphery. Each holder bar carries spring-loaded plungers with printing pads and a program-bar across the width of the cylinder. The program-bars are moved mechanically to determine if the printing pads print or not, thus, producing a pattern effect. The combination of the pattern and color produced at each printing station gives the final design. The Craford Pickering system belongs to this class.

Package Impregnation Process.   A cross-wound package of yarn is injected with dye solutions at predetermined points using a series of hypodermic needles. Regular and controlled wetting of the yarn is achieved by suitable auxiliary agents and dyes. In another method, the dye solutions are injected at various points within a package that is rotating about its central axis. Penetration is achieved by centrifugal action. In the latter method, the colors are applied sequentially and not simultaneously.

Miscellaneous Space-Dyeing Methods.   In the Japanese Daito Eastern Color Dyeing Machine, the yarn is passed by the side of rotating plates that cause the yarn to be spattered with spots of dye solution of controllable size. In the Vald Henrikesen VH Syno Flow Type 60 machine, the hanks are stacked around a slotted stainless-steel drum and the dye is injected through the slots.

16.3   PIECE DYEING OF CARPETS

Carpet material is produced from undyed yarns and is then colored in piece form to get multicolored effects by using either differentially dyeable fibers or special dyeing techniques. Thus, the coloring can be manipulated per market demand with a minimal inventory and a short delivery time. Dyeing is carried out either as a batch process or a continuous process.

Beck Dyeing.   Nylon carpets are dyed by this old technique, particularly when differentially dyeable fibers are used.[27-29] The process involves rotating an endless carpet belt in a dye-liquor in a beck. An arrangement for the uniform heating of the dye-liquor and a winch with a speed roller (which keeps the carpet moving in open width without any creases) are provided for uniform dyeing. The success of dyeing depends largely on the control and uniformity of the temperature in the dye beck and the selection and addition of dyes across the beck. For PET carpets, high-pressure winch-beck units are used. After dyeing, the exhaust liquor is dropped and the dyed carpet is rinsed, soaped, washed, and drained in the beck. The carpet is in a tension-less condition during beck-dyeing. Thus, the feel of carpet is not adversely affected during dyeing. Energy consumption and pollution due to exhaust liquor are major problems. Low-temperature dyeing of nylon carpets,[30-34] repeated use of the exhaust dyebath,[35,36] and modified winch-beck units working with low liquor-to-goods ratios[37] are described to save energy and minimize pollution.

Open-width dyeing has the advantage over rope dyeing of uniform bulk, easy loading and unloading, and freedom from lengthwise creases. However, the tendency to side-to-center shading, expensive special winches, and high labor costs for open-width dyeing are such large disadvantages that 95% of all carpet-becks in the United States are operated using the rope-dyeing technique.

Beam Dyeing.   Tufted carpets in which the piles do not crush during piece dyeing are dyed on beam dyeing machines with air jets.[38] Air at about 10 kg/cm^2 pressure is injected into the suction side of the centrifugal pump by two jet nozzles. This forms a milky "foam" of air bubbles that keeps the layers of fabric apart, maintains the fabric bulk, and prevents water marking of the piles. Another important objective is to achieve a high rate of liquor circulation, thus, permitting a shortened dyeing cycle.

Continuous Dyeing Techniques.   Continuous dyeing machines have been developed to color carpets in open widths. These machines have achieved high productivity; low costs of labor, water, and energy; and low water pollution compared to machines for batch dyeing.[39,40] The good surface appearance of the carpet piles because of relatively mild mechanical action, the possibility of dyeing any type of carpet including easily damaged acrylic carpets, no mismatching of shades, and the capability of producing long runs of a uniform single color are responsible for the phenomenal growth of continuous dyeing techniques. The main limitations of these techniques are the high cost of machinery, uneconomical process for dyeing small lengths of carpet, and the necessity for careful selection of dyes, auxiliaries, carriers, and conditions of steaming.[41,42]

In continuous dyeing machines, the dye-liquor is applied to the dry or prewetted carpet in an open width and the color is fixed by steaming. The unfixed residual dye and the pad-liquor additives are washed out on an open-width washer connected in line. The carpet is then vacuum-extracted and dried. The dye-liquor is applied by slop-padding, nip-padding, or using the lick-roller–doctor-blade coating method. The slop-dye process depends on the equilibrium set up between the padded carpet moving upwards from the impregnated liquor and the downward gravitational draining of the picked-up liquor (Figure 16.2). The impregnated carpet is taken at least 3 m upwards before its direction is changed so that it can be introduced into a steamer. Alternatively, the quetch roller in the trough is vertically adjusted to control the dip length.

In the Kusters or Fleissner process, the impregnation is carried out on a prewetted carpet (100% wet pickup) by applying a closely controlled volume of dye-liquor (300–400%) from a doctor blade running against a lick roller (Figure 16.3). The level of the dye-liquor in the trough is controlled to get uniform application. The dye is fixed without intermediate drying by steaming in a horizontal, festoon, or suction-drum ager. Frostiness at the surface of the fiber is caused by the condensation of steam on the surface of the cold carpet as it enters the steamer. Frostiness is minimized by preheating the

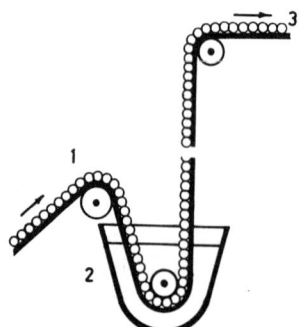

FIGURE 16.2.   BDA slop-dye process. 1: carpet, 2: trough with dye-liquor, 3: to steamer.

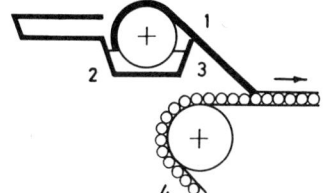

FIGURE 16.3. Kusters dye-applicator. 1: bowl; 2: dye-liquor trough; 3: doctor blade; 4: carpet.

carpet. The steaming time is decided by the class of dyes and the type of fiber.[43] After steaming, the carpets are thoroughly washed to remove all unfixed dyes, gums, thickeners, and auxiliary products. Special washing units are developed for washing the system, which consists of a vacuum washer or vacuum-mesh washer and a jet washer.[44] As the carpet leaves the steamer, it is guided down with its pile outward over the vacuum washer. In this position, water is then sprayed onto the carpet back and is sucked off immediately afterwards by a vacuum slot. In a jet washer, the water is sprayed vertically from a wide slot-jet over the entire width of the carpet. The water strikes the surface of the carpet and penetrates right down to the backing. Inadequate washing results in color bleeding or crocking. After washing, the water is vacuum-extracted through the back of the carpet. The carpet is then dried by forced hot air.

Low Liquor-Volume Processes. In the continuous foam-dyeing method,[45–48] the carpet is padded with a foam containing the dye and other auxiliaries. The amount of applied dye-liquor is only 130–150% of the weight of the carpet compared to 300–500% in the case of padding in an aqueous medium. Because of this, a great savings in energy in drying and steaming is realized. Furthermore, the concentration of the auxiliary chemicals can be reduced, the fixation of the dyes is much faster, and higher color yields are realized. The migration of chemicals and dyes is significantly reduced due to the low liquor add-on. The pile is more open, softer, and uniform. The washing of the carpets after dyeing is easier since less dye and chemicals need to be removed (see Chapter 20).

The Artos Fluid-O'therm System is based on a continuous exhaust dyeing in which a preheated carpet enters the dyeing chamber. A metered volume of dye-liquor is applied continuously on the advancing carpet. The submerged carpet moves through a very long narrow dyebath. Both carpet and dye-liquor move in the same direction in order to complete the dyeing in a few minutes. The carpet is then washed in a rinsing tunnel similar to the one used for the dyeing process, vacuum-extracted, and stenter-dried. Thus, this process is a combination of a batch beck-dyeing and continuous pad-dyeing techniques. It has the following advantages: (*a*) no thicking or foaming agents; (*b*) carpet is not crushed, creased, or folded; (*c*) no special formulations of dye-liquor; (*d*) low water consumption; and (*e*) uniform dyeing.

Backing Coloration.   The dyeing of the carpet is not complete until the backing of the carpet is dyed to roughly match the shade on the pile yarn. The dyes are selected to suit the chemical nature of the fiber in the backing. For jute and polypropylene, selected disperse dyes are used. Cationic dyes are also used for jute backings. Direct dyes are used for cellulosic fibers and acid dyes are the most suitable for nylon. These dyes are incorporated in the dye-liquor used for dyeing the carpet piles. Care must be taken that they do not adversely alter the main shade of the carpet.

16.4   MULTICOLORED EFFECTS BY DYEING

Space-dyed yarns have gained considerable importance in the production of random multicolored effects in tufted carpets. These yarns are, however, quite expensive and attempts have been made to develop methods of producing space-dyed effects by piece dyeing techniques using differentially dyeable yarns. Nylon is marketed as (*a*) regular dyeing, (*b*) deep dyeing, (*c*) light dyeing, and (*d*) cationic dyeing. Similarly, PET is available as regular dyeable, carrier-free dyeable, and cationic dyeable (see Chapter 12). By a careful selection of dyes and fibers, it is possible to dye multicolors in a single dyeing process. It is also possible to resist dye, thus leaving whites in an otherwise colored carpet. Differentially dyeable yarns are also used to get a unibroken design that gives the appearance of a single shade.

Kustur TAK Processes.   The TAK process consists essentially of a device for sprinkling droplets of dye solution on to a predyed or prepadded carpet.[44,49] Thickened dye-liquor is removed from the furnishing roller by an oscillating, serrated doctor blade. The resulting dye stream is broken into droplets by first falling on to a laterally running wire brattice and then on to reciprocating hooked comb-bars.[44] The color effect depends on the speed of these components together with the carpet speed and dye-liquor viscosity. Each machine can produce two colors plus the base shade and can be incorporated into a Kuster continuous dyeing range. The Multi TAK machine differs from the TAK machine in that it is equipped with additional counterdoctors that move in opposite directions to one another. This feature enables the machine to apply the dye-liquor to specific areas of the carpet, thereby, creating patterns. Further developments in the machine give a random pattern capability.[48] In the Gum TAK process, a thick layer of gum is applied on the surface of the carpet before the TAK drops are applied. This layer of gum allows the TAK drops to go beneath the surface of the carpet in order to yield a softer, more sophisticated appearance. The gum layer protects the fiber tips from being dyed and thereby creates a multilayer color effect. By combining a roller or rotary-screen printing machine with a TAK or multi TAK machine, it is possible to get highly attractive designs on carpets.

Stalwart–Pickering Spectra Color Process.   In this process,[50] a ground-color liquor with low viscosity is applied first, followed by two or more color pastes to produce the design. The latter tend to displace the ground-color liquor. Color or blank liquor applied at the spotting stage develops color or almost white spots without mixing with the ground color. To produce the design, two or more lick rollers transfer dye-liquor to stationary doctor blades. The streams of liquor obtained are then broken into droplets and are randomly distributed over the carpet surface. The quantity of dye-liquor delivered to the carpet surface is controlled by the running speed of the lick roller and the viscosity of the liquor.

Polychromatic Dyeing.   This process[51–54] can produce a virtually infinite variety of patterns on any fabric that can be processed on a continous pad-steam system. The principle involves feeding of streams of meterized volumes of dye-liquor to the carpet which is then immediately passed through the nip of a mangle. The jets or streams of liquor can be traversed to give wave lines or to run intermittently to give dots, blobs, or streaks in a pro-grammed manner. The penetration of the color into the carpet is excellent. The prewetted carpet resists the rapid uptake of further dye solutions and the first dye predominates in the final pattern emerging from the nip. The color is fixed by steaming and the carpet is then washed, vacuum-extracted, and dried in the standard manner. Polychromatic dyeing gives free-flow rough designs with approximate repeats. It is a cheap process for producing multicolored effects in a continuous manner.

16.5   PRINTING OF CARPETS

Handmade carpets produce rich color effects and designs. However, they are very expensive and the market demands attractively priced carpets that are truly patterned and multicolored. Differential dyeing and the use of space-dyed yarns were the initial methods of meeting this demand. Multicol-ored piece-dyeing techniques brought down the price of multicolored effects but the sophistication of design, color, and pattern of handmade carpets has yet to be matched. The true potential of patterned carpets is only recognized when one sees the growth in production of printed carpets (Table 16.1). Because it might be possible to produce sophisticated designs on carpets, printing is the area of carpet coloration that has received maximum attention and has experienced rapid development in the past few years.[2,55,56] Carpet materials of simple construction, produced on high-speed machines, and using one type of yarn can be printed to produce a range of qualities and patterns. This entails a savings in the yarn inventory and a rapid changeover to new patterns and designs that will reach customers and markets rapidly. Inferior quality carpets can be printed to cover up defects. Printed carpets have better soil-hiding properties. However, the printing process is not without limitations. Most of the printing machines are expensive. The penetration of the printing paste down the pile is not always satisfactory.[57] The selection of a thickener is important. It should assist in getting enough color on the carpet and should hold the pattern definition on the carpet surface, for which modified guar gum, locust bean gum, emulsion thickening, and synthetic thickeners are commonly used.[58,59] Most of the printed carpets are tufted nylon carpets. Carpets are printed by roller or relief printing, flat-bed screen printing, rotary printing, jet-spray printing, and heat-transfer printing (see Chapters 13–15).

Roller Printing.   The roller printing machines operate on a principle similar to the relief printing machine.[60] The raised pattern is produced by combining hard and soft rubber materials such as natural sponge rubber or appropriate neoprene brands and mounting them on a wooden roller. These materials take up slightly thickened dye-solutions from the trough and apply them to the carpet. The machine can print 3–4 colors on a carpet of about 3–4-m width. The penetration of the color is good and the speed is around 9 m/min. A typical such machine is the Stalwart Roller Printer.[60]

Flat-Bed Screen Printing.   This least-expensive carpet-printing process gives good penetration of the color in the carpet. The method can produce a large number of patterns with free-flow or abstract designs. However, fine lines, geometric and oriental designs requiring accurate registration, and printing on fine-gauge low-pile carpets cannot be produced precisely by the screen printing process.

In the Peter Zimmer TPA flat-bed printing machine, the usual squeegee is replaced by a pair of metal rollers.[60] The print paste is pumped into the space between the rollers. Powerful electromagnets are arranged under the screen so that they move across the screen in a longitudinal direction and pull the rollers along, which results in good pile penetration.[57] The screen is either polyester or metal with mesh sizes specially designed for the carpet. The dye paste is fed in from a series of sets of nozzles positioned above the screen—one set for each roller. The carpet is transported on an endless belt with small needles that penetrate into the back of the carpet, thus ensuring that the carpet is firmly positioned. The machine can print 12 colors of 5-m width at a speed of 4–5 m/min. After printing, the carpet goes through a long horizontal steamer, multiple wash-boxes with pressure water sprays, and multiple nips to remove residual dye and to extract as much water as possible. This is followed by drying.

In the BDA unit,[60] the color is applied by a sponge or twin squeegees. The print paste is sucked into the pile from below by a vacuum in order to ensure good penetration. The carpet must be backed to allow free suction through the carpet. The Mitter Semimatic flat-bed screen printing machine is suitable for printing bordered carpets with large repeats.[60] The unit uses a twin squeegee and the application of the print paste is assisted by a suction slot with high suction. The printing screens are stacked over the area to be printed, each screen is lowered in turn automatically on to the carpet, moved along up, and returned to the top of the pile. This makes the unit compact and space-saving.

Rotary-Screen Printing.   Flat-bed machines require larger screens for printing and occupy more floor space than rotary-screen machines. Furthermore, the screens may cause marks and carpet displacement, thus, spoiling the design. These defects are absent in rotary-screen printing.[61] In rotary-screen printing, the carpet moves continuously rather than intermittently, which enables this method to have faster linear speeds.[7] A disadvantage of rotary-screen machines is low penetration of the color compared with flat-bed machines. Rotary-screen machines for carpets employ the same basic principles as for textile printing but operate for larger widths. Three special squeegee systems are available: the Mitter machines has a positively driven squeegee, the Peter Zimmer machine has a friction-driven magnetic roller squeegee, and the Peter Zimmer hydroslot system has the print paste under hydrostatic pressure.[57,62] In a Johnnes Zimmer three-stage printing unit known as a A + Z (Artos Zimmer),[63] only a part of the print paste is applied in the initial stage. The printed side of the carpet is exposed to IR radiation in order to partially fix the dye. In the last stage, blank print paste is applied to bring about complete penetration of the piles. The dye is fixed by steaming in the normal way. A final washing-off and drying complete the process.[64] Other machines are described by Abramson.[13]

16.6   JET-SPRAY PRINTING MACHINES

The typical jet-spray machines are (*a*) the Millitron machine, (*b*) the Titan machine, (*c*) the Chromotronic machine, and (*d*) the Coloromatic machine.[65]

The Millitron Machine.   The Millitron machine[13,60,66] is a computer-controlled, micro-jet, color-injection dyeing machine. It can reproduce any color or complicated printed pattern at the relatively high speed of 6–8 m/min without the aid of screens or pattern rollers. The design to be duplicated is carefully photographed, corrected, and transferred onto a magnetic tape with the help of computers. After the dye-liquors are fed to their storage tanks and the carpet to be dyed is threaded into the machine, the operations become automatic and are controlled by the magnetic tape.

The carpet is passed under a boxlike unit that is over 5 m long and approximately 60 cm wide. Within this area a series of air jets force the dye-liquors into the carpet. Each jet is individually computer-controlled, the dye penetration and pattern definition are outstanding, and the patterning capabilities are almost unlimited. During the dyeing process nothing but dye-liquor is in contact with the face of the carpet and, consequently, there is no pile distortion or crushing. This results in a very attractive surface appearance and pattern clarity. After tearing the dye head, the carpet is steamed, washed, and dried in the standard manner.

The Millitron machine has a number of outstanding advantages. It is a universal dye machine capable of producing both solid colors and intricate patterns. There are no screens or pattern rollers. This dramatically reduces the cost of screens, space, and clean-up time, both on and off the machine. The machine has the flexibility of changing both pattern and/or size of repeat within a couple of minutes. Relatively short runs of highly specialized contract or oriental designs can be produced quite economically.

The Titan Machine.   This machine was developed by an Australian firm, Godfrey Hirst and Co. Ltd.[67,68] It consists of various types and specifications of color-application heads which can be added if necessary to the existing installations such as rotary or flat-bed or continuous-dye lines. The Titan Mark 4B/124/156/RR is a computerized color-applicator jet head. Each head delivers a single color and is controlled by a computerized Electron Pattern Controller (EPC). The Titan Jet printer can produce multicolored random or symmetrical designs. Changeover at high speed (15 m/min) from one design to another is possible without stopping the machine and with barely 3 cm of wastage. The unit can be programmed in advance to cover a full day of production. Digital Drawing Monitoring (DDM) equipment is supplied to provide the pattern input for the EPC which controls the Mark 4B heads. The head can produce either 3-m or 5-m wide carpets or bordered rugs. The claimed advantages of the Titan Jet Printer are competitive price, good efficiency, good flexibility, and simplicity in operation.

The Chromotronic Machine.   The Peter Zimmer Chromotronic machine[69–71] jet-spray system has a fast opening and a closing nozzle that is moved close to the substrate. The jets are self-cleaning because of the high pressure created during the closing of the jet. In a jet plate with eight colors, one carriage machine is equipped with 384 jets/color. An electronic control system is provided for both the mechanical machine parts such as the blanket advance, the carrier controls and so on, and for the time and position precise impulses for the jets and the relevant jet amplifiers. On an eight-carriage chromotronic machine, the computer has a 15 megabyte magnetic-disc memory and a magnetic station responsible for the input of the pattern data. Thus, the machine gives sharp pattern outlines, sharp definition throughout the depth of piles, sufficient penetration without any frosty effect, low color consumption, and trouble-free operation.

The Colormatic Machine.   The machine consists of preparation units, jet heads, an evenflow applicator, a steamer, a washer, and a dryer. The main parts of this jet-printer[65] preparation unit are sew-on machines for carpet-piece sewing, a double *J*-box for scouring, a back beater for cleaning and opening the carpet surface, a presteamer for yarn conditioning, a padder for applying the ground dye and chemicals, a center-line guider, and an evenflow applicator to apply gum and/or dye in a solid sheet to the surface of the carpet. The next units on the line are three-to-eight jet-printing heads—one for each color—with nozzles mounted directly above the carpet. When the valve is opened by a signal from the computer, the dye-liquor is forced under pressure on to the carpet. The carpet passes below at a speed of up to 20 m/min. The pressure on the dye-liquor can be adjusted as required for the pattern or the base material. The jet heads fire the dye-liquors one after another in order to give proper registration to the design on the moving carpet. Minor pattern shifts can be done electronically for registration or correction. The computer senses the speed of the carpet and controls the jet operations accordingly. The jet head can be purged, washed, reloaded, and blended in about 18 min while other heads in the line are operating. Pattern changes without color changes are made almost instantaneously without any loss of carpet material. After the last jet head, the carpet on the transport belt passes to another evenflow applicator which applies a solid sheet of gum and/or dye to the carpet in order to give a low viscosity wash or a gum layer or to obtain a particular effect. Solid shades can also be dyed using the foam technique. After the carpet is taken to a steamer with a horizontal section (18–20 m) followed by a vertical loop section (60–65 m capacity), it is washed and dried on a stenter and a mesh conveyor.

Advantages of Jet-Spray Printing.   The different types of machines described above have the following advantages.[65] They can produce almost any look currently offered in the carpet trade. They are very fast and can reach speeds up to 15–20 m/min. Computerized operations eliminate the use of screens, their maintenance and repair, downtime, space, cost, and time for their manufacture and storage and so on. The loss and seconds of carpet material during jet-spray printing are very low.

16.7 TRANSFER PRINTING OF CARPETS

This technique can print extremely intricate design details and very short runs. However, full-width carpets are not printed by the transfer-printing technique; only small pieces and floor tiles are printed. These carpet tiles are then assembled on the floor to get the desired design. The principle of transfer printing of flat textile materials (see Chapter 15) is to transfer the dyes from a special type of paper on to the surface of a carpet by the application of heat and pressure and/or vacuum. The dyes are transferred by the sublimation process[72,73] and both nylon and polyester can be readily colored. However, it is not suitable for acrylic, polypropylene, cellulosic, or animal fiber.

The major problem of transfer printing is the penetration of dye into the body of the carpet.[74] In thick materials, such as carpets, the dye has to penetrate the piles and diffuse from the tips of the exposed fiber piles to their roots. This can take place through the fiber phase or can move conveniently and rapidly along the length of the fiber pile (Figure 16.4).[74] The amount of dye per unit area on the substrate has to be higher than in flat material in order to color the thick material. Drastic conditions of time and temperature may be needed to vaporize this large amount of dye from the paper, which may impair the appeal of the thick carpet materials. Diffusion of dye in the fiber phase has to be rapid and the affinity of the dye has to be low, so that the dye vapor will fully penetrate into the fiber as well as the body of the thick material. Datye[75] has shown that the rate of arrival (condensation) of dye on the fiber surface is always higher than the rate of dye transfer to the fiber which is controlled by the slow diffusion of dye in the fiber phase. Thus, the dye vapor is available for migration outside the fiber phase along the length of the carpet piles. The penetration of dye into the body of the thick material is therefore dependent on the vapor pressure of the dye. The carpet industry has not accepted the transfer-printing process under vacuum, even though penetration of the dyes under vacuum is claimed to be deeper.[76] The main objection to vacuum-transfer printing of carpets is that a permanent deformation of the piles can occur when they are in the thermoplastic state.

In a typical tufted carpet material, about 20–25 percent of the silver strand length is exposed in the surface of the carpet pile and is easily accessible to the dye vapor without any penetration process. Under conventional-transfer

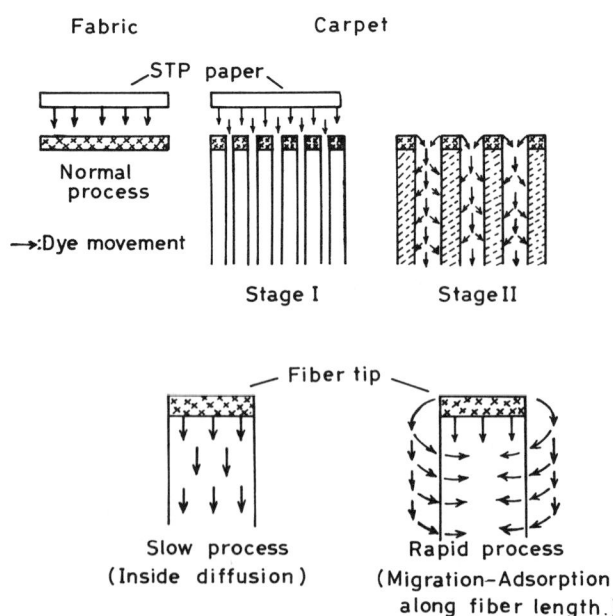

FIGURE 16.4. The mechanism of dye transfer to flat and carpet material by the sublimation transfer process.[74]

printing conditions, these exposed surfaces receive the dye. On the other hand, the remaining 70–80% of the carpet material remains uncolored under these conditions. For practical coloring, it is necessary to color at least 85–90% of the silver strand length (10–15% percent in the back of the carpet may be neglected). Datye[74] showed that the penetration of the dye into the body of the carpet is not influenced by the load on the carpet–paper assembly, concentration of dye, formulation of dye, repetition of a transfer process of short duration, heating after transfer, and temperature. The penetration is found to be a function of the vapor pressure of the dye. If the vapor pressure is the same, the penetration of the dye remains the same, irrespective of temperature of transfer printing. The vapor pressure of the dye can be increased only by increasing the temperature. However, the transfer cannot be carried out at very high temperatures because it is detrimental to the appearance of the carpet. At elevated temperatures, synthetic fibers soften, get crushed, and are permanently pressed in deshaped form. Thus, alternative techniques must be developed in order to get better dye transfer without an increase in temperature.

Even though the vapor pressure of a dye at a given temperature cannot be increased, the rate of vaporization can easily be increased by blowing away the dye vapor with hot gases. This process is called *the transpiration process*. The directional flow of hot gases carrying the dye vapor away from the back of the carpet material improves the penetration. Even commercial dyes of high sublimation fastness penetrate the carpet materials almost completely (85–90%) within 20–30 sec at 190–200°C. All types of carpet materials containing polyester and nylon are fully colored without any crushing of the material by the transpiration–transfer-print process.[74] Heating under similar conditions but without the directional flow of air, that is, by the thermofixation process, produces only a marginal improvement in the penetration.

The transfer–transpiration process can be carried out by taking the dye on a perforated substrate such as filter paper or lightweight cloth such as cotton voiles.[74,77] Lightweight cloth as a substrate for nonionic dyes and hot air suction through the carpet give brilliant uniform coloring of any depth throughout the carpet within 60 sec at 180–200°C. A process has been reported in which cotton voile is padded with a primary color such as yellow, red, and blue and is used with a stencil for the transfer–transpiration process.[77] Mixed colors are obtained using two dyes on two layers of voile (substrate). With the right combination of the three primary colors, any hue or color can be easily produced in any depth without crushing the carpet or causing it to lose its fresh appearance or handle in the process. After use, the voiles are washed, repadded, dried, and reused. The transfer–transpiration process can be carried out in one or two stages. In the two-stage process, the dye is deposited on the tips of the fiber piles by either printing on the surface or the conventional transfer-printing process. Hot gas (air) is then forced through the carpet to obtain uniform penetration. In the one-stage process, the dye may be taken up on a porous substrate and hot gas is then sucked out or forced through the substrate, the carpet, and out of the carpet to complete the transfer–penetration phase in a single transpiration step as described above. These two processes give satisfactory coloring without damage to carpetlike materials within a very short time of heating.

16.8 FINISHING OF CARPETS

Drying, shearing, steaming, back-coating, and application of special finishes are the operations involved in the finishing of carpets.

Drying. The wet carpet is vacuum-extracted and given a rapid vibration and back-beating on a series of rapidly rotating serrated rollers or on back-beater beltings to open up the crushed pile and to remove soft creases and distortions in the carpet. The water content in the carpet must be uniform before it enters a continuous drum tenter dryer.[78] This dryer design embodies the advantages of the principle of flow-through on a suction drum combined with a carpet transport or a stenter chain. This achieves high specific evaporation; uniform drying in a relaxed undistorted state; absence of turbulence, thus no fading of the pile; low energy consumption. Drying is done very carefully because it affects the handle and surface appearance of the carpet. The carpet is cooled after drying in order to avoid fiber deformation during rolling.

Shearing. A number of pills and protruding fibers are generally present on the carpet surface after dyeing. They spoil the appearance of the carpet.[79,80] It is therefore necessary to lightly shear the carpet in order to remove the pills and fibers. The shear is a unit that heavily brushes the carpet pile in order to make it erect and uniform. The carpet then passes over a series of rotary knives or blades that shear or cut off the fiber tips at a precise, adjustable height. In carpet shearing, it is common to have either two or four shear blades operating in tandem.

Steaming. The carpet is steamed to improve its appearance. The conditions of steaming depend on the type of fiber present in the carpet.

Back Coating. Back-coating is a must for carpets because the tufts inserted by needling into a comparatively light prewoven backing fabric of jute or polypropylene are only loosely attached.[81,82] This treatment is also given to some woven carpets. The main objectives of a back-coating treatment[1,13] are (a) to bind the pile tufts securely in position thereby stabilizing the carpet and making it durable, (b) to give dimensional stability, (c) to confer a firm but flexible hand, (d) to produce nonfray properties, and (e) to improve sound and thermal insulation.

Typical back-coating materials are starches, glue dispersions, polyvinyl acetate copolymers, polyvinyl chloride, natural and synthetic rubber latex, polyacrylates, polypropylene, polyethylene and so on. In addition to the main ingredients, other additives are also incorporated into the back-coating latex. These include fillers such as clay and calcium carbonate and stabilizers such as zinc oxide, color, pigments, antioxidants, thickeners and so on.

The usual back-coating method is by pickup from a lick roller or by a spreading technique using a doctor blade. The viscosity of the coating material should ensure adequate penetration without danger of wicking through to the pile side of the carpet. After a back-coating material is applied, the carpet is dried at a temperature of 140°C. High-solid latex systems contain only 17–18% moisture compared with 23–24% moisture in normal latex. Because of this, the energy required for drying this latex system is low.

In certain tufted carpets, a secondary backing of jute or synthetic fiber is also used. The carpet is first treated with back-coating chemicals. A separate roll of secondary backing (jute or synthetic material) is positioned onto the latex coating and the two materials are pressed together. This laminate is then dried and cured at 110–150°C for 4–5 min.

The polyurathane in the back-coating of carpets has a number of advantages such as excellent tuft bind, high resilience, low smoke emissions on burning, and a low flame-spread rating. Some of the processes for polyurathane back-coating of floor coverings are described by Fleissner.[78]

Special Finishes.   Static accumulation in carpets is responsible for the shocks that a person receives while walking on a carpet. It is also responsible for attracting more soil to the carpet. In order to minimize this problem, antistatic synthetic fibers as well as antistatic finishing agents for carpets were developed. Fluorochemical soil-retardants provide protection against oily stains, waterborne stains, and dry soiling[83] (see Chapter 19).

Carpets made from natural and synthetic fibers are flammable. In a number of countries, regulations have been enacted prohibiting the use of flammable carpets in public buildings. Therefore, efforts have been made to develop flame-retardant carpets (see Chapter 18). The backing as well as the finishes applied to the carpet should be flame-retardant.

Antibacterial finishes are often applied to carpets. The principle types of chemicals used as antibacterial finishes are chlorinated and brominated phenols and salicylanilides, organic tin compounds, and quaternary ammonium compounds.[84,85]

REFERENCES

1. Robinson, G., *Carpets and other Textile Floor Coverings,* Textile Book Service, U.S.A. (1972).
2. Dunkerley, K., *Rev. Prog. Color.,* **11** (1981), 74.
3. Budding, J., *Rev. Prog. Color.,* **6** (1975), 29.

4. Crawshaw, G. H., and Ince, J., *Text. Prog.*, **4**(2) (1972), 1.

5. Arbuckle, A. W., *Dyer*, **155** (1976), 254.

6. Crawshaw, G. H., *Text. Prog.*, **9**(2) (1977), 34.

7. Turner, G. R., *TCC*, **13** (1981), 51.

8. Ridgway, B., *Text. Month* (May 1980), 23.

9. Dawson, T. L., *JSDC*, **91** (1975), 289.

10. Kroh, K. D., *Chemiefasern/Textilindustrie*, **29/81** (1979), 599.

11. Anon, *Dyer*, **164**(3) (1980), Carpet Dyer, 9.

12. Dowson, T. L., *Rev. Prog. Color.*, **2** (1971), 3.

13. Abramson, C. C., *Practical Carpet Dyeing, Printing and Finishing*, Reg. Burnett Inc., (1978), p. 6.

14. Anon, *Dyer*, **164**(3) (1980), Carpet Dyer, 16.

15. Mayer, U., and Riechert, M. A., *ADR*, **57** (1968), 1104.

16. Dowson, T. L., *TCC*, **1** (1969), 336.

17. Burley, R. W., Flower, J. R., and Rattee, I. D., *JSDC*, **85** (1969), 187; **87** (1971), 278.

18. Raymont, J., *JSDC*, **88** (1972), 122.

19. Philpott, G. C., *Dyer*, **153** (1975), 298.

20. Limbert, K., *JSDC*, **91** (1975), 299.

21. Park, J., *Dyer*, **158** (1977), 481.

22. Badertscher, P., *Textilveredlung*, **2** (1967), 864.

23. Pfizenmaier, C., *ADR*, **56** (1967), 221.

24. Cratchley, G., *Text. Month*, (Dec. 1971), 61.

25. Beal, W., and Warwick, J. J., *JSDC*, **90** (1974), 425.

26. Roberts, B. P., *Rev. Prog. Color.*, **9** (1978), 13.

27. Park, J., *JSDC*, **84** (1968), 601.

28. Crowder, E., *ADR*, **60**(11) (1971), 28.

29. Budding, J., *JSDC*, **90** (1974), 232.

30. Beiertz, H., *ADR*, **68**(6) (1979), 22.

31. Vavala, L., *Chemiefasern/Textilindustrie*, **29/81** (1979), 193.

32. Toon, J. J., *ADR*, **69**(6) (1980), 26.

33. Stakelbeck, H. P., and Engeler, E., *MTB*, **62** (1981), 579.

34. Petty, J. B., *ADR*, **70**(6) (1981), 34.

35. Tincher, W. C., *ADR*, **66**(5) (1977), 36.

36. Cook, F. L., and Tincher, W. C., *TCC*, **10** (1978), 1.

37. Rudolf, T., *Dyer*, **162** (1979), 441.

38. *Tufting and Needling News*, (May 1973), 8; (Feb. 1974), 1.

39. Hobson, P. H., *TCC*, **4** (1972), 232.

40. Fryer, S. E., *JSDC*, **90** (1974), 229.

41. Stern, H., and Kaiser, L. E., *JSDC*, **85** (1969), 653.

42. Stewart, N. D., *JSDC*, **89** (1973), 258.

43. Skelly, J. K., *Text. J. of Australia*, **48** (Sept. 1973), 48 and **48** (Oct. 1973), 22.

44. Zimmerli, K., *ADR*, **67**(6) (1978), 19.

45. Nambodri, C. G., and Gregorian, R. S., *ADR*, **67**(6) (1978), 27.

46. Weber, R., and Tuschen, J., *Chemiefasern/Textilindustrie*, **30/82** (1980), 745.

47. Hartman, W., *ADR*, **69**(6) (1980), 21.

48. Dawson, T. L., *JSDC,* **97** (1981), 261.

49. Dawson, T. L., *JSDC,* **90** (1974), 235.

50. Hoechst, A. G., Ger. Pat. 133,541, 154,941 (1971).

51. ICI, *Polychromatic Pattern Dyeing by Dye Weave and Flow Form Techniques,* Tech. Information D 1140 (July 1970).

52. Newton, C., *MTB,* **52** (1971), 728.

53. Stetson, G. R., *ADR,* **60**(2) (1971), 20.

54. Fox, M. R., *JSDC,* **89** (1973), 17.

55. May, J., *ADR,* **62**(11) (1973), 42.

56. Bradertscher, W., *MTB,* **54** (1973), 513.

57. Kramrisch, B., *Dyer,* **157** (1977), 23.

58. Barnhardt, G., *TCC,* **11** (1979), 224.

59. Perkins, T. G., *TCC,* **11** (1979), 213.

60. Wirtz, J., *Chemiefasern/Textilindustrie,* **29/81** (1979), 195.

61. Mitter Co., *Textilveredlung,* **6** (1971), 393.

62. Dunkerley, K., and Hughes, J. A., *JSDC,* **91** (1975), 265.

63. Zimmer, J., *Textilbetrieb,* **93** (April 1975), 53.

64. Bodenhorst, P. B., *TCC,* **11** (1979), 161.

65. Harrel, B. L., *ADR,* **69**(6) (1980), 19.

66. Dawson, T. L., and Roberts, P., *JSDC,* **93** (1977), 439.

67. McKendrick, G. A. R., *ADR,* **67**(6) (1978), 42; **69**(6) (1980), 28.

68. Anon, *ADR,* **68**(6) (1979), 39.

69. Kudlich, H., Proceedings of a Symposium on Advances in Preparation, Coloration and Finishing, SDC, Sheffield, U.K. (Sept. 1979).

70. Zimmer, P., *ADR,* **69**(6) (1980), 32.

71. Eible, J., *MTB,* **62** (1981), 565.

72. Datye, K. V., Textile Printing 2nd Annual Symposium, IIT, New Delhi, India, Sept. 1979, p. 15.

73. Datye, K. V., *Colourage,* **27**(20) (1980), 3.

74. Datye, K. V., *JSDC,* **96** (1980), 434.

75. Datye, K. V., *JSDC,* **94** (1978), 415.

76. Anon, *Dyer,* **161**(6) (1979), Carpet Dyer, 12.

77. Datye, K. V., *Advances in Textile Chem. Processing,* IIT, New Delhi, India (May 20–June 5, 1971), p. 327.

78. Fleissner Co., *Dyer,* **164**(3) (1980), 6.

79. Anon, *Dyer,* **161**(6) (1979), Carpet Dyer, 24.

80. Bottom, K., *Dyer,* **164**(3) (1980), 11.

81. Phipps, J. N., and Porter, D., *TII,* **12** (1974), 68.

82. Dinsdale, R., *TII,* **14** (1976), 321.

83. Bierbraur, C. J., Goebel, K. D., and Landucci, D. P., *ADR,* **68**(6) (1979), 19.

84. Klesper, H., *Text. Prax.,* **30** (1975), 101.

85. Anon, *Carpet Review Weekly* (Jan. 6, 1977), 42.

17 | FINISHING OF SYNTHETIC FIBER MATERIALS

Finishing aims at increasing the aesthetic value, serviceability, and comfort of textiles of synthetic fibers and their blends. It also covers or overcomes defects, or imparts certain desirable properties. The ultimate aim of a finishing process is to develop and/or modify the fibers or finishes for conventional synthetic fibers so that they give the comfort of natural fibers, the richness of animal fibers, and all the advantages of synthetic fibers—durability, plasticity, attractive appearance, wash-and-wear characteristics and so on.

Many of the finishes are temporary because they are not fast to washing. Softeners, stiffening agents, calendering and so on fall into this category. Some finishes last longer and therefore are called *permanent finishes*. An antistatic finish can be permanent or temporary depending on the durability of the finish. Wash-and-wear, permanent-press, flame-proofing, and antipilling finishes are permanent finishes.

Synthetic fibers have many features that make finishing these fibers and their blends quite different from finishing natural or animal fibers. Most of the synthetic fibers are hydrophobic in nature and have a very low moisture content. This creates the following problems. (*a*) Synthetic fibers accumulate static electricity and attract soil from the atmosphere. The removal of the soil during aqueous laundering is difficult. (*b*) Synthetic fibers are uncomfortable to wear in humid weather. (*c*) The application of finishes from aqueous liquors becomes difficult.

Synthetic fibers are stronger and tougher and their fine structure is more compact than natural fibers. Therefore, they may need drastic temperature and time conditions to impart a finishing effect. Because of their high strength, the synthetic fibers wear slower than the natural fibers. This creates many problems; the most important is pilling during wear. Most of the synthetic fibers and their blends have a very serious flammability problem. Many synthetic fibers melt easily and stick to the body of a user during burning. Advantage can be taken of their thermoplastic nature to permanently impart dimensional stability, crease resistance, and pleating.

It is possible to incorporate finishing agents into synthetic fibers during their manufacture. Thus, finishes on synthetic fibers can be broadly divided into two types: (a) A finishing agent incorporated into the polymer prior to the spinning of the filaments; (b) A finishing agent applied to the filaments after spinning. Many times, the polymer is modified by using a comonomer or another polymer to form a polymer alloy so that the desired properties are imparted to the filaments. This enlarges the scope of finishing synthetic fibers to an almost unlimited extent. Furthermore, such effects may be permanent.

Various finishing effects are described in the present and following two chapters. Both ways of achieving the desired finish, namely, before and after the spinning of the filaments, are given so that all possible ways of finishing are available for comparative study. Some of these methods may not fall under the category of finishing. In fact, many times a modification in the fiber material during polymerization gives the desired effect. A few methods of testing the efficiency of the finish are also given.

17.1 OPTICAL WHITENING

White is a color of great subjective brightness—chromatic but with such a low saturation as to be almost achromatic. While greater subjective brightness always has a positive effect, the hue may have a positive or a negative influence on the perceived whiteness; the hue with a blue or bluish-violet tone will always appear whiter than the one with a yellow cast. Textile fibers do not appear perfectly white because of the presence of colored impurities that absorb some of the incident light in the blue end of the spectrum (400–480 nm). In fact, the textile fibers show a pronounced yellowish tinge. Application of a tinting dye or blueing agent (dominant wavelengths lie around 460 nm) will increase the whiteness but will reduce the subjective brightness. It is only if saturation and subjective brightness are increased simultaneously [for example, with optical brighteners or fluorescent whitening agents (FWA)], that the whiteness level continues to increase. Although blueish-white is always assessed better than yellowish-white, a considerable difference of opinion may arise when comparing blueish-violet–white with more neutral blue or blueish-green–white.[1]

17.1.1 Mechanism of Whitening Action

The yellow tinge on the fiber can be reduced but not completely eliminated by chemical bleaching.[2] Blue tinting compensates for the yellow appearance:

the blue dye absorbs the excess yellow portion. This makes the textile appear whiter. The correction, however, is only obtained at the expense of the reflected light. Fluorescent whitening agents can provide the blue-violet light needed to compensate for the yellow without absorbing any visible light.

When a specimen transforms a part of the absorbed light into light of another wavelength instead of into heat (as is the case with normal dyed specimens), it is called a *fluorescent specimen*. The fluorescent light has a longer wavelength than the absorbed light. The fluorescent brightening agents absorb ultraviolet light (λ 300–400 nm) from daylight and emit it in the visible range (λ 400–500 nm) at the blue–violet end of the spectrum (Figure 17.1). As a result of this light conversion, the textile reflects more visible light in the range of blue light and the spectral reflectance may increase to compensate for the yellow tinge (Figure 17.2). There is an increasing trend to enhance the degree of whiteness by fluorescent brightening agents (FBA). These fluorescent brighteners compensate for the absorbed blue component of the light by their blue fluorescent light which is excited by the ultraviolet component of daylight. At the same time, they also increase the luminosity of the goods. Thus, the brilliant bluish-white shades that appear whiter, more brilliant, and brighter to the eye than ideal white are produced by increasing the spectral reflectance to >100%.

The reflectance curves in Figure 17.2 are for magnesium oxide (neutral white standard for 100% reflectance) and cotton specimens unbleached (1); chemically bleached (2); chemically bleached and tinted with a blueing agent (3); chemically bleached and treated with FBA (4); chemically bleached, tinted with a blueing agent and FBA (5). The blue tint cancels out the yellow tinge in the bleached fabric (substractive process). The cloth appears more neutral in tone because the indigo-blue complements the orange-yellow. However, since there is less reflection than before, this white appears dull. Tinting colors are used in low quantities along with FBAs in order to get the desired white tint in daylight and artificial light (Table 17.1). Optical brightening agents are effective in daylight, which contains some ultraviolet light, but not in tungsten or other artificial light which does not contain ultraviolet

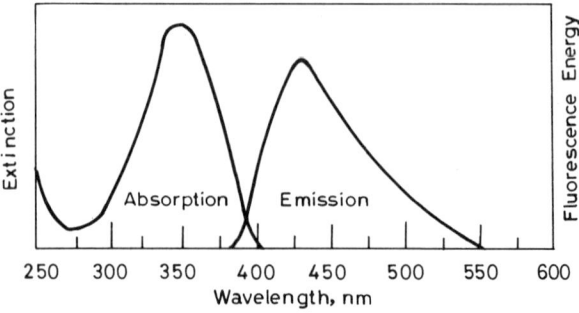

FIGURE 17.1.   Absorption and emission spectra of a solution of a fluorescent whitening agent.

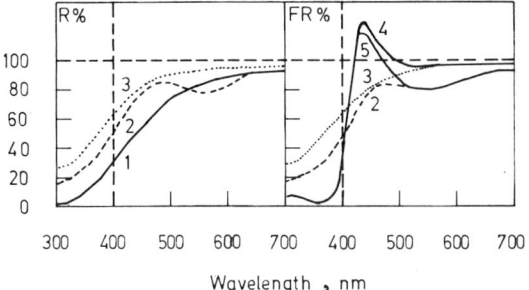

FIGURE 17.2. The reflectance and fluorescent reflectance of magnesium oxide (– – –) and white fabric under various conditions.[1] (For details, see text.)

light. The whitening action of FBAs on cotton is also valid for synthetic fibers.

Whitening Process. Fluorescent whitening agents (also called *optical brighteners*) resemble dyes in all respects except that they have no visible color and are thus called *colorless dyes.* The first FBA found [IG-Farben (1940)] was 4,4'-diamino stilbene-2,2'-disulfonic acid. Various stilbene derivatives, naphthalimides, benzimidazolyl and benzoxazolyl derivatives, amino cumarins, and pyrazolins were developed.[1] FBAs suitable for addition during polymerization, drying of chips, filament-spinning, and spin-finishes are now used to produce white synthetic fibers. The FBAs must be thermally stable so that they can withstand the high temperature necessary for polymerization, polycondensation, and melt-spinning without any noticeable degradation. They must not be volatile or sublime from the melt. Interaction between the polymer chain and the FBA is possible but must not lead to changes in the polymerization process or the finished fiber material

TABLE 17.1 Effect of Tints and Fluorescent Brightening Agents on White Produced in Daylight and Artificial Light[1]

FBA Fluorescence	Blueing Tint (Tinge)	Appearance	
		Daylight	Artificial Light
Violet	Nil	Blueish tinge	No effect
Green	Nil	Blueish tinge	No effect
Violet	Green	Blueish tinge	Green tinge
Green	Violet	Blueish tinge	Violet tinge
Violet	Violet	Violet tinge	Violet tinge
Green	Green	Green tinge	Green tinge

(see Section 17.1.7). Generally, the material to be dyed may be produced without any FBA-type whitening agent since the latter may affect the color adversely. The whitening agents are preferably applied after the fabric is woven or knitted, by the conventional dyeing techniques[3] or, preferably, during the bleaching process. FBA are usually applied by the exhaustion method or the pad-dry-cure methods. For blend fabrics, a mixture of two FBAs that are suitable for both the component fibers is employed. When the carrier method is used for PET brightening, the choice of a suitable carrier is considerably restricted; the best carrier is the chlorinated benzene type. For polyamide, the bleaching agents can be of the reducing type, and the brightening agent, if applied during bleaching, must withstand the reducing action of hydrosulfite. Polyamide fibers give excellent and permanent whiteness if they are treated with sodium borohydride[1] prior to fluorescent brightening. It is important to know the concentration of the brightening agent on the fiber that gives the maximum whitening effect. Higher concentrations cause a noticeable drop in the whitening effect while lower concentrations fail to give a full whitening effect. Many fluorescent brighteners exhibit excellent fastness properties on one fiber and not-so-good properties on another fiber. The effect of the softener, the antistatic agent, and the resins on the properties of the FBA must be evaluated before selecting the FBA. Thus, the brightening agent, its concentration, and the method of application must be properly selected for each fiber in order to achieve the best fastness properties. The liquors of the brightening agents are not stable to light and cannot be stored for a long time or exposed to light. The pH of the FBA solution is properly maintained in order to get the desired final tone of the white material. An optical brightener should not be used on dyed materials because it flattens the shade. A bactericide (e.g., Sanitized of Sanitized Inc., Lexington, N.Y.) that produces a sanitary finish on polyamide and blend fabrics can be incorporated in the fluorescent brightening bath.

The method for stripping the brightening agent should be properly selected so that the fabric does not develop a latent fault. For example, if stilbene-based brightening agents on polyamide are destroyed by a chlorite bleach, they give permanent yellow patches when exposed to light in a wet state after domestic washing. The best method of correcting the faulty application of a brightening agent is to brighten the goods again instead of stripping the faulty material.

Tone of a Fluorescent Brightening Agent. It has already been mentioned that the fluorescent brightening agents absorb the invisible (UV) portion of sunlight. They reemit the same at a longer wavelength, which is visible light comprising spectral colors from red to blue. Since blue is in excess, the reemitted light may be considered a mixture of white and blue light; this is observed as blueish fluorescence. During the process of neutralizing the yellowish–off-white appearance of the fabric, a proportionate amount of grey is formed. For the complete conversion of fabric yellowness to grey (black and white), it is necessary for red, yellow, and blue to be present. A reddish tinge is usually present in the fabric. However, if it is not present, the fluorescent brightening agent should have a slight reddish component. Different fluorescent brightening agents have different tones and, in practice, it is not possible to get absolute white by use of fluorescent brightening agents alone. The desired white is usually obtained by combining a fluorescent brightening agent and a shading color in the finishing bath. The shading colors of red, violet, blueish-red, and green may be selected from a wide range of disperse, cationic, and acid dyes. The concentration of shading colors is in the range of a few mg per kg of the fabric. The shading color must be stable under the conditions of application and use. It will however show its color in artificial light (Table 17.1).

17.1.2   Application to Polyester

The fluorescent brightening agents suitable for application to polyester fabrics are insoluble in water and are usually supplied as greenish-yellow dispersions or in liquid form. They produce a good white effect on polyester materials and have satisfactory wash, light, and sublimation fastness.[4-6] The trade products most commonly used are Tinopal ERT, Uvitex ERN, Palanil White RR, Mikawhite ATN and so on.

Carrier Method.   The bath in the jigger is set with

0.1–1.2%	Fluorescent brightener
$X\%$	Carrier
0.1–0.5%	Acetic acid to get pH 6

The material is entered at 40°C, run for 15–30 min, the temperature is raised to boiling in 30 min, and is maintained at boiling for 30–60 min. The goods are then rinsed with hot water and dried. The degree of whiteness for a given FBA is decided by the carrier and its concentration (Figure 17.3).[7]

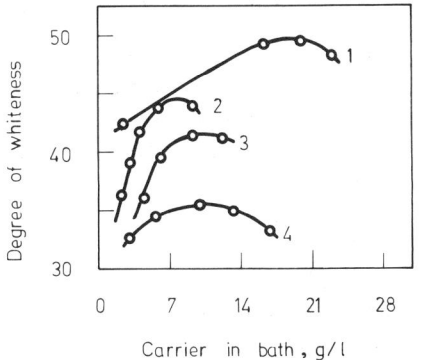

FIGURE 17.3. Degree of whiteness of PET fabric with different carriers and their concentration.[7] 1: Palanil Carrier A; 2: Palanil Carrier B; 3: Levagol PE; 4: Palanil Carrier PE. FWA: Palanil White R (0.6%). Brightening: 95°C/90 min.

HT Method. The fabric is treated with FBA in a HT-beam dyeing machine or in a jet dyeing machine at 120–130°C using 0.5–1.5% FBA, and 0.5–1% dispersing agent (optional) at pH 5 to 6. The goods are entered at 40°C and the temperature is rapidly raised to 120–130°C in 30 min. The temperature is maintained for 30–45 min before the bath is dropped. Goods are rinsed with hot water. The temperature of application greatly influences the degree of whiteness.[7]

Thermofixation Method. This is a continuous application method of FBA to polyester fabric. The bleached fabric is padded with 5–25 g/liter fluorescent brightener at room temperature. It is then carefully dried at 100–110°C. This is followed by thermofixation at 180–200°C for 30–60 sec. Final washing and drying complete the process.

Pad-Steam Method. The bleached fabric is padded with a dispersion of FBA at room temperature, dried or the wet goods are run continuously through the steamer. The goods are steamed at 100–101°C for 2–3 h or under pressure (1–1.5 kg/cm^2) for 25–30 min. Alternatively, the goods are thermofixed by superheated steam at 150–170°C for 3–5 min. Textured goods are brightened by the steaming process because the handle of the fabric is not adversely affected by the treatment.

17.1.3 Application to Polyester–Cotton Blend Fabrics

The FBA for the PET in the blend fabric is applied by the methods described above. The goods are then either bleached with hydrogen peroxide or given a reduction-clear treatment with sodium hydrosulfite to remove the brightening agent deposited mechanically on the cotton portion of the blend. This step is essential because the effect of the fluorescent brightener for cotton is adversely affected by the presence of the FBA for polyester. If the thermofixation method is used for the FBA on the polyester portion, the bleaching or reducing aftertreatment is also useful in removing the yellowness of the fiber that might have occurred during thermofixation. After the aftertreatment, the cotton portion of the blend is treated with a fluorescent brightener for the cellulosic fiber. In certain cases, the fluorescent brightener for cotton is applied along with resin finishing.

Combined Chlorite Bleaching and Optical Whitening. The FBAs for polyester are stable to chlorite bleaching. A method for simultaneous chlorite bleaching and application of fluorescent brightening is shown below.

The bath in the jigger or winch is set with

1–5 g/liter	Sodium chlorite
0.5–1.5%	Fluorescent brightener (on the weight of the fabric)
	Acetic acid to pH 4

The goods are treated at 40°C for 15 min. The temperature is raised to boiling in 30 min and treatment at boiling is carried out for 30–60 min. The bath is drained and the goods are given a treatment with an FBA for cotton.

Combined Resin Finishing and Optical Brightening.   In this process, the goods are padded with a liquor consisting of

X g/liter	Cross-linking agent
Y g/liter	Catalyst (such as magnesium chloride)
Z g/liter	Softener
5–15 g/liter	Fluorescent brightener for polyester
1–2 g/liter	Fluorescent brightener for cotton
0–1 g/liter	Wetting agent

dried, and cured at 140–150°C for 4–5 min, where the fluorescent brightener is fixed on polyester and the resin on cotton. A final scouring followed by washing completes the process. Although the process is economical and gives very high production, the brightening effects are not very intense. This is because the yellowish tinge resulting from setting is not removed by bleaching and the cotton fibers are not cleared of the polyester FBA.

17.1.4 Application to Nylon Fabrics

A variety of optical brightening agents are available for application to nylon by different techniques.[8-12] Some of the optical brightening agents for cellulosic fibers which resemble acid dyes can be applied to nylon fiber under acidic conditions.[13] However, special products such as Tinopal WG, Uvitex WGS, Blanchphor WT, Tinopal WHN and so on were developed for nylon. The FBA are applied as below.
    The bath is set with

0.05–0.5%	FBA
1–2%	Formic acid to pH 3.5–4

The goods are treated at 40°C for 10 min. The temperature is then raised to boiling in 15 min and the treatment is continued for 30 min. The bath is then drained and the goods are rinsed. The light fastness of these anionic products is not all that could be desired. The wash fastness is about 3.
    Tinopal WHN and similar products are applied to nylon, preferably by the HT method at 110–120°C.
    The bath is set with

0.2–2%	FBA
5 g/liter	Sodium hydrosulfite
0.1–0.5 g/liter	Metal chelating agent

The bleached goods are entered at 40°C and are treated at this temperature for 10 min. The temperature is raised to 110–120°C in 30 min. The goods are treated at this temperature for 30 min. The bath is cooled to 80°C and then drained. The light fastness of the product is around 3 and wash fastness about 5.

In addition to acid-dye type, FBAs, a few disperse-dye type FBAs (which are normally employed for polyester) are used for nylon. They exhibit better light fastness (3–4) and wash fastness (4–5 at 60°C) than acid-dye type FBAs.

The application method consists of setting the bath with

0.1–1%	Fluorescent brightening agent
0.5–1%	Dispersing agent
	Acetic acid to pH 5–6

and treating at 100°C as described earlier.

FBAs can be applied to nylon by continuous methods.[14] In the acid-shock process, the fabric is padded with a cold solution of FBA. The goods are then treated in a boiling bath containing formic acid in an open-width washer, washed, and dried. The thermosol process of application involves padding with a solution of FBA and acetic acid, drying, and thermofixing at a temperature of 170–190°C for 30–45 sec. A final rinsing with hot water completes the process.

17.1.5 Application to Acrylic Fabrics

Acrylonitrile fiber can be mass-colored with FBA by adding the agent directly in the solution (dope) of the polymer or in the spin-finish bath in the wet-spinning process. The FBA must be readily soluble in the solvent and insoluble in the water–solvent mixture used in the coagulating bath.

The nature of the copolymer in the acrylic fiber determines the affinity of the fiber for various fluorescent brighteners. The cationic-type brighteners are used for acrylic fibers containing anionic copolymer,[15] whereas fibers containing cationic copolymer are brightened with anionic FBAs, similar to those used for nylon.

The method of application of cationic FBAs is as below.

The bath is set with

0.5–1%	Fluorescent brightening agent
3–5%	Formic acid to get pH 3–4

The treatment is started at 50°C and the temperature is raised to boiling in 15 min. The treatment is continued at boiling for 30 min. This is followed by washing and drying. The fastness properties of these agents are quite good. A continuous process of application of cationic-type brightener is suggested.[16,17] The fabric is padded with a solution of brightener in water under acidic conditions and steamed for 10 min. Special types of FBAs have been suggested for the continuous process.

Some of the disperse-type optical brighteners used for polyester can also be applied under acidic conditions (pH 3–4) to acrylic fibers. The acidic conditions are employed not for promoting exhaustion of the agent but to avoid discoloration of the acrylic fiber under boiling alkaline conditions. The agent is applied at boiling for 30–40 min. The exhaustion is slow, but can be accelerated in pressurized machines by raising the temperature to 105°C.

Acrylic–Cotton Blends.   The acrylic component is first optically brightened with a cationic-type FBA. This is followed by the application of FBA to the cotton component.

Acrylic–Wool Blends.   A cationic-type FBA such as Leucophor EFR liquid (Sandoz) is first applied to the acrylic fiber by treatment at boiling for 1 h at pH 4.5–5. The wool component is then bleached with hydrogen peroxide in the second bath and brightened in the third bath with Arostit BL and Leucophor PAF powder or Leucophor BSB liquid.[18]

17.1.6 Application to Cellulose Acetates

Disperse-type FBAs are applied to cellulose acetate and cellulose triacetate. For example, Uvitex ER (Ciba–Geigy) can be applied by the exhaustion method, the padding method, from scouring and bleaching liquors, in discharge printing, and in resin finish. The FBA concentration is 0.1–1%, the dyeing temperature is 50–70°C for cellulose acetate, and is above 75°C for cellulose triacetate. FBA can be added to the dope before the spinning of the filaments.

17.1.7 Spun-Whitened Fibers

Incorporation of FBA to get extra-white material during the manufacture of filaments has been attempted with some success, even though the FBA has to meet stringent specifications.[19–21] Some of these FBA are also suitable for other application techniques, for example, Uvitex 551 (Ciba–Geigy) can be incorporated before or during the polycondensation reaction or dry-blended with polyester chips prior to spinning.[22] It can be added during polymerization as a dispersion in ethylene glycol and imparts a reddish to neutral blueish tinge to the fiber. The whiteness of the fiber increases with the concentration of the FBA (Figure 17.4).[19,22] Fastness properties are excellent. For nylon and acrylic polymer, similar experimental products are described.[22] Uvitex OB (Ciba–Geigy) in minute concentrations (0.005–0.05 g/kg) makes polypropylene fiber extra-white by removing the yellowish tinge.[23] The resultant fiber has a brilliant white appearance of moderate-to-good fastness to light and washing. Various FBA in the fiber manufacturing industry are discussed by Hefti.[24]

In Figure 17.4, the whitening effects obtained with FBAs applied by the pad-thermofix process at 185° and 210°C are compared with those given by FBAs incorporated in the polymer before extrusion.[19] The best results are obtained with a high concentration of FBA incorporated before extrusion. The fastness properties of all three FBA are good; light fastness is 6–8. A reactive FBA for PET has the following structure:[19]

MeOOC—⟨ring⟩—C—R—C—⟨ring⟩—COOMe

It takes part in the condensation reaction that produces polyethylene terephthalate.

It is possible to distinguish between mass-brightening and textile-brightening effects by examining the cross-section of a fiber under a fluorescent microscope. Mass-brightened material gives equal fluorescence intensity throughout the fiber while textile-brightening effects have a ring of bright fluorescence with little intensity in the center of the cross-section. If the FBA is applied by the pad-dry thermofixation process, the outer layer of the cross-section will fluorescence less strongly than the inner part. In the thermofixation process, a part of the FBA is lost by sublimation. This does not occur with FBAs used for mass-brightening polyester because they must be stable to withstand the polymerization–melt-spinning conditions without subliming.[19]

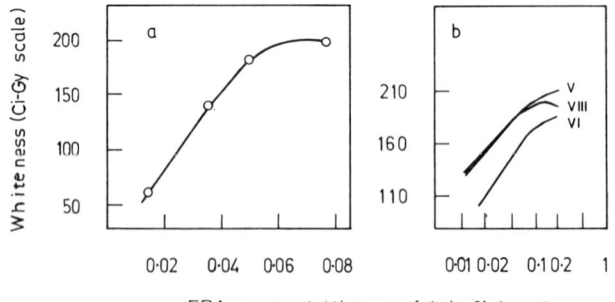

FBA   concentration   on   fabric,%   by   wt

FIGURE 17.4. Brightening effects with different concentrations of FBA.[19,22] (*a*); Uvitex 551 Ciba–Geigy (For the formula of FBA, see Ref. 19.); (*b*): V: Mass brightening; VI and VIII: Pad-thermofixing.

17.1.8 Efficiency of Fluorescent Brightening Agents

The whitening effect of FBA depends on various factors.[25,26] Its intrinsic effectiveness as a fluorescent compound, spectral absorption and emission characteristics, self-color, concentration, and physical form on the textiles are some of these factors. For example, the reflectance of PET material in the UV-range falls with increasing concentration of FBA (Uvitex ERN Ciba–Geigy) until a certain proportion of the total UV-light is absorbed. If the FBA concentration is increased further, the absorption shifts into the visible region and the substrate turns yellow.[27] It is essential that the FBA penetrates the fiber. Mere deposition on the surface of the fiber will result in yellowish-green discoloration and poor light fastness. If used to print a white ground, the effect of the FBA on other dyes must be checked since it may adversely influence the fastness properties of printed portions.

The efficiency of the FBA is evaluated after it is applied to the fiber by various methods.[1,2,28–31] Daylight as a source of light may give erroneous results because of variations in spectral energy. An examination under UV light is more precise in comparing FBA efficiency but the method may give results that differ from those observed in normal daylight by a human eye. Differences in the spectral energy distribution of UV light from various sources and UV light in normal daylight may give wrong results.

The visual assessment of a white object is based on comparison of a test sample with a standard sample of the same size, structure, and gloss.[29–31] The transparency of the samples must be the same since the light is lost (or reflected back) that passes through the fabric. The samples are folded often enough to eliminate the transparency effect. The white effect is largely dependent on the lighting. The degree of radiation in the UV range should not be too low compared to that of the visible range. Colorful surroundings or a colored background have a disturbing influence on the comparative assessment of a white sample. A nearby tree can interfere with the assessment because of the extra green light reflected from the leaves of the tree. Light grey, neutral tiles or paint, unwhitened paper or cotton fabric is a suitable background. The samples are viewed at a 45° angle and the positions of the samples are changed several times during the test. Daylight facing North (or South in the Southern Hemisphere) behind a window glass that absorbs as little UV light as possible between 10 AM to 3 PM is the best condition. A white scale with increasing perceptible whiteness has been suggested by Ciba–Geigy.[1,31]

Instrumental assessment of whiteness is based essentially on the principles used for (see Chapter 3) determining the x, y, and Y values in colorimetry. The color of the fluorescent sample can be described as a combination of a non-self-luminous color and a primary source.[1] In addition to the reflectance of the nonfluorescent substrate, there is an emission (fluorescence). The L^*, a^*, b^* values for the standard reference substrate and the sample are determined. The former values are substracted from the latter to get ΔL^*, Δa^*, and Δb^* values. This determines the whiteness and tonal differences. The light source is an important factor: too much or too little UV light will give unacceptable results.

A tristimulus filter photometer and a spectrophotometer are used for color measurement. The absolute accuracy of tristimulus values measured by filter photometers is generally poorer than values obtained with the spectrophotometer.[31] However, this may not be so important for finding out differences between the samples. Some instruments measure the L^*, a^*, b^* coordinates and directly indicate the ΔL^*, Δa^*, and Δb^* values with respect to the standard.

17.2   ANTIPILLING FINISHES

Pills are bundles of entangled fibers formed on the surface of a fabric during wear. They usually occur in the form of balls, but occasionally as rolls or ridges. On clothing made of wool, cotton, and other low-tenacity fibers, the pills are often rubbed or washed off during wear and create no problems. Pills of high-tenacity fibers such as nylon and polyester do not break off at the surface of the fabric and thus, spoil the appearance of the garment. The tendency to pill is a common fault in blends of high and low-strength fibers such as polyester and wool.[32] The tendency of a fabric to pill depends on a variety of factors including the fiber, the yarn, and the fabric.

Fiber Properties and Pilling.   The pilling tendency depends on the fiber in the fabric.[33] Polyester has the most severe pilling problem; acrylic, wool, and acetate fibers have a less severe pilling problem; cotton and viscose have a minimal pilling problem (Figure 17.5).[33] Pills rub off from the surface of the fabric with increasing abrasion, particularly when the tenacity of the fiber is low.

The ease of entanglement or deformation of a fiber with a circular cross-

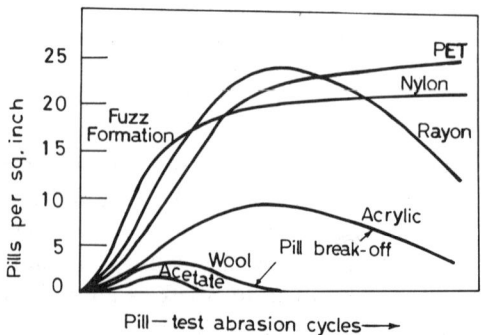

FIGURE 17.5.   Typical pill curves for common textile fibers.[33]

section depends on the fiber diameter or the denier of the fiber. As the fiber denier increases, the resistance to deformation increases and the pilling decreases. The relationship is not a linear one since the fiber rigidity increases as the square of the fiber denier. Additionally, when the fiber denier is high, there are fewer fibers in a given denier yarn to entangle.[34] Fibers with a circular cross-section and a smooth surface have a greater pilling tendency than those with a dumbbell-shaped cross-section. Fibers with a serrated cross-section have the least pilling tendency. The behavior of polyester fibers with circular, trilobal, and pentalobal cross-sections under identical fabric conditions shows differences in pilling tendencies (Figure 17.6).[34] Fibers with longer staple length are difficult to work loose from the yarn and give lower pilling[35,36] (Figure 17.7). As the crimp in the fiber increases, it is anchored in the fabric to a greater extent and gives less pilling (Figure 17.8).[37] The electrostatic properties of a fiber do not play a direct role in

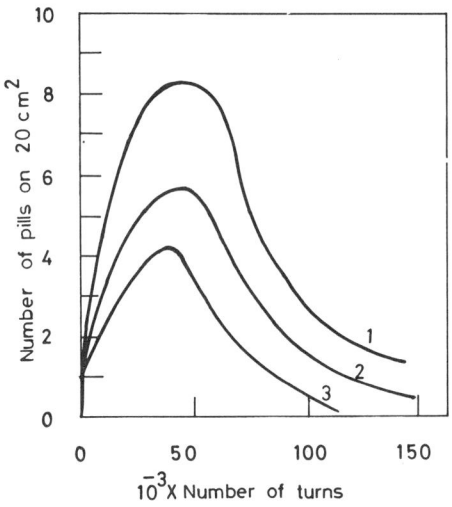

FIGURE 17.6. Relationship of pilling and fiber cross-section.[34] Polyester Fabric 1: Circular; 2: Trilobal; 3: Pentalobal.

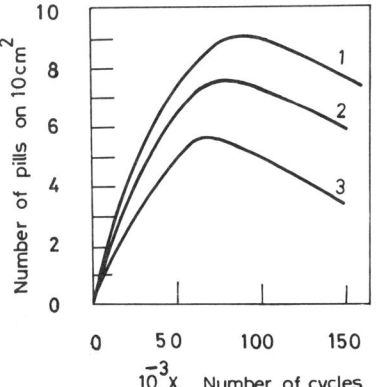

FIGURE 17.7. Pilling tendency of PET fabric with different staple lengths.[35,36] 1: 60 mm; 2: 80 mm; 3: 120 mm.

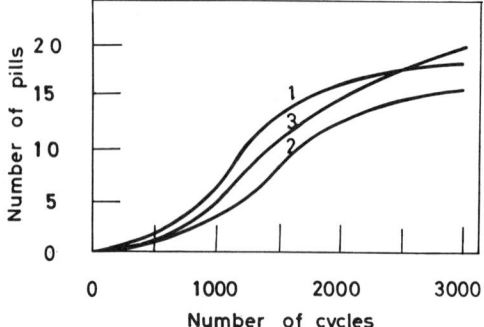

FIGURE 17.8. Relationship of pilling and crimps per inch (cpi).[37] Yarns spun on worsted system—1: 100% nylon; 3 denier, 10 CPI; 2: 50:50 wool:nylon; 3: 100% nylon, 3 denier, 15 CPI.

increasing the pilling tendency, although pilling may be greatly aggravated by electrostatic pickup of tint, dirt, or other foreign matter.

Yarn Properties and Pilling. Yarn properties influence the pilling behavior of a fabric, but manipulation of the yarn construction can minimize the pilling problem. Very effective pilling control can be achieved by using yarn of the finest count, consistent with the fabric requirements. With finer count yarn, the number of twists per meter is large for a given twist factor and the number of yarn intersections per unit area is also large for a given cover factor. This helps to decrease pilling. Twist is inserted into the yarn in order to bind the fibers together so that no slippage occurs.[37] Hence, higher twist in the yarn will lower the pilling tendency of the fabric[38,39] (Figure 17.9). In a single yarn, adequate twist levels are very important in controlling pilling and must be as high as the fabric character will permit; twist multipliers in

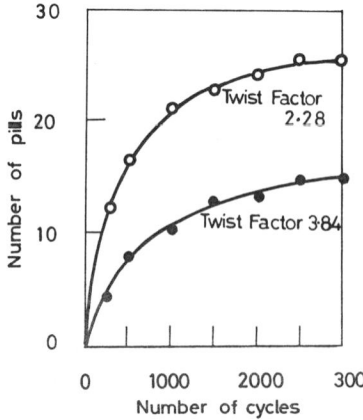

FIGURE 17.9. Effect of twist on pilling of 20/80 Fibro/Nylon blend fabric.[38,39]

the range of 4.0–4.5 do a good job. In the piled yarns, it is more practical to keep the single-yarn twist at a normal level, while increasing the ply twist to provide the necessary binding of the constituent fibers. Ply yarns are found to give more satisfactory results than single yarns in controlling pilling because the protruding fibers in a single yarn are locked into the twofold structure.

Pilling can be severe because of thick and thin places in the yarn. The spinning twist will concentrate itself in the finer portions of an irregular yarn. Thick portions will contain less twist and the fibers that are loosely held in such heavy places can be teased out to form pills in wear. Hairy yarn results in a fabric with protruding fibers and encourages pilling.

Blend Composition. Although the relationship is not a simple one, the extent of pilling in a blended fabric increases with an increase in the content of a fiber known to be prone to pill formation.[40] The components of the blend have to be properly balanced in order to get minimum pilling.

Fabric Construction and Pilling. *Structure.* Fabrics with loose open structures are far more prone to pilling than those with tight constructions. Because of this, knitted fabrics made from spun-yarn pill much more compared to woven fabrics.

Float Length. Greater float lengths in weaves such as sateen give a long uninterrupted length of yarn and so facilitate the ease with which a fiber is teased to the surface during wear. Thus, plain weaves (where the warp and filling yarns are more closely bound in the fabric) are more pill-resistant than twill or other types of weaves.

Sett. The fabric sett or cover factor is the most important parameter that contributes to the restriction of fiber movements. This prevents the fibers from moving to the fabric surface and thereby, controls pilling. The firmness of the sett is most important and pilling can be reduced appreciably by increasing the pickage. At the same time, overconstructing the fabric can impart harshness and stiff feel to the fabric. Two approaches to tackling the pilling problem are (*a*) preparing a modified fiber with less of a pilling tendency and (*b*) applying antipilling finishes.

17.2.1 Modified Fibers

Most of the research to produce antipilling fibers is confined to polyester. Polyester is blended with cellulosic or wool fiber and such blends give large pilling problems. Visible pills on the surface of the fabric are the result of two opposing processes; formation of pills and wearing-off of pills. By using modified fibers with slightly lower tenacities, the wearing-off of pills is facilitated and the pilling problem is minimized (Figure 17.10).[33] The number of

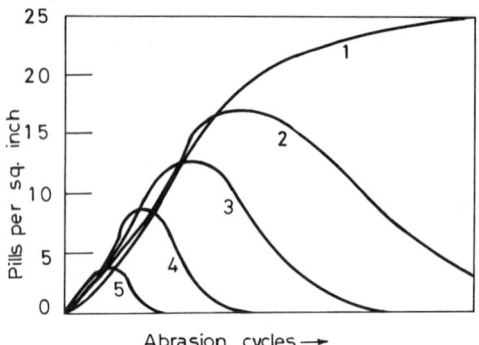

FIGURE 17.10. Pilling curves for several modified polyester fibers.[33]

Item	Tenacity	Flex Life
1	3.9	13,000
2	3.8	4200
3	3.6	2900
4	2.5	1800
5	1.8	800

pills formed as well as those retained on the fabric decrease with a decrease in tenacity. Vaidya[41] has reviewed the literature on various methods of producing low-molecular-weight polyester which gives low tenacity and, in turn, low pilling. By blending the spinning mass with 1–5% of a terephthalate of barium, calcium, or zinc, it is possible to manufacture polyester fiber with a smaller pilling tendency. The use of esters of boron or silicon alters the relationship between the melt viscosity and the solution viscosity and thus, permits direct spinning of a low-molecular-weight polymer. The incidence of pilling in polyester fiber is reduced by the addition of a boron compound, an organic antimony, chromium, iron, tin, titanium, vanadium, zirconium compound, or sebacic or adipic acid to the reaction mixture before esterification.[41] Pill-resistant polyester fiber (Grisuten) is produced by incorporating an organosilicon modifer during the polycondensation reaction.[42] The molecular weight distribution of the polymer is changed and the resultant fiber has a low-average-molecular-weight compared to normal polyester.

Low-molecular-weight and low tenacity polyester fibers such as Tergal T 900 and T 906 (Rhone-Poulene)[43] Hydrophilic WD-2 polyester (Monsanto)[44] and cationic-dyeable polyester[45] exhibit low-pilling tendencies. A polyester with 3.5–5.0 g/d tenacity and antipilling properties have been described by Toyo Boseki Co.[46] To overcome the pilling problem in acrylic fiber, DuPont has claimed the use of an acrylonitrile-styrene sulfonate copolymer in composite filaments and yarn with acrylonitrile-alkyl vinyl pyridine-alkyl acrylate or low-strength acrylic fiber.[47] Although modified polyester and acrylic fibers are useful in reducing the pilling problem, these fibers have certain drawbacks such as low wear-life of the garments, oligomer formation in dyeing, and low crease recovery and dimensional stability; hence, these fibers have not been widely commercially accepted. At present, the only approaches used commercially to minimize pilling in synthetic-fiber fabrics are standardizing mechanical and chemical processing operations and applying antipilling finishes.

17.2.2 Antipilling Finishing

Finishing treatments play an important part in reducing pilling in textile fabrics. The following operations are very important.[40,48]

Shearing and Cropping. In this operation, the protruding fibers are cut off from the fabric and the pilling is reduced.

Singeing. This is one of the most important operations for reducing pilling in PET/CO blend fabrics. The effect of singeing is to soften and melt the protruding fiber-ends on the fabric. It is preferable to carry out the singeing operation twice, using a less severe flame and at a speed of 100–150 m/min. A washing operation is indispensable after singeing to remove the smell and dust of burnt-off fibers from the fabric. During singeing, tiny melt beads are formed on the surface that have a tendency to dye more deeply than the rest of the fabric. This gives a speckled appearance. Furthermore, the dyeability of the fabric is changed in various places if the size of the flame is not the same. Because of these reasons, it is preferable to carry out singeing after the dyeing operation, particularly when the carrier dyeing method is used.

Heat Setting. This is also useful in minimizing pilling. During heat setting, the fibers are set in a wavy entangled form within the fabric.

Chemical Treatments. Considerable control of pilling is possible by weakening the fibers that anchor pills to the surface of the fabric and by causing them to break off ideally as the pills are being formed. This can be conveniently achieved in fabric form rather than during fiber or yarn manufacture, thus avoiding possible difficulties that may arise during spinning, weaving, or knitting. One proven technique is to subject the fabric to a caustic-soda solution treatment that weakens the polyester fibers by preferential attack at the crimps where maximum deformation has taken place. In a typical process, polyester fabric is treated with caustic soda (2 g/liter) at 60°C for 30 min, after which it is washed, dried, and heat set.[49] The process is not expensive and gives fabric with a smaller pilling problem. Increasing the caustic soda concentration or processing time does not significantly reduce the pilling tendency. An increase in temperature from 60 to 100°C is useful in getting fabric with lower pilling tendency, but it causes a reduction in the abrasion resistance of the fabric. Addition of 0.3 g/liter of a quaternary ammonium compound such as Matexil SC-ASO (ICI) or Vantoc CL (IC) to the caustic bath is also useful in getting better pill resistance of the fabric with some acceptable loss in abrasion resistance. The effect of the caustic soda treatment on the pilling tendency is thoroughly examined by Timmis[49] using standard and low-pilling polyester and soda scour (60°C/30 min), caustic scour (60°C/30 min) without and with Cirrasol AC-RT or Matexil SC-A 50(3 g/liter), and the combination of soda scour–jet dyeing–caustic scour. He found that the treatment is more effective with higher denier fibers, with low-pilling polyester, and with additives. Other chemical treatments have been reviewed by Vaidya.[41]

Pilling of fabrics made from polyester, polyamide, and acrylic fibers can be reduced by heating the fabric quickly by passing it through an organic liquid at 180–190°C, or by weakening the fabrics by a short high-temperature heating while in contact with a fiber-swelling agent such as terephthalic acid dibutylester. Polyamide and polyester fibers may be weakened at closely spaced intervals along their length by first crimping them (12–20 crimps/10 cm) and then treating them with hydrogen peroxide or sodium hydroxide solution. Treating polyester with steam, ammonia, or methylamine weakens the fiber and reduces pilling. A process for a wide range of fabrics containing polyester and its cotton blends involves padding the fabric with polyethylene glycol, drying, and curing at 220°C for 30 sec.[50] Treating polyester with chlorine lowers the viscosity of individual fibers and reduces pilling. The treatment can be given after dyeing or finishing.

Preventing Fiber Migration. The commercially available antipilling finishes are based on preventing fiber migration by some form of adhesion at points of contact and/or modifying the interfiber friction. Some of these finishes are Dispersion 505W (ICI Mond Division), Pillmaster TF (Tenatex Chemical Co.), Turpex KM (Ciba–Geigy),[49] silicone-elastomer-resin-based Dicrylan WK (Ciba–Geigy), and Q2-4011B (Dow Corning Co.) for acrylic fabric.

The application of a cross-linking resin-emulsion by the exhaustion technique and a heat treatment can limit fiber migration and minimize pilling. A typical example of this technique is the application of Sovatex SRT (Standard Chemical Co., Cheshire U.K.) by exhaustion during the rinse cycle of the perchloroethylene dry-cleaning process. Fabrics containing polyester and other high-tenacity fibers may be treated with a thermoplastic vinyl resin or, alternatively, the inside of the fabric is given a shallow coating of butadiene or butadiene-acrylonitrile latex followed by vulcanization. In the case of polyester/wool blends, a treatment with a 2–6% aqueous emulsion of polyvinyl chloride by the pad-dry-cure process is useful in reducing pilling. DuPont has suggested a colloidal-alumina product known as Baymal which when applied at low concentrations in size or finish on acrylic fabrics reduces their pilling tendency.[51] Nylon or nylon/wool blend fabrics can be treated with an aqueous liquor containing a condensate of formaldehyde and an aromatic sulfonic acid, a nonionic dispersing agent or water-repellent liquid polysiloxane, or an aqueous dispersion of polyvinyl acetate, polyvinyl butyrate, or a modified acrylic disperson.

With the present state of technology, pilling is a phenomena with which we have to live. If a completely nonpilling synthetic fiber is made, it is almost certainly not suitable for apparel because of unacceptable aesthetics or poor wear performance.

17.2.3 Evaluation of Pilling

The pilling resistance of a specific fabric in actual wear varies with the individual wearer and the general conditions of use. ASTM has recommended the following test methods for determining pilling resistance and other surface effects such as fuzzing.[52]

Appearance Retention Test. The specimens are subjected to simulated wear against an abradant that moves in a uniform circular direction for a specific period. The resultant pills are mechanically colored with ink and are evaluated by comparison with visual standards.

Brush and Sponge-Pilling Test. The specimens are subjected to simulated wear conditions by brushing the fabric to form free fiber-ends and then subjecting the fabric to a circular rubbing action with a sponge that rolls the fiber-ends into pills. The appearance of the fabric is then evaluated by comparison with visual standards.

Random Tumble Procedure. Fabrics are caused to form typical pills by a random rubbing motion produced by tumbling specimens in a cylindrical chamber lined with a mildly abrasive material. In order to form pills that resemble wear pills in appearance and structure, small amounts of short-length cotton fibers (grey in color) are added to each chamber. Besides ASTM tests, there are two other tests for pilling that are in extensive use: the ICI Pill-Box Test and the Atlas Random-Tumble Pill Test.[49]

17.3 DURABLE-PRESS FINISHES

Anticrease or crease-resistant finishes, wash-and-wear finishes, and durable-press finishes are achieved by similar techniques; the textile material is treated with a precondensate of a resin or a cross-linking agent in the presence of a catalyst followed by a heat treatment to set the resin or to form cross-links or both. Wash-and-wear and durable-press finishes involve the use of cross-linking agents (which may be resinous), a softener, wetting agent, and a catalyst or a combination of catalysts.

The crease-recovery properties of blend fabrics are better than those of 100% cellulosic fabrics. However, these fabrics still must be ironed, particularly after washing. The wash-and-wear finish is well-suited for textiles where pressed-in creases are not required. On the other hand, where pressed-in creases are required for a neat and fashionable appearance, a fabric with the wash-and-wear finish is totally unsuitable since the built-in wrinkle resistance hinders the formation of sharp creases during ironing. Furthermore, the inserted creases quickly disappear during wear and thus, the garment has a poor appearance. For such an end-use, the fabric is given a durable-press finish so that the garment can be shaped permanently.

Cross-linking agents are of two types:[53] (a) nitrogenous or (b) nonnitrogenous. Nitrogenous cross-linking agents are urea-formaldehyde precondensates, cyclic urea-formaldehyde products, melamine-formaldehyde precondensates, carbamates, and triazones. Urea and cyclic urea react with formaldehyde under slightly alkaline conditions to give mono and di-methylol derivatives and their mixture. The methylolurea finish exhibits poor wash fastness and susceptibility to chlorine during bleaching with hypochlorite and is therefore suitable only for unbleachable colored goods. It is not suitable for durable-press finishes because the precondensate of urea and formaldehyde cures rapidly and is very difficult to keep in an uncured state for a long time.

Dimethylolethylene urea (DMEU) produces high wrinkle-resistance levels at low add-on and without excessive loss of useful fabric properties. It exhibits a slight susceptibility to chlorine, and the finish is sensitive to acid. Therefore, the treated fabric needs a thorough afterwash to remove the acid catalyst. It is useful for durable-press garments. Dimethylolpropylene urea (DMPU) is not susceptible to chlorine retention and is therefore used for white goods. Its resistance to acid is also better than DMEU's. Dimethyloldihydroxyethylene urea (DMDHEU) finishes give excellent crease recovery and wash-and-wear properties with high fastness to laundering. However, it exhibits a chlorine-retention problem and therefore is not used for white goods. It is an excellent agent for the durable-press finish. Triazones are prepared by reacting urea, formaldehyde, and aliphatic amines. Thus, an addition of ethylamine (or other aliphatic amine) to dimethylol urea gives the triazone:

Dimethylolethyltriazone

The main advantage of the triazones is their resistance to damage because of chlorine absorption during bleaching. The tertiary nitrogen blocks the hy-

drochloric acid liberated by the action of heat and chlorine on the cotton in
the blend fabric. The triazones give a white finish that can be bleached
without danger of yellowing. However, they give poor crease recovery,
impart a fishy odor, and are not suitable for deferred-cure durable-press
finishes.

Melamine–formaldehyde precondensates have good storage stability.
Their esterified products give a finish with good durability. They do not
absorb chlorine during hypochlorite bleaching and thus, the problem of degradation of cotton after bleaching is absent. Nevertheless, the cloth finished
with these precondensates becomes yellowish. These products, like similar
urea–formaldehyde precondensates, also cure very easily and are unsuitable
for durable-press finishes.

Carbamates or urethanes are used for precure durable-press finishing of
white goods. A carbamate can react with two molecules of formaldehyde in
an alkaline medium to give a corresponding N,N dimethylol derivative.

$$ROOCNH_2 + 2HCHO \longrightarrow$$

$$ROOCN(CH_2OH)_2$$

The substituent R has a profound effect on the properties of the treated
fabric. When R is a lower allkyl group, the resin imparts crease-resistance
and shrink-resistance to the treated fabric. If R is a long-chain fatty alkyl
group, it acts as a lubricant, improves abrasion resistance and tear strength,
and may impart water-repellent properties. If R is a halogenated group like
perfluoro-alkyl, it may impart oil repellency and soil-release properties. Carbamates give good wash-and-water properties. The finish is stable during
laundering and is free from damage because of chlorine retention—if cured
properly. The drawbacks of carbamates are a high release of formaldehyde
and toxicity. The finished fabric suffers a considerable loss of tear and tensile strengths and abrasion resistance.

Nonnitrogenous Agents.   All the aminoaldehyde agents have the potential
drawbacks of chlorine retention, fishy odor, and liberation of free formaldehyde. Nitrogen-free cross-linking agents may not have such limitations.
Such agents are (a) formaldehyde, (b) acetals, (c) epoxides and epichlorohydrin, and (d) sulfones. Formaldehyde in the presence of an acid catalyst
gives wet or dry crease recovery depending on the swelling of cotton. Formaldehyde gives a durable-press finish, but the loss of strength in the finished fabric is too high to be acceptable. Acetals, the condensation products
of aldehydes with alcohols in acidic mediums; for example, diethylene gly-
col-formaldehyde, pentaerythritol-formaldehyde can be used as a cross-linking agent. Epoxides and epichlorhydrin finishes are stable to acid or alkaline-
hydrolyzing conditions. Butadene diepoxide gives good wash-and-wear

properties.[53,54] Divinylsulfone adduct and bis (hydroxy ethyl) sulfone are also used as cross-linking agents.

Additives.   When PET/CO blend fabrics are treated with cross-linking agents for the preparation of wash-and-wear or durable-press textiles, there is a considerable loss in abrasion resistance, tear strength, and tensile strength of these materials. In order to minimize the loss in abrasion resistance and tear strength of textiles, the addition of ionic or nonionic softeners or other chemicals to the resin-finishing bath is recommended. The softeners are also useful in getting a better handle. Addition of an emulsion of a polymer such as polyethylene, polyvinyl chloride or acetate, polyacrylate, polyethers, silicones, dispersions of polyester, carbamates based on ethylene-oxide adducts, and polypropylene glycols to the resin bath or pretreatment with polyurathanes or methylol derivatives of fatty alkyl carbamates prior to resin finishing improves abrasion resistance of the finished fabric. The catalyst is either an acid such as tartaric, citric, or sulfamic acid, acidliberating ammonium compounds, metallic salts such as magnesium and zinc haldies, nitrates, phosphate, or fluoroborates or their mixtures with the acid catalysts.

Wash-and-Wear Finish.   The heat set PET/CO blend fabric is padded through a liquor containing a cross-linking agent (9–12% of the weight of the cotton component), a catalyst, a wetting agent (0.5–1 g/kg), and an additive (1–10 g/liter) to improve abrasion resistance. It is then dried on a hot-flue drier or a stenter at 100–110°C. Drying is an important operation when the migration of the resin has to be controlled. The Remaflam drying process[54] and drying with electromagnetic (radio) frequencies[55,56] are two of the methods used to avoid migration. The amount of liquor on the fabric can be minimized by using foam techniques, vacuum impregnation and so on. The dried fabric is cured at 140–160°C for 3–5 min, preferably by the moist-cure process, or at 125–140°C for 5–10 min by superheated steam. The fabric is washed in an open soaper and dried.

Durable-Press Finish.   PET/CO blend fabrics are given a durable-press (DP) finish by various techniques.[57,58] The fabric is padded with the liquor and dried as in the wash-and-wear process. The dried fabric is then cut and made into garments within 4–6 months; creasing and shaping are done very carefully with a hot-iron press. Finally, the garment is oven-cured at 150–160°C for 10–15 min, washed, and dried. Alternatively, the stitched garment may be soaked with the liquor; centrifuged; given the desired crease, pleats, flat areas and so on using a hot iron; and then cured, washed, and dried.

In a two-stage fixation process, such as the wet-fixation process or the polyset process, the reactant or resin or their mixture is fixed in the first stage. In the wet-fixation process, the padded fabric is kept in a batch and rotated in a pad-roll system at 70°C for 2 h. The fabric is then washed, dried, repadded with a catalyst (zinc nitrate), and dried. After stitching the garments and pressing the shape and creases, the fabric is cured at 165°C for 5–10 min, washed, and dried. In the polyset process, zirconium acetate is used as the polymerization catalyst and magnesium chloride or zinc nitrate as the cross-linking catalyst.[59] The padded and dried fabric is cured at 150–160°C for 3–5 min, while the shaped garment is cured for 10–15 min.

Durable-Press Finish on Dyed Fabric. Disperse dyes on polyester migrate during curing at elevated temperatures which can cause a color change, staining of lining, a loss in color fastness, and frosting in cross-dyed PET/CO garments.[60–62] Addition of Mykon 39 (Sun Chemical Corp) in the resin bath minimizes frosting.[63] Selecting disperse dyes with high sublimation fastness and testing the suitability of the resin composition prior to use are two ways to minimize this problem.

Durable-Press Material without Resin Finish. A blend fabric containing a low-temperature dyeable polyester (80%) and cellulosic fiber (20%) is dyed with disperse and reactive or vat dyes. Garments from this fabric are then made, shaped, and stabilized in a special press at 170–190°C. After pressing, the head is raised and a vacuum is applied to cool the garments to a nonplastic state.[64]

17.4 WATER-REPELLENT FINISHES

Water-repellent synthetic-fiber fabrics are required for a number of end-uses such as raincoats, umbrellas, and tents. Although synthetic fibers are essentially hydrophobic in nature, they get wet on exposure to water for a short duration. It is therefore necessary to treat them with water-repellent finishes to make them suitable for the above applications.[65–67]

A good water-repellent for synthetic-fiber fabrics is compatible with other finishing agents, easy to apply, inexpensive, readily available, does not cause pollution of water, and is not toxic. The water-repellency of the finished fabric must be fast to washing and dry-cleaning, and the air porosity of the fabric must not be affected by the finish. The important water-repellent agents are (*a*) wax dispersions, (*b*) pyridinium compounds, (*c*) waxy thermosetting resins, (*d*) silicone, and (*e*) fluorochemicals.

Wax Dispersions. These products are anionic alkaline dispersions used on nylon, polyester, and cellulose-acetate taffetas. They are usually applied with melamine-formaldehyde or urea-formaldehyde resins to give a slightly crisp hand. A typical recipe consists of

10%	Wax dispersion
3%	Melamine-formaldehyde resin
1%	Ammonia
0.5%	Ammonium sulfate

The fabric is padded, dried at 100°C, and cured at 150–160°C for 3–4 min. The aqueous dispersions of wax containing aluminium acetate or other salts (Dipsanil V, Waxol W, Waxol P etc.) are applied at pH 3.5–5.5 by padding, followed by hard drying at 110–125°C. An advantage of these products is that the feel of the fabric is not adversely affected. The zirconium salts–wax emulsion is applied in conjunction with the urea-formaldehyde resins by the pad-dry-cure process.

Pyridinium Compounds. These are cationic products such as

$$R—\overset{+}{N}\diagup\!\!\!\bigcirc \quad Cl^-$$

where $R = C_{17}H_{33}CONHCH_2$ for Zelan; $R = C_{18}H_{37}OCH_2$ for Velan PF; $R = C_{17}H_{35}COCH_2$ for Norane. Pyridinium compounds impart a good water-repellent effect and a soft feel to the fabric.

Waxy Thermosetting Resins. These are based on modified thermosetting resins containing long-chain fatty alkyl groups. Resins of melamine-formal-dehyde, triazine, and carbamate are generally used in their preparation. Phobotex water-repellents (Ciba–Geigy) belong to this group and consist of fatty-acid-substituted melamine resin.

Silicones. These chemicals differ from other water repellents in that long chains of hydrocarbon or fluorocarbon are not present. Water repellency is provided by methyl groups oriented and attached to the fiber surface by silicone links. The silicones are of one of the following two types:

$$\left[\begin{array}{c} CH_3 \\ | \\ —O—Si— \\ | \\ H \end{array}\right]_n \qquad \left[\begin{array}{c} CH_3 \\ | \\ O—Si— \\ | \\ CH_3 \end{array}\right]_n$$

$$\text{I} \qquad\qquad \text{II}$$

Type I is reactive and is generally used as a water repellent along with type II. Low-molecular-weight polysiloxanes are liable to undergo further poly-merization by lengthening and by cross-linking of adjacent Si–O chains. This is undesirable in textile applications. It is prevented by replacing the end hydrogen atoms by inert substituents like methyl groups. Organometallic compounds of titanium, zirconium, tin, zinc, and other metals, for example, triethanolamine titanates or triethanolamine-lead acetate complexes, act as catalysts.[67] The fabric is padded, dried at 100–110°C, and cured at 150–160°C for 2–3 min. The silicones give a good water-repellent finish that is moderately fast to washing and dry cleaning. The handle of the fabric is smooth and soft.

A copolymer of ethylene and vinyl triethoxy silane gives good water repellency for polyester and PET/CO blends. The copolymer can be applied from a solvent or an aqueous solution. The treated fabric is then steamed.[68] Finishing with sodium ethylsiliconate is recommended. Water-repellent finishing can be combined with crease-resistant finishing for PET/CO blend fabrics using DMEU, Texil MP (silicone-oil emulsion), and a catalyst.[69,70]

Fluorochemicals. Organic compounds in which a high proportion of hydrogen atoms are replaced by fluorine are called *fluorochemicals*. The fluorinated carbon chain that is evenly distributed on the fiber with proper orientation presents an essentially fluorinated surface to water or oil, which imparts water and oil repellency. One of the most successful ways of obtaining this condition is the incorporation of the fluorinated molecule into a polymer molecule in which the perfluoro group constitutes the side chains; for example, acrylic acid–1 : 1 dihydroperfluoro octanol ($C_7F_{15}CH_2OH$) copolymer (I) will impart water and oil repellency. The fluorochemicals are supplied as

emulsions and are applied by the pad-dry-cure process. Fluorochemicals give water-repellent and soil-release finishes in conjunction with other water repellents (called *extenders*)[71,72] with good resistance to washing and moderate resistance to dry cleaning.

Oil and water repellency are imparted to polyester, PET/CO blend, nylon, acrylics, wool, or cotton fabrics by impregnating the fabric with a dioxan solution of a perfluoro-alkyl-carboxylic acid ester.[73] The fabric is dried well and cured at 150°C for 4–5 min. A polymethacrylate derivative of fluorinated alcohols,[74] perfluoroalkyl alkyl nitriles,[75] fluoroalkyl acrylate or methacrylate in tetrachloroethane polymer[76] with urathane phosphate, and maleic anhydride resin impart similar finishes on polyester and nylon. Alternatively, fluorocarbon is grafted by a free radical initiator on to a polymer.[77]

17.4.1 Testing of Water Repellency

The large number of tests for water repellency may be divided into two types: (*a*) those which judge the resistance of the fabric to penetration by rain drops (e.g., Rain test, Impact Penetration test) and (*b*) those which measure the resistance of the fabric to surface wetting or penetration into but not through the fabric (e.g., Spray test, Hydrostatic Pressure test).

Rain Test (AATCC Test Method 35: 1974.)[78]   The test specimen is backed by a blotting paper of known weight and is clamped in the specimen holder in a vertical position. A horizontal water spray is directed against the specimen at a fixed distance from the face of the spray nozzle. The water is sprayed for 5 min. The blotting paper is then removed and weighed quickly. The penetration of water is indicated by an increase in the weight of the blotting paper. The apparatus is provided with an arrangement whereby the height from which water falls onto the specimen can be altered.

Impact Penetration Test (AATCC Test Method 42: 1974).[78]   One end of the test specimen is clamped under a spring clamp at the top of the inclined stand. Another clamp of standard weight is clamped to the free end of the test specimen. 500 ml of distilled water at $27 \pm 1°C$ is poured into the funnel of the tester and is allowed to fall on to the test specimen from a fixed height. The specimen is then carefully lifted, the blotter is removed, and is weighed quickly. The increase in weight of the blotter is calculated.

Spray Test (AATCC Test Method 22: 1974).[78]   This test consists essentially of allowing a spray of water to fall onto the fabric under test and comparing the effect produced against a standard chart. It measures the resistance of the fabric to surface wetting; no account is taken of penetration (rain resistance). A sample is mounted in an embroidery loop and placed on a frame at a 45° angle. Water (250 ml) is sprayed onto the sample from a fixed distance. The sample is then removed and the degree of spotting is compared with the chart and graded in one of six categories.

Hydrostatic Pressure Test (AATCC Test Method 127: 1974).[74]   This test is a measure of the combined effect of the waterproofing agent and the fabric structure. It is particularly suitable for testing the resistance of heavy fabrics such as tent cloths, heavy ducks, and tarpaulins, which in actual use may be expected to resist the penetration of pools of water accumulated in folds. The instrument consists essentially of a hollow cylinder to which the fabric to be tested is clamped with the face side in contact with water. The cylinder is provided with a side tube attached to a graduated vertical tube through which water is allowed to enter the cylinder at a constant rate and to build-up a hydrostatic pressure on the fabric. The height of the water is measured on the graduated tube. When sufficient pressure has been built up, water penetrates the fabric and the height of the water is noted when three drops have come through.

Bundesmann Test.   This test measures the resistance of a fabric to penetration by rain as well as resistance to surface wetting.[79,80] The specimens to be tested are mounted over special cups and are subjected to a shower of water at pH 6–8 at 18–20°C from a multinozzle drop producer for 10 min. The rate of flow is adjusted to 62–68 ml/min/cup. The cups are rotated at 5 revolutions/min during spraying. Special wipers inside the cup rub the undersides of the specimens to simulate the rubbing action that occurs when a raincoat is worn. The water that penetrates the fabric is collected in the cups and is measured, as is the water retained by the fabric specimen. The former gives the value of water penetration, while the latter gives the value of water absorption. In addition to these tests, static absorption tests (AATCC 21: 1975) and tumble-jar dynamic absorption tests (AATCC 70: 1975) are available for evaluating water-repellent finishes.[78]

17.5   SILKLIKE POLYESTER

Garments of natural silk are greatly valued for their softness, drapey suppleness, warm and sensuous feeling, and comfort properties.[81] If polyester fabric is constructed from very fine denier yarn, the feel of the fabric is close to natural silk. However, it is uneconomical to weave or knit fabrics from very fine denier yarn. Therefore, a finish is given to PET fabrics to make them silklike. This finishing treatment is based on controlled hydrolysis of polyester with sodium hydroxide. It gives some loss of weight of the fabric which reduces the denier of the fiber and thus, gives a thin fabric.

The hydrolysis of polyester with caustic soda is an old technology.[82] In a 1952 Patent,[82] ICI disclosed a treatment for polyester fabrics made from flat yarn to get a soft hand and low synthetic feel. DuPont developed a process (calender–heat set–caustic soda hydrolysis) to get fabric with high luster and better feel.[83] The treatment was of occasional interest during the 1960s and early 1970s.[84] During all these years, weight loss in hydrolysis was kept low at about 2–5%. Since 1977, Japanese PET manufacturers have extensively used the controlled hydrolysis technique to get silklike polyester.

Saponification Reaction.   Hydroxyl ions in a solution of sodium hydroxide attack the carbonyl group in the polymer, which results in the formation of disodium terephthalate and ehtylene glycol.[84] Disodium terephthalate is soluble in alkaline solutions (pH above 8) up to 13–14%. Free terephthalic acid is formed below pH 8 which is insoluble in water and may deposit on the surface of the fabric. It is therefore essential to wash out all disodium terephthalate with alkaline water before neutralizing the fabric.[84] The loss in weight of PET depends on the alkali concentration, temperature, and time of treatment[8,84,85] (see Chapter 5). The relation between time of treatment, weight loss, and thickness of the fabric is shown in Figure 17.11.[82] It can be seen that weight loss increases whereas thickness decreases with an increase in time of treatment. The addition of quaternary ammonium compounds accelerates the hydrolysis reaction (Table 17.2).[81] The weight loss

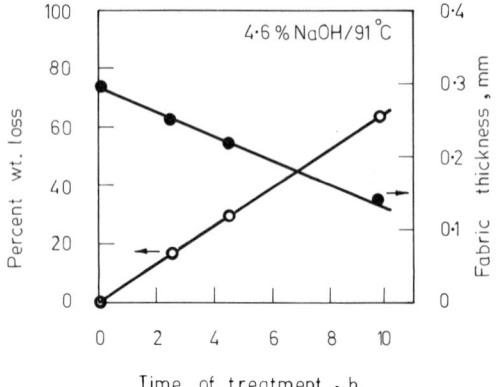

FIGURE 17.11.   Treatment of polyester fabric with sodium hydroxide.[82]

decreases with an increase in the $M:L$ ratio in the range of $1:10$ to $1:30$.[81] The behavior of flat and textured yarns is the same but may vary in magnitude of response.[84] The relation between weight loss and breaking load as well as fabric thickness is shown in Figure 17.12.[82] Both breaking load and fabric thickness decrease with an increase in weight loss. It is thus necessary to control the weight loss of the fabric in order to get a fabric with the desired strength and thickness. It is not possible to recommend the exact conditions of the treatment because different PET fibers react differently. Before dyeing, the caustisized polyester is brought to pH 5–6 so that dyeing with disperse dyes is not disturbed.[86] There is no significant difference in dye affinity between normal and caustisized polyester.[87]

TABLE 17.2   Effect of Quaternary Ammonium Salts on Weight Loss of PET (7.5% NaOH, 95°C, 30 min, $M:L$ ratio $1:10$)[81]

Compound	Concentration (owf)	Weight Loss (%)
1. Alkyl (C_{12}–C_{14}) dimethyl benzyl ammonium chloride	0.33	6.8
	0.67	11.5
	1.00	13.7
2. Substituted imidazoline compound	0.60	3.54
	1.20	3.75
	1.80	4.03
3. Amphoteric compound	0.45	5.28
	0.90	5.49
	1.35	5.56
4. Sodium hydroxide (control)	0	4.3

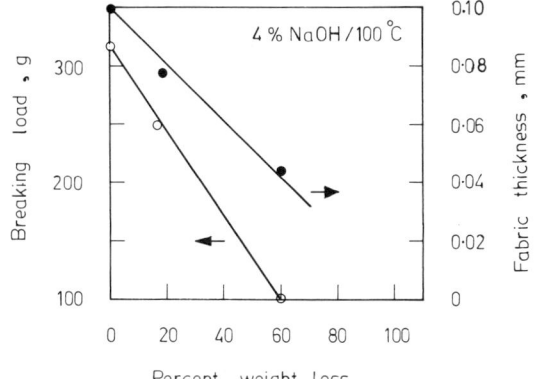

FIGURE 17.12. Effect of percentage weight loss by sodium hydroxide treatment on fabric thickness and breaking load.[82]

REFERENCES

1. Ciba-Geigy, *Ciba-Review*, **140** (1973/1), p. 26.
2. Allen, E., *ADR*, **46** (1957), 425.
3. Narkar, R. K., and Narkar, A. K., *Text. Dyer Printer*, **2** (Nov. 1969), 12.
4. Scheller, M., and Gartner, R., *Deut. Textiltech.*, **19** (1969), 628.
5. Liddiard, A., *Textilveredlung*, **5** (1970), 93.
6. Williamson, R., *Dyer*, **157** (1977), 408.
7. Technical Information, *Bleaching of Man Made Fibers*, BASF (1973), 43.
8. Weihsbach, J., *Text. Prax.*, **23** (1968), 825.
9. Nick, H., *Bayer Farben Review*, **15** (1969), 44.
10. Scheller, M., *Deut. Textiltech.*, **20** (1970), 443.
11. Weihsbach, J., *Dyer*, **144** (1970), 179.
12. Williamson, R., *Dyer*, **157** (1977), 359.
13. Williamson, R., *Man Made Text.*, **39** (1962), 40.
14. Jakobs, K., and Reiner, M., *Text. Prax.*, **21** (1966), 27.
15. Blume, W., *Textilveredlung*, **4** (1969), 88.
16. Ciba, BP, 889,642 (1962).
17. Rosch, G., *MTB*, **50** (1969), 1199.
18. Sandoz, Sandocryl B dyes 1470/4075/76.
19. Echkhardt, C., and Hefti, H., *JSDC*, **87** (1971), 365.
20. Barten, D., and Davidson, H., *Rev. Prog. Color.*, **5** (1974), 3.
21. Ackyroyd, P., *Rev. Prog. Color.*, **5** (1974), 86.
22. Ciba-Geigy, Preliminary Product Information, Experimental Fluorescent Whitener G 13-127, and G-14-551 (1974); Uvitex MD.
23. Lanter, J., *Ciba-Review* (1964/3) 42.
24. Hefti, H., *Textilveredlung*, **4** (1969), 94.
25. Williamson, R., *Dyer*, **128** (1962), 709.
26. Vaidya, A. A. *Colour and Chemicals Weekly, India* (Nov. 9 1977), 125.

27. Lanter, J., *Textilveredlung,* **19** (1964), 474.
28. Taylor, G. G., *JSDC,* **71** (1955), 697.
29. Anders, G., *JSDC,* **84** (1968), 125.
30. Crum, F., *TCC,* **13** (1981), 187.
31. Griesser, R., *Rev. Prog. Color.,* **11** (1981), 25.
32. Nims, N. B., *Man Made Text.,* **53** (April 1957), 46.
33. Gintis, D., and Mead, E. J., *TRJ,* **29** (1959), 578.
34. Kulkarni, G. G. and Trivedi, S. S., *Processing of Polyester Cotton Blends,* ATIRA, 1967, p. 55.
35. Timm, K., *MTB,* **51** (1970), 177.
36. Rao, K. S. S., and Phalgumani, G. R., *Text. Highlights India,* **3**(1) (1976), 31.
37. Hurten, J., *Text. Prax., Inter.,* **33** (1978), 832.
38. Baird, M. E., Hatfield, P., and Morris, G. J., *JTI,* **47** (1956), T181.
39. Turnball, D., *Man Made Text.,* **39** (1962), 30.
40. Modi, J. R., and Mali, N. C., *Colourage,* **26**(21) (1979), 31.
41. Vaidya, A. A., *Colourage,* **24**(16) (1977), 135.
42. Schmidt, W., and Prengel, H. C., *Textiltechnick,* **26**(8) (1976), 483.
43. Crosjean, P., *Industrie Text.,* (July/Aug 1973), 429.
44. Burnthall, C. V., and Lomartire, J., *TCC,* **2** (1970), 218.
45. Thimm, J. K., *TCC,* **21** (1970), 69.
46. Toyo Boseki, K. K., BP, 1,424,931 (1976).
47. DuPont, USP, 1,279,163 (1964).
48. Vaidya, A. A., and Shah, R. C., *Text. Dyer Printer,* **7**(2) (1973), 49.
49. Timmis, J. B., *Dyer Inter,* **156** (1976), 355, 410; *Textiles* **5**(3) (1976), 73.
50. DuPont, *Mod. Text.,* **38**(9) (1957), 82.
51. DuPont, *Dyer,* **126** (1961), 23.
52. ASTM, D 1375–67 *Annual Book of ASTM Standards* (1971), 231.
53. Cook, T. F., and Weigmann, H., *TCC,* **14** (1982), 100.
54. Ulrich, H., *MTB,* **54** (1973), 1206.
55. Ulrich, H., *Chemiefasern/Textilindustrie,* **23/75** (1973), 1212.
56. Gibson, M., *Text. Prax,* **29** (1974), 348.
57. Kopacz, B. M., and Perkins, R. M., *TCC,* **1** (1969), 80.
58. Pandey, S. N., *Text. Ind. Trade J.* (*India*) (Jan.–Feb. 1976), 2.
59. Valik, S. L., Verburg, G. B., Young, A. H. P., and Parikh, D. V., *TRJ,* **39** (1969), 505.
60. Morris, M. A., and Young, M. A., *TCC,* **4** (1972), 221.
61. Urbanik, A., *TCC,* **6** (1974), 78.
62. Patton, J. P., *TCC,* **6** (1974), 90.
63. Nye, S. H. W., *ADR,* **60**(11) (1971), 37.
64. Mousalli, F. S., and Browne, C. L., *TCC,* **3** (1971), 202.
65. Caldwell, J. R., and Dannely, C. C., *ADR,* **56** (1967), 77.
66. Davis, C. A., *ADR,* **56** (1967), 555.
67. May, J. M., *ADR,* **58**(20) (1969), 15.
68. Hoechst, BP 1,372,453 (1974).
69. Kolyadenko, S. S., Ivanova, G. V., and Strepetova, N. V., *Tekno Prom. No.* **109** (1976), 22; *vide World Text. Abstract,* 1976/4445.

70. Warno, E., *Bull. Inf. Bar Srod. Prom.*, **18**(3) (1974), 274; *vide World Text. Abstract*, 1975/3802.

71. Grajeck, E. J., and Peterson, W. H., *TRJ*, **32** (1962) 320.

72. Roth, P. B., *ADR*, **60**(7) (1971), 34.

73. Ciba-Geigy, BP, 1,439,432 (1976).

74. Pennwalt, Corp., BP, 1,437,255 (1976).

75. Ciba-Geigy, BP, 1,429,279 (1976).

76. Hoechst, A. G., BP, 1,406,866 (1975).

77. Gisen Co. Ltd., USP, 3,903,330 (1975); BP, 1,444,942 (1976).

78. AATCC Technical Manual, **51** (1975), 227–235.

79. Bundesmann, G., *MTB*, **16** (1935), 128.

80. Sanderson, S., *JTI*, **44** (1953), 550.

81. De Maria, A., *ADR*, **68**(10) (1979), 30.

82. ICI, USP, 2,590,402 (1952).

83. DuPont, USP, 2,828,528 (1958).

84. Gorrafa, A. A. M., *TCC*, **12** (1980), 83.

85. Namboori, C. G., *TCC*, **1** (1969), 50.

86. Rotogers, A., *Chemiefasern/Textilindustrie*, **29/81** (1979), 135.

87. Maekawa, S., *Japan Text. News*, **294** (1979), 87.

18 | FLAME-RETARDANT FIBERS AND FINISHES

The flammability of textiles is a major factor in fire accidents and loss of life and property. Typical statistics for fires in the United States in 1977–1978 are given in Table 18.1.[1] A number of countries have introduced a flammable-fabric act by which the production and sale of flammable textiles are prohibited for certain end-uses. This has generated considerable interest in flame-retardant fabrics. A flame-retardant fabric may be defined as a fabric that does not propagate flame, although it may burn or char when subjected to sufficient heat. Most flame-retardant textiles have inferior textile properties. Since flame-retardant fabrics have no special aesthetic appeal or added properties that may catch markets, they are produced only when statutory obligations demand them. The flame-retardant textiles are required for the following:[2]

1. Defense Application: Uniforms for soldiers and air-force pilots; fabrics for tents, ropes, baggage, parachutes; and seat covers for tanks, jeeps, airplanes and so on.

2. Use in Industry: Uniforms for operators in ammunition factories, oil refineries, coal mines, steel factories, and similar industries.

3. Use in Space Research: Uniforms for astronauts and fabrics for specialized uses in spacecraft.

4. Civil Uses: Nightwear for children, elderly persons, and the physically handicapped; fabrics for hospital uses; carpets; interior decorations in hotels; and uniforms for fire-fighters.

The hazards associated with the combustion of cellulosic textiles have long been recognized.[3] The need for developing flame-retardant synthetic

470

TABLE 18.1   The Fire Statistics in the United States (1977–78)[1]

Fires	
Reported	3,030,000
Unreported	30,000,000
Deaths	
Civilian	8,460
Firefighters	170
Injuries	
Civilians, reported	33,220
Civilians, unreported	200,000
Firefighters	58,650
Property Loss	5.31×10^9
Indirect Loss	0.45×10^9
Fire-related expenditures	15.2×10^9

Source: USFA national estimates, 1977–78 (These totals include reports from nonpublic fire services.)

fibers has only been recognized during the last 15 years. Perhaps the reason for this is that these fibers are produced from thermoplastic polymers that melt upon ignition and carry away the flame.[2] It has now been established that synthetic fibers are equally dangerous and can cause serious injuries by melting and dripping when ignited. The fire hazard is increased to a large extent when synthetic fibers are blended with cellulosic fibers.[4] In such blends, a carbonaceous gridwork, that is, a scaffolding effect, is formed from the combustion of cellulosic or nonmelting material.[5] Since the use of blends is increasing, the problem of flame-retardant synthetic fibers assumes a significant importance.

18.1   MECHANISM OF THERMAL DECOMPOSITION OF FIBERS

Heat from an ignition source raises the temperature of a polymer above its decomposition point yielding flammable volatile polymer fragments that mix with the air and are transported to the flame front. This is followed by combustion. Heat from combustion continues to generate flammable gases and burning proceeds.[6] Table 18.2 lists the burning behavior of textile fibers.[7,8]

Polyester.   When subjected to an open flame, polyester fabric shrinks and then catches fire. The fiber melts due to heat, the melt drips and carries away

TABLE 18.2   Flammability of Textile Fibers[7,8]

Fiber	Ignition Temperature (°C)	Flammability
Cotton	400	Burns readily with char formation, afterglow
Viscose rayon	420	Burns very rapidly with char formation, no afterglow
Acetate	—	Burns, melts ahead of flame
Triacetate	540	Burns readily, melts ahead of flame
Nylon 6 and Nylon 66	530	Supports combustion with difficulty, melts
Polyester	450	Burns readily with melting and soot
Acrylic	560	Burns readily with melting and sputtering
Modacrylic	450	Melts, burns very slowly
Polypropylene	570	Burns slowly
Wool	600	Supports combustion with difficulty
PVC and Polyvinylidene chloride	—	Does not support combustion

most of the flamming portion.[8] Although polyester fiber cannot be regarded as highly flammable, the extent of its flammability is governed by factors such as the fabric structure, dyes, and finishes present on the fiber.[9,10] The limiting oxygen index (LOI) is a test for assessing the flammability of textile fibers (see Section 18.10). Polyester fiber is considered to be flammable because it has a LOI value of 21. A LOI value of over 26% is required for flame-retardant fibers.[8] The major product of pyrolysis of PET is acetaldehyde.[11–13] The constitution of the gaseous product of pyrolysis is shown in Table 18.3.[14] The burning of carbon monoxide, ethylene, methane, and benzene is highly exothermic, and thus, the pyrolysis in burning polyester continues with this reinforcement of heat. At the initial stage, there is a rupture of the alkyl–oxygen bond. Random chain breaking has also been postulated from the type of compounds in the pyrolysis product[11] and by kinetic studies.[13]

Polyamide.   Nylon 6 and nylon 66 support combustion with difficulty but melt on burning.[6] The molten material retracts from the flame and drips away in the form of globules. Quite often, the flame is carried away with the droplet, making the material self-extinguishing. The molten droplets, however, can be dangerous if they continue to burn.[15] When a flame is applied to the lower edge of a vertical strip of fabric, the nylon fiber material tends to

TABLE 18.3   Composition of the
Pyrolysis Product of PET[14]

Compound	Mole %
CO	8.0
CO_2	8.7
H_2O	0.8
CH_3CHO	80.0
C_2H_4	2.0
2-Methyl dioxalan	0.4
CH_4	0.4
C_6H_6	0.4

melt and burn with a small flame as long as the ignition source is kept in contact with the strip. Within a few seconds after the removal of the ignition source, the burning ceases because the molten polymer droplets carry away the flame. If, however, the molten nylon is prevented from receding or dripping, it burns upon reaching the ignition temperature with a hot flame until consumption. The presence of chrome dyes and coatings of melamine resins considerably increase flammability of nylon fabrics.[16] The flammability of nylon may be increased by finishes in two ways. First, the finish itself can be flammable and may burn before the nylon has time to melt and drop away. Second, the finish can strengthen the fabric and support the burning nylon, thereby ensuring continued flame propagation. Melamine and urea-formaldehyde resins (used to stiffen nylon nets) are examples of the second effect. These resins have a high melting point and are considered nonflammable. In conjunction with nylon fabric, however, they cause the finished product to be extremely flammable. Blends of nylon with nonmelting fibers such as cotton and rayon burn rapidly. These blends are highly flammable and dangerous because they can cause injuries both by burning and melting.[17]

Acrylics.   Acrylic fibers burn readily with melting and sputtering and therefore are considered flammable.[6] Modacrylic fibers containing 35–85% polyacrylonitrile along with vinyl chloride or vinylidene chloride have greater flame resistance than acrylic fibers.

Methods of Producing Flame-Retardant Textiles.   Flame-retardant synthetic-fiber textiles can be produced by three different ways: (*a*) incorporation of a flame-retardant additive during or immediately after the polymerization reaction to produce modified fibers; (*b*) application of a flame-retardant finish to fiber, yarn, or fabric; and (*c*) development of new flame-retardant fibers.

18.2 FLAME-RETARDANT POLYESTER FIBER

Incorporation of a flame-retardant additive during the transesterification or polycondensation reaction of PET or mixing the additive with the polymer chips before melt-spinning gives a flame-retardant polyester fiber. The additive should be stable at 270–290°C, effective at a low concentration, cheap, nontoxic, readily available, and should give a finish fast to laundering and dry cleaning. It should not adversely affect the rate of polymerization, spinnability, stretchability of filaments, and the mechanical properties of the fiber. The majority of flame-retardant additives to polyester are based on antimony trioxide, phosphorous, bromine, and their mixture.[18] The increasing order of efficiency is $Sb_2O_3 <$ Br $<$ P. Bromine compounds are efficient, flame-retardant additives but their fastness to light is not always satisfactory. Chlorinated aralkyl hydrocarbons and bis-(2,4,6-trichlorophenyl) phthalate have been suggested. Some of the commercially available flame-retardant polyester fibers are described below.

Heim (Toyobo Co. Ltd.).[19] A flame-retardant polyester (under the trade name *Heim*) is produced by incorporating a phosphorous compound during polymerization. Heim was introduced in 1972 as a staple as well as continuous filaments. The flame-retardant property of the fiber is fast to laundering and the feel is similar to normal polyester. The fiber exhibits improved dyeability and can also be dyed with cationic dyes. Other finishes such as antistatic or water-repellent finishes can be applied without adversely affecting the flame retardancy. Heim can be blended with acrylics, normal polyester, nylon, triacetate, and wool to provide blends with acceptable flame retardancy.

Spectran Polyester (Monsanto Textiles Co.).[20] Spectran polyester fiber is particularly suitable for blending with modacrylic fiber to give flame-retardant fabrics. The fiber can be dyed at atmospheric pressure (100°C) and has an excellent soil-release property, improved hand, drape, and general fabric appearance.

Dacron Type 900 F (DuPont). Fabrics made from polyester fiber containing more than 6% covalently bonded bromine as part of the polymer backbone meet the standards for children's sleepwear. Dacron Type 900 F is very similar to normal polyester but it must be heat set at a lower temperature (120–140°C). It dyes more readily and carriers are required only for deep shades.[20]

Tetoron Extar (Teijin Ltd.). This phosphorous-containing flame-retardant polyester has the excellent fiber qualities of regular polyester and meets flammability standards.[21] The whiteness of this fiber is better than regular fiber and is best suited for lace curtains. It can be dyed to any shade at temperatures below 120°C. The resistance to weathering and yellowing on exposure to light is excellent and the flame-retardant property is fast to laundering and dry cleaning.

Trivera 270 and 271 (Hoechst AG). These fibers are flame-retardant and this property is retained after laundering. Their dyeability is better than that of the normal fiber and they give dyeings with equal fastness properties.[22] Some other commercial flame-retardant polyester fibers are GH fibers (Toyobo, Co.), Wistel FR (Snia Viscose, Ltd.), and Diolen FR (Enka).[23]

18.3 FLAME-RETARDANT FINISHES FOR POLYESTER

Most of the flame-retardant finishes are based on halogen compounds, often in conjugation with chemicals containing phosphorous or antimony to lower the requirement of the former.[24] According to Stepniezka[25] who scanned 78 flame-retarding agents, 33 are halogen compounds, 17 are phosphorous compounds, 20 are phosphorous–halogen compounds, and 8 are nitrogen–halogen compounds. The most widely used flame-retardant for polyester is tris(2,3 dibromopropyl) phosphate[26]

$$(BrCH_2BrCHCH_2O)_3P{=}O$$

which contains 68.7% bromine (by weight) which makes it a highly efficient flame-retarding agent.[24] The compound has a very good affinity for polyester so that it is absorbed and fixed by almost the same procedure as that used for dyes. The pad-dry-thermofixation process is found to be particularly effective. The flame-retarding effect of tris (2,3 dibromopropyl) phosphate is fast to laundering. About 14×10^{10} flame-retarding garments for children's sleepwear had been manufactured in the United States by early 1977 and about 40% of these were estimated to have been treated with tris(2,3 dibromopropyl) phosphate.[27] Recently, tris(2,3 dibromopropyl) phosphate and fabric treated with this chemical have been found to cause cancer in animals.[27] Garments treated with this compound are now banned in the United States. Many other fire-retarding finishing agents for polyester are described in the literature[18,24] and are now commercially available, such as Flamex MM (Guardian Laboratories, U.S.A.), Fire Master LV (Michigan Chemical Corporation, U.S.A.), and Tanotard PNZ (Chas S. Tanner Co., U.S.A.).[9]

Tetoron Unfla 3 (Toray Industries Ltd., Japan) is a normal polyester filament yarn finished with a flame-retarding agent.[28] The fiber is not discolored or degraded during the application of the finish and has a permanent flame-retardant property. It exhibits high soil resistance, good antistatic property, and can be dyed by conventional methods. It is used in lace, curtains, draperies, babies' sleepwear, nightdresses and so on.

18.4   FLAME-RETARDANT POLYESTER–COTTON BLENDS

Polyester–cotton blend fabric is extremely flammable.[4,29] When the cotton component comes in contact with a flame, it burns very rapidly and propagates the flame. The polyester portion starts melting along with the burning cotton; these two simultaneous actions can cause considerable injuries or even death to the wearer. The relationship between the LOI and the blend composition is shown in Figure 18.1.[30] The blends, which range in cotton content from 15% to 85%, have a lower LOI (higher flammability) than either 100% polyester or 100% cotton. The blend fabric ignites sooner, burns faster, thermally decomposes faster, and evolves more volatile hydrocarbons than the individual components.[31]

It is thus highly desirable to reduce the flammability of a PET/CO blend fabric. In blends, the presence of flame-retardant polyester has little effect on the combustion of the cotton portion. The addition of a flame retardant that is especially effective for one component does not necessarily result in decreasing the flammability of the blend.[17,32] This makes the task of finishing more difficult. Two approaches for the production of flame-retardant PET/CO blend fabrics are (*a*) to use flame-retardant polyester and to apply a flame-retardant finish to the cotton and (*b*) to apply flame-retardant finishes to both the components.

For 100% cotton fabrics, flame-retardant finishes based on organophosphorous compounds such as tetrakis (hydroxymethyl) phosphonium salts are most effective, whereas for 100% polyester, flame-retardant finishes containing halogen (particularly bromine) are most effective.[33] A combination of organophosphorous and bromine-containing compounds is likely to give good results for PET/CO blends.[34]

The LOI for polyester varies inversely with the carboxyl end-group content of PET. The mode of incorporation of a bromine flame-retardant into polyester is critical. Bromine from tetrabromobisphenol is more effective in reducing total heat out of the burning blends than bromine from polyvinyl

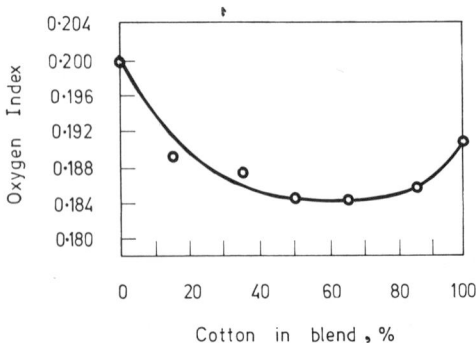

FIGURE 18.1.   Relationship of LOI to blend composition for polyester–cotton felts.[30]

bromide. For finishes in which phosphorous is not present, it is preferable to use a combination of bromine with antimony oxide,[34] when a bromine level of 7–8% is usually sufficient to pass the FF-3 sleepwear test.

Some of the finishing agents and processes are now mentioned. Tris(dibromopropyl) phosphate and tetrakis(hydroxymethyl) phosphonium chloride (THPC) are applied by the pad-dry-cure process.[35–37] By the same process, a mixture of tetramethylol-2-4-diamino-6(3,3,3 tribomo-1-propyl) 1-3-5 triazine and colloidal antimony oxide gives a good flame-retardant property. However, in view of its toxicity, "tris" is not used at present.[27]

THPC is used in combination with other phosphorous and/or bromine compounds[38] or as THPC-Urea and poly(vinyl bromide)–polyvinylchloride (PVB–PVC) copolymer.[39] A typical process consists of padding the fabric with a solution containing 30% THPC-Urea, 4% disodium hydrogen phosphate, 6.4% PVB–PVC copolymer, and about 2% NaOH (50% solution) to adjust the pH to about 5.7, drying at 85°C, and curing at 160°C for 1.5 min. The treatment imparts good flame retardancy to a 50:50 PET/CO blend fabric, which can withstand up to 50 laundry cycles.

Sandoz[18] has claimed a compound of the following general structure. This phosphazine-based compound, FR 1030-190, is applied by the pad-dry-cure

process. Phosphonium salt oligomer, Pyrovate 3762 (Ciba–Geigy), gives a finish fast to laundering.[40] Many other phosphoric acid-quaternary ammonium hydroxide condensates have been described.[18] Toyobo Co., Ltd. markets Taien-TPD-V and Taien-TPD-100 with a phosphorous-based agent. Fyrol 76 (Stauffer Chemical Co.) is a water-soluble oligomeric vinyl phosphate with 22.5% phosphorous by weight. It can be applied with a coreactant such as N-methylol acrylamide. Its requirement for effective flame retardancy increases with increasing polyester content. The possibility of applying of Fyrol 76 to 50:50 polyester/cotton as well as 50:50 Dacron T 900 F (flame-retardant polyester)/cotton by electron beams has been explored.[41] Fyrol 76 along with Triton X-100 (a wetting agent) is padded on the fabric. This is followed by drying at 105°C for 15 min. Fixation of the flame retardant is carried out by exposing fabric samples placed in a self-sealing polyethylene bag to a 500-key electron accelerator. The samples are then washed and dried. The fabric samples are found to have good flame resistance. A fabric sample made from 50:50 Dacron T 900 F and cotton requires a smaller amount of Fyrol 76 than fabric made from 50:50 normal polyester and cotton for getting similar nonflammability characteristics. The flame-retardant effect is durable to washing.

The LRC-15 (LeBlanc Research Corporation) used for PET/CO blend fabrics is a condensation product of THP-sulfate-trimethyl phosphoramide. It is applied with trimethylol melamine and urea by the pad-dry-cure process to get a wash-fast flame-retardant finish on a 50:50 blend fabric.[42]

Caliban FRP-44 (White Chemical Corporation), a dispersion of 46% decabromo diphenyl oxide and 22.5% antimony trioxide, (or THPC) is applied in conjunction with 15–20% soft acrylic binder.[43,44]

A flame retardant based on poly-2,3-dibromopropyl acrylate (PDBPA) can be applied to 50:50 PET/CO blend fabrics in two ways.[45] In the first method, PDBPA is applied with THPC-urea system, whereas in the second process, PDBPA is applied with THPS condensates and melamine resins and urea to produce a durable finish with a stiffened hand. It can also be applied in combination with melamine, glyoxal, and antimony oxide. This results in a durable flame-retardant finish that has good hand, good strength, and durable-press properties. The poly-2,3-dibromopropyl acrylate has shown some promise as a flame-resistant hand builder for 100% polyester fabrics. PDBPA does not cause any discoloration during curing. It does not increase the photodegradation of fabrics or photofading of dyes. The soil-release properties of the fabric are not affected by the presence of PDBPA. Although toxicological tests have not been done so far, PDBPA is unlikely to cause toxicological problems since its monomer has low toxicity.[45]

In a novel method, the cotton portion of the blend is first acetylated and then the usual flame-retardant for PET, tris(2,3 dibromopropyl) phosphate, is applied to the system.[46,47] Grafting of flame-retarding monomer on the cellulosic component, followed by application of a flame retardant for polyester is also suggested.[48]

It appears from the foregoing that the best treatment for flame-retardant finishing a PET/CO blend fabric is the one that combines phosphorous for cotton and bromine for polyester. The THPC-Urea, vinyl bromide system developed by Southern Regional Research Center and Ethyl Corp. has come close to meeting the demands for such a treatment.[34] Although a large number of flame-retardants have been developed for blends, no one is ideally suitable for different end-uses. The majority of flame retardants are quite costly and adversely affect the feel of the fabric. Thus, there is a wide scope for developing new and better flame retardants for PET/CO blends.

18.5 FLAME-RETARDANT NYLON FIBER

The additive may be incorporated during the polymerization reaction or preferably coated on to the chips of the polymer before melt-spinning. The additives can be grouped into two categories, similar to those for polyester:

(a) halogen-containing compounds and (b) organophosphorous compounds of the oligomer-type.

Some of the important flame-retardant additives are[18] (a) halogenated hydrocarbons and their dicarboxyl, anhydride, or imine derivatives as such or in conjunction with Sb_2O_3, PbO, SnO_2, sodium borate, or sodium antimonates; (b) phosphorous-containing compound such as ethylene glycolphosphorous trichloride condensate, bis(2 chloroethyl) vinyl phosphate, aliphatic and aromatic phosphorous compounds, alkyl phosphonium phosphate, O-halophenyl phosphites, phosphates, and phosphonates; and (c) triacetyl melamine.

Flame-retardant nylon fiber is not produced on a commercial scale. Polyamides are melt-spun at temperatures ranging from 250–300°C and only a few flame retardants are sufficiently stable to such high temperatures (if they are present in the polymer melt). The carbonamide linkages in nylons are considerably more reactive than many other polymer linkages. Because of the reaction between polymer and potential flame retardants at high processing temperatures, the polymer gets degraded, usually by the release of hydrochloric or hydrobromic acid from the flame retarder. The high amount of additive (up to 30%) required for effective retarding impairs the fine structure (crystallinity) of the fiber which is necessary for the desired fiber properties. Another effect that restricts the use of additives in polyamides is the tendency of the additive to migrate from the interior to the surface of fibers. This effect causes a serious soiling problem and impairs the hand of the fabric.

18.6   FLAME-RETARDANT FINISHES FOR NYLON

Many flame retardants that have been successfully applied to cellulosic and natural protein fibers tend to increase the flammability of nylon, with the exception of a few sulfur and/or chlorine-containing compounds.[49] The latter compounds prevent flame propagation of molten nylons because they produce acidic sulfur compounds and hydrogen halides on pyrolysis. These, in turn, decompose the ammonium carbonate produced in heated nylon under the generation of carbon dioxide.[49] The only flame retardants useful for nylon fabrics are those which lower the melting point of nylon. Flame retardants that are effective on natural fibers but ineffective on polyamides do not significantly lower the melting point of nylon.[50]

Many compounds containing antimony trioxide, sulfides, halogenated compounds and PVA-polyvinylidene chloride, resins containing formaldehyde, thiourea, cresol, and resorcinol, many phosphorous compounds, phosphorous-halogen with or without nitrogen compounds and so on have been described as flame-retardant finishes for nylon.[18,24] THPC and THPO are often reacted with an amine to form a resin on the fibers, which renders them flame-retardant. A typical recipe for a nylon flame-retardant consists of

18.5%	THPC
8.5%	Trimethylol melamine
8.8%	Urea
0.1%	Surfactant

This is applied by the pad-dry-cure process. All the above-mentioned flame-retardant finishes give improved flame resistance but none of them is ideally suitable to give a satisfactory flame-retardant finish on polyamide fabrics. The present flame-retardant finishes on nylon may have the following drawbacks:[9] (a) the finish is not fast to laundering, (b) the fabric hand is adversely affected because of the presence of a large amount of finishing agent, (c) the strength of the fabric is reduced, and (d) the shade of the dyed fabric is adversely affected.

18.7   FLAME-RETARDANT ACRYLIC FIBER

The flame-retarding additive can be dissolved with the polymer (or in its dope) before the wet or dry-spinning of the fiber. The stability of the additive at high temperatures is not a problem. Although the method appears to be quite simple, the resultant fibers may have low sticking temperatures and high shrinkage at elevated temperatures.[51] Most of the additives are based on esters of phosphorous containing halogen in the alkyl side chain. Typical additives recommend for preparing flame-retardant acrylic fibers are:[18] (a) tris (2,3 dibromopropyl) phosphate, allyl or vinyl esters of phosphoric or phosphonic acid, brominated halomethanes, and unsaturated phosphates; (b) vinylidine chloride, vinyl bromide, chloroacrylonitrile, and a mixture of vinylidene chloride and acrylamide; and (c) antimony oxide-brominated compound.[52]

Flame-Retardant Modacryl Fibers.   Typical commercial modacryl fibers are described below.

Dynel (Union Carbide).   It contains 60% vinyl chloride and 40% acrylonitrile.[20] The inherent self-extinguishing properties of garments made from Dynel are not affected by the normal variables found in laundering practices. Dynel is of particular use as tent fabric, draperies, bed sheets, and work clothes. The 65:35 blends of Dynel and polyester have adequate flame retardance and good physical and chemical properties.

SEF Fiber (Monsanto Textiles).   This fiber has much lower dry-heat shrinkage than other modacrylic fibers and does not appreciably shrink at temperatures below 190°C. The flame-retardant property of SEF fiber is not impaired by repeated laundering with phosphate or nonphosphate detergents. Blends of 65 : 35 SEF and Spectran polyester fibers have good flame retardant property, abrasion resistance, wash-and-wear properties, and dyeability.[20]

Lufnen (Kanebo Ltd.).   This fiber consists of more than 50% acrylonitrile, another monomer for the normal textile properties, and a flame-resistant monomer, and is manufactured by the standard wet-spinning techniques for the PAN fiber. The fiber properties are fast to laundering and dry cleaning, and the fiber is not toxic. It does not shrink at 120°C in wet conditions and shrinks only 4% at 160°C in dry state. It can be dyed by conventional methods using cationic dyes and has excellent bulk and hand. Lufnen has a LOI value of 29 and can be used for general apparel, interior furnishings, bedding, and, particularly, children's sleepwear. Blends of Lufnen (40%) with normal acrylic fiber are suitable for use in carpets.

Flame-Retardant Finishes for Acrylic Textiles.   Since flame-retardant modacrylic fibers are easily produced, development of a finish for acrylic textiles is limited to the following chemicals:[18,24]

Ammonium sulfide or polysulfide
Ammonium phosphates
Urea
Urea derivative of esterified melamines
$TiCl_4$–Sb_2O_3 reaction product
Urea-formaldehyde and ammonium bromide
Hydroxylamine salt with methylol melamine
Ammonium salt of phosphoric or metal phosphoric acid in higher paraffins

However, the commercial processes have not yet been developed.

18.8   MISCELLANEOUS FIBERS

Cellulose Acetate and Triacetate.   The most practical and economical way of producing flame-retardant cellulose acetate and triacetate fibers is to add flame-retardant chemicals to the spin-dope. These flame-retardant cellulose acetate fibers have excellent textile properties such as strength, softness, absorbancy, and warmth. Typical flame-retardant acetate fibers are Arnel triacetate (Celanese Co.),[53] FLR acetate (DuPont), Acele acetate (DuPont),[54] and Estron R-14 (Eastman Chem. Products Inc.).[55]

Polypropylene.   Attempts have been made to prepare flame-retardant polypropylene by incorporating suitable additives before melt-spinning.[7,56] A combination of chlorinated paraffin, tris(2,3 dibromopropyl) phosphate, chlorendic anhydride, and stearic acid is suggested to get a fiber with excellent textile properties. Chlorophosphorylation of polypropylene fiber by a treatment with ozone in carbon tetrachloride and then with phosphorous trichloride yielded polypropylene with improved receptivity to dyes with hydroxy and amino groups.[57,58] A fiber containing more than 4.7% phosphorous is flame retardant.

Polyvinyl Alcohol. PVA fibers can be made flame retardant by converting a portion of hydroxyl groups to acetate using chlorine, bromine-containing acetaldehydes, and benzaldehydes.[59] Phosphorylation of PVA fibers by a number of ways has been tried. PVA fibers in which 43% of the OH groups were acetalated are reacted with phosphorous chloride in the presence of tertiary amines to yield partially phosphorylated fibers of reduced flammability.[60] Another method of phosphorylation is to treat with urea and aqueous phosphoric acid at elevated temperatures.[61] Phosphorylation can also be done with phosphonates.[62]

Polyvinyl Chloride Fiber. Polyvinyl chloride fibers have low flammability and burn with a slow rate and a smoky flame, which gives hydrochloric acid, carbon monoxide, and carbon dioxide apart from some aliphatic, aromatic, and olefinic hydrocarbons as the combustion products at 600°C. Although much smoke is produced during burning, it does not produce poisonous gases such as chlorine, phosgene and so on. The main problem with PVC fibers is their high thermal shrinkage. They start shrinking at 65–70°C and shrink up to 40% of their length. A modified product, Leavil (Montedison U.S.A.), produced by a novel spinning process exhibits better resistance to heat and dry-cleaning solvents.[63] It is dimensionally stable up to 130°C and can be dyed or printed with disperse dyes. Improved properties such as crease resistance, weathering resistance, and easy washing are exploited in draperies, in apparel as such, or as a blend with polyester or cellulose acetate. Because of its acceptable flame-retarding properties, it is used in home furnishings and children's sleepwear. Teviron and Valren (Teijin Ltd.) are fibers that are extensively used in furnishings and draperies.[64] They exhibit good resistance to chemicals, mildew, rotting, weather, and fire. Valren staples are blended with other fibers for apparel use. Their properties are summarized in Table 18.4.[64]

Cordelan (Kohjin Co., Japan). Developed in 1967, this fiber consists of three components—polyvinylacetal, polyvinyl chloride, and polyvinylalcohol grafted with vinyl chloride.[65] It is available in two forms: high-flame-resistant-type (FBCH) and low-flame-resistant-type (FBC). Very few toxic fumes are produced when Cordelan is burned. Cordelan has a specific gravity of 1.32 and moisture regain 3% with good abrasion resistance. Flame-resistance and self-extinguishing characteristics of Cordelan are built into the

TABLE 18.4   Properties of Polyvinyl Chloride Fibers[64]

Polyvinyl chloride	100%
Chlorine	56–57%
Limiting oxygen index	35
Flammability	Self-extinguishing
Ignition point	503°C
Heat shrinkage up to 100°C	
Teviron	Considerable
Valren	Almost nil

fiber. This means that garments made of Cordelan can be laundered or dry cleaned repeatedly with no loss of flame resistance. Cordelan has a relatively low melting temperature and hence, it has to be finished below 90°C. It has a good resistance to chemicals like caustic soda, sulfuric acid, and solvents. The fiber accumulates less static electricity than any other synthetic fiber, absorbs more water than polyester, and therefore has a comfortable feel. It is a very soft fiber. It can be dyed with disperse, cationic, premetalized, and vat dyes. It is used in children's sleepwear, apparel, curtains, draperies, bedspreads, blankets, and auto/aeronautical fabrics.[28] Cordelan can be blended with cotton or polyester, Cordelan 11 fiber has a LOI of about 3 units higher than conventional Cordelan. Cordelan 11 maintains all the desirable properties of Cordelan.[42]

18.9   NEW FLAME-RETARDANT SYNTHETIC FIBERS

Although some of the modified fibers such as Heim polyester, Spectran polyester, Dacron Type 900 F polyester, Dynel, Verel, and SSF modacrylic fibers have some commercial success, they are not able to give entirely satisfactory performances in different end uses. Attempts have been made to produce altogether new synthetic fibers with an inherent flame-retarding property.[66] These fibers are found to be very useful in applications where stability to exposure to extreme heat is required.[66,67] Some such fibers are described below:

Aromatic Polyamides.   *Nomex.*   A DuPont fiber introduced in 1967 under the trade name *Nomex* is extensively used in protective clothing. As shown in Scheme 18.2, Nomex is an aromatic polyamide-poly (*m*-phenylene isophthalamide) prepared from *m*-phenylenediamine and isophthaloyl chloride.[67,68] A solution of *m*-phenylenediamine and sodium carbonate (as an acid binder) in water is vigorously blended with another solution containing isophthaloyl chloride in tetrahydrofuran to complete the condensation polymerization within 10 min. The polymer is collected by filtration, washing, and drying to get a 94% yield. The polymer is dissolved in dimethyl formamide containing 4.5% lithium chloride (to improve the solvent power). A twenty percent dope is dry-spun in hot air at 200–210°C. The fiber is then extracted with cold water and drawn in stream at a draw ratio of 5.5. The fiber is flame-resistant, has a melting point of 371°C, and sticking temperature over 300°C. If a single thread of Nomex is lit with a match, it flames brightly for a moment but extinguishes immediately. If a flame is held underneath a piece of Nomex fabric, the fabric hardens as it starts melting, discolors, and chars. The fire risk is negligible and this is of great value in spacesuits. Nomex exhibits better resistance to ionizing radiation than nylon 6. Chemical resistance is good but is less than Teflon. The dimensional stability is good. It shrinks only up to 2% in boiling water or in dry air at 260°C. Dry tenacity is 5.3 g/d at 65% RH/20°C and wet tenacity is 4.1 g/den. Specific gravity is 1.38. Nomex is off-white in color and its dyeability is poor.[69]

m-Phenylenediamine   Isophthaloyl chloride   Nomex

Nomex was initially used as an industrial fiber in dryer belts, laundry-press fabrics, and for high-temperature filtration. As protective clothing, it finds use in drivers' overalls, refinery operators' garments, fire-fighters' coats and so on. Being non-flame-propagating, self-extinguishing, and very low in smoke generation, it is used in carpets and upholstery, uniforms of military forces, for example, uniforms of tank crews, aviation pilots and so on. The thermal resistance of Nomex is further increased by a treatment with thionyl chloride at elevated temperatures or with a reactive organophosphorous compound, a nitrogen-containing binder, and a curing catalyst by the pad-dry-cure technique.[70]

In conditions where total immersion in flames for several seconds is likely to take place, garments made from heavier weight or double-layer Nomex fabrics offer good protection. However, uniforms made from lighter weight fabrics break open as a result of high overall shrinkage on total immersion in flames. A HT-4 fiber (DuPont) withstands such conditions even in lighter weight fabrics.[71,72] A comparison of Nomex and HT-4 is given in Table 18.5.[71] HT-4 fabric produces a very low temperature rise due to heat transfer and has higher strength retention under dynamic heat exposure conditions than Nomex. The long-term thermal stability to temperatures below the onset of rapid pyrolytic degradation is better for Nomex than for HT-4. Similarly, Nomex's hydrolytic stability and stability to acids and alkalies is better than HT-4. HT-4 is preferred where total immersion in flames for

TABLE 18.5 Properties of Nomex and HT-4[71]

Property	Nomex	HT-4
Fabric weight (oz/yd^2)	4.5	4.4
Break open time (sec)	1.2	60
High-temperature free shrinkage (%)		
2 sec	55	0
4 sec	60	0
6 sec	60	5
Afterflame time (sec)	11	2–3
Limiting oxygen index %	29	40

several seconds is likely to occur. Nomex is halogenated in the benzene ring to produce Durette 400 (Monsanto Co.).[71] For example, a light gold colored fabric is obtained by a treatment with chlorine gas at 300°C for 15 min in an inert atmosphere. Durette 400 fiber contains 9–10% chlorine and exhibits a high LOI (35–37). It can be used for protective garments in severe fire hazards.

PBI (DuPont and Celanese Corporation).[73,74] PBI fiber is dry-spun from the polymer dope (formed by melt condensation of diphenyl isophthalate and diamino benzidine) in dimethylacetamide. It has a tenacity of 5 g/d and elongation of 15–20%. The retention of strength after exposure for 1 min at 450°C is about 40%. PBI fiber has thermal properties that exhibit high resistance to ignition and poor burning rates.

POD-Z and BBB (Air Force Materials Laboratory, U.S.A.). The POD-Z fiber is dry-spun from a dope in dimethyl sulfoxide of a polymer-poly(1,3/1,4-phenylene-2,5-(1,3,4 oxadiazole)) made from terephthaloyl chloride and isophthaloyl hydrazide. Its short-term temperature response is similar to PBI fiber, but its long-term stability is better.

BBB fiber consists of poly(bisbenzimidazobenzophenanthroline) produced by condensation of 1,4,5,8 tetracarboxylic acid and 3,3'-diamino benzidine. The fiber is produced by wet-spinning a solution in concentrated sulfuric acid into a dilute (70%) acid bath. The fiber is drawn at 550–575°C so that it has a tenacity of 4.5 g/d and elongation of about 4%. The strength retention at high temperatures is superior to PBI and POD-5 fibers. Even at 575°C, the strength is about 60%. It has excellent stability to light and has a natural deep-olive color suitable for military purpose.

Kynol. (Carborundam Co.). Kynol is a phenol-formaldehyde cross-linking polymer with excellent flame-resistant properties.[73,74] Depending on the denier, Kynol has a moisture regain of 4–8% which results in comfort in wear. It is available in staple form and only in gold color. The flame-resistance properties are:

Limiting Oxygen Index %	32–36
Vertical flame test (5–6 oz/yd²)	
Char length (in.)	0.1
Glow time (sec)	2.8
Time to increase the back-face temperature by 25°C	
(5–6 oz/yd²) fabric (sec)	7.0

The resistance to abrasion and the fastness properties of Kynol are inferior to those of cotton. The fiber is used as an insulator in jet aircraft and in protective clothing for furnace workers.

Kevlar.   Kevlar is produced by DuPont mainly for use as a tire cord. Its chemical constitution is poly(p-phenylene terephthalamide) (PPT) and is prepared by the reaction of terephthalolyl chloride and N-methyl pyrrolidine.[66] The concentrated sulfuric acid solution of the polymer is spun into water or dilute sulfuric acid. After washing, a further heat treatment under tension is applied. The fibers are not drawn. Depending on the draw-down in spinning and take-up, fibers with varied tensile properties are obtained. Kevlars are both a flame and high-temperature-resistant fibers.[66,75,76] These fibers are used in ropes and cables for various outdoor end-uses.

Inorganic Textile Fibers.   Some of the promising flame-retardant inorganic fibers are A-150,[60] a glass fiber developed by the Air Force Materials Laboratory (U.S.A.) with good flame-retardant properties and suitable for wearing.

18.10   EVALUATION OF FLAMMABILITY

The measurement of the relative flammability hazard of textile materials poses many difficulties:

1. Flammability is a dynamic and complex process. The way an article burns is influenced by surrounding conditions such as temperature, relative humidity, oxygen content, and rate of air flow.
2. Textile materials are diverse and include apparel, floor coverings, and home furnishing.
3. Within the same category of textiles, there are many factors that influence the flammability characteristics drastically. In apparel, for example, flammability depends on the fiber, fabric construction and weight, and the design of garment among other things.

4. Human reactions during fire is a complicating factor and is in many cases unpredictable.

It is not possible to standardize a single method of testing flammability of all textiles.[77-79] It is necessary to base the test method on the specific end-use requirements.

AATCC Test Method 33-1962. This test is designed to indicate the textiles that ignite easily and, once ignited, burn with sufficient intensity and rapidity to be hazardous when worn.[80] A test specimen is cut from the fabric to be tested and if it has a raised surface, it is then brushed. It is then held in a flammability tester at a 45° angle, and ignited by applying a standardized flame for one second. The time required for the flame to proceed 127-mm up the fabric is recorded. Three classes of flammability are recognized: normal, intermediate, and rapid intense burning.

Normal Flammability: More than $3\frac{1}{2}$ sec (7 sec for textiles with raised surface) for the flame spread indicates no unusual burning characteristics.

Rapid and Intense Burning: Time less than $3\frac{1}{2}$ sec (less than 4 sec for raised surface fabrics) for flame spread.

Textile Institute (U.K.) Methods.[81] *Method A.* This method measures the rate of flame propagation, duration and spread of the afterflame and afterglow, and the char length. It can be applied to all textile fabrics except floor coverings. In this method, a strip of fabric is held vertically in a box with an open front. The flame is ignited in a standard manner and the time to travel between two markers is noted. The rate of propagation per min is calculated. The time that elapses between the removal of the flame and the flame extinction is called the *duration of the afterflame.* The time between the flame extinction and the end of any glowing is termed the *afterglow.* The extent of charring is given by the *char length,* that is, the difference between the original length and the remaining undamaged length of the specimen.

Method B. Method B is intended as a confirmatory test for fabrics that do not burn when tested by Method A because they melt, shrink, or curl away from the flame. In this method, the test specimen is hung in sheet form in a rectangular box (as used in Method A) and the igniting flame is applied at right angles to the sheet and near the bottom edge for 5 sec. The damage and the time taken are observed.

Method C. Method C is intended solely for determining whether the piles will promote the rapid spread of a flame with which it is in transient contact. A sheet of fabric is held vertically in a substantially draft-free enclosure and a flame is traversed across the surface of the pile at a known speed to determine whether ignition of the pile fibers occurs.

Limiting Oxygen Index Method. This method[82] is used widely for accessing the flammability of textiles. The method yields a value known as the limiting oxygen index (LOI) which is defined as the minimum percent oxygen concentration that permits the entire specimen to burn. This device determines the relative flammability of materials through precise and reproducible measurement of the minimum concentration of oxygen required to just support the burning of a specimen. Oxygen and nitrogen are fed through a highly accurate gas-metering unit to the bottom of a glass-combination chamber, mixed, and then flow past the specimen. Fibers with a LOI between 18–20 are considered flammable. Fibers with a LOI above 27 are nonflammable. The LOI of important fibers are given in Table 18.6.[82]

TRI Flammability Analyzer. The Textile Research Institute (U.S.A.) flammability analyzer is a wheel-feed apparatus that works on the principle of maintaining the flame in a fixed position by feeding the specimen into it at a rate equivalent to the flame-propagation rate. Thus, this test differs from the earlier ones which measure the time of progress of a consuming flame along a stationary specimen. The method is applicable to all types of flexible, single, and multilayer sheet materials.[83] The specimen is mounted on the rims of a pair of linked wheels that rotate about a horizontal axis at a speed that is continuously adjusted until the flame remains stationary at a desired point. The wheel is designed to operate within an encloser fed at the bottom by a gas-mixing system so that the gaseous environment, particularly the

TABLE 18.6 Limiting Oxygen Index of Textiles[82]

Textile Material	% LOI
Acrilan	18.2
Acetate	18.6
Cotton	20.1
Dynel	26.7
Heim	27.5
HT-4	40.0
Kynol	33.0
Lufnan	29.0
Nomex	28.2
Nylon	20.1
Polyester	20.6
Polypropylene	18.6
Rayon	19.7
Teviron	35.0
Triacetate	18.4
Valren	35.0
Wool	25.2

oxygen content, can be controlled and varied systematically. The top cover of the encloser has a port that can be positioned for igniting the sample. This same port affords a means of adopting the unit to measure smoke density, flame temperature, and heat generation. The TRI flammability analyzer has simplified evaluation of the effects of fabric weight, density, and construction on flame characteristics, and has improved the precision of such measurements. Another novel feature of this technique is its ability to deal with the tendency of fusible synthetic fibers to drip and shrink away from the flame.

Apex Test.   For 100% thermoplastic textile materials,[84] the specimen is supported on a lightweight glass fabric in a carefully designed manner in order to overcome the dripping away of the melt. Otherwise, the method is similar to the AATCC-33-1962 test, and the length of the char-melt area and burning time are used as measures of flammability.

Test on Mannequin.[71]   This test for protective garments involves clothing a life-sized polyester/fiberglass mannequin in a uniform and carrying it through a pit of flaming aircraft fuel. The rate of travel of the mannequin is adjusted to expose the uniform for varying amounts of time. Temperatures and thermal load on the mannequin are determined by temperature-sensitive tapes and copper wedge calorimeters. While this test is an excellent simulation, it is extremely expensive (135 liters fuel/run, 30 runs/test), time-consuming, and not very reproducible.

REFERENCES

1.   Tovey, H., and Katz, R. G., *TCC,* **13** (1981), 118.

2.   Vaidya, A. A., Chattopadhey, S., and Chipalkatti, V. B., *Text. Dyer Printer,* **8** (1974), 21.

3.   Gay-Lussac, J. L., *Ann. Chim.,* **18**(2) (1921), 211.

4.   Kruse, W., *MTB,* **49** (1968), 203.

5.   Kruse, W., and Segall, W. M., *ADR,* **58**(6) (1969), 22.

6.   Nametz, R. C., *Ind. Eng. Chem.,* **62**(3) (1970), 41.

7.   Press, J. J., *Man Made Textile Encyclopedia,* InterScience, New York (1959), p. 143.

8.   Chattopadhayay, S., and Ravishankar, S., *Colourage,* **23**(22) (1976), 29.

8.   Vaidya, A. A., Chattopadhyay, S., and Ravishnkar, S., *Text. Dyer Printer,* **9**(10) (1976), 37; **10**(6) (1977), 49.

10.   Delaware Valley Section, *TCC,* **10** (1978), 252.

11.   Pohl, H. A., *J. Am. Chem. Soc.,* **73** (1951), 5660.

12.   Goddings, E. P., *Soc. Chem. Ind. Monograph No.* **13** (1961), 211.

13.   Marshall, T., and Todd, A., *Trans. Faraday Soc.,* **49** (1963), 67.

14.   Modorsky, I. S. L., *Thermal Degradation of Organic Polymers,* Wiley–Interscience, New York (1964), 272.

15.   Stepnickza, H. E., *Ind. En. Chem., Product R and D,* (12 March, 1973), 29.

16. Aenishaenslin, R., *MTB,* **49** (1968), 1210.

17. Kruse, W., *MTB,* **50** (1969), 460.

18. Datye, K. V., Jitendra, K., and Vaidya, A. A., *Man Made Text. India,* **23** (1980), 617; **24** (1981), 23.

19. Masai, Y., *Japan Text. News,* **214** (1972), 66.

20. Leblanc Research Symposium, *TCC,* **6** (1974), 125.

21. Takayama, M., *Japan Text. News,* **267** (1977), 74.

22. Zimmermann, H., *Chemiefasern/Textilindustrie,* **28/80** (Dec. 1978), 1054.

23. Holker, J. R., *Text. Month,* (May 1979), 61.

24. Yehaskel, A., *Fire and Flame Retardant Polymers,* Noyes Data Corporation, U.S.A. (1979).

25. Stepniezka, H. E., *Textilveredlung,* **10** (1975), 180; **10** (1975), 234.

26. Nametz, R. C., *TRJ,* **41** (1971), 593.

27. Metropolitan Section, AATCC, *TCC,* **10** (1978), 6.

28. Anon, *Japan Text. News,* **267** (1977), 76.

29. Tesero, G. C., *TCC,* **5** (1973), 235.

30. Tesero, G. C., and Meiser, C. H., *TRJ,* **40** (1970), 430.

31. Miller, B., Martin, J., Meiser, C. H., and Garginllo, M., *TRJ,* **46** (1976), 530.

32. Tesero, G. C., and Rivlin, J., *TCC,* **3** (1971), 156.

33. Reeves, W. A., Perkins, R. M., and Drake, G. L., *Fire Resistant Textiles Handbook,* Technomic Pub. Co., Westport, Conn. (1974).

34. Barker, R. H., and Drews, M. J., *Report on Experimental Technology Incentive Programme,* National Bureau of Standards, U.S. (Sept. 1976).

35. Piedmont Section, *ADR,* **57**(10) (1968), 373.

36. Sharma, V. N., *Colourage,* **26**(7) (1979), 27.

37. Timpa, J. D., Chance, L. H., and Goynes, W. R., *ADR,* **68**(6) (1979), 60.

38. Reeves, W. A., *TCC,* **6** (1974), 125.

39. Donaldson, D. J., Normand, F. L., and Drake, G. L., *ADR,* **64**(9) (1975), 30.

40. Geigy, German P., 1,150,044 (1963).

41. Bittencourt, E., Ennis, J., and Walsh, W. K., *ADR,* **67**(1) (1978), 32.

42. LeBlanc, R. B., Dicarlo, J. P., and LeBlanc, D. A., *TCC,* **10** (1978), 207.

43. LeBlanc, R. B., Textile Industries (Feb. 1977), 29.

44. McMackin, L. V., *TCC,* **9** (1977), 202.

45. Palmeto Section, AATCC, *TCC,* **9** (1977), 421.

46. Mehta, R. D., *TRJ,* **45** (1975), 637.

47. Mehta, R. D., Loughlin, J. E., and Sheets, D., *ADR,* **65**(1) (1976), 30.

48. Tesero, G. C., *TCC,* **5** (1973), 235.

49. Hasselstrom, T., Coles, H. W., Balmer, C. E., Hannigan, M., Keller, M. M., and Brown, R. J., *TRJ,* **22** (1952), 742.

50. Douglas, D. O., *JSDC,* **73** (1957), 258.

51. Masuda, K., *Japan Text. News,* **267** (1977), 76.

52. Masuda, K., *Japan Text. News,* **268** (1977), 106.

53. Vandermaas, J. K., Holmes, T. L., and Zybko, W. C., *TRJ,* **44** (1974), 561.

54. Shealy, D. L., Lynch, J. A., and Arnold, H. W., *Mod. Text.,* **55**(10) (1974), 50.

55. Casey, D. A., *ADR,* **64**(9) (1975), 33.

56. Knopfel, H. P., and Mitterhofer, F., *Chemiefasern/Textilindustrie,* **30/82** (1980), 808.

57. Vsiansky, J., Czeck Patent, 11,352 (1966).
58. Bellus, D., Czeck Patent, 111,995 (1964).
59. Matsubayaski, K., *Sen-i Gakkaishi,* **19**(1) (1963), 27.
60. Tseitlina, L. A., *Khim Volokna,* **4** (1965), 16.
61. Joshnson, J. H., USP, 3,210,147 (1961).
62. Orlov, N. F., USSR Patent, 1,89,515 (1966).
63. Susani, P. L., *TCC,* **6** (1974), 128.
64. Tsuzuki, R., *TCC,* **6** (1974), 128.
65. Koshiro, T., and Goldfarb, A., *Mod. Text.* (Nov. 1973) 40.
66. Hughes, A. J., McIntyre, J. E., et al., *Text. Prog.,* **8** (No. 1), (1976), 1.
67. Van Krevelen, D. W., *Appl. Poly. Symposium,* **31** (1977), 269.
68. DuPont, USP, 3,006,899 (1961).
69. Ulrey, H. E., *JSDC,* **90** (1974), 401.
70. Hathaway, C. E., and Early, C. E., *J. Appl., Polymer Symp.,* **21** (1973), 101.
71. Shivers, J. C., and Hentschal, R. A. A., *TRJ,* **44** (1974), 665.
72. Gloor, W. H., *ADR,* **57** (1968), 482.
73. Economy, J., *Fire and Flammability,* **3** (1972), 114.
74. Economy, J., *Textilveredlung,* **88**(12) (1972), 114.
75. *Text. Prog.* **8**(1) (1976), 25.
76. Sucheki, S. M., *Text. Industries,* **142**(2) (1978), 29.
77. Lebalnc, R. B., and Weaver, J. W., *TCC,* **8** (1976), 144.
78. Weaver, J. W., *TCC,* **8**(11) (1976), 176.
79. Galil, F., and Lomartire, J., *TCC,* **8** (1976), 121.
80. AATCC Test Method 33-1962, AATCC Technical Manual, (1975), 210.
81. Anon, *JTI,* **59** (1968), 47.
82. Fenimore, C. P., and Martin, F. J., *Modern Plastics,* **44**(3) (1966), 141.
83. Miller, B., and Meiser, C. H., *TCC,* **3** (1971), 118.
84. Endler, A. S., and Hurwitz, M. L., *ADR,* **56** (1967), 694.

19 | ANTISTATIC AND SOIL-RELEASE FIBERS AND FINISHES

Synthetic fibers such as polyester, polyamide, and acrylics exhibit two undesirable properties that can be attributed to their hydrophobic nature: (*a*) they accumulate static charges on their surfaces and (*b*) they attract soil easily. They are finished with antistatic agents and soil-releasing compounds to overcome these limitations; a common finishing treatment may achieve both effects. Such finishes are described in the present chapter. Since synthetic fibers can be easily modified during their manufacture, attempts have been made to produce a fiber with antistatic and soil-releasing properties. Such modification processes and modified fibers are also discussed in this chapter.

19.1 STATIC CHARGE ON FIBER

During spinning, weaving, and finishing, textile fibers, yarns, and fabrics are subjected to friction. Static electricity is thus generated on the fiber. Cellulose and protein fibers have sufficient electrical conductivity in their normal-air dry-state to conduct the static electricity to the earth. However, synthetic fibers and cellulose acetates have much lower conductance and, therefore, accumulate static electricity.

The static electricity on the fiber gives rise to a number of problems.[1,2] During melt-spinning of polyester and nylon filaments, static attraction causes breakages. The yarn may balloon during the winding processes as a result of similarly charged individual filaments repelling each other, and fabrics may stick to the rollers over which they are being drawn during finishing. Operators at the delivery end of a stenter or a Sanforizing machine may get electric shocks because of static electricity. Carpets made of synthetic fibers accumulate excessive static charge and give unpleasant electric shocks to anyone walking over them. Garments made of synthetic fibers attract soil during normal wear, which is difficult to remove. They also have a tendency to cling to the body of the wearer or other garments. Sparks created by the garments because of excessive accumulation of static electricity are particularly dangerous for workers in coal mines, ammunition factories, oil refineries and so on. It is essential to minimize static build-up on synthetic fibers. Two approaches that are likely to solve the problem of static accumulation on synthetic textile materials are: (a) avoiding the generation of static charge and (b) increasing the rate of dissipation of the generated charge. The reduction of the rate of dissipation or the total avoidance of the generation of static charges is rather difficult. The relationship between the structure of the fiber and the generation of the static charge is not fully understood. The ultimate surface of the fiber represents one or two molecular layers that are responsible for the generation of the static charge. The control of the surface structure is not at all easy and feasible since the trace impurities are invariably present on the surface. These impurities not only influence the surface structure of the fiber but may significantly alter its electrical characteristics. All the available methods of minimizing problems of static electricity on textiles are therefore based on increasing the rate of dissipation of the static charge. The electrical conductance (reciprocal resistance) is taken as a measure of ability to dissipate static charge.

The role of water in reducing accumulation of static charge has been studied by Valko et al.[3] They observed that electrolytic conductance requires the presence of a medium of high dielectric constant and that water is the only medium in this class that is replenished from the atmosphere when it is lost. Thus, the moisture regain of the fiber plays an important role in static dissipation, as is seen from the inverse relationship of moisture regain (%) with the log specific area resistivity (log SAR) (Table 19.1).[4] The higher the moisture regain, the lower is the log SAR and the lower is the static charge accumulation. Cellulosic fabrics have a low log SAR value and the electrical conductance of cotton is about 10,000 times greater than that of

TABLE 19.1 Moisture Regain and Fiber Resistivity[4]

Fiber	Moisture Regain (%)	Log of Specific Area Resistivity
Viscose rayon	12	7
Cotton	8	7
Acetate	6	12
Polyamide	4	12
Polyacylic	1	14
Polyester	0.4	14
Polyvinyl acetate	0	15

polyester and acrylic fibers. Since cotton just escapes giving static trouble under proper humidity conditions, it is essential to lower the log SAR values of synthetic fibers to about that that characterizes cotton. Of equal importance in static dissipation is the presence of ions in the fiber. It has been observed that if all the other factors are constant, the conductance of textiles is roughly proportional to the equivalent concentration of electrolytes present on the fiber. Antistatic fibers are obtained by increasing these two factors.

19.2   ANTISTATIC FIBER

Modified polyester, nylon, or acrylic fibers are produced by incorporating an antistatic agent during or immediately after the polymerization reaction. Incorporation of an antistatic agent after polymerization by the chips-coating method is preferred because the antistatic agent is not exposed to the rigor of the polymerization reaction for a long time. Furthermore, the course of polymerization is not adversely affected by the additive. Another way to tackle the static problem is to develop entirely new high-performance antistatic fibers and to blend a small proportion of these fibers with conventional synthetic fibers.

Preparation of Antistatic Fiber.   An antistatic agent is incorporated during or immediately after the polymerization reaction so that the fiber's surface characteristics are modified. The antistatic agent has to meet many stringent specifications.[5] It should give a permanent antistatic property at low add-on without creating any pollution or toxicity problems. It should be thermally stable, compatible with the ingredients in the polymerization reactions, and should not have any adverse effect on the fiber properties. It should get uniformly distributed in the polymerizing mass and should have a very small particle size (less than 2 μm). It is usually not possible to get an ideal antistatic agent that meets all the above specifications. Often, one has to make a compromise and select an antistatic agent that will affect the fiber properties to the least extent. The majority of the compounds used as antistatic agents are based on ethylene oxide condensates which meet many of the above-mentioned specifications and are readily available.

Antistatic Polyester Fiber.   The antistatic agent is added either during the transesterification or the polycondensation reaction. Alternatively, a polymer containing the antistatic agent is separately prepared as a master batch and added to the normal polyester. Some of the additives are sodium sulfate, sulfonated condensation products of ethyleneoxide, epichlorohydrin, polyether urathane, metal chlorides or pyrophosphates, antimony stearyl phosphate, sodium benzene sulfonate-ethyleneoxide adduct, sodium naphthalene sulfonic acids, polyethylene glycol dibenzoate or diacetate, and compounds containing ethyleneoxide units.[6]

Antistatic Nylon Fiber.   Nylon with antistatic properties is prepared by incorporating (*a*) a variety of products obtained by condensation of ethyleneoxide, ethylene glycol sulfonates, tetrol and so on during the polymerization of caprolactam,[1,5] or (*b*) electrically conductive particles of carbon black, silver, aluminum, or bronze into the fiber. Similarly, spinning bicomponent filaments with polyamide as a sheath and a conductive material as a core, for example, nonfilament-forming polyetherester, aliphatic polyester, *N*-alkylpolycarbonamide, or a core of carbon black dispersed in a polymer and polyamide as a sheath give antistatic nylon.

Antistatic nylon is commercially marketed under various trade names, for example, Antron 111 (DuPont), Ultron and Cadon (Monsanto), Stataway (Celanese), and Bodyfree and Auso-X (Allied Chemical Corp.). Most antistatic nylon is used in the carpet industry.

Antistatic Acrylic Fiber.   The antistatic agent is added during polymerization to form part of the polymer chain or in the polymer dope before spinning. Thus, polytetrahydrofuran is uniformly mixed in a spinning solution of a fiber-forming polymer of acrylonitrile–vinyl–acetate (PAN) or any other ethylenically unsaturated monomer.[7] A polymer consisting predominantly of a polyether or polyester block copolymer and a polymer consisting of polyacrylonitrile-vinyl monomer are wet-spun from a dope.[8] Oxyalkylated linear polyamide (about 1–10%) is mixed in the PAN–polymer dope and wet-spun into fiber.[9] Hydroxyalkylated polyvinyl alcohol (7–15%) obtained from polyvinyl-alcohol–ethylene oxide is mixed in the PAN–polymer dope and spun into fiber.[10]

High-Performance Antistatic Fibers.   Stainless steel, carbon, or metal-coated fine fibers are good conductors of electricity. By blending such high-performance fibers with normal fibers, it is possible to impart antistatic properties to yarn or fabric. Typical fibers are listed in Table 19.2.[11] An advantage of high-performance antistatic fibers is that their effectiveness does not depend on the relative humidity of the atmosphere. Furthermore, the antistatic effect of the fabric containing high-performance antistatic fibers is permanent. Although the cost of high-performance fibers is quite high, the quantities used for the preparation of the blend are so small that the overall cost of the fabric is not altered.

19.3   ANTISTATIC FINISHING TREATMENTS

The antistatic agents are applied at various stages of processing such as immediately after melt-spinning to the filaments (as spin-finishes), or during

TABLE 19.2 High-Performance Antistatic Fibers[11]

Product	Manufacturer	Type of Fiber
Selguard AFD	Allied Chemical Corp.	Slit aluminum nylon film laminate
Bekinox, Bekitex	NV Bekaert SA	Stainless steel blended with a nylon staple fiber
Brunsmet	Brunswick Corp.	Stainless steel blended with a nylon 6 staple fiber
Zefstat, Zefstat 901	Dow Badische Co.	Nylon 6 with carbon embedded in the surface
Conducting core hetero filament	DuPont	Nylon 66 sheath with conducting carbon in polyethylene core
Antron 111 bulked filament yarn	DuPont	Blend of above with Antron 11
Epitropic filament yarn	ICI	Nylon 66 hetero filament with carbon embedded in the surface
First Safety	Kuraray Co., Ltd.	Metal-organic fiber composite
Nylfrance, no-static filament yarn and staple fiber	Rhone-Poulene Co.	Nylon 66 with silver embedded in the surface
X-Static	Rohm and Hass Co.	Silver-coated nylon filament yarn
Metalian	Teijin Ltd.	Carbon or silver-loaded resin on nylon filament yarn

cone-winding (as coning oils) to the yarn or fabric. The antistatic agents are of two types: (*a*) nondurable antistatic agents and (*b*) durable antistatic agents. The durability of an antistatic agent to laundering is of importance to consumers. The nondurable agents are adequate for use during spinning and weaving operations.

Nondurable antistatic agents are usually hygroscopic surface-active materials, closely allied in composition to softeners and wetting agents. Anionic, cationic, and nonionic materials have all been proposed, for example, sulfated fatty acids or alcohols, quaternary ammonium compounds, and polyethylene glycols, respectively. The cationic and nonionic types are preferred since they are generally more hygroscopic, and being more oil-soluble, are more compatible with the fiber itself. These agents are often applied with other additives, such as fiber lubricants, with which they must also be compatible. Because of their hygroscopic properties, they help in maintaining a layer of moisture on the surface of the fiber, which is responsible for dissipating the static electricity. The drawback of these nondurable agents is that they impart a sticky hand to textiles and increase the soil pickup. They have been shown to create problems in soil removal during washing.[12]

The development of durable antistatic agents was difficult because the two main requirements of antistatic agents, namely, hygroscopicity and an ionized state, make them soluble in water and therefore removable during laundering.[13] The only answer to the problem is deposition of a hygroscopic and ionizable polymer on the surface of the fiber; the polymer is insolubilized by the cross-linkages between macromolecules. Some of the durable antistatic agents and their application processes reviewed in the literature[14-16] are listed below.

1. When a resin precondensate made by reacting melamine with formaldehyde and polyethylene oxide is applied to polyamide, polyester, polyvinyl alcohol, or cellulose acetate in the presence of ammonium nitrate (as a catalyst) by the pad-dry-cure process, these fibers do not give static troubles. Valko et al.[3] have used polyamine resins as antistatic agents for polyester fibers.

2. A condensation product of fatty acids such as oleic acid, stearic acid, lauric acid, glycerine, and ethylene oxide is useful for rendering polyester and nylon material antistatic.[4] Another antistatic agent is produced by reacting a sodium salt of diaminostilbene disulfonic acid with diglycidyl ether of glycene.

3. When stearamide-propyl-dimethyl nonyl-etheneoxy ammonium chloride is applied as an aqueous dispersion by the pad-dry-cure process, it gives a wash-fast antistatic finish on hydrophobic fibers.[7]

4. A product based on the ammonium or sodium salt of a copolymer of maleic acid and styrene is recommended as an antistatic agent for polyamide and polyester fibers. It is applied as an aqueous liquor in the presence of polyglycol and ammonium chloride by the pad-dry-cure process.[4]

5. When aqueous liquor containing a diglycidyl ether of polyethylene glycol and an amine such as ethylene diamine or triethylene tetramine is applied by the pad-dry-cure process, it makes the synthetic fibers antistatic.

6. A condensation product of ethylene pentamine with ethylene oxide and dichloride of polyethylene glycol is a useful antistatic agent for synthetic fibers.

7. Products based on polyethylene glycol derivatives of polyamides or a combination of polyamine with epichlorohydrin or polyepoxides and quaternary ammonium derivatives of such products have been suggested.

8. Complex acrylic polymers have also been suggested including both copolymers with ethylene oxide actually formed in the fibers and terpolymers of glycidyl methacrylate and cellulose methacrylate as antistatic agents. A reaction product of polyethylene glycol with vinyl acrylic copolymer is also used as an antistatic agent.

9. A combination of silicones and fluorochemicals with poly-aziridinyl derivatives and polyethoxy-based polyelectrolytes, and block copolymers of siloxanes and polyglycols have been claimed as antistatic agents for synthetic fibers.

10. Compounds based on amidophosphoric acid esters such as dieth-

ylstearylamido phosphate, diethyl oleylamidophosphate, polyammonium quaternary salt derivatives from dimethylamine acrylate, diethylamino-ethyl methacrylate, and morpholine N-ethyl methacrylate are recommended as antistatic agents.

11. A polycondensation of an aliphatic dicarboxylic acid, a hydroxy oxyalkylene, and an amino acid or diamine are useful antistatic agents for polyamide fibers. Treatment of polyamide fibers with polycaprolactam ethylene oxide condensate in the presence of dimethylol ethylene urea and magnesium chloride (as a catalyst) gives an antistatic finish. A similar formulation may be the hydroxy alkyl derivative of cellulose or a starch with a cross-linking resin and a catalyst.

12. For acrylic fibers, the use of polyethylene glycol acrylate prepared by reacting polyethylene glycol with acrylic acid in molecular proportions of 1 : 1.5 in a benzene solution using sulfuric acid as a catalyst is suggested.

Some of the commercial antistatic finishing agents are given in Table 19.3.

19.4 EVALUATION OF STATIC CHARGE ON FIBERS

Three typical standard methods for the evaluation of static charges on textiles are described below:

Measurement of Electrical Resistivity.[17,18]   The yarn or fabric specimen is conditioned for 4–24 h at 24°C and 40% R.H. The residual surface charge on the specimen is removed by passing a radioactive metal bar over the sides of the sample. The electrical resistance of the specimen is then measured using an electrical resistance meter. The electrical resistivity (R) is then computed. In the case of yarn, the resistance values of *n* specimens are mea-

TABLE 19.3 Antistatic Agents for Synthetic Fibers

Trade Name	Manufacturers
Antista Oil, Antista D, and Antista M	Ahura Chemical Products, India
Antistatic Oil, Antista D	Hico Products Ltd., India
Antistatin C, D and M	BASF
Ceranine R NAS	Sandoz
Cirrasol PT	ICI
Emcol OC-9	Witco Chemical Co.
Lektrostat VVG	Dexter Chemical Corp., U.S.A.
XT-4-Oil 2	Dow Corning Co., U.S.A.

sured and resistivity is computed in n ohms (Ω)/cm/strand of yarn by using the formula

$$R = \frac{S}{D} \times \frac{r_1 + r_2 + \cdots + r_n}{n}$$

where R = resistivity in Ω/cm/strand; S = number of strands/specimen; D = distance of individual specimens; n = number of specimens tested; r = resistance of individual specimens.

Electrostatic Clinging of Fabrics: Fabric-to-Metal Test.[19]   A charged fabric clings to the human body because of the instantaneous induction of an equal and opposite charge on the surface of the body when the charged fabric is brought close to it and oppositely charged materials attract each other. A metal plate exhibits the same phenomenon of instantaneous charge induction as the human body. Therefore, a metal plate can be used to simulate the conditions of clinging observed between charged garments and a human body. Some individuals are more static-prone than others and a given individual may be more static-prone at one time than at another time. Therefore, fabric–to–metal cling times cannot be related directly to fabric–to–body clinging for different individuals. This test method integrates the effect of the fabric weight, stiffness, construction, surface character, finish application, and other fabric parameters that affect the tendency of fabrics to cling. The test method is not intended to determine suitability of a fabric for use in hazardous areas where the electrostatic generation of sparks might result in fires or explosions. In the test method, the conditioned fabric sample is kept on a metal plate and is rubbed 12× with a rubbing block with an attached rubbing nylon or polyester fabric. The tendency of the fabric to cling to the metal plate is then determined.

Dirt Pickup Test.[5]   This test is useful for quickly evaluating the development of static charge on a fabric. The test specimen is rubbed 10× with a woolen cloth and is held 2.54-cm above an open container of synthetic soil for 10 sec. The amount of attracted dirt is estimated visually and grades are given. Synthetic soil is prepared by mixing 20% coal ash, 40% pigment-grade carbon black, and 40% bone charcoal, each of which is ground fine enough to pass through a 200-mesh screen before mixing.

19.5   SOILING OF TEXTILES

Removing accumulated soil from garments or fabrics and maintaining them in a satisfactorily clean condition are age-old problems. The problem of soiling was aggravated considerably by the introduction of synthetic fibers, since these hydrophobic fibers attract soil to a greater extent than natural

fibers. Moreover, synthetic fibers do not easily release soil during laundering. The absorption and retention of soil results in graying, yellowing, and
deterioration in whiteness and brightness of the fabric.[20] The longer wear-life
of synthetic fabrics exposes them to soiling media longer, and the need for
further research to elucidate soiling phenomenon becomes apparent. In February 1967, a soil-release finish was introduced in the United States. The
soil-release (SR) finish makes for easier, more thorough removal of all types
of soil. Easier care is the hallmark of all soil-release finished textiles. The
soil-release finish may have the following functions:[21]

Antisoiling or Soil Repellency.   This includes protection against dry soiling
matter and oil-repellency during wear. The antisoil finish envelopes the fiber
so that soiling matter adheres less tenaciously. Interfacial properties are
modified to resist greasy soiling matter.

Soil Release.   Soil-release finishes have special functional groups that remove soil from fabric and transfer it to detergents. Carboxyl groups in the
fiber facilitates soil removal. The soil-release finish is hydrophilic in character and may allow soil to adhere to the fabric during wear.

Antisoil-Deposition.   It is the property that prevents a redeposition of the
soil that has already been dissolved or dispersed in the wash liquor. It is
obvious that the functions of the detergents and the finishing chemicals must
supplement each other in this process. Although the process is often confused with soil release, it is not the same thing; in fact, there is very little
direct connection between the two.

Antistatic Effect.   The antistatic effect prevents soil and dust from being
attracted and held by electrical forces on the fabric surface.

Nature of Soil.   The soil on textiles comes from two different sources: (*a*)
the body of the wearer and (*b*) the surrounding atmosphere. Bille *et al.*[21]
have analyzed the composition of fatty substances secreted by the human
skin. Most of the secretions consist of free fatty acid, wax, triglycerides, and
other fatty substances (Table 19.4). They have also examined street dirt
which contains a large proportion of inorganic matter (Table 19.5). The ash
of a typical dirt contains 21.5% SiO_2, 11.1% Fe_2O_3, 1.7% MgO, and 6.4%
CaO. More than 50% of the dirt has particle sizes below 5 μ; particles range
in size from 2–40 μ.[22–24]

An analysis of domestically soiled personal items such as shirts, linen
collars, pillow cases, hand towels, and wool socks showed that soil content
varies 0.25% (for shirts) to 1.2% (for collars) based on the weight of the
fabric.[25] In view of the considerable differences in climatic conditions, industrial developments, and composition of dirt, it is very difficult to draw any
general conclusions about the nature of soil.

TABLE 19.4 Composition of Fatty Substances Secreted by Human Skin[21]

Component	% Content
Free fatty acid	22–27
Wax and stearol ester	20–22
Triglyceride	20–25
Diglyceride	6–10
Squalene	10–15
Stearol	2–5
Paraffin	0.5–1.5

TABLE 19.5 Typical Analysis of City Dirt[21]

Component	Percentage
Water-soluble	15.4
Ether-soluble	7.7
Moisture	2.1
Total carbon	26.9
Inorganic matter (as ash)	50.5

19.6 MECHANISM OF SOILING

There are two important forces responsible for soiling:[26] (*a*) forces of contact and (*b*) forces of retention.

Forces of Contact. Fabric and dirt come together either by direct contact with a soiled surface or by contact with airborne substances. Simple mechanical forces act through direct contact to transfer oily and/or solid materials directly from a soiled surface to the uneven surface of a fabric. Fabrics also collect soil from the air or pick up dissolved or suspended substances from liquids. Electrostatic forces also play an important role in the attraction of dirt.[27,28] Hydrophobic synthetic fibers accumulate static charges during processing and in use and, thereby, attract soil particles from the air. Scheme 19.1 shows how the hydrophobic nature of synthetic fiber is responsible for its greater soiling.

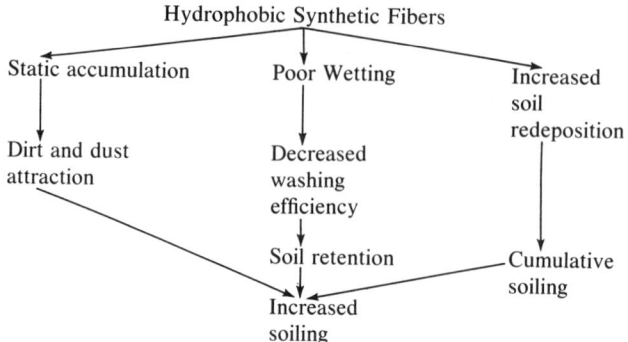

SCHEME 19.1. Soiling of synthetic fibers.

Forces of Retention. There are two different types of forces of retention: (*a*) those dependent on some form of bond energy and (*b*) those dependent on mechanical entrapment. Thus, dirt is retained in a fabric by mechanical and electrostatic forces and by soil bonding.[29] The soil particles also adhere to fibers by occlusion in pits and crevices on fiber surfaces,[8] entrapment in intra or interyarn spaces, in the irregularities of the fiber surface, and by sorption by van der Waals or Coulombic forces.[30]

The possible mechanisms of soiling[25] for various textiles are given in Table 19.6. The parts of the garments that are often rubbed against the skin are soiled quickly.

19.7   FACTORS INFLUENCING SOILING

Particle Size of Soil.   The smaller the size of the soil particles, the greater is the soil retention by the fabric.

Textile Structure.   Mechanical entrapment of the soil is closely related to the fiber, yarn, and fabric structure.[31] The soil retention increases with decreasing filament denier and is a linear function of the gross surface area of the filament, that is, the finer the filament, the higher the soil retention per unit weight of filament. For any type of fiber and for a given denier, a circular cross-sectional fiber retains less soil than one with a serrated or lobal cross-section. The mechanical soiling is proportional to the degree of roughness of fiber, yarn, or fabric. The chemical structure of the fiber has a greater influence on soil retention than the physical characteristics. Thus, polyester fiber with favorable physical shape of filaments exhibits greater soil retention than other fibers. With increasing twist in the yarn, the soil retention increases to maximum and then decreases to values below that of untwisted yarn.[31] Fabric with a tightly woven structure prevents the entrance of all but the smallest particles. However, the removal of soil by washing is difficult. In the case

TABLE 19.6 Probable Soiling Mechanism for Textiles[25]

Item	Mechanism of Contact	Mechanism of Retention	Normal Service Time for Soiling
Light curtains	Interception and internal effect (at open windows) and diffusion and depositions (at closed windows)	Occlusion, oil bonding by finishing	Few weeks to few months
Suits	Sleeves, seat, and cuffs—direct transfer	Occlusion	Few days
	Top of shoulders—deposition		Few weeks
	Body—direct interception and diffusion		Few weeks
Women's stockings	Interception, direct transfer	Occlusion, oil bonding by secretions	Few hours to few days
Men's shirts	Sleeves, collars, and cuffs—direct transfer	Occlusion, oil bonding by secretions	Few hours
	Top of shoulders—deposition	Occlusion	Few days
	Body—interception and diffusion	Occlusion	Few days
Carpets	Direct transfer—deposition	Occlusion, oil bonding	Few months
Working cloths	Direct transfer—interception	Occlusion, oil bonding	Few hours–few days
Tents	Deposition, direct transfer interception	Occlusion, oil bonding by fireproof finish, if any	Few weeks–few months
Gloves	Direct transfer	Occlusion	Few hours–few days

of loosely woven fabric, penetration of soil in the fabric is easy but so is removal of soil during washing.

Electrostatic Charge.   Synthetic fibers have a tendency to accumulate static charges during wear. The soiling problem is considerably aggravated because the charged fibers attract the soil from the atmosphere.[12] Positively charged fabric is soiled more than negatively charged fabric.

Moisture Regain.   Soiling increases rapidly when the moisture content of the fibers drop below 4% (Figure 19.1).[32] Polyester has the least moisture regain among the synthetic fibers studied and displays maximum soiling.

Wet Soiling Behavior.   In aqueous media, hydrophobic fibers generally soil more than hydrophilic fibers; in dry-cleaning media, the opposite is true. Polyester has a very hydrophobic surface because the surface is made up of ether oxygen (C—O—C) linkages while hydrophilic ester oxygen (C=O) is facing towards the core of the fiber. The soiling of polyester is therefore very high. If the fabric is not washed frequently, greasy soil penetrates into the fiber and the polyester material becomes extremely dirty. Polyester fabric can collect dirt from aqueous liquors, even in the presence of a dispersing–sequestering agent.

19.8   ANTISOIL AND SOIL-RELEASING FIBERS

The moisture regain of the synthetic fiber is usually improved by adding a variety of compounds during or after polymerization.[1,6] For example, addition of sodium sulfate (5–10%), sulfonic-acid-containing compounds, polyethylene oxide ether (0.4–5%) and a salt (0.3–3%), polyethylene dibenzoate

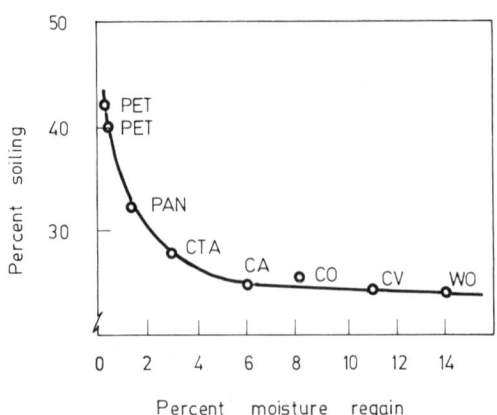

FIGURE 19.1.   Soiling of different fibers.[32]

or diacetate to polyester, and polyether urathane, and ethylene oxide to polyester and polyamide during polymerization improves the soil-release properties. Incorporation of polyether esteramide (adipic acid-polyethylene glycol-caprolactam or hexamethylene diamine condensation polymer) into polyamide improves the soil-releasing properties of nylon 6 or nylon 66. Alternatively, the polymer chips are blended with polyethylene oxide (molecular weight 500, terminal group esterified), block polyether caproamide, a mixture of sodium dodecyl benzene sulfonate and polyethylene glycol and are melt-spun to get improved soil-release properties. The soiling tendency of modified fiber is quite low, as is seen from Table 19.7.[32] Most of the additives suggested for incorporation in the dope are useful in making the fiber soil-releasing.

19.9 SOIL-RELEASE FINISHING TREATMENTS

A large number of soil-release finishing agents have been described and patented.[33-39] Polyethylene terephthalate lacks hydrogen-bonding sites on the fiber surface because the ether oxygen atoms and not the ester oxygen atoms are outermost on the surface. Because of this, polyester is very hydrophobic in nature.[40] Nylon is less hydrophobic than PET because hydrogen bonds are more readily formed in polyamides. The soiling problem is therefore much smaller in nylon than in PET. The PET/CO blend fabric, which is subsequently treated for permanent press or durable-press effect, soils to a great extent. Soil-release finishes are therefore of great use in the case of PET and its blend fabrics.[41]

19.9.1 Polyester and Its Blends

A temporary finish is given using epoxides, polyethylene oxide derivatives, quaternary ammonium compounds, or silicones with polar side chains.[6,42] When polyester fabric is padded with polyethylene glycol and sodium hydroxide, dried, and baked, some ester interchange between terephthalate and sodium polyglycol oxide takes place and hydrophilic grafts— $COO(CH_2CH_2O)_nH$—are formed on the fiber. The polyester fabric shows an

TABLE 19.7 Soiling of Normal and Modified Polyester[32]

Fiber	Moisture Regain (%)	Soiling (%)
Normal PET	0.4	38.5
Modified PET	1.1	19.2

improvement in soil release and soil redeposition properties as well as a drop in static charges even after repeated washings.[43] It is likely that the polyester fabric gets degraded during the alkaline curing. ICI has developed a special finish known as Cirrasol PT in United States and Permalose T in other countries.[44-48] It is a nonionic polymer and is applied to PET/CO blend fabrics by the pad-dry-cure process. Usually 1–3% of finish is applied to get the following benefits:

1. Reduction in the gradual greying that results from redeposition of soil during washing.
2. Help in the removal of soil stains during washing.
3. Prevention of static build-up and reduction in clinging and attraction of airborne dust.

The finish can be used in conjunction with selected wash-and-wear resins, such as hydroxyethyl carbamate or methoxy ethyl carbamate, for preparing permanent-press PET/CO garments. Softeners and other additives to be used with Permalose T should be screened before use. The finish is fast to household washing and dry cleaning, but highly alkaline washes, particularly at high temperatures, will gradually remove it. Ammonium or sodium salts of styrene-maleic acid copolymer or of methyl vinyl ester-maleic acid copolymer can be applied along with urea, melamine, or other wash-and-wear resins on polyester or PET/CO blends to impart soil-release properties to the fabrics.[6] Wet soiling of polyester and other hydrophobic fibers is minimized by treatment with a polymer containing only carboxyl groups or sulfonic or phosphoric acid groups adhered to the fiber with polyol-polyisocynates reaction product.[49] The finish is cured to cause a reaction between the two polymers. Typical polymeric acid components include polyacrylic and polymethacrylic acids, copolymers of acrylic acid with acrylic esters, copolymers of maleic anhydride with styrene or sulfonated styrene and so on. Rohm and Haas Co. has developed a soil-release finish called *Rhoplex SR-488,* which is a hydrophilic acrylic copolymer. It can be applied to polyester, nylon, rayon, and cotton fabrics and their blends and is particularly effective as a top finish (e.g., aftertreatment) on precured durable-press fabrics. If Rhoplex SR-488 is applied by a one-bath treatment, the best results are generally obtained with dimethylol ethylene urea or methoxyethyl carbamate. Careful selection of softener and other additives is important.

Polyethylene oxide-terephthalate polymer and specific *S*-triazine or piperazine derivatives are suitable for antistatic, hydrophilic, soil releasing, and soil-resistant finishes for polyester and its blends.[6,38] A copolymer of terephthalic acid and polyethylene glycol is applied to polyester fabric by the pad-dry-cure process. The product imparts hydrophilicity to the fiber because of the presence of polyglycol groups in the copolymer. The durability of finish results from the hydrophobic portion of the copolymer forming a cosolution with the polyester fiber substrate.

A treatment of polyester fabric with sodium hydroxide solution (e.g., 10% at 60°C for 10 min) makes the fabric soft as silk, hydrophilic, and more washable.[50–53] Treatment with inorganic acid such as phosphoric, phosphorous, or perchloric acid and heating is recommended to improve the soil-releasing property of PET. Because of polar groups formed on the fiber, PET becomes hydrophilic.[6]

Ethoxylated alkyl phenols are applied to both blend fabrics and cross-linking agents by the pad-dry-cure process.[6] A finish based on a monomer containing a fluoroalkyl group, an ester of the formula $CH_2{=}CH{-}COO$ $(CH_2CH_2O)_nR$, and acrylonitrile is applied in emulsion, solution, or aerosol form to PET/cotton blend fabrics.[6,38]

To overcome the severe soiling of polyester, the various polyvalent finishes described above are used to impart an antistatic finish resistant to washing and dry cleaning. Such a finish has a neutral handle and gives no loss of crease-resistance properties. The moisture absorbent effect of the finish promotes the removal of perspiration from apparel. A catalyst and a soil-releasing agent are usually applied by the pad-dry-cure process at 150°C for 20 sec. Polyester/wool blends are also treated with such compounds, for example, Migafor 5763 (Ciby–Geigy) and Nonax 1166 (Bohme Fettchemie). Helizarine Delustrant (BASF) modified with Perapret D and applied with Helizarine Binder F are suitable for PET and PET/cotton blend fabrics. Acramin (By) and Melustral Delustrant (Hoechst) are also suitable for these fabrics. A typical recipe for a soil-release finish for PET net materials is:

50 parts	Antisoiling agent (Migafar 5763)
5 parts	Magnesium chloride (catalyst)
10–40 parts	Fluorescent brightener
0.005–0.020 parts	Shading color
	Water to make
1000 parts	Total

Pad (60% pickup), dry, and cure for 20 sec at 180–210°C.

19.9.2 Polyamides and Acrylic Fibers

Nylon. Compared with polyester, nylon has only a small soiling problem and hence, only a few soil-release agents have been developed so far. Some of the important agents and processes are:[38]

1. Treatment of nylon with a hydroxyalkyl ether of cellulose or starch along with a cross-linking agent is suggested to reduce soil pickup by nylon from dirty washwater.[54]
2. By impregnating nylon fabric with a liquor containing 0.5% *N*-methoxy methyl–nylon 6, 3% polyoxyethylene sorbitol ether, and an

ammonium tartrate catalyst, followed by drying and curing, it is possible to get fabric with soil release and antistatic properties.[55]

3. 3M Co. has suggested a fluorochemical that has long chain molecules of very high-molecular-weight and consists of two types of segments: hydrophobic and hydrophilic. The product is claimed to be suitable for all types of natural and synthetic fibers.[56]

Acrylics.   A nonoily soil-resisting agent is suggested for acrylic fibers.[57] It is a cationic surface-active agent that is dispersable in hot water. At a concentration of 1.5–6.0%, it gives a good soil-resisting effect on acrylic carpets. The finish is also said to be somewhat water-repellent.

19.9.3   Grafting

Grafting hydrophilic monomers onto nylon and polyester fibers using irradiation techniques has been a subject of investigation in recent years.[58–64] Nylon has been grafted from solutions containing acrylic acid, sodium styrene sulfonate, acrylamide, N-methylol acrylamide, polyethylene oxide, and water. This imparts a soil-releasing property to fibers and films.[58,60,61] For imparting a soil-releasing property to polyester, grafting with a solution containing sodium styrene sulfonate, dimethyl sulfoxide, acrylic acid, methylol acrylamide, polyethylene oxide, and water is suggested.[59,62] With acrylic acid grafting, the moisture regain of polyester increases from 0.4% to 4.46%.

Mares et al.[63] have grafted methacrylic acid on to PET/CO blend fabric. They have observed that the soil-removal efficiency of the grafted fabric is about 88% compared to the soil-removal efficiency of the untreated fabric (50%). The moisture regain of the grafted fabric is increased from 3.11% to 4.84%. They have concluded that higher moisture regain, the presence of electron-donating groups to reduce charge on the fabric, and increased water penetration into crevices and intrayarn and interyarn spaces are essential for an efficient soil-release process.

Walsh[65] has carried out an electron-beam-induced polymerization of a series of cross-linked hydrophilic finishes on polyester fabrics. The monomers used in the study are Carbowax 550 acrylate, Tergitol 15-5-12 acrylate, 2 acrylamido-2-methyl-propanesulfonic acid, and N(1,1-dimethyl-3-dimethylamino-propyl) acrylamide. The monomers are applied by padding and drying at 70°C. The fabric is then irradiated with electrons at a rate of 1–3M rad/sec and a total dose of 10M rad. The samples are washed thoroughly with water and tested for soiling, soil release and so on by the standard AATCC methods. All the treated samples give a rating of 5 compared with a rating of 2–3 for untreated samples. Thus, the process is useful in improving the oily soil-release property of polyester. Although grafting techniques for imparting hydrophilic and soil-release properties on nylon and polyester are now established on a laboratory scale, commercial activity is negligible.

19.10 EVALUATION METHODS

19.10.1 Standard Soiling

A number of attempts have been made to soil fabric artificially in a manner similar to natural soiling. However, this task is difficult because the nature of soil, the surrounding atmosphere, and soil secreted from person to person vary considerably.[66] Many types of artificial soils have been proposed. Several contain carbon black because of its powerful effect on light reflectance. In general, it is essential to have oil or grease present in the soil because this can influence the binding of the soil to the fabric.[67,68] The carbon black particle size should be uniform. Tsunoda et al.[69] have formulated an artificial soil by mixing carbon black and clay to form the inorganic soil and a liquid mixture to form the organic soil. Salsbury et al.[70] have used Juvenon R for testing their antisoiling agent. Its composition is given in Table 19.8.

The defects of carbon black are avoided by taking a mixture of kaolinite and a six-component synthetic sebum. Reproducibility of the results is recorded by using wool grease in carbon tetrachloride that contains road dirt.[71] Ferric oxinate of the uniform particle size is preferred to carbon black as a standard soil for experimental work.[72] In this case, estimation of ferric oxinate removed by various detergents is made colorimetrically.

A large number of methods have been developed for applying soil to the textiles. The tumbler test is one of the oldest and most commonly used.[28] Fabric samples are tumbled end-over-end in a jar in a Launderometer together with dry (pigment) soil and steel balls. After a short run, test fabrics are removed, given a light shake, and placed in a perforated metal dusting cage with steel balls. They are then tumbled again in a Launderometer to remove excess soil. The amount of soil retained by the fabric is estimated gravimetrically.

The Blower Test consists of passing equal volumes of air directly through the fabric samples being tested until the degree of soiling produced by the natural airborne dust is enough to be evaluated visually or photometri-

TABLE 19.8 Composition of Artificial Soil (Juvenon R)[70]

Ingredients	Amount %
Peat moss	38
Cement	17
Kaolin clay	17
Silica	17
Carbon black	1.75
Red iron-oxide pigment	0.50
Mineral oil (Nujol)	8.75

cally.[28,73] The fabric to be tested is mounted over the intake end of one of the tubes, each of which contains a motor and fan assembly.

Hemmendinger and Lambert[74] and AATCC[75] have suggested a collar test. As the name implies, the method employs a collar made from the fabric to be tested which is worn underneath the regular shirt. Soiling is measured by the change in reflectance and color.

Compton and Hart[30] have developed the chopped Fiber Method in which the fibers are cut into 1 mm or less lengths and soiled with an aqueous dispersion of carbon black in water. The extent of soiling is measured by reflectance.

Berch et al.[76] have suggested a method that consists of preparing soiled cubes with a mixture of oily soil and dispersed solid soil. The fabric samples to be tested are then tumbled with these soiled cubes and soiling is evaluated. It is claimed that the soiling of the test specimen is more uniform.

Soiling by the spray method was also suggested.[77] In this method, a spray gun moved by a motor at right angles to the fabric web delivers 200 ml/min. of a soil dispersion on to the fabric. Simultaneously, a second spray gun sprays 10 ml/min of an aqueous sodium chloride solution on to the fabric. The fabric is squeezed and the soiling is evaluated by reflectance measurement.

19.10.2   Soil Release

AATCC has developed a test for measuring the soil-release property of the fabric.[78] Staining is produced on a test specimen by using a weight to force a given amount of the staining substance (Nujol refined oil) into the fabric. The stained fabric is then laundered in a prescribed manner and the residual stain is rated on a scale from 5 to 1 by comparing it with a standard stain release replica showing a gradual series of stains. In spite of the fact that so many methods have been developed for evaluating the soiling behavior of fabrics, the ideal method has not yet been found.

REFERENCES

1. Redston, J. P., Burnholz, W. F., and Schaltter, C., *TRJ,* **43** (1973), 325.
2. Postman, W., *TRJ,* **50** (1980), 444.
3. Valko, E. I., Tesero, G. C., and Ginilewicz, W., *ADR,* **47** (1958), 403.
4. Hall, A. J., *Textile Finishing,* Haywood Books, London (1966), p. 435.
5. Nahta, R., *ADR,* **64**(4) (1975), 41.
6. Vaidya, A. A., and Nigam, J. K., *Man Made Text. India,* **22** (1979), 363.
7. American Cyanamide Co., USP, 3,976,738 (1976).
8. Toray Ind., USP, 3,963,803 (1976).
9. BASF, BP, 1,263,936 (1972).
10. American Cyanamide Co., BP, 801,721 (1958).

11. Hughes, A. J., and McIntyre, J. E., *Text. Prog.*, **8**(1) (1976), 24.

12. Wilson, D., *JTI*, **53** (1962), T1.

13. Valko, E. I., and Tesoro, G. C., *TRJ*, **29** (1959), 21.

14. Johnson, K., *Antistatic Compositions for Textile and Plastics*, Noyes Data Corp., U.S.A. (1976).

15. Vaidya, A. A., and Trivedi, S. S., *Textile Auxiliaries and Finishing Chemicals*, ATIRA, India (1976), p. 134.

16. Palmer, J. W., *Textile Processing and Finishing Aids*, Noyes Data Corporation, U.S.A. (1977).

17. AATCC Test Method 76-1975.

18. AATCC Test Method 84-1973.

19. AATCC Test Method 115-1973.

20. Lewis, H. M., *ADR*, **57** (1968), 132.

21. Bille, H. A., Eckell, A., and Schmidt, G. A., *TCC*, **1** (1969), 600.

22. Sanders, H. L., and Lambert, J. M., *TRJ*, **21** (1951), 680.

23. Masland, C. N., *Rayon Text. Monthly*, **20** (1939), 573.

24. Rodman, C. A., and Pansey, A. W., *TRJ*, **20** (1950), 873.

25. Alurkar, R. H., ATIRA, Ahmedabad, India Report (1971).

26. Getchell, N. F., *TRJ*, **25** (1955), 150.

27. Hearle, J. W. S., *JTI*, **48** (1957), 40.

28. Reeves, W. A., Beninate, J. V., Perkins, R. M., and Drake, G. L., *ADR*, **57** (1968), 1053.

29. Snell, F. D., Snell, C. T., and Reich, I., *J. Am. Oil Chemists Soc.*, **27** (1950), 62.

30. Cromption, J., and Hart W., *Ind. Eng. Chem.*, **43** (1951), 1564.

31. Weatherburn, A. S., and Baylley, C. H., *TRJ*, **27** (1957), 199, 358.

32. Bowers, C. A., and Chantrey, G., *TRJ*, **39** (1969), 1.

33. Marsh, J. T., *Text. Mfr.*, **94** (1968), 465.

34. Tsuzuki, R., and Yabuuchi, N., *ADR*, **57** (1968), 472.

35. Dorset, B. C. M., *Text. Mfr.*, **94** (1968), 248.

36. Bille, H., and Schmidt, G., *MTB*, **50** (1969), 1481.

37. Haug, E., and Hell, M., *Text. Prax.*, **24** (1969), 741.

38. Ranney, M. W., *Soil Resistant Textiles*, Noyes Data Corporation, New Jersey, USP (1970).

39. Gagarine, D. M., *TCC*, **10** (1978), 247.

40. Ellison, A. M., and Zisman, W. A., *J. Phys. Chem.*, **58** (1954), 503.

41. Burkit, F. H., and Heap, S. A., *Rev. Prog. Color.*, **2** (1971), 51.

42. Campbell, J. K., *ADR*, **57** (1968), 75.

43. Sheard, D. R., Paper Presented at Gorden Research Conference on Textiles, 1965 (through *Rev. Textile Progress*, **17** (1966), 466).

44. Garett, D. A., and Hartley, P. N., *JSDC*, **82** (1966), 252.

45. Ghionis, C. A., and Brown, C. L., *ADR*, **57** (1968), 254.

46. Perry, E. M., *ADR*, **57** (1968), 405.

47. Aliman, W. T., Dunlap, R. K., and Zybko, W. C., *ADR*, **57** (1968), 1086.

48. Moyse, J. A., *TII*, **8** (1970), 43.

49. Eastman Kodak, USP, 3,152,920 (1969).

50. Liljmark, N. T., and Asnes, H., *TRJ*, **41** (1971), 733.

51. Rotgers, A., *Chemiefasern/Textilinder*, **28/81** (1979), 135.

52. Hiedenann, G., Stein, G., and Valk, G., *MTB,* **60** (1979), 350.
53. Demaria, A., *ADR,* **68**(10) (1979), 30.
54. ICI, BP, 1,005,839 (1965).
55. Teijin Ltd., BP, 1,214,839 (1970).
56. 3M Co., BP, 1,215,861 (1970).
57. Menin, B., USP, 3,159,502 (1964).
58. Hofmann, A. S., and Berbeco, G. R., *TRJ,* **40** (1970), 975.
59. Tealdo, G. C., Mundaris, S., and Calgari, S., *CA,* **73** (1971), 99861.
60. Byrne, G. A., and Jones, D. M., *JSDC,* **87** (1971), 496.
61. Mehta, P. C., and Trivedi, I. M., *JAPS,* **19** (1975), 1.
62. Lokhande, H. T., Kale, P. D., Rao, K. M., and Rao, M. H., *JAPS,* **19** (1975), 461.
63. Mares, T., Arthur, J. C., and Harris, J. A., *TRJ,* **46** (1976), 563.
64. Needles, H. L., Adgar, K. W., and Tai, A., *ADR,* **67**(6) (1978), 36.
65. Walsh, W. K., Northern Piedmont Section, *TCC,* **9** (1977), 19.
66. Venkatesh, G. M., Dweltz, N. E., Madan, G. L., and Alurkar, R. H., *TRJ,* **44** (1974), 352.
67. Berch, J., and Peper, H., *TRJ,* **33** (1963), 137.
68. Kennedy, J. M., and Stout, E. E., *ADR,* **57** (1968), 2.
69. Tsunoda, T., Oba, Y., and Kashiwa, I., *CA,* **71** (1969), 71496.
70. Salsbury, J. M., Cooke, T. E., Pierce, E. S., and Roth, P. S., *ADR,* **45** (1956), 190.
71. Ilg, J., *Text. Prax.,* **19** (1964), 520.
72. Okuyama, H., Fujji, T., Tsujikawa, M., and Morita, H., *CA,* **63** (1965), 13582b.
73. Henno, J., and Jouhet, R., *JTI,* **41** (1950), A431.
74. Hermendinger, H., and Lambert, J. M., *J. Am. Oil Chem. Soc.,* **30** (1953), 163.
75. Washington Section, AATCC, *ADR,* **43** (1954), 751.
76. Berch, J., Peper, H., Ross, J., and Drake, G. L., *ADR,* **56** (1967), 167.
77. Ilg, J., *Text. Prax. Inter.,* **19** (1964), 520.
78. AATCC, Test Method 130-1969.

20 | ENERGY AND WET PROCESSING

The textile industry ranks among the ten top energy-consuming industries[1] and the spiraling rise in crude-oil prices has hit the textile industry squarely and severely. During the last decade, the price of crude oil has increased more than $20\times$ (Figure 20.1),[2] followed by a similar increase in the cost of alternative sources of energy. The cost of energy is continuously increasing and is becoming a significant share of the total cost of processing textiles. For example, in 1973, energy accounted for 5–6% of the total cost of dyeing and finishing of textiles. This increased to 12–14% by 1975.[3,4] The industry has recognized this trend and a considerable effort to minimize energy consumption, to conserve energy, and to recover energy is being expended. Thus, the Department of Energy Govt. of USA set a goal of 22% reduction in energy consumption by 1980 and conserved 17% by the year 1978 (Base year: 1973).[5] About 60% of the total energy used by the textile industry is consumed in energy-inefficient wet-processing.[6] There is a lot of room to make wet processing energy efficient and, thus, to conserve energy.

Table 20.1 gives a rough estimate of energy consumption in various wet-processing operations.[7] Of the total energy consumed in wet-processing, heating process–water consumes 35–65%; drying, heat setting, and baking, 25–60%; liquor circulation, 10%; the rest is consumed in moving and handling fabric and so on.[8] Thus, in order to conserve energy, it is necessary to concentrate on operations that involve the heating of water, drying, or baking of textiles. Any reduction in water consumption will result in the reduction of consumption of energy in the form of steam. According to Sucheki,[9] the amount of water used by the industry can be reduced by at least 25%. This may be achieved by formulating processing sequences that need a minimum number of fresh baths and by keeping the liquor ratio as small as

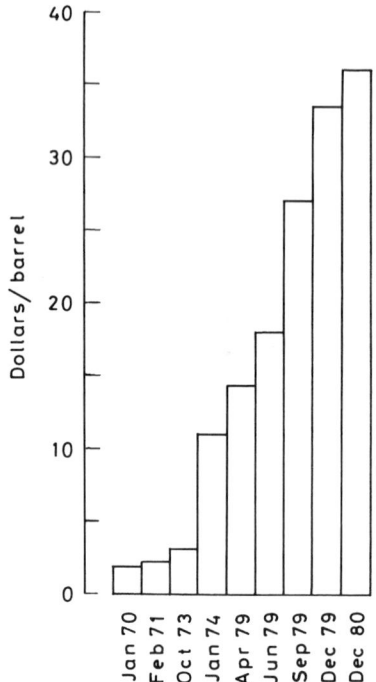

FIGURE 20.1. Crude-oil price in the years 1970–1980.[2]

TABLE 20.1 **Energy Requirements for Wet Processes**[7]

Fabric	Process	Therms/100 kg Fabric
PET/Cotton	Preparation-bleaching	14.8
Shirtings	Dyeing (thermosol, pad-steam)	21.4
	Resin finishing	11.5
	Screen printing (Disperse/vat)	21.2
Textured PET	Jet-scour-dye	19.4
jersey	Dry-heat set (stenter)	6.2
Knitting Nylon	Heat set	4.0
shirting	Scour-beam dyeing	8.8
	Dry-heat set	6.0
Nylon	Heat set	4.0
Taffeta	Scour-beam dyeing	8.8
	Dry-heat set (stenter)	6.4

1 Therm = 10^5 Btu = roughly 50 kg dry saturated steam at 2.5 bar pressure when injected in a dyebath at about 50°C. Thus, 1 Therm/100 kg fabric is equal to about 0.5 kg steam/1 kg fabric.

514

possible in each step. Similarly, the number of drying operations and the volume of water to be removed by drying should be minimal. Since most of the energy supplied to the processing units is in the form of steam, generation, transport, and utilization of the steam must be properly controlled.[10] The idling time and heating during the idling time of a machine should be avoided or kept to a minimum. Other approaches to conserving energy during processing are described below.

20.1 PRETREATMENT

A smaller volume of water, lower temperature, and a shorter time cycle are the main approaches for conserving energy in pretreatments.[11] Scouring at 40°C instead of at 80–90°C is made possible by special surfactants.[12] ATIRA has developed a cold scouring and bleaching process.[13,14] Better utilization of water and steam and improved washing efficiency were achieved by the development of new machines in which the principle of countercurrent flow of liquor-to-goods movement with good expression between successive compartments is applied.[15,16] A major source of heat loss in open-width washing machines is the evaporation of water. In modified machines, a cover is provided to minimize the water evaporation and to eliminate high humidity and possible condensation of water in the area around the machine.

20.2 COLORING

The use of mass-colored fibers (see Chapter 4) avoids a dyeing process and thus, saves energy. The coloring matter is introduced into the fiber during its production and hence, mass coloration does not entail substantial consumption of additional energy.

The replacement of conventional normal fibers by modified fibers brings down the energy requirements. For example, a carrier-free dyeable-type modified polyester fiber requires 20% less time for dyeing at 100°C than normal polyester by the HT dyeing method at 130°C.[17] Similarly, the thermofixation temperature is 20–30°C lower and the fixation time is about 50% less than that for normal polyester. The cross-staining of cotton or wool with disperse dyes under mild conditions is quite low and thus, a rigorous aftertreatment to remove the stain from the cotton or wool in the modified polyester blend is not required. Because of all these factors, considerable energy is saved in dyeing carrier-free dyeable polyester and its blends with cotton or wool. Similarly, Courtelle CP2 acrylic fiber, which has a slightly lower glass-transition temperature, can be dyed at 80°C instead of 100°C.[3]

Yarn Dyeing. The rapid dyeing machines were developed for shorter dyeing times. This was achieved by increasing the rate of liquor flow and reducing the liquor-to-goods ratio.[18-23] The powerful pumps and motors required for higher flow-rates consume extra energy which is, however, more than offset by the energy saved in shorter dyeing cycles. The heat generated by the friction in the pump is used for heating the liquor and thus, some savings in steam may be achieved.[11] The excess of liquor in the yarn package is removed after dyeing by centrifuging and vacuum extraction[24] so that the energy for drying is minimal.

Fabric Dyeing. The dyeing machine decides the liquor-to-goods ($L:G$) ratio and, in turn, the energy consumption[3] (Table 20.2). Modified machines were developed to bring down the liquor ratios, for example, the Rudolf Then Winch[25] works with a liquor ratio of 15:1. Squire[4] has suggested fitting to a winch an inner chamber shaped to facilitate the natural movement of liquor with the aid of a small flow-inducer pump (Figure 20.2).[4] The liquor ratio is thus brought down from 25:1 to 17:1.

In a beam dyeing machine, a thin layer of fabric gives good penetration of the liquor. For this purpose, a large diameter perforated beam is preferred. However, this leaves a large central volume inside the beam ineffective and thus increases the liquor ratio. The introduction of an adapter "sausage," which is either a detachable or a permanent part of the machine can bring down the liquor ratio from 15:1 to 12.5:1. A further reduction in the liquor ratio can be achieved by adding a filling insert[4] (as shown in Figure 20.3).

The jet dyeing machine design has been modified so that a liquor ratio of as low as 5:1 can be achieved.[26,27] However, the need for an even distribu-

TABLE 20.2 Dyeing of Polyester on Different Machines[3]

Machine	$L:G$ Ratio	Steam kg/100 kg Goods	Relative Energy Consumption
Winch	25:1	1375	100
Venturi Jet	15:1	723	53
Geston County Jet	10:1	482	35

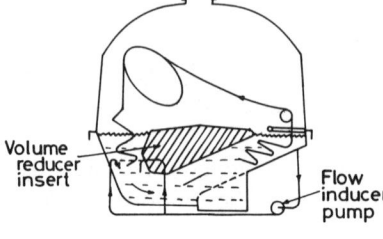

Volume reducer insert

Flow inducer pump

FIGURE 20.2. Winch with volume reducer insert.[4]

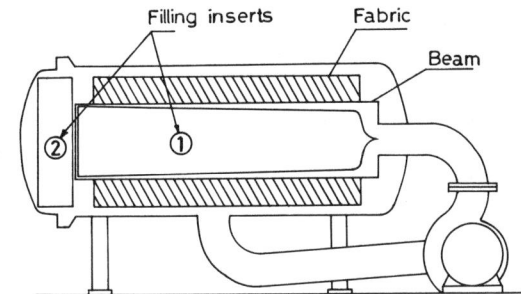

FIGURE 20.3. Beam dyeing machine with volume reducer insert.[4]

tion of the dye-liquor and for sufficient lubrication of the fabric is a limiting factor in using low liquor ratios.

20.3 FOAM PROCESSING

Foam containing auxiliaries, dyes, and finishing agents offers potential savings in materials and energy. The foam processes bring down the liquor ratios required for pretreatment, dyeing, and finishing by producing a uniform foam with the required characteristics in terms of viscosity, stability, blow ratio (i.e., volume of foam to volume of liquor), bubble size distribution and so on.[28–34] The potential advantages[35] of foam dyeing are (a) savings of water, chemicals, dyes and so on, (b) better color yield, (c) superior levelness and reproducibility of shades, (d) no frostines or color-flushing effect since the volume of condensed steam during steaming is low, (e) a savings in the energy required for heating, drying, thermofixing, steaming and so on because the water content is very low (Table 20.3), (f) greater throughput rates because both the heating up and dye-fixation rates are faster, (g) major reduction in thickening agent usage in printing, particularly, of carpets, and (h) no wash-off after steaming in certain cases.

TABLE 20.3 Steam Usage in the Dye-Fixation[35]

Process	% Picked Up	% Increase (Due to Steam Condensation)	Steam Usage (kg/h)
Continuous dyeing	500	75	2250
Carpet printing	300	45	1700
Foam application	50[a]	7.5	1100

[a] 100–250% volume pick-up.
Carpet (800 g/m²), Running Speed: 8 m/min.

Garments and fabrics are foam-dyed in rotating drum-type machines.[35] The dry textile goods are placed in the drum and rotated. The dye-liquor is sprayed on the material and rotation is continued for 15 min. The temperature is then raised rapidly by injecting hot air, steam, or both. Dyeing at boiling is carried out for 30 min. The goods are then removed, washed, soaped if necessary, and dried. The liquor for the foam process is prepared as follows:

Nylon

X g/liter	Acid dye
20–40 g/liter	Foaming agent
0.1–0.5 g/liter	Acetic acid
	Water to get a liquor-to-goods ratio of 1.5 : 1.

Polyester

X g/liter	Disperse dye
20–50 g/liter	Foaming agent
45 g/liter	Benzyl alcohol
0.5–1 g/liter	Acetic acid (to pH 5.5)
	Water to get a liquor-to-goods ratio of 2 : 1.

Acrylic Fiber

X g/liter	Cationic dye
30 g/liter	Foaming agent
0.3–0.5 g/liter	Acetic acid (to pH 5.0)
	Water to get a liquor-to-goods ratio of 2 : 1.

Blends. In fiber blends, the dye deposits on the hydrophilic fiber in the initial stages. It subsequently migrates to the appropriate fiber only if there is sufficient liquor. High quality dyeing cannot be achieved using a low ratio of foam to fabric. A higher liquor ratio (4 : 1) gives sufficient liquor for the dye to migrate from the hydrophilic fiber to the hydrophobic fiber and may be used for dyeing blend fabrics. A low liquor ratio without the foam technique is recommended when the design of the dyeing equipment permits it.[36,37] Under such a dyeing process, the hand and appearance of the fabric should not be affected.

Pretreatment and Finishing. Desizing, bleaching, and finishing as well as fluorescent brightening of goods can be done using a foam technique such as the Sancowad process (Sandoz)[38] (see Section 20.9).

20.4 DYEING RECIPE

It is normal practice to take samples to check the shade during dyeing and to add shading dyes to match the tone. The dyebath is cooled before the addition of the shading dye and is then heated again to the maximum dyeing temperature. This shading process consumes a large amount of energy, as can be seen from the data in Table 20.4.[3] The energy requirement is particularly high in HT and rapid dyeing processes. The need for shading can be minimized by using an automated dye-cycle controller and by developing correct dye-recipes. The shade should never be produced directly by the trial-and-error technique on bulk dyeing machines. Pilot dyeing experiments for matching the shade should be done first in a laboratory or on a pilot dyeing machine. Datye and Mishra[39] have developed a simple technique to find out the suitability of dyes for mixed shades. A solution of the dye mixture is prepared that is supposed to give the desired shade. Samples of the fabric are dyed with this solution for different lengths of time, and for full lengths of time using different percentages of shade. The ratio of dyes is kept the same. The shade developed on these samples is then evaluated as *CIELAB* color coordinates L^*, a^*, b^*. These coordinates are plotted as a^* vs. b^* and L^* vs. $(a^{*2} + b^{*2})^{1/2}$ for the two sets of samples. These plots clearly show the process of adsorption of dyes, the build-up of shades, and the time to take out a sample for a final check of the matching. Since, the fabric and dyes as well as the dyeing process (temperature, $L:G$ ratio, time, stirring, etc) in these pilot experiments are the same as those used in bulk

TABLE 20.4   Energy Requirements for Shading[3]

Process	Number of Shading Additions	Steam kg/100 kg Goods	% Increase in Energy Consumption
HT-package dyeing	0	482	0
(conventional)	1	607	26
	2	732	52
	3	857	78
HT-package dyeing	0	343	0
(rapid)	1	686	100
	2	1029	200
	3	1372	300
HT-beam dyeing	0	487	0
(with HT sampling facilities)	1	529	8.6
	2	571	17.2
	3	613	25.9

Fiber: Polyester
Liquor: Goods Ratio = 10:1

dyeing, the desired shade is produced on a large-scale machine with minimum or no shading.

20.5 RECOVERY OF HEAT FROM EFFLUENTS

Liquors from scouring, bleaching, and dyeing baths are drained hot. The heat in these liquors can be recovered and used for heating fresh process water.[6] Some of the arrangements used for this purpose are (a) heat exchanger, shell and tube type or spiral type; (b) coil in tank; (c) flash vessel. The hot liquor flows through the tubes which are placed in a tank filled with fresh water wherein the heat exchange takes place. Alternatively, fresh water passes through a coil immersed in an effluent tank. The cleaning of the latter is easier. The spiral heat-exchangers are easy to dismantle and convenient to clean. The effluent from a HT dyeing machine is hot (130°C) and under pressure. It has to be handled carefully to avoid the hazards of flash steaming. A typical way of handling such an affluent is to have a flash tank and heat exchanger to handle the steam, as is shown in Figure 20.4.[4]

The effluent is dumped through the high-pressure drain into a flash chamber where the steam generates with a drop in pressure. The flash steam then passes through a heat exchanger. It is followed by the effluent which is at atmospheric pressure. Most of the heat is recovered in the countercurrent heat-exchanger before the liquor is discharged. The flash-recovery system is suitable for liquors from kiers, boiler blowdowns, and high-pressure steam condensates. The condensed water from the flash steam is clean enough to use for dyeing. The recovered heat is used to warm the incoming water and thus saves energy. Typical savings, if preheated water is used for making the dye solutions, are shown in Table 20.5.[3]

20.6 REUSE OF EXHAUSTED DYEBATH

The energy required for heating a fresh dyebath can be saved by using a hot exhausted dyebath for dyeing a fresh lot without any intermediate purifica-

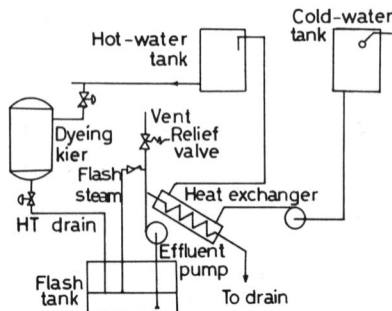

FIGURE 20.4. Heat recovery from effluents under pressure.[4]

TABLE 20.5 Energy Saving in Using Preheated Water[3]
(Polyester Dyeing)

Number of Shading Additions	Steam Consumption (kg/100 kg Fabric)		% Saving
	Cold Water (10°C)	Preheated Water (60°C)	
0	482	213	56
1	607	333	45
2	732	453	38
3	857	573	33

tion. The required amount of dye and 10% of the auxiliaries (to make up losses during the earlier dyeing) are added to the exhausted dyebath and a fresh lot is dyed in this liquor.[40] In the dyeing of nylon carpets, the bath is reused for up to five times before it is dumped.[6] The variation of shade in these five lots is within acceptable limits. There are savings in chemicals (33%), water (58%), and energy (67%). The build-up of impurities in the dyebath is minimized by replenishing 15–25% of the original volume with fresh water in each run; the bath can be reused for a number of dyeing cycles.[41] The use of a standing bath is recommended for the Irgasolvent method of dyeing wool and nylon loose stocks (Ciby–Geigy).

A novel method of purifying the exhausted dyebath is the *dynamic membrane hyperfiltration method,* also known as *reverse osmosis.*[42,43] In this method, the exhausted dyebath is passed through a dynamic membrane hyperfiltration unit, where the chemicals and dyes are removed and the water obtained is used for dyeing a fresh lot. Since the performance of this process improves at high temperatures, it is possible to recover hot pure water. The dyes and chemicals are also recovered. When effluents are passed through the unit, the resultant purified hot water is suitable for use in dyeing the next lot. The process has great potential for energy savings, dyes recovery, and effluent treatment.

20.7 LOW-TEMPERATURE DYEING METHODS

Lowering the temperature of dyeing brings down the energy requirement. Attempts have been made to give polyester a pretreatment so that it can be dyed at 100°C. The polyester fiber is treated with a carrier solution (biphenyl in methylene chloride) at room temperature, followed by dyeing with disperse dyes at boiling for a short time.[44,45] The dyeing cycle for this method and for the conventional HT dyeing method is shown in Figure 20.5.[44] There can be a considerable savings in energy and time. Disperse dyes of low,

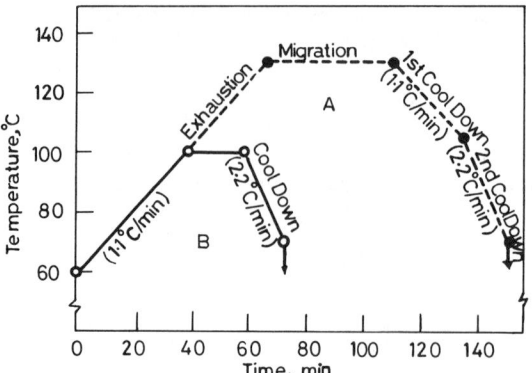

FIGURE 20.5. Long and short dyeing cycles for PET.[44]

medium, and high sublimation fastness exhaust at almost the same rate and hence, can be readily mixed. The pretreatment does not influence the mechanical properties of the fiber or fastness properties of the dyeings. The elimination of the carrier from the dyebath minimizes water pollution. The only snag in this process is the use of methylene chloride, an extremely toxic, habit-forming solvent with a very low-boiling point. In fact, polyester can be dyed in methylene chloride dyebaths at low temperatures but the process cannot be commercially exploited. Methylene chloride brings down the glass-transition temperature of polyester thereby, facilitating the rapid diffusion of dye in the fiber, even at 30–40°C. The use of a special carrier, for example, Tinosol ST (Ciba–Geigy) in the HT dyeing of polyester can bring down the dyeing temperature from 130°C to 114°C to 120°C without adversely affecting the build-up of the shade.

Dyeing of nylon carpets at 70–80°C is suggested to save energy and to minimize the distortion of fibers.[46] However, the migration of acid dyes on nylon at 70°C is about 50% of that at boiling. It is therefore necessary to select dyes from a range that gives level dyeing and to maintain the uniformity of temperature throughout the dyeing vessel. The pH of the dyebath is maintained at 7–8 for pale shades and at 6–7 for medium and deep shades.

20.8 MISCELLANEOUS COLORING TECHNIQUES

Rapid-Inverse Dyeing of Polyester–Cotton Blends. The conventional exhaust dyeing of PET/CO blends with disperse and reactive dyes consists of dyeing the polyester component with disperse dyes, reduction-clearing, and then by dyeing the cotton component with reactive dyes and soaping. This process is inefficient in terms of energy, chemicals, water, and time. A rapid-inverse dyeing (RID) method is suggested, where the cotton is dyed first (with reactive dyes) and then the polyester is dyed with disperse dyes.[47–49] The goods are rinsed in-between and the vigorous soaping required for cleansing the cotton of unreacted dye is eliminated. Highly reactive Procion MX dyes (ICI) or Remazol dyes (Hoechst) with good stability under acidic conditions[50] are particularly suitable for the reverse dyeing method. The method gives dyeings comparable to those obtained by the conventional method with a considerable savings in energy, water, and time.

Combining Two Operations.   It is possible to increase productivity and save energy, water, and labor[3,4] by combining different processes with dyeing operations or converting two-bath dyeing processes into single-bath processes. For example, scouring and dyeing knitted goods using suitable dispersing/emulsifying agents, dyeing and softening acrylic materials with cationic dyes and softening agents, and a one-bath dyeing process for nylon–acrylic blends with acid and basic dyes using an antiprecipitant may be mentioned. The dyeing, scouring, or finishing additives to the bath must not interfere in the dyeing operation.

Automatic Controls.   The use of automatic controls in batch-processing machines can save a considerable amount of energy.[51,52] They minimize the shading additions to the dyebath and help to maintain the correct dyebath temperatures and conditions. Wherever possible, automatic controls should be fitted to save energy (see Chapter 6).

Continuous Dyeing Methods.   Most of the energy required by a continuous dyeing process is used in removing moisture from the fabric and for heating the fabric to fix the dye. Reducing the wet pickup in the padding lowers the volume of water to be dried and thus, conserves energy.[5] This can be achieved by various techniques (see the chapters on finishing). Partial replacement of water with methanol, which is later burnt to generate heat for drying the padded fabrics, is claimed to save energy in Remaflam process[53] (see Chapter 7).

Printing.   Special binders and catalysts in pigment printing combine the drying and curing operations into a one instead of a two-stage process.[54–57] Modified steamers need less steam.[11,54,58,59] The use of synthetic thickeners in place of emulsion thickeners provides a savings in energy (and kerosene). Sublimation-transfer printing processes offer a considerable savings in energy for the textile printer. However, the energy and labor used by the manufacturer and the printer of the paper must also be considered while calculating the total energy input into the transfer printing process. Repeated use of the paper (up to 10×) is therefore suggested.[60]

20.9 FINISHING

Most of the finishing operations involve four steps: (*a*) padding, (*b*) drying, (*c*) curing, and (*d*) washing-off. The major amount of energy in finishing is consumed in removing water from the textiles and in heating the fabric. Removal of water by mechanical means is far cheaper than using heat and a considerable amount of energy can be conserved by reducing the wet pickup during padding. There are two principal ways of reducing wet pickup: (*a*) the expression methods and (*b*) the topical methods.

Expression Methods.   In these methods, the fabric is thoroughly saturated with finishing liquor. The surplus liquor is then removed by squeezing or other means. The wet pickup can be reduced considerably during squeezing by using the right combination of pad-roll hardness, roll diameter, nip pressure, and fabric speed (Figure 20.6).[61] The wet pickup decreases with an increase in the hardness of the pad-roll and the nip pressure. Decreasing the diameter of the pad-roll is also useful in getting lower wet pickup. Wet pickup can also be reduced by using a supplementary device to extract or eject the surplus liquor in conjunction with a padder. A vacuum slot extractor was found to be particularly effective in reducing the wet pickup of fabrics.[62–65] As an alternative to the vacuum, compressed air is also suggested for ejecting the surplus liquor after padding.[62–66] While using these supplementary devices, care must be taken to see that the applied finish is uniformly distributed throughout the width and length of the fabric.

Topical Methods.   In the topical methods for finishing fabrics, wet pickup is controlled simply by applying less liquor. One topical method receiving much attention at present is the application of finish in the form of foam.[67–69] By adding an appropriate surfactant to the finishing bath, a foam is generated either by injecting air or by mechanical means; thus, the specific gravity of the pad bath is lowered to below 0.5. The viscosity of the foam bath is very high and the bath does not flow at all. Equipment designed for applying highly viscous pastes are suitable for applying such foamed finishes.[61] Some of the methods for applying foam are listed below:

Knife Over Roll Method.   In this process (Figure 20.7), the knife spreads the

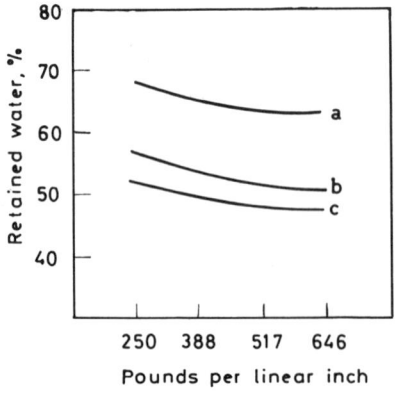

FIGURE 20.6. Expression of the padding mangle with top roller of different hardness.[61] Top roller: rubber. (*a*): 83 shore A, (*b*): 88 shore A, (*c*): 98 shore A. Bottom roller: steel.

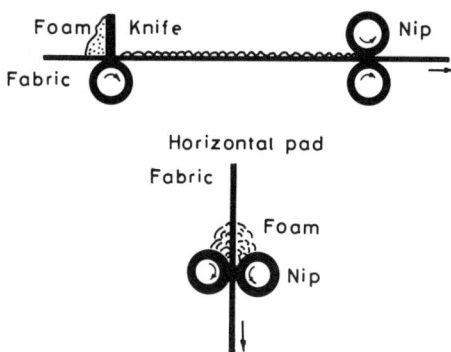

FIGURE 20.7. Foam padding systems.[69]

foam across the fabric.[69] The viscous material sits on top of the web until it is collapsed by the nip-rolls. The collapsed foam becomes a thin liquid and penetrates the fiber.

Horizontal Pad Method. In this method, the fabric passes through the pad-rolls vertically instead of horizontally (Figure 20.7).[69] The foam is supported above the rolls so that it can be applied to both sides of the fabric. The pad-rolls meter the foam and cause its collapse. The collapse of the foam is critical to the success of the process and can be accomplished mechanically as well as by applying a vacuum.

The Gaston County Dyeing Machine Co. has developed a foam finishing technology (FFT) applicator (Figure 20.8 and 20.9) that can be inserted into an existing finishing range.[70] It gives wet pickup as low as 10%.[71] The applicator nozzle is designed to apply foam to the fabric under static pressure within the applicator and dynamic forces in the foam mixture. In the FFT process, the foam is contained under pressure at the point of application. The collapse of the foam takes place within the fabric, where the air is

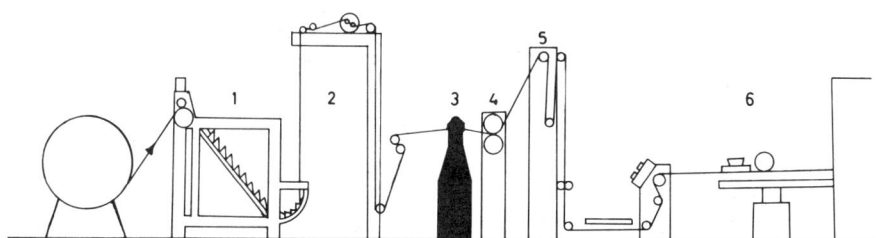

FIGURE 20.8. Gaston County FFT applicator, schematic drawing.[70,71] 1: Let-off scray; 2: Entry gantry with guiders; 3: FFT applicator; 4: Squeeze rolls; 5: Compensator; 6: Tenter frame.

FIGURE 20.9.   Photograph of FFT unit. (Courtesy of M/s Gaston County Dyeing Machine Co.)

separated from the liquid and the liquid is absorbed by the fibers. Force is required to do it. Hence, the more air in the foam, the greater is the pressure required to force the foam through the physical resistance of the substrate to reach a satisfactory level of penetration. The pressure in the applicator then increases.

The advantages of applying a foam finish are similar to those for applying foam for dyeing (see Section 20.3). They are (*a*) savings in energy, water, and chemicals; (*b*) uniformity of pickup of solids; (*c*) no migration during drying; (*d*) high productivity; (*e*) ability to process delicate, easily damaged fabrics under low tension; (*f*) improvement in aesthetics with reduction or possible elimination of softeners; (*g*) reduction in shrinkage in width and the need for stretching back to width because of low liquor pickup. A drawback of foam processing is that foam is a troublesome pollutant and can create problems in the aeration basins of water-treatment plants.[72]

Other Topical Methods.   The kiss-roll method can be used for applying a controlled amount of finishing liquor to the fabric.[73] Spin-finishes and coning oils are applied by this technique to synthetic filaments during their manufacture. A spraying method is also suggested for controlling wet pickup at low levels. However, the uniformity of the finish application can be a problem.

20.10 DRYING AND HEATING OF FABRICS

In the textile industry, water is added and removed many times during pretreatment, dyeing, printing, and finishing. Several different types of equipment are involved in removing the water. In any drying system, free water, as opposed to moisture, is removed by a suitable mechanical means, for example, by squeezing, vacuum, or centrifuging. These devices are electrically driven and are the most efficient method for removing water in bulk. In centrifuging packages of fibers, the centrifugal force varies from the side of the basket to the center of the centrifuge and invariably, the material near the center retains more water than at the side. In a typical case of dyed nylon muffs, the water retained after centrifuging was 15% and 10%, respectively. The rate of water removal increases as the speed is picked up by the centrifuging basket and then drops within a short time. No more water will be removed after this time. This optimum time of centrifuging depends on the size of the basket and the material to be centrifuged. The initial water in the material does not influence the optimum time of centrifuging or the final moisture content of the goods. The textiles are dried a number of times during wet processing, usually on cylinders (Figure 20.10), stenters (Figure

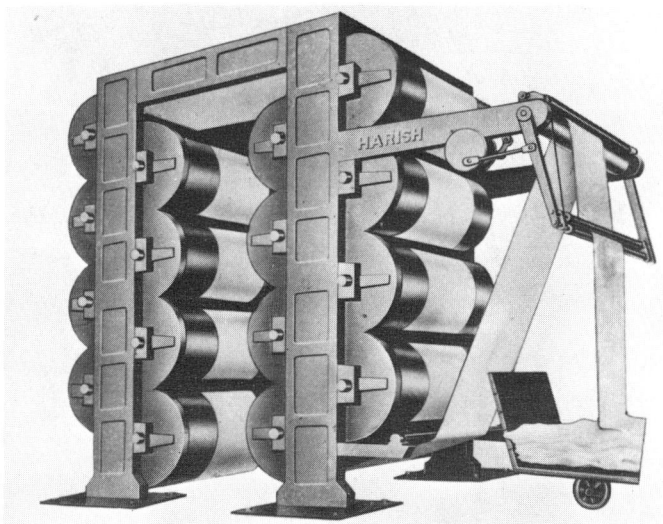

FIGURE 20.10. Drying cylinder range. (Courtesy of Harish Engineering Works, Bombay, India).

20.11), or curing ovens.[74] Synthetic fiber textiles also require heat for the thermofixation of dyes and finishes as well as for setting.

Cylinder Drying. Cylinder drying ranges are inefficient in utilizing steam energy for drying. A weak convection current arises out of rotating cylinders that is not strong enough to drive away the steam formed on the fiber surface because of the evaporation of water. Thus, a stagnant boundary layer of moisture (steam) forms over the evaporating surface and hampers the evaporation process and the diffusion of steam to the ambient atmosphere. If there are good convection currents in the vicinity of the evaporating surface, the thickness of the boundary layer decreases, the removal of moisture is rapid, and the drying rate is high.[75–77] Furthermore, there is no overheating of steam leaving the fiber surface. Devices such as the *Shirley Hood*[75] and the *ATIRA Rapidry system*[78–80] produce useful convection currents to disperse the stagnant boundary layer and thus, improve the drying efficiency. There are other dryers, for example, dual dryers, which utilize convection currents for heat and mass transfer.[81–84]

Stenters. Stenters have a high-intensity convection drying system that consists of high-velocity air jets.[85–93] The air is heated to about 150°C by steam

FIGURE 20.11. Harish Supra 3 Stenter. (Courtesy of Harish Engineering Works, Bombay, India).

heaters. A certain volume of air is continuously removed from the system through one or more exhaust fans to prevent the build-up of excessive humidity. The rest of the air is recirculated. The energy consumed in the stenter can be divided into five portions: (a) to heat water, evaporate it, and heat the vapor to reach the exhaust temperature; (b) to heat the substrate; (c) to heat fresh air; (d) to circulate the air and to drive the machine; (e) to heat the surrounding through radiation and convection from the housing.

The amount of water evaporated during drying is decided by the difference between the water content of the incoming fabric and the moisture content of the outgoing dry fabric. The lower the water content of the incoming fabric, the lower is the heat required for drying. Drying over the normal moisture regain of the fabric is a waste of energy and the outgoing dry fabric should have at least that much moisture. Typical data on the stenter drying of two fabrics is shown in Table 20.6.[10]

Heating and mass-transfer rates at the evaporating surface of the fabric are determined by the air jet velocity in the stenter, which are higher at higher velocities. The air jet velocities are therefore adjusted to optimum levels. The unsaturation or humidity of the jet air affects both the production speed and the specific steam consumption of a stenter. The drier (less saturated with moisture) the jet air, the faster is the drying. However, for keeping the jet air unsaturated, it is essential to withdraw hot saturated air from the stenter and to take in fresh cold unsaturated air. More heat is required for heating the fresh air; thus, the gains in energy by faster drying are lost. In order to satisfy these conflicting trends, the humidity of the jet air is adjusted to the optimum level for the machine. During the idling period when wet fabric is not being fed to the stenter, the amount of air exhausted should be as low as possible. Automatically controlled dampers provide sensitive control for this purpose.

Hot-air ovens are similar to stenters. Many of the approaches discussed above can be extended to such ovens.

TABLE 20.6   Drying on a Stenter[10]

Operational Details	I	II
Moisture		
Before drying %	120	80
After drying %	2	6
Steam consumption during		
drying (kg/h)	894	870
kg/kg of fabric	2.33	2.07

Fabric width: 103 cm; Running speed: 80 m/min; Average chamber temperature: 149°C; Hot air: Moisture ratio: 7–6; Average pressure at heater: 7.8 kg/cm².

Dryers and Heat-Setting Units. Heat can be recovered from the hot exhaust air from stenters and ovens used for drying, curing, thermofixing, and setting synthetic fibers.[3,10,11,94] Two types of heat exchangers are available for heat recovery: (*a*) air/air exchangers where energy is transferred through a conducting surface to an incoming fresh air stream, and (*b*) air/water exchangers that result in the production of hot water. An important consideration in deciding between the two is the convenience of utilizing the clean hot-air stream and the clean hot water.

Use of Solar Energy in Dyeing and Finishing. Solar energy can be used for dyeing and finishing. At Pali, Rajasthan, India, dyed and finished goods are dried in festoons in the open air where the fabric is heated by the sun. Solar energy is easily absorbed by black bodies and converted into heat energy. The heating of the water may thus be easily achieved, if the roof is made of corrugated galvanized sheets painted black and the water can flow underneath. The hot water thus formed can be used for dyeing, washing and so on.

A plant was established for utilizing solar energy for dyeing below 90°C at the Riegel Textile Corporation in La France, South Carolina.[95,96] The system was successful, but it is not yet competitive in price with the conventional system. However, with effective cost reductions for solar systems and an increase in the prices of other energy sources, the system will become economical in 1980s.

New Techniques. Dielectric heating, infrared drying, and heat pumps should be regarded not only as replacements for but also additions to existing processes for achieving greater throughput, saving energy, and controlling quality. For example, the infrared dryer may be fitted prior to the regular stenter or hot flue to partially dry the padded PET/CO blend fabrics so that the migration of the dye during drying is minimized. Infrared drying is done only to remove free water so that the movement of disperse dye particles is stopped.

Dielectric Heating.[97-100] *Microwave Heating.* The property of microwaves that makes them attractive for dye fixation and other uses is their ability, under suitable conditions, to produce rapid and uniform heating throughout the material exposed to them. The materials, for example, water that absorbs microwaves, are described as *lossy.* The ability to absorb radiation is related to the electrical polarization properties of its molecule. In the presence of a high-frequency electromagnetic field, the molecules oscillate at an amplitude that depends on the material and causes heat to be generated through intermolecular friction.

The heating effect is completely uniform when the absorbing species is evenly distributed. Thus, microwave heating is ideal for textiles. Because heating is uniform, there should be no migration of dye or finish during the treatment and the resultant dyeing should show uniformity. Completely dry fibers appear to absorb microwaves to only a slight extent. Materials such as glass or polypropylene are useful in constructing the microwave application while metals are avoided. A similar form of heating, which operates at lower frequencies (around 30 MHz), is called *dielectric heating*. Howeve, sparking is a problem in such cases. Nevertheless, the use of higher frequencies (896–2450 MHz) has not been used much in textile dyeing and drying. Radio-frequency (rf) heating systems have gained greater acceptance than microwave heating because of their lower capital cost and ease of application.[97]

Radio-Frequency Heating. Approximately 35 conveyer-type rf dryers with 40–60 kW output have been installed in the United Kingdom. Rf heating is carried out in its simplest form for systems consisting essentially of a top and bottom plate or an electrode system plus a conveyor belt to pass the textile materials through the rf field. The effect of the rf is to generate heat in the moisture within the product to be dried. However, the speed at which the moisture will be liberated to the surface is controlled by the structure of the material. At the surface, the moisture will diffuse away, which is facilitated by an air flow without which the water vapor may condense, and the moisture build-up may create arcing between the electrodes. Applying rf energy too rapidly to the textile packages may result in package distortion because of excessive build-up of steam pressure within.

A typical 50-kW output continuous rf drying unit operating at 27.12 MHz resulted in a dramatic cut in drying time for PET, acrylic, cotton, wool, and blend fibers. The moisture content of the fiber is maintained to an accuracy of ±5% of that required. The energy cost for drying was reduced by 65%; a significant part of this savings is due to the self-regulating characteristics of rf. Thus, rf energy is consumed in proportion to the moisture content of the material to be dried. Products are conveyed through the equipment on a polyester belt, a material chosen because of its lack of reaction to heating when in an rf field. The electrode gap can be adjusted to cope with a wide range of product forms such as muffs, hanks, tops, loose stock, and packages. Varying the electrode gap and belt speed enables the operator to accurately match the energy requirements to the load. Convected air is available from the generator's cooling system for the removal of surface moisture. The rf dryer can be ready for full production within 3 min of switching compared with the 20 min or so required with conventional dryers. The rf unit can be a self-contained dryer, which can be used in isolation without steam. There is a substantial savings in energy, labor, and capital cost. A combined radio-frequency and vacuum dryer is used for packages of yarns. The wet packages on vertical spindles are mounted on a carousel that transports the spindles into the chamber fitted with concentric electrodes. A vacuum is applied to the package via the spindles, which are rotated on their axes in order to give uniform penetration of rf. Rf techniques have been developed by Dawson International Ltd. for transfer printing and loose-stock dye-fixation. There is considerable savings in energy with a manyfold increase in production.[97]

REFERENCES

1. Cooper, S. C., *The Textile Industry–Environmental Control and Energy Conservation,* Noyes Data Corporation (1978), p. 70.
2. Vaidya, A. A., and Datye, K. V., *Synthetic Fibers, India,* **10** (Oct./Dec. 1981), 6.
3. Skelly, J. K., *JSDC,* **92** (1976), 117.
4. Squire, D. H., *JSDC,* **92** (1976), 109.
5. Wygand, W., *TCC,* **10** (1978), 60; Wallenberger, F. T., and Holfeld, W. T., *TCC,* **13** (1981), 173.
6. Tincher, W. C., *ADR,* **66**(5) (1977), 36.
7. Jones, F., Shirley Institute Conference Publication, **S13** (1974).
8. Scarce Resources Committee of SDC, *JSDC,* **95** (1979), 401.
9. Sucheki, S. M., *Text. Industries,* **140** (Oct. 1976), 25.
10. Ratna-Prabhu, M., Parajia, J. S., Vyas, M. M., and Subramaniam, K., *Heat Economy in Textile Mills,* ATIRA, Ahmedabad, India (1974), p. 76.
11. Wyles, D. H., *Rev. Prog. Color.,* **9** (1978), 35.
12. Zika, H. T., and DiPasqua, R. A., *ADR,* **65**(9) (1976), 68.
13. Mehta, H. U., and Mashruwala, M. N., *Colourage,* **29**(6) (1982), 9.
14. Kothari, B. C., *Colourage,* **29**(6) (1982), 17.
15. Schiffer, K., *MTB,* **56** (1975), 820.
16. Weber, H., AATCC Conference, Atlanta, Georgia (1977).
17. Anon, *Text. Industries,* (Oct. 1977), 49.
18. Dautricourt, M., *Teintex* **36** (1971), 769.
19. Anon, *Chemiefasern,* **5** (1974), 412.
20. Ulrich, H., and Kunz, W., *Text. Prax.,* **29** (1974), 645.
21. Ulrich, H., Reuther, A., and Wassmuth, J. J., *MTB,* **55** (1974), 549.
22. Bremhurst, G., *Chemiefasern/Textil Industrie,* **24/76** (1974), 747.
23. Carbonell, J., Egli, H., Hasler, R., and Walliser, R., *Textilveredlung,* **10** (1975), 3.
24. Camp, J. G., *ADR,* **66**(5) (1977), 52.
25. Anon, *Dyer,* **162**(9) (1979), 441.
26. Patterson, D., *Rev. Prog. Color.,* **4** (1973), 80.
27. Ratcliffe, J. D., *Rev. Prog. Color.,* **9** (1978), 58.
28. Lister, G. H., *Textilveredlung,* **6** (1971), 708.
29. Lister, G. H., *JSDC,* **88** (1972), 9.
30. Tosa, M., and Wade, H., *Japan Text. News* (March 1974), 74.
31. Carbonell, J., Hasler, R., and Walliser, R., *JSDC,* **92** (1976), 100.
32. Carbonell, J., Egli, H., and Perrigg, M., *ADR,* **66**(8) (1977), 44.
33. Davenport, R., *ADR,* **66**(11) (1977), 56.
34. Skoufis, J., *ADR,* **68**(7) (1979), 20.
35. Dawson, T. L., *JSDC,* **97** (1981), 262.
36. Carbonell, J., Hasler, R., and Walliser, R., *TCC,* **8** (1976), 86.
37. Sandoz, *ADR,* **66**(9) (1977), 72.
38. Aqualuft, Peace Dyeing by Sancowad Method, AATCC Natl. Tech. Conf., Atlantic City, New Jersey (1973).
39. Datye, K. V., and Mishra, S., *Textilveredlung,* **18** (1983), 211.
40. Cook, F. L., and Tincher, W. C., *TCC,* **10** (1978), 1.

41. Carr, W. W., and Cook, F. L., *TCC,* **12** (1980), 106.
42. Bradman, C. A., Energy and Pollution Control Conference, Hilton Head, South Carolina Professional Development, Clemson University, South Carolina, 1976.
43. Olsen, E. S., *TCC,* **9** (1977), 34.
44. Matkowsky, R. D., Weigmann, H. D., and Scott, M. C., *TCC,* **12** (1980), 55.
45. Moore, R. A. F., and Weigmann, H. D., *TCC,* **13** (1981), 70.
46. Brownewell, R., and Roochetti, A., *ADR,* **65**(6) (1976), 14.
47. Southeastern Section, AATCC, *TCC,* **11** (1979), 246.
48. ICI Tech. Bull. 655, ICI, UK, Rapid Inverse Dyeing of Polyester/Cotton Blends.
49. Stetson, G. R., and Thompson, C. W., *ADR,* **68**(3) (1979), 28.
50. Kenyon, G. H., *ADR,* **68**(3) (1979), 19.
51. Gailey, I., *Dyer,* **152** (1974), 634.
52. Gailey, I., *JDC,* **91** (1975), 165.
53. Hoverath, A., *TCC,* **13**(2) (1981), 41.
54. Anon, *Int. Text. Bull* (1974/3), 255; (1975/3), 304.
55. Modi, J. R., Palekar, A. W., and Trivedi, S. S., *Text. Dyer Printer,* **9**(8) (1976), 53.
56. Parikh, D. V., *TCC,* **10** (1978), 58.
57. Khanna, S. R., *TCC,* **11** (1979), 158.
58. Tischbein, C., *Int. Text. Bull.,* **3** (1975), 325.
59. Artos Research, *Chemiefasern,* **6** (1975), 547.
60. Glover, R. D., BP, 1,556,705 (October 10, 1975).
61. Goldstein, H. B., and Smith, H. W., *TCC,* **12** (1980), 49.
62. Fox, M. R., Marshall, W. J., and Stewart, N. D., *JSDC,* **83** (1967), 493.
63. Fox, M. R., *Rev. Prog. Color.,* **4** (1973), 18.
64. Fox, M. R., *JSDC,* **89** (1973), 46.
65. Goldstein, H. B., *TCC,* **11** (1979), 148.
66. Rudnick, E. S., *ADR,* **63**(8) (1974), 49.
67. Carter, T. W., *ADR,* **67**(12) (1978), 48.
68. Anon, *ADR,* **68**(7) (1979), 28.
69. Turner, G. R., *TCC,* **13** (1981), 13; **14** (1982), 76.
70. Chifford, G. F., *ADR,* **68**(1) (1979), 32.
71. Clifford, G. F., *TCC,* **12** (1980), 46.
72. Bahorsky, M. S., *ADR,* **68**(10) (1979), 26.
73. Schwemmer, M. H., Bros, H., and Gotg, A., *Textilveredlung,* **10** (1975), 15.
74. Slater, K., *Text. Prog.,* **8** (1976), 3.
75. Jones, E. H., *J. Inst. Fuel,* **24** (1951), 1.
76. Anon, *Textile Weekly,* **62** (1962), 1315.
77. Ratna Prabhu, M., and Subramaniam, K., Proceedings 6th Jt. Tech. Conf. BTRA, Bombay, India, (1965), 117.
78. Ratna Prabhu, M., Proceedings 5th ATIRA Tech. Conf., Ahmedabad, India (Sept. 13, 1967), p. 1.
79. Ratna Prabhu, M., and Subramaniam, K., Ind. Patent, 1,14,822 (1968).
80. Ratna Prabhu, M., and Subramaniam, K., *Colourage,* **16**(10) (1969), 54.
81. Anon, *Dyer,* **117** (1957), 949.
82. Gandhi, K. S., Indian Patent, 105,240 (1966).

83. ATIRA (Ahmedabad) Indian Patent, 129,192 (1970).

84. Anon, *ATIRA Tech. Dig.,* **6**(3) (1972), 8.

85. Wadsworth, P., *JTI,* **51** (1960), 552.

86. Lyons, D. W., and Vollers, C. T., *TRJ,* **41** (1971), 661.

87. Vernazza, J., *MTB,* **52** (1971), 579.

88. Glaser, D., and Klenn, T., *Deutsche Textiltechnik,* **22** (1972), 110; **22** (1972), 171.

89. Kuechler, W. E., *ADR,* **61**(7) (1972), 32.

90. Hebrank, W. H., *ADR,* **63**(4) (1975), 32; **63**(5) (1975), 44; AATCC Conf., Atlanta, Georgia, 1977.

91. Parish, G. J., *Dyer,* **155** (1976), 385.

92. Dambroth, J. P., *Int. Text. Bull.,* **3** (1976), 221.

93. Beard, J. N., *TCC,* **8** (1976), 47.

94. Richter, F., *MTB,* **60** (1979), 864.

95. Trice, J. B., Spera, R. J., Haas, S. A., Konig, A. A., and McCarthy, R. L., *ADR,* **66**(5) (1977), 24.

96. Trice, J. B., and Chen, A. D., *ADR,* **67**(12) (1978), 43.

97. Hulls, P. J., *JSDC,* **98** (1982), 251.

98. Jones, P. L., *JSDC,* **98** (1982), 248.

99. Henderson, K., McAulay, T., Young, R., and Smith, G., *JSDC,* **98** (1982), 303.

100. Harrison, D. H., *JSDC,* **98** (1982), 305.

21 | POLLUTION AND WET PROCESSING

Pollution is the discharge of material residues and energy into the environment. Some of these residues are unconverted raw materials, some are unrecovered or not fully recovered products, and some are byproducts. Energy consumption and pollution are often interrelated. By conserving energy, it is possible to minimize pollution to a considerable extent.[1]

Large volumes of water and thousands of chemicals are used for pretreatments, dyeing, printing, and finishing of textiles. The Color Index lists more than 8000 products associated with the dyestuff industry alone. The diversity of chemical structure for compounds used by the textile industry spans from inorganic compounds and elements to biochemicals, polymers, and organic products.[2] A certain proportion of these chemicals always end up in the environment, which leads to pollution. The pollution associated with the chemicals is divided into (a) water pollution, (b) air pollution, and (c) solid-waste pollution. Apart from this, pollution due to energy waste such as noise pollution, heat pollution, or radiation pollution may be mentioned.

21.1 WATER POLLUTION

The textile industry is a leading consumer of water,[3] which is mainly used in the chemical processing of fibers. Consumption varies from fiber to fiber[4] (Table 21.1). The quantities of pollutants that form the pollution load going into the wastewater depend on the chemical process[5] (Table 21.2). Thus, although synthetic fibers do not contain any natural impurities, they produce pollutants during wet processing, the composition of which is highly variable and difficult to generalize.[6] The agents on synthetic fibers that cause water

TABLE 21.1 Water Consumption for Chemical Processing[6]

Substrate	Water Consumption (kg/kg Fabric)
Cotton	250–350
Wool	200–300
Polyamide	125–150
Polyester	100–200
Acrylic	100–220

TABLE 21.2 Pollution Load from Wet Processing (as Percent Weight of Fabric)[5]

	Cotton	Wool	Acrylic	Polyester	Nylon[a]
Natural impurities	3–5	20–30	. . . . .nil.		
Size, oil antistats	0.5–10	0.2–9	. . . . .0.5–6		
Scouring	0.5–6	1–15	. . . . .0.5–6		
Dyeing	0.2–8	0.5–10	0.5–10	0.3–60	0.2–5
Special finishes	. . . . .0.2–8.				
Total load	4–37	21–72	2–30	4–80	1.5–25

[a] Rayon and acetate fibers have similar values as those of nylon.

TABLE 21.3 Water Pollutants in Wet Processing[5,6]

Fiber	Process	Pollutants
Acetate	Scour and dye	Oil, dye, detergent, antistatic lubricants
	Scour and bleach	Detergent, hydrogen peroxide
Nylon	Scour, dye, and finish	Antistatic lubricants, detergents, tetra-sodium pyrophosphate, soda-ash, dye, oligomers, leveling agent, softener
	Bleach and finish	Peracetic acid, softener.
Polyester	Scour, dye, and finish	Oil, antistatic lubricant, detergent, size such as PVA, dyes, carriers, oligomers caustic soda, hydrosulfite, finishes.
	Scour, bleach, and finish	Oil, antistatic lubricant, detergent, size (PVA), chlorite, sodium nitrite, acid, bisulfite, finishes.
Acrylic and Modacrylic	Scour, dye, and finish	Oil, antistatic lubricant, detergent, dyes, finishes
	Scour, bleach, and finish	Oil, antistatic lubricant, detergent, chlorite, sodium nitrite, acid, bisulfite, finishes

pollution in wet processing are listed in Table 21.3.[5,6] A rough estimate of the pH, biological oxygen demand (BOD) (See Section 21.5), and total solids in wastewater is given in Table 21.4.[5] The water effluent discharged into the local sewage system has to satisfy certain stipulations that vary from place to place. Typical conditions are shown in Table 21.5.[7]

Such regulations are imposed to facilitate further purification of sewage water, which can then be discharged into a river or other major water source. This is essential since this source supplies water for domestic use, which has to meet very rigid specifications such as those given in Table

TABLE 21.4   Characteristics of Wastewater from Wet Processing of Synthetic Fibers[5]

Process	Fiber	pH	BOD-5 (ppm)	Solids (ppm)
Scour	Nylon	10.4	1360	1882
	Acrylic/modacrylic	9.7	2190	1874
	Polyester	9–10.5	500–800	NA
Scour and dye	Acetate	9.5	2000	1778
	Nylon	8.4	368	641
Dye	Acrylic/modacrylic	1.5–3.7	175–2000	833–1968
	Polyester	5–6.5	480–27,000	NA
Final scour	Acrylic/modacrylic	7	668	1191
	Polyester	7	650	NA

NA: Not available.

TABLE 21.5   Typical Limits for Discharge into Sewage Systems[7]

Effluent Water	Maximum Limit
Temperature	45°C
pH	6–8
Suspended solids	300 mg/liter (No gross material allowed)
Oil and grease	Not visibly detectable
Compounds containing chromium, copper, nickel, cadmium, zinc, lead, tin	20 mg/liter (in total)
Above metals in soluble form	10 mg/liter (in total)
Iron	150 mg/liter
Cyanide and cyanogen	5 mg/liter
Sulfate (as S)	300 mg/liter
Five-day biochemical oxygen demand at 20°C	600 mg/liter
Synthetic detergents	10 mg/liter

21.6.[8] If the effluents are not treated in the plant, then municipal authorities charge for their treatment at their end. These charges are increasing at a rapid rate (Table 21.7).[9] The rates vary from unit to unit depending on its municipal locality. In order to reduce the pollution load of wastewater from the processing of synthetic fibers, attention has to be paid to every operation in the wet processing.

TABLE 21.6 Specifications of Water for Domestic Use[8]

Substance or Parameter	Highest Desirable Level (mg/l)	Maximum Permissible Level (mg/l)
Anionic surfactant	0.2	1.0
Mineral Oil	0.01	0.3
Phenolic compounds (as phenol)	0.001	0.002
Total hardness (as $CaCO_3$)	100	500
Calcium	75	200
Chloride	200	600
Copper	0.05	1.5
Iron	0.1	1.0
Magnesium	30	150
Manganese	0.05	0.5
Sulfate (as SO_4)	200	400
Zinc	5	15
Fluoride	0.8	1.4
Arsenic	nil	0.05
Cadmium	nil	0.01
Cyanide	nil	0.05
Lead	nil	0.1
Mercury	nil	0.001
Selenium	nil	0.01
pH value	7.0–8.5	6.5–9.2

TABLE 21.7 Effluent Charges in the United Kingdom

Unit	1973	1975	1977	1979	1980
A	100	182	248	183	333
B	100	217	867	1567	1733
C	100	243	268	336	339
D	100	275	467	651	742
E	100	125	1750	2500	2250

Desizing. The most common sizing agents are polyvinyl alcohol (PVA) and carboxymethyl cellulose (CMC). Both are not readily biodegraded.[10,11] A treatment of effluent with calcium oxide and carbon dioxide removes CMC by precipitation.[12] Methods of adsorption or molecular filtration are suggested for removing PVA.[13] Wheatley and Baines[14] have reported that it is possible to biodegrade PVA by using acelimated organisms in properly designed and operated systems in waste-treatment plants. The desizing waste is recovered for its reuse by hyperfiltration or reverse osmosis.[15] Recovery of PVA is as high as 96%.[16] There is, thus, a considerable savings in desizing material and in the cost of an effluent treatment.

Scouring. Synthetic fibers are scoured with nonionic or anionic surfactants in the presence of mild alkali. Significant factors that contribute to water pollution are detergents, oils, solvents, and antistatic agents removed from the fiber. The pollution load in wastewater because of anionic surfactants is high if branched-chain nonbiodegradable detergents are used and becomes significantly less by replacing them by the readily biodegradable straight chain compounds.[17] Nonionic ethoxylated alkyl phenols are difficult to degrade at low temperatures.[18,19] Apart from its ability to degrade, it is also essential that a detergent has a low foaming property since excessive foam seriously affects the efficiency of a water-treatment plant for biodegradable waste. The coning oils, solvents, and antistatic agents extracted by the scouring liquor are generally inert to biological treatment.[10,20] Coagulation with calcium chloride or any other coagulant is useful for removing coning oil. Alternatively, the flotation method can be used.

Bleaching. The important bleaching agents used for synthetic fibers are hydrogen peroxide, sodium chlorite, sodium hypochlorite, and peracetic acid. Hydrogen peroxide decomposes into water and oxygen and does not leave any objectional residue except a small amount of sodium silicate and alkali added to the bleach-liquor. In fact, it can raise the concentration of dissolved oxygen in waste streams, thus decreasing the BOD. A biodegradable surfactant has to be used in peroxide bleaching. Sodium hypochlorite also does not cause any water pollution. Sodium nitrite added to the bleach-liquors of sodium chlorite to avoid corrosion causes water pollution. Similarly, sodium sulfite used as an antichor is a pollutant. Peracetic acid also causes water pollution. In fluorescent bleaching, most of the fluorescent brightening agent applied by the pad-dry-cure-wash process is taken up by the fiber. Such a low concentration of FBA in the washwater does not create a serious pollution problem, even though these agents are not biodegradable. They are broken down to harmless products slowly over a long period.[21]

Dyeing. Dyes generally impart only a small fraction of the total organic load in wastewater. However, they produce a high degree of color that is readily detectable and that detracts from the aesthetic value of streams, rivers and so on.[21-23] The removal of color from the wastewater is often more important than the removal of the soluble colorless organic substances that usually contribute to the major fraction of the biochemical oxygen demand (BOD). In a typical year, about 640,000 tons of dyes were used by textile and other industries, of which roughly 94,000 tons were lost in processing and added to the effluent water (Table 21.8).[21] Most of the dyes with very stable aromatic molecular structures are not destroyed by biological treatments and their removal from the effluent waters remain a major problem.[24-26] Metal-complex dyes and catalysts with heavy metals such as arsenic, cadmium, cobalt, mercury, chromium, lead, and zinc have to be eliminated to a very large extent since they are toxic to useful microorganisms present in biological effluent-treating tanks. Dyes free from heavy metals are therefore preferred.

It is essential to treat the wastewater so that the color is removed. Biological treatments are not efficient in completely removing the color, and mechanical, physical, and chemical treatments either alone or in addition to a biological treatment are usually necessary. To start with, all wastewater is mixed for equalization. The different liquors neutralize each other and the pH can be easily adjusted to neutral to prevent shock loadings to the treatment plant. The waste-liquor is then treated further.

21.2   WASTE-LIQUOR TREATMENT

Sedimentation, Flotation, and Flocculation.   Grit or solid material is made to settle or float out of the waste-liquors. Screening is used to remove solids such as undissolved chemicals, dirt, grit, and fibers.[27,28] The natural force of gravity attracts heavy solids to the bottom of the tank where they form watery sludge, which is mechanically removed and disposed of by incineration or dumping.[29,30] The separation and concentration of suspended solids as well as sludge are handled by the flotation technique.[31,32] The wastewater

TABLE 21.8   World Dye Production in 1975 and Losses in Processing[21]

Organic Dye (Active Substance)	Production		Losses in Processing	
	(%)	(Tons)	(%)	(Tons)
Textile	56	360,000	Ca 10–20	72,000
Paper/leather	14	90,000	Ca 10	9,000
Pigments	23.5	150,000	Ca 1–2	3,000
Others	6.5	40,000	Ca 25	10,000
Total	100	640,000		94,000

or a portion of clarified effluents is pressurized in the presence of air to 3–5 bar to approach saturation with the dissolved air. It is then released in the flotation unit where dissolved air forms minute bubbles that attach themselves to and become enmeshed in the floc or suspended particles. The suspended solids or the sludge flocs float because of these minute air bubbles. They then rise to the surface where they are skimmed off. Flocculation uses polyelectrolytes to cause rapid sedimentation, flotation, or as a solids reduction technique.[22,33] The solids in the waste and the flocculating agent are held together by molecular ionic forces, thus increasing particle size to give gelatinous particles known as floc which are removed by sedimentation, flotation, or filtration.

Coagulation. Chemical coagulants that are useful in the removal of color include lime, aluminum sulfate, ferrous and ferric sulfate, ferric chloride, and polyelectrolytes.[22,33–35] Sometimes two coagulants are used in conjunction with each other. The coagulation method uses the action of chemical adsorption or bonding to separate dissolved contaminants from the effluents. This economically feasible method has two drawbacks: (*a*) it is time-consuming—special tests are required to establish which coagulating agent or combination of agents is most suitable for each type of an effluent and (*b*) it produces a sludge, which creates a further problem.

Oxidation. Powerful oxidizing agents destroy the color of many organic materials.[8] Chlorine is the cheapest oxidizing agent but hydrogen peroxide or ozone are also used.[36]

Biological Treatment. A biological process, required to reduce the BOD of the wastewaters,[37,38] may not always be effective in completely destroying the coloring matter. The biological treatment of industrial wastewater involves contacting the water with a mixed culture of microorganisms (bacteria being the most important species) under favorable conditions.[6] The microorganisms metabolize the wastewater components for energy and synthesis of their cells. The microorganisms use dissolved oxygen, produce carbon dioxide, and synthesize new cells. These reactions continually occur in nature but are slow. In biological treatment plants, the rate of reaction is enhanced by providing a high concentration of microorganisms, substantial mixing, adequate dissolved oxygen, and good agitation for frequent contact between the microorganisms and their food. Methods of biological treatments are aerobic or anaerobic. The aerobic system use the free oxygen in the wastewater to convert wastes in the presence of microorganisms into more microorganisms, energy, and carbon dioxide. The anaerobic process occurs in the absence of free oxygen and converts the waste into methane and carbon dioxide. The process of aerobic biological treatment is carried out in stabilization ponds, aerated lagoons, and trickling filters. A stabilization pond usually has a water depth of 1–2 m. Oxygen is supplied by surface entrainment or algae. The BOD loading must be kept low and the retention time ranges between 5–25 days.[39] Aerated lagoons are designed with a 2–5 m liquid depth, depending on the available aeration system, which dramatically increases the ability of the system to remove organic matter. The retention time varies from 2–10 days. Bioaeration alone is ineffective in removing the color but bioaeration with activated carbon catalyst can remove the color.[40] In the activated sludge process, active microorganisms are settled and returned to the process in order to increase their concentration in the aeration basin. This allows for a higher BOD loading and reduces the required retention time.

Figure 21.1 shows a schematic flow diagram of an activated sludge treatment.[23] The incoming waste water is mixed with a biologically active sludge or a suspension of microorganisms. The mixture is aerated with compressed air or mechanical aerators for the desired length of time and is then transferred to another tank where the activated sludge is separated by sedimentation. The isolated sludge is disposed of or returned to the process, if needed. In these systems, high-purity oxygen instead of air may be used to oxygenate the biological mass.[41]

Trickling filters are generally cylindrical tanks packed either with stone or a synthetic medium.[42] A typical system is schematically shown in Figure 21.2.[23] The effluent flows onto the filter media by means of a rotating arm that distributes the wasteload uniformly over the circular bed. The effluent trickles through the filter over a slime of bacteria that adheres to the filter media. As bacteria die, they fall off the filter and are removed from the effluent through the secondary settling stage. The removal efficiency depends upon the type of media used, the organic loading ratio, the ratio of waste to recycled wastewater, and the operating temperature.[43]

Adsorption.   Adsorption is a physico–chemical process that is used to produce quality effluents that are low in concentrations of dissolved organic matter. Dissolved molecules are attracted to the surface of the adsorbent by physical–chemical forces. Activated carbon is the most commonly used adsorbent for the removal of soluble dyes from the wastewater.[44–48] However, activated carbon is expensive. It can be reactivated after use. A variety

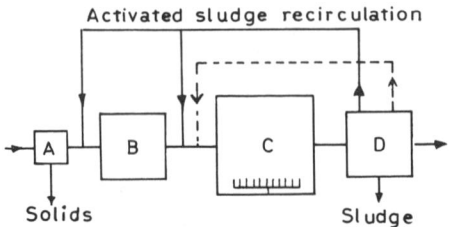

FIGURE 21.1.   Flowchart of activated sludge plant.[23] A: Preliminary treatment, B: Primary treatment, C: Aeration tank, D: Secondary settling. – – –: Effluent recirculation.

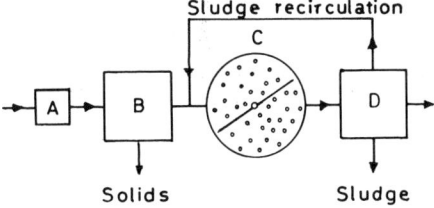

FIGURE 21.2. Flowchart of trickling filter plant.[23] A: Preliminary treatment; B: Primary treatment; C: Trickling filter (dosing arm and media); D: Secondary settling.

of other substances such as silica gel,[49] peat bauxite, Fuller's earth (clay), and wood have also been recommended as adsorbents.[22]

Membrane Filtration. Dynamic membrane hyperfiltration or reverse osmosis can remove all types of dyes from the waste-liquor.[44,50] The process leaves a concentrated dye solution which must be disposed of by some other method such as evaporation and incineration. High capital costs and the possibility of clogging the membrane with dyes after long usage are the two disadvantages of this system.[51] The possibility of reusing the dyes after recovery is being explored.

Ion-Exchange Method. In this method, the wastewater is passed over the ion-exchange resin until the available exchange sites are saturated and the contaminant appears in the effluent. At this point, the treatment is stopped, the bed is backwashed, and the resin is regenerated using a strong acid or base solution. After regeneration, the bed is rinsed with water to wash out residual regenerant. The bed is then ready for the next treatment cycle. This process removes anions and cations from the wastewater.[52]

Distillation. This is the oldest demineralization process. It separates wastes from water through the conversion of liquid into vapor and the subsequent removal of vapor. It gives water in a very pure form. However, the process is not economical and is not recommended as a wastewater treatment.

21.3 REDUCTION IN POLLUTION LOAD

Dyeing. Acetic acid is extensively used for adjusting the pH of the dye-liquor. Since acetic acid is difficult to biodegrade, it should be replaced by the readily biodegradable formic acid.[53] Furthermore, a lower concentration of formic acid gives the desired pH and hence, the pollution load decreases further.

The use of boron compounds should be restricted as far as possible since boron is detrimental in effluents. Sodium perborate, which is used for oxidation of leuco-vat dyes to parent vat dyes, can be easily replaced with hydrogen peroxide. Minimum concentrations of bichromate in chrome dyeing and copper sulfate for the aftertreatment of direct dyes should be selected so that most of the metallic compound goes on to the fiber and very little goes into the effluent.[53]

Many carriers used for dyeing polyester (e.g., 2 hydroxydiphenyl) are not biodegradable[54,55] and increase the pollution load. Biodegradable carriers such as biphenyl, trichlorobenzenes, or methyl biphenyl are therefore preferred.

Printing.   Most of the wastewater in printing is generated during the washing-off of printed goods. This washwater contains surfactants, dyes or pigments, thickeners, and other ingredients present in the printing paste. The selection of ingredients from the viewpoint of their biodegradability is important in order to reduce the pollution load. Thickeners constitute the major portion of impurities in the wash-liquor. Starch thickening is readily biodegradable.[10] Carboxymethyl cellulose, polyvinyl alcohol, and synthetic thickeners are not biodegradable.[10,56–58] Depending on the ingredients present in the printing paste, the wastewater can be given a biological treatment alone or in combination with the physical and chemical processes.

Finishing.   A large number of chemicals are employed for finishing the synthetic fibers. Depending on the type of chemicals used, a suitable effluent treatment is selected. Most of the polymer emulsions used for textile finishing are sensitive to pH, salt, or agitation and may coagulate when they enter the wastewater streams.[10] The sewer lines may then be clogged with these inert solid materials which have to be removed physically. While bulk of the polymer emulsion can be coagulated and removed in a treatment plant, some of it remains emulsified and is not removed by the biological treatment. For the complete removal of a polymer emulsion, a chemical treatment is sometimes necessary.

Many finishing agents used for wash-and-wear and permanent-press fabrics are manufactured from urea, melamine, formaldehyde, and glyoxal. Some of these products are biodegradable but others are not.[9] Formaldehyde derivatives can react with themselves or other chemicals in the waste stream to form insoluble products that are removed by the sedimentation.

Flame-retardant finishing agents are compounds generally containing phosphorous and nitrogen. Two important flame-retarding agents, tetrakish (hydroxy methyl) phosphonium chloride (THPC), and trisaziridinyl phosphine oxide (APO), create a serious problem if they are allowed into a natural stream.[59,60] The soil-release finishing agents such as acrylate and methacrylate copolymers containing free carboxyl groups are resistant to biological treatment and are removed by either physical or chemical methods.[61] Mothproofing agents such as DDT, Eulan, and Mitin FF are nonbiodegradable.[8] Wherever possible attempts should be made to replace them by suitable biodegradable products.

21.4 OTHER METHODS

Recycling of Water. Repeated use of the process water, that is, recycling, achieved by a suitable treatment given to an effluent stream, lowers the consumption of fresh water.[62,63] The proportion and properties of the treated water to be recycled are decided by the process for which it is to be used. The process may not have the rigid specifications which the discharged effluent has to meet. For example, wastewater after filtration and active carbon treatment is found to be suitable for dyeing hosiery.[64] Similarly, a sequence of treatments provided by an active-sludge plant, flocculation, adsorption on active carbon, and ion-exchange resin give water suitable for recycling.[65] Reverse osmosis either alone or in combination with other treatments also gives water suitable for recycling.[26,66]

Foam Processing. Foam is used as a medium for the application of dyes, pigments, and finishes to textiles. Foam processing requires less water and very low energy. However, foam is troublesome from the viewpoint of water pollution. Foaming of drinking water is not acceptable, even from an aesthetic viewpoint. Foam creates problems in aeration basins of biological treatment plants.[67] The surfactants required to generate foam has to be biodegradable, a quality that does not contribute to the formation and functioning of the foam. The overall load of pollutants in foam processes is very low and thus, is easy to handle.

Solvent Processing. A decade ago, nonaqueous solvents and their mixtures with water were used for pretreatment, dyeing, and finishing textiles. One of the claims of solvent processing is the recovery of solvent and thus, little, if any, water pollution. However, a part of the solvent is always lost to the atmosphere during processing and solvent recovery. Heavy molecules of solvents such as chlorinated hydrocarbons tend to settle down around the processing units. This generates a dangerous level of solvent–vapor concentration in the ambient atmosphere and creates serious air pollution. Solvent processing has now lost its commercial importance.

21.5 MEASUREMENT OF WATER POLLUTION

Various government and municipal agencies specify methods of measuring water pollution.[3,62,68–71] The effluent water is treated for dissolved and suspended solids, pH, metallic impurities, sulfate, phenolic compounds, color, oxygen content and so on.[72] The oxygen demand of the effluent is measured by determining (*a*) the biological oxygen demand after storage for five days (BOD-5), and (*b*) the chemical oxygen demand (COD).

BOD-5. BOD-5 measures the amount of oxygen required by microorganisms present in the test water for breaking down the carbon (organic) compounds. BOD-5 is thus a measure of the quantity of organic material present in water and which is accessible to bacterial attack.

The liquor to be tested is diluted to a suitable level of pollutants and a known volume added to oxygen-saturated water containing trace elements and an inoculum of organisms. Closed bottles are incubated precisely at 20°C for five days, and the residual dissolved oxygen is compared with that on duplicates measured before incubation. The difference gives the consumption of oxygen due solely to the biodegradable organic matter present. Trials have to be made with different dilutions of the test solution, both to observe the effect of dilution and to arrive at a level whereby about half the dissolved oxygen is consumed in the test.

The drawback of the BOD-5 test is that it is a time-consuming method and does not give any idea about the amount of organic pollutant present in the water that is not degraded by bacteria. A BOD-5 value often indicates a lower degree of pollution than is actual one.

COD.   Two techniques used for measuring the chemical oxygen demand (COD) are (*a*) permanganate value and (*b*) dichromate value.

Permanganate Value.   The consumption of potassium permanganate under acidic conditions by a known quantity of effluent at 27°C in 4 h is measured. It gives a quick indication of the expected BOD value if there is a moderately reasonable relationship between the two values. However, for textile-processing effluents, these two values have no correlationship, the permanganate values fail to indicate the expected BOD values.

Dichromate Value.   The consumption of potassium dichromate under acidic conditions by a known quantity of effluent at boiling in 2 h is measured. The value indicates total organic matter, both biodegradable and nonbiodegradable. Thus, these values are generally higher than those of BOD.

21.6   AIR POLLUTION

Air pollution from textile mills consists of low levels of particulates such as lint and a small amount of organic chemicals.[6] Organic chemicals come from spin-finishes, coning oil, pigment printing (kerosene), stain-removing solvents and so on. These chemicals become airborne during heat setting, curing, and drying operations. Stenters, polymerizing ovens, and cylinder dryers are the machines from which organic chemicals are exhausted into the atmosphere. Pigment printing using kerosene is a major source of air pollution. Attempts have been made to recover kerosene from the exhaust of polymerizing units used for drying and curing of pigment prints. During the last few years, alternative thickeners were developed to replace emulsion thickeners.[58,73,74] These synthetic thickeners based on long chain polymers derived from substituted vinyl compounds, polyacrylic acid, or polymerized maleic acid are useful in reducing air pollution. Various methods for removing organic chemicals from exhaust gases of stenters and polymerization ovens are under development. These are based on incineration, scrubbing, and electrostatic precipitation.

Air pollution due to boiler-house exhaust is a common factor to all industries that use steam for heating or the generation of electricity. Coal-fired boilers create more pollution than oil-fired boilers, but the cost of steam by the latter is almost three times higher.

21.7 SOLID WASTE

Every effluent treatment plant produces solid waste, which has to be disposed of properly because it contains hazardous chemicals.[6] Some of these materials are highly toxic, others are corrosive, flammable, irritants, explosive, infectious and so on. Disposal of this solid waste always presents a problem. Common waste-disposal systems are landfill, incineration, ocean-dumping, spray irrigation, or hauling to a municipal treatment plant. While dumping the waste for land fill, proper precautions have to be taken to avoid further water pollution because of the seepage of water during the rainy season.[10] Land-destined hazardous waste may be physically encapsulated in impervious materials such as concrete, asphalt, or plastics prior to disposal. The incineration of textile waste is usually not carried out for two reasons: (*a*) it causes air pollution and (*b*) the ash produced from incineration has a high concentration of heavy metals. Ocean-dumping, spray, and irrigation methods are also not very safe if the solid waste contains a large proportion of hazardous chemicals.

21.8 TOXICITY

The term *toxicity* is defined as the ability of a chemical to produce injury once it reaches a susceptible site in or on the body.[75] Since the textile industry uses a large number of chemicals, the evaluation of the toxicity of these chemicals is very important to protect the health of the workers. Certain terms used in the evaluation of toxicity are defined below.

Threshold Limit Value (TLV).   The TLV is the level of atmospheric concentration of potentially hazardous gases, vapors, and dusts to which workers may be exposed during working hours throughout the year without adverse effect on their health or efficiency.[75] A low TLV value indicated a high hazard. TLV is expressed in ppm or mg/m^3.

LD_{50} Test. LD_{50} is the dose level (administered as a single dose and expressed in terms of weight of chemical compound per unit weight of the

animal, e.g., mg/kg) that kills 50% of the experimental animals thus treated.[75] An LD_{50} of less than 1 mg/kg represents an extremely toxic substance.

LC_{50} Test.[21,76]   This test is used for measuring the toxicity of chemicals in effluents to fish.

Toxicity of Pretreatment Chemicals.   No health hazards have ever been reported from the use of enzymatic desizing agents. During the singeing of fabrics, dust and fly are produced that can cause some irritation. However, by arranging a proper exhaust system, this hazard can be almost completely eliminated. Sodium hydroxide used in scouring and mercerizing is poisonous when ingested or breathed as fumes or dust. It is corrosive and causes burns. Ammonia is also corrosive and causes burns to the eyes and skin. Its TLV is 50 ppm and 35 mg/m^3. Anhydrous ammonia constitutes a fire hazard and is explosive. All the bleaching agents, namely, hydrogen peroxide, sodium hypochlorite, and sodium chlorite are corrosive. Hydrogen peroxide (35% w/w) causes blistering of the skin and irritation to the mucous membrane. It also constitutes a fire and explosion hazard. Sodium chlorite solutions are corrosive and chlorine dioxide evolved during its decomposition is a highly toxic gas. By comparison, sodium hypochlorite solutions are less toxic. A sodium hypochlorite solution is often used by dyehouse workers for cleaning their hands. This can damage the skin. If it is essential to use sodium hypochlorite for removing dye from hands, it is advisable to use only dilute solutions and the treatment time has to be very short, say, 15–20 sec.[77] The hands are immediately rinsed with plenty of cold water followed by a dilute solution of sodium bisulfite, and finally a soap solution and cold water. Organic solvents like trichloroethylene and perchloroethylene act as an anesthetic and, if breathed in high concentrations, can cause death. The TLV of perchloroethylene and trichloroethylene are 100 ppm, 41 ml/100 m^3 and 100 ppm, 37 ml/100 m^3, respectively.[78]

Toxicity of Dyes.   The problem of toxicity of dyes attracted considerable attention after the discovery of the carcenogenicity of 2-naphthylamine and benzidine which were used as intermediates in the manufacture of dyes. The use of these chemicals is now banned in most countries.[78] As far as dyes are concerned, the toxicity hazards are negligible, since the carcenogens are destroyed in the manufacture of the final dye or pigment and the finished compounds used in dyeing and printing are inactive.[75,76,79–81] Manufacturers now test the toxicity of their newly developed dyes and only those dyes without any health hazard are produced on a commercial scale. The majority of dyes on the market have been tested for toxicity to fish and have been found to be only slightly poisonous; the LC_{50} (48-h trout) is mostly above 100 ppm, and, in some cases, even above 1000 ppm.[21] Values below 1 ppm are very rare. Little and Lamb[82] have observed that only two dyes *viz.* Methyl

Violet and Malachite Green have LC_{50} values of 0.05 ppm and 0.12 ppm, respectively. These two basic dyes constitute the bulk of dyes toxic to fish.

Toxicity of Finishing Chemicals. Solutions and vapors of formaldehyde represent both an acute and a chronic irritant.[83] Frequent or prolonged exposure can cause hypersensitivity. If swallowed, it causes violent vomiting and diarrhea and can lead to collapse. Formaldehyde produces a chronic inflammatory reaction to the respiratory tract. It causes a hardening or tannin effect on skin. If formaldehyde vapors are inhaled for a long time, they cause bronchitis or bronchial pneumonia. Studies on inhalation of formaldehyde by animals have produced evidence of lung cancer.[75] Exposure to formaldehyde is suspected to be responsible for a number of abortions during the early stages of pregnancy. Free formaldehyde is always present in small quantities in the commercial urea-formaldehyde resins. Free formaldehyde also evolves during the finishing process.[84] Dimethylol-dihydroxy-ethylene-urea gives the lowest level of free formaldehyde during finishing, while urea-formaldehyde, melamine-formaldehyde, and carbamate give the highest levels of evolved formaldehyde.[85] The TLV limit for formaldehyde is 2 ppm and is likely to be lowered to 1 ppm.

The resin-finishing of textiles using magnesium chloride (as a catalyst) was considered to be a safe process. However, the reaction of hydrochloric acid and formaldehyde in air produces bis-chloromethyl ether (BCME) at a very rapid rate.[86] At room temperature and 40% relative humidity, equilibrium is reached in less than a minute to yield parts per billion (ppb) of BCME from ppm of reactants. BCME is known to be a highly potent inducer of tumors in the respiratory tract in mice and rats.[87] BCME has been assigned a TLV of 0.001 ppm (1 ppb). There is a controversy regarding the formation of BCME during resin-finishing. Davies[88] and Ellgehausen[89] have reported that BCME is not formed during resin-finishing. Hurwitz[90] also observed that in the presence of cellulose-substrate BCME is not formed, whereas Zollinger[91] advises caution in certain cross-linking systems where in the presence of hydrogen chloride, acetic acid, or dioxane, a measurable amount of BCME is formed. However, he carried out experiments in the absence of cellulose-substrate. Thus, no definite conclusion has yet been drawn regarding the formation of BCME during the resin-finishing of textiles using magnesium chloride as a catalyst.

Tris(2,3 dibromopropyl) phosphate, (Tris), was extensively used as a flame-retarding finishing agent for polyester, cellulose acetate, and triacetate fibers. Initially, Tris was shown to be a nonhazardous chemical.[92] However, Tris was found to produce cancer in laboratory animals and it is probably carcinogenic to human beings.[72] There is no evidence, however, from any source to establish that Tris has, in fact, caused cancer in human beings. In April 1977, the U.S. Consumer Product Safety Commission acting on a report from the National Cancer Institute banned the application of Tris to children's clothing.[75]

REFERENCES

1. Royston, M., *Pollution Prevention Pays,* Pergamon Press, London (1980).
2. Kolber, A. R., *TCC,* **12** (1980), 134.
3. Durig, G., *Rev. Prog. Color.,* **7** (1976), 70.
4. Carbonell, J., Egli, H., Perrig, M., *ADR,* **66**(8) (1977), 44.
5. U.S. Environmental Protection Agency (2) Waste Water Treatment Systems: Upgrading Textile Operations to Reduce Pollution, Report EPA-625/3-74-004, Washington D.C. (October 1974).
6. Cooper, S. G., *The Textile Industry, Environmental Control and Energy Conservation,* Noyes Data Corporation, U.S.A. (1978), p. 210.
7. Best, G. A., *JSDC,* **90** (1974), 389.
8. Gardiner, D. K., and Borne, B. J., *JSDC,* **94** (1978), 339.
9. Dwek, J. C., *JSDC,* **97** (1981), 390.
10. Souther, R. H., *ADR,* **51** (1962), 363; **58**(15) (1969), 13.
11. Porter, J. J., Nolan, W. F., and Aberanthy, A. R., *Chemical Eng. Symposium Series, Water* **67**(107) (1970).
12. Pangle, J. C., USP 3,419,493 (Dec. 1968).
13. Robert, A., Taft Water Res. Centre, Report No. TWRC (1969).
14. Wheatley, Q. D., and Baines, F. C., *TCC,* **8** (1976), 23.
15. Cacho, J. W., *TCC,* **12** (1980), 78.
16. Sibley, W. A. L., *ADR,* **68**(10) (1979), 23.
17. Barnes, W. V., and Dobson, S., *JSDC,* **83** (1967), 313.
18. Mann, A. H., and Reid, V. W., *J. Am. Oil Chem. Soc.,* **48** (1971), 588, 794, 798.
19. Stiff, M. J., and Rootham, R. C., *Wat. Res.,* **7** (1973), 1407.
20. LeBlanc, R. B., *ADR,* **56** (1967), 623.
21. Anliker, R., *Rev. Prog. Color.,* **8** (1977), 60.
22. Porter, J. J., *Pollution Engineering,* (Oct. 1973), 27.
23. Mckay, G., *ADR,* **68**(4) (1979), 29; **69**(3) (1980), 38.
24. Adams, A. D., *ADR,* **65**(4) (1976), 32.
25. Horning, R. H., *TCC,* **9** (1977), 73.
26. Porter, J. J., and Sargent, T. N., *TCC,* **9** (1977), 269.
27. Katz, W. J., and Geinopolis, A., *Chem. Eng.* **75**(10) (1968), 78.
28. Berger, O., *Chem. and Ind.* (Jan. 19, 1974), 50.
29. Fitch, B., *Ind. Eng. Chem.,* **54**(10) (1962), 44.
30. Shannon, P. T., and Tory, E. M., *Ind. Eng. Chem.,* **57**(2) (1965), 19.
31. Rohlick, G. A., *Sewage Ind. Wastes,* **26** (1954), 1056.
32. Sessler, R. E., *Sewage Ind. Wastes,* **27** (1955), 1178.
33. Datye, K. V., and Sharma, N. D., *Popular Plastics and Rubber,* **XXVI**(4) (April 1981), 3–6.
34. Summers, T. H., *JSDC,* **83** (1967), 373.
35. Dehme, C., and Martinola, F., *Chem. and Ind.* (Sept. 1, 1973), 823.
36. Ogden, M., *Ind. Water Eng.* (June 1970), 36.
37. Alspaugh, T. A., *TCC,* **5** (1973), 255.
38. Purvis, M. R., *ADR,* **63**(8) (1974), 19.
39. Evers, D., *Dyer,* **143**(1) (1970), 56.

40. Porter, J. J., *Water and Wastes Engineering* (Jan. 1972), A8.

41. Matsch, L. C., and Dedeke, W. C., *Chem. Eng. Progress,* **69**(8) (1973), 75.

42. Purchas, D. B., *Chem. and Ind.* (Jan. 19, 1974), 53.

43. Flood, J. E., *Chem. Eng.,* **73**(13) (1966), 63.

44. Porter, J. J., *ADR,* **61**(8) (1972), 24.

45. Davis, R. A., Kaempf, H. J., and Clemens, M. M., *Chem. and Ind.* (Sept. 1, 1973), 827.

46. Leslie, M. E., *ADR,* **63**(18) (1974), 15.

47. Digiano, F. A., Frye, W. H., and Natter, A. S., *ADR,* **64**(8) (1975), 15.

48. Roy, C., and Volesky, B., *TCC,* **10** (1978), 94.

49. Mckay, G., Poots, V. J., and Alexander, F., *Ind. Eng. Chem. Process Res. Dev.,* **17** (1978), 406.

50. Brandon, C. A., Johnson, J. S., Minturn, R. E., and Porter, J. J., *TCC,* **5** (1973), 134.

51. Rhode Island Section, AATCC, *TCC,* **3** (1971), 239.

52. Arden, T. V., *Water Purification by Ion Exchange,* Butterworths, London (1968).

53. Durig, G., and Hausmann, J. P., *JSDC,* **94** (1978), 331.

54. Haas, J. M., Earhart, H. W., and Todd, A. S., *ADR,* **64**(3) (1975), 34.

55. Simmons, P. B., Branson, D. R., Moolenaar, R. J., and Bailey, R. E., *ADR,* **66**(8) (1977), 21.

56. Beersma, P. J. A., *Water and Waste Management,* **19** (June 1976), 26.

57. Bayerlein, F., *MTB,* **58** (1977), 1017.

58. Habereder, P., *MTB,* **61** (1980), 165.

59. Beroza, M., and Borkoveu, A. D., *J. Med. Chem.,* **7** (1964), 44.

60. LeBlanc, R. B., *Text. Ind.,* **132**(10) (1968), 274.

61. Parsons, W. N., *Mod. Text. Mag.,* **48**(12) (1967), 52.

62. Parish, G. J., *Rev. Prog. Color.,* **7** (1976), 55.

63. Goodman, G. A., and Porter, J. J., *ADR,* **69**(10) (1980), 33.

64. Pardue, E. E., Stark, M. M., and Lalli, B. M., Book of Papers 1974 National Technical Conference, AATCC, New Orleans, USA (1974), p. 422.

65. Lohmann, J., *Chemiefasern,* **22** (1972), 687.

66. Dittrich, V., *MTB,* **54** (1973), 853.

67. Bahorsky, M. S., *ADR,* **68**(10) (1979), 26.

68. Fearn, R. J., Hetherington, W. H., Jackal, S. M., and Ward, C. D., *JSDC,* **83** (1967), 146.

69. Hinge, D. C., *Chem. and Ind.* (Aug. 4, 1973), 727.

70. Williams, B. R., *Chem. and Ind.* (Feb. 2, 1974), 89.

71. Little, A. H., *Water Supplies and the Treatment and Disposal of Effluents,* The Textile Institute (1975), 56.

72. Vaidya, A. A., and Datye, K. V., *Colourage,* **29**(1) (1982), 3.

73. Thomas Holst, L., *TCC,* **11** (1979), 53.

74. Hughes, D. W., *JSDC,* **95** (1979), 381.

75. Madaras, G. W., *ADR,* **69**(5) (1980), 28.

76. Tooby, T. E., Hursey, P. A., and Alabaster, J. S., *Chem. and Ind.* (June 21, 1975), 523.

77. Gadian, T., *Rev. Prog. Color.,* **7** (1976), 85.

78. Fabric Care Research Ass., Tech. Circular **157** (June 1974).

79. British Association of Urological Surgeons, *British J. Urology,* **23** (1961), 1.

80. Anthony, H. M., and Thomas, G. M., *J. Nat. Cancer Inst.,* **45** (1970), 879.

81. Veys, C. A., *British J. Ind. Med.*, **31** (1974), 65.

82. Little, L. W., and Lamb, J. C., *Dyes and Environment*, Vol 1, New York ADME (1973), Chapter 6.

83. Sax, N. I., *Dangerous Properties of Industrial Materials*, 3rd Ed., Reinhold Publishing Corp., New York (1968).

84. Cook, T. F., and Weigmann, H. D., *TCC*, **14** (1982), 100.

85. Lund, G., *Shirley Inst. Bull.*, **48**(1) (March 1975), 17.

86. Rohm and Haas Co., *News Release*, Dec. 27, 1972.

87. Kuschner, M., *Archives of Environmental Health*, **23** (1971), 135.

88. Davies, D. C., *Chemistry in Britain*, **10** (1974), 359.

89. Ellgehausen, D., *Chimia*, **30**(8) (1976), 338.

90. Hurwitz, M. D., *ADR*, **63**(3) (1974), 60.

91. Zollinger, H., *TCC*, **9** (1977), 96.

92. Caniher, F. A., Proceedings of the 1976 Symposium on Textiles, LeBlanc Research Corp., U.S.A. (1976).

BIBLIOGRAPHY

References for further reading are given at the end of each chapter. Books of general interest and utility are as follows.

1. Abramson, C. C., *Practical Carpet Dyeing, Printing, and Finishing,* Reg. Burnett Inc. U.S.A. (1978).
2. Ahmed, M., *Polypropylene Fibers Science and Technology,* Elsevier Scientific Publishing Co., Amsterdam (1982).
3. Bhatnagar, V. M., Ed., *Advances in Fire Retardant Textiles,* Technomic Publications, U.S.A. (1975).
4. Bird, C. L., and Boston, W. S., Eds., *The Theory of Coloration of Textiles,* Dyers Company Publications Trust, U.K. (1975).
5. Chakravertty, R. R., and Trivedi, S. S., *Technology of Bleaching and Dyeing of Textile Fibers,* Mahajan Brothers, India (1979).
6. Chavan, R. B., Ed., *Textile Printing,* IIT, Delhi, India (Sept. 1979).
7. Cheetham, R. C., *Dyeing Fiber Blends,* D. Van Nostrand Co. Ltd., London (1966).
8. Clarke, W., *An Introduction to Textile Printing,* 4th Ed., Newness Butterworths, London (1974).
9. Cook, J. G., *Handbook of Textile Fibers,* 4th Ed., Merrow Publishing Co. Ltd., U.K. (1968).
10. Cooper, S. G., *The Textile Industry-Environmental Control and Energy Consumption,* Noyes Data Corp., U.S.A. (1978).
11. Crockett, S. R., and Hilton, K. A., *Dyeing of Cellulosic Fibers and Related Processes,* Leonard Hill, London (1961).
12. Diserens, L., *Chemical Technology of Dyeing and Printing,* Rheinhold Publication Corp., New York (1951).
13. Gulrajani, M. L., Ed., *Texturising,* IIT, Delhi, India (1977); *Polyester Textiles,* The Textile Association, India (1980); and *Blended Textiles,* The Textile Association, India (1981).
14. Hall, A. J., *Textile Finishing,* Heywood Books, London (1966).

15. Happey, F., Ed., *Applied Fiber Science*, Academic Press, New York (1978).

16. Hearle, J. W. S., and Peters, R. H., Eds., *Fiber Structure*, The Textile Institute and Butterworths, London (1962).

17. Johnson, K., *Antistatic Composition for Textile and Plastics*, Noyes Data Corporation, U.S.A. (1976).

18. Kale, D. G., *Principles of Cotton Printing*, Mahajan Brothers, India (1976).

19. Kulkarni, G. G., and Trivedi, S. S., *Wet Processing of Polyester/Cotton Blends*, ATIRA, India (1967).

20. Ludwig, H., *Polyester Fibers*, Wiley-Interscience, New York (1971) (English Translation).

21. Mark, H. F., Atlas, S. M., and Cernia, E., Eds., *Man Made Fibers, Science and Technology*, Interscience, N.Y., Vol. I (1967), Vol. II and III (1968).

22. Mark, H. F., Wooding, N. S., and Atlas, S. M., Eds., *Chemical After Treatments of Textiles*, Wiley Interscience, New York (1971).

23. Marsh, J. T., *Self Smoothing Fabrics*, Chapman and Hall Ltd., London (1962).

24. Moilliet, J. L., Ed., *Water Proofing and Water Repellency*, Elsevier Publishing Co., Amsterdam (1963).

25. Moncrieff, R. W., *Man Made Fibers*, Newness Buttersworths, London (1975).

26. Mortan, W. E., and Hearle, J. W. S., *Physical Properties of Textile Fibers*, The Textile Institute and Butterworths, London (1962).

27. Nunn, D. M., Ed., *The Dyeing of Synthetic Polymer and Acetate Fibers*, Dyers Company Publications Trust, U.K. (1979).

28. Peters, R. H., *Textile Chemistry*, Elsevier Scientific Publishing Co., Amsterdam, Vol. I, (1963), Vol. II, (1967), and Vol. III (1975).

29. Piller, B., *Bulked Yarn, Production, Processing Application*, The Textile Trade Press, Manchester, UK, (1973).

30. Ranney, M. W., *Soil Resistant Textiles*, Noyes Data Corporation, U.S.A. (1970).

31. Robinson, G., *Carpets and Other Textile Floor Coverings*, Textile Book Service, U.S.A. (1972).

32. Schmidlin, H. U., *Preparation and Dyeing of Synthetic Fibers*, Chapman and Hall Ltd., London (1963).

33. Shenai, V. A., *Technology of Textile Processing*, Vols. I–V, (1971–1976) Sevak Publications, India.

34. Trotman, E. R., *Textile Scouring and Bleaching*, Griffin, London (1968); *Dyeing and Chemical Technology of Textile Fibers*, 5th Ed., Charles Griffin and Co. Ltd., London (1975).

35. Vaidya, A. A., and Trivedi, S. S., *Textile Auxiliaries and Finishing Chemicals*, ATIRA, India (1975).

36. Venkataraman, K., Ed., *Chemistry of Synthetic Dyes*, Vols. I–VIII, (1952–1978) Academic, N.Y.

37. Wilcock, C. L., and Ashworth, J. C., *Dyeing with Coal Tar Dyestuffs*, 6th Ed., Bailliere Tindall and Cox, London (1964).

38. Ziabiki, A., *Fundamentals of Fiber Formation*, John Wiley and Sons, New York (1976).

INDEX

949.404

1488